ENGINEERING PRACTICES FOR MANAGEMENT OF SOIL SALINITY

Agricultural, Physiological, and Adaptive Approaches

Innovations in Agricultural and Biological Engineering

ENGINEERING PRACTICES FOR MANAGEMENT OF SOIL SALINITY

Agricultural, Physiological, and Adaptive Approaches

Edited by
S. K. Gupta, PhD
Megh R. Goyal, PhD
Anshuman Singh, PhD

APPLE
ACADEMIC
PRESS

Apple Academic Press Inc. | Apple Academic Press Inc.
3333 Mistwell Crescent | 9 Spinnaker Way
Oakville, ON L6L 0A2 Canada | Waretown, NJ 08758 USA

© 2019 by Apple Academic Press, Inc.
First issued in paperback 2021

Exclusive worldwide distribution by CRC Press, a member of Taylor & Francis Group
No claim to original U.S. Government works

ISBN-13: 978-1-77463-162-1 (pbk)
ISBN-13: 978-1-77188-676-5 (hbk)

Library and Archives Canada Cataloguing in Publication

Engineering practices for management of soil salinity: agricultural, physiological, and adaptive approaches / edited by S.K. Gupta, PhD, Megh R. Goyal, PhD, Anshuman Singh, PhD.

(Innovations in agricultural and biological engineering)

Includes bibliographical references and index.

Issued in print and electronic formats.

ISBN 978-1-77188-676-5 (hardcover).--ISBN 978-1-351-17108-3 ((PDF)

1. Soils, Salts in. 2. Soil management. 3. Soil mechanics.
I. Goyal, Megh Raj, editor II. Gupta, S. K. (Suresh Kumar), 1949-, editor III. Singh, Anshuman, editor
IV. Series: Innovations in agricultural and biological engineering

S595.E54 2018	631.4	C2018-903694-X	C2018-903695-8

Library of Congress Cataloging-in-Publication Data

Names: Gupta, S. K. (Suresh Kumar), 1949- editor. | Goyal, Megh Raj, editor. | Singh, Anshuman, editor.
Title: Engineering practices for management of soil salinity : agricultural, physiological, and adaptive approaches / editors: S.K. Gupta, Megh R. Goyal, Anshuman Singh.
Description: Waretown, NJ : Apple Academic Press, 2018. | Includes bibliographical references and index.
Identifiers: LCCN 2018027319 (print) | LCCN 2018028697 (ebook) | ISBN 9781351171083 (ebook) | ISBN 9781771886765 (hardcover : alk. paper)
Subjects: LCSH: Soils, Salts in. | Soil management. | Soil mechanics.
Classification: LCC S595 (ebook) | LCC S595 .E54 2018 (print) | DDC 631.4--dc23
LC record available at https://lccn.loc.gov/2018027319

Apple Academic Press also publishes its books in a variety of electronic formats. Some content that appears in print may not be available in electronic format. For information about Apple Academic Press products, visit our website at **www.appleacademicpress.com** and the CRC Press website at **www.crcpress.com**

OTHER BOOKS ON AGRICULTURAL & BIOLOGICAL ENGINEERING BY APPLE ACADEMIC PRESS, INC.

Management of Drip/Trickle or Micro Irrigation
Megh R. Goyal, PhD, PE, Senior Editor-in-Chief

Evapotranspiration: Principles and Applications for Water Management
Megh R. Goyal, PhD, PE and Eric W. Harmsen, Editors

Book Series: Research Advances in Sustainable Micro Irrigation
Senior Editor-in-Chief: Megh R. Goyal, PhD, PE

Volume 1: Sustainable Micro Irrigation: Principles and Practices
Volume 2: Sustainable Practices in Surface and Subsurface Micro Irrigation
Volume 3: Sustainable Micro Irrigation Management for Trees and Vines
Volume 4: Management, Performance, and Applications of Micro Irrigation Systems
Volume 5: Applications of Furrow and Micro Irrigation in Arid and Semi-Arid Regions
Volume 6: Best Management Practices for Drip Irrigated Crops
Volume 7: Closed Circuit Micro Irrigation Design: Theory and Applications
Volume 8: Wastewater Management for Irrigation: Principles and Practices
Volume 9: Water and Fertigation Management in Micro Irrigation
Volume 10: Innovation in Micro Irrigation Technology

Book Series: Innovations and Challenges in Micro Irrigation
Senior Editor-in-Chief: Megh R. Goyal, PhD, PE

- Micro Irrigation Engineering for Horticultural Crops: Policy Options, Scheduling and Design
- Micro Irrigation Management: Technological Advances and Their Applications
- Micro Irrigation Scheduling and Practices

- Performance Evaluation of Micro Irrigation Management: Principles and Practices
- Potential of Solar Energy and Emerging Technologies in Sustainable Micro Irrigation
- Principles and Management of Clogging in Micro Irrigation
- Sustainable Micro Irrigation Design Systems for Agricultural Crops: Methods and Practices
- Engineering Interventions in Sustainable Trickle Irrigation: Water Requirements, Uniformity, Fertigation, and Crop Performance

Book Series: Innovations in Agricultural & Biological Engineering
Senior Editor-in-Chief: Megh R. Goyal, PhD, PE

- Dairy Engineering: Advanced Technologies and Their Applications
- Developing Technologies in Food Science: Status, Applications, and Challenges
- Engineering Interventions in Agricultural Processing
- Engineering Practices for Agricultural Production and Water Conservation: An Inter-disciplinary Approach
- Emerging Technologies in Agricultural Engineering
- Flood Assessment: Modeling and Parameterization
- Food Engineering: Emerging Issues, Modeling, and Applications
- Food Process Engineering: Emerging Trends in Research and Their Applications
- Food Technology: Applied Research and Production Techniques
- Modeling Methods and Practices in Soil and Water Engineering
- Processing Technologies for Milk and Dairy Products: Methods Application and Energy Usage
- Soil and Water Engineering: Principles and Applications of Modeling
- Soil Salinity Management in Agriculture: Technological Advances and Applications
- Technological Interventions in the Processing of Fruits and Vegetables
- Technological Interventions in Management of Irrigated Agriculture
- Engineering Interventions in Foods and Plants
- Technological Interventions in Dairy Science: Innovative Approaches in Processing, Preservation, and Analysis of Milk Products

- Novel Dairy Processing Technologies: Techniques, Management, and Energy Conservation
- Sustainable Biological Systems for Agriculture: Emerging Issues in Nanotechnology, Biofertilizers, Wastewater, and Farm Machines
- State-of-the-Art Technologies in Food Science: Human Health, Emerging Issues and Specialty Topics
- Scientific and Technical Terms in Bioengineering and Biological Engineering
- Engineering Practices for Management of Soil Salinity: Agricultural, Physiological, and Adaptive Approaches
- Processing of Fruits and Vegetables: From Farm to Fork
- Technological Processes for Marine Foods, from Water to Fork: Bioactive Compounds, Industrial Applications, and Genomics

CONTENTS

About the Lead Editor: S. K. Gupta..*xi*

About the Senior Editor-in-Chief: Megh R. Goyal ....*xiii*

About the Editor: Anshuman Singh... *xv*

List of Contributors..*xvii*

List of Abbreviations ....*xxi*

Foreword 1 by K. P. Singh .... *xxix*

Foreword 2 by S. K. Chaudhari .... *xxxi*

Preface 1 by S. K. Gupta ....*xxxiii*

Preface 2 by Megh R. Goyal .... *xxxvii*

Preface 3 by Anshuman Singh .... *xxxix*

PART I: Management of Saline/Sodic Stress: Field Practices 1

**1. Nomenclature and Reclamation of Sodic (Alkali)
Soils Using Gypsum: A Review of Historical Perspective** 3

S. K. Gupta

**2. Soil Salinity Management in Fruit Crops: A Review of
Options and Challenges**..39

Anshuman Singh, D. K. Sharma, Raj Kumar, Ashwani Kumar,
R. K. Yadav, and S. K. Gupta

**3. Role of Conservation Agriculture in Mitigating Soil Salinity
in Indo-Gangetic Plains of India**...87

Ajay Kumar Mishra, Athuman Juma Mahinda, Hitoshi Shinjo, Mangi Lal Jat,
Anshuman Singh, and Shinya Funakawa

**PART II: Physiological and Molecular Innovations to
Enhance Salt Tolerance** 115

**4. Physiological and Biochemical Changes in Plants Under Soil
Salinity Stress: A Review** 117

Vikramjit Kaur Zhawar and Kamaljit Kaur

5.   Biochemical, Physiological, and Molecular Approaches for
     Improving Salt Tolerance in Crop Plants: A Review 159

     Archana Singh, Veda Krishnan, Vinutha Thimmegowda, and Suresh Kumar

6.   Genomics Technologies for Improving Salt Tolerance in Wheat 209

     Amit Kumar Singh, Rakesh Singh, Rajesh Kumar, Shiksha Chaurasia, Sheel Yadav,
     Sundeep Kumar, and Dharmmaprakash P. Wankhede

PART III: Adaptations and Screening of Plants Under Water
Logging and Salinity Stresses 255

7.   Morpho-Biochemical and Molecular Markers for Screening and
     Assessing Plant Response to Salinity 257

     Md. Nasim Ali, Lucina Yeasmin, Vibha Singh, and Bhaswati Ghosh

8.   Plants Under Waterlogged Conditions: An Overview 285

     Anuj Kumar Singh, Pandurangam Vijai, and J. P. Srivastava

PART IV: Non-Conventional and High-Value Crops for
Salt-Affected Lands 327

9.   Potential and Role of Halophyte Crops in Saline Environments...........329

     Ashwani Kumar, Anita Mann, Arvind Kumar, Sarita Devi,
     and Prabodh Chander Sharma

10. Approaches for Enhancing Salt Tolerance in Seed Spices 367

     Arvind Kumar Verma, Anshuman Singh, Rameshwar Lal Meena, and Balraj Singh

Index...403

ABOUT THE LEAD EDITOR: S. K. GUPTA

 S. K. Gupta, PhD, is an agricultural/civil engineer with specialization in soil and water resources engineering. He is an INAE Distinguished Professor at the Central Soil Salinity Research Institute, Karnal, Haryana, India. He has served as Head of the Division of Drainage and Water Management; Head of the Division of Irrigation and Drainage Engineering; Head of the Indo-Dutch Network Project, and Project Coordinator of the All India Network Project on Use of Saline Water in Agriculture at CSSRI, Karnal, India. He was also a visiting scientist at the US Salinity Laboratory, Riverside, California and the International Institute for Land Reclamation and Improvement, The Netherlands.

Dr. Gupta has published about 130 peer-reviewed papers in various international and national journals and has also published 100 technical, popular articles, book chapters, and bulletins. In addition, he has published 15 books and written an unpublished manual.

Dr. Gupta is currently the Chief Editor of the *Journal of Water Management* and has been on the editorial boards of the *Journal of Agricultural Engineering* and the *Indian Journal of Soil Salinity and Water Quality.* The Indian Council of Agricultural Research (ICAR) awarded Dr. Gupta its most prestigious Rafi Ahmad Kidwai Award in 2000. In addition, he has received many other awards from many other academies, societies, and national organizations, including a Gold Medal each from the Indian Society of Agricultural Engineers and Institution of Engineers (India). He is the Fellow of several organizations.

Dr. Gupta has acted as a consultant to the United Nations Development Programme through WAPCOS, the Ministry of Water Resources, and the Government of Ethiopia. He has been the chairman/member of a number of committees and has been the chairman of the committee to prepare BIS (Bureau of Indian Standards) standards on water quality as well as a member of course development and a contributor in agriculture and environment of the IG National Open University. Dr. Gupta has more than 40 years of research experience.

He obtained his BTech and Master's in Agricultural Engineering from Punjab Agricultural University (PAU), Ludhiana, Punjab, India. He earned his PhD in Civil Engineering from Jawaharlal Nehru Technological University, Hyderabad, Telangana, India.

Readers may contact him at drskg1949@yahoo.com.

ABOUT THE SENIOR EDITOR-IN-CHIEF: MEGH R. GOYAL

Megh R. Goyal, PhD, PE, is a Retired Professor in Agricultural and Biomedical Engineering from the General Engineering Department in the College of Engineering at the University of Puerto Rico–Mayaguez Campus; and Senior Acquisitions Editor and Senior Technical Editor-in-Chief in Agriculture and Biomedical Engineering for Apple Academic Press, Inc. He has worked as a Soil Conservation Inspector and as a Research Assistant at Haryana Agricultural University and Ohio State University.

During his professional career of 45 years, Dr. Goyal has received many prestigious awards and honors. He was the first agricultural engineer to receive the professional license in Agricultural Engineering in 1986 from the College of Engineers and Surveyors of Puerto Rico. In 2005, he was proclaimed as "Father of Irrigation Engineering in Puerto Rico for the Twentieth Century" by the American Society of Agricultural and Biological Engineers (ASABE), Puerto Rico Section, for his pioneering work on micro irrigation, evapotranspiration, agroclimatology, and soil and water engineering. The Water Technology Centre of Tamil Nadu Agricultural University in Coimbatore, India, recognized Dr. Goyal as one of the experts "who rendered meritorious service for the development of micro irrigation sector in India" by bestowing the Award of Outstanding Contribution in Micro Irrigation. This award was presented to Dr. Goyal during the inaugural session of the National Congress on "New Challenges and Advances in Sustainable Micro Irrigation" on March 1, 2017, held at Tamil Nadu Agricultural University. Dr. Goyal is slated to receive the Netafim Award for Advancements in Microirrigation: 2018 from the American Society of Agricultural Engineers at the ASABE International Meeting in August 2018.

A prolific author and editor, he has written more than 200 journal articles and textbooks and has edited over 55 books. He is the editor of several book series, including *Innovations in Agricultural & Biological Engineering, Innovations and Challenges in Micro Irrigation, and Research Advances in Sustainable Micro Irrigation.*

He received his BSc degree in engineering from Punjab Agricultural University, Ludhiana, India; his MSc and PhD degrees from Ohio State University, Columbus; and his Master of Divinity degree from Puerto Rico Evangelical Seminary, Hato Rey, Puerto Rico, USA.

Readers may contact him at goyalmegh@gmail.com.

ABOUT THE EDITOR: ANSHUMAN SINGH

 Anshuman Singh, PhD, is an agricultural scientist with specialization in fruit science (pomology). He is currently working as a scientist at the ICAR-Central Soil Salinity Research Institute, Karnal, Haryana, India. He has been actively engaged in research and extension activities. His research interests include understanding the physiological basis of salt tolerance in crops, salinity management techniques, and agro-biodiversity management. The main objective of his research is to develop technologies to green salt-affected soils and improve the salt tolerance of fruit and vegetable crops through varietal selection and agronomic manipulations.

Dr. Singh has to date published 19 research papers in refereed international and national journals. In addition, he has contributed seven book chapters; written 12 technical/popular articles; and brought out four e-publications. He has made several presentations at international and national conferences and has made significant editorial contributions to key R&D documents of the Central Soil Salinity Research Institute, such as annual reports, the *Salinity Newsletter*, and CSSRI Vision 2050.

Dr. Singh was awarded a University Silver Medal and the Dr. Kirtikar Memorial Gold Medal during graduation; and a Chancellor's Gold Medal and a Vice-Chancellor's Gold Medal during post-graduation. In recognition of his research in greening salty lands through horticulture, Dr. Singh received the Young Scientist Award of the Society for the Upliftment of Rural Economy, Varanasi, India. Recently, the Government of Australia bestowed on him the Australia Awards Fellowship for attending an international training program in Australia.

Readers may contact him at anshumaniari@gmail.com.

LIST OF CONTRIBUTORS

Md. Nasim Ali, PhD
Associate Professor, Department of Agricultural Biotechnology, Bidhan Chandra Krishi
Viswavidyalaya, Mohanpur, Nadia 741252, West Bengal, India. E-mail: nasimali2007@gmail.com

S. K. Chaudhari, PhD
Assistant Director General, (Soil and Water), Indian Council of Agricultural Research, KAB II,
New Delhi 110012, India. E-mail: surchaudhari@hotmail.com

Shiksha Chaurasia, MSc
Junior Research Fellow, ICAR-National Bureau of Plant Genetic Resources, New Delhi 110012, India.
E-mail: chaurasia.shiksha785@gmail.com

Sarita Devi, PhD
Assistant Scientist, Department of Botany and Plant Physiology, Chaudhary Charan Singh Haryana
Agricultural University, Hisar 125004, Haryana, India. E-mail: devisaritaa@gmail.com

Shinya Funakawa, PhD
Professor (Soil Science), Soil Science and Terrestrial Ecosystem Management Laboratory,
Graduate School of Global Environmental Studies, Kyoto University, Kyoto 606-8502, Japan.
E-mail: funakawa@kais.kyoto-u.ac.jp

Bhaswati Ghosh, MSc
School of Agriculture and Rural Development, IRDM Faculty Centre, Ramakrishna Mission
Vivekananda University, Ramakrishna Mission Ashrama, Narendrapur, Kolkata 700103, India.
E-mail: bhaswatig2@gmail.com

Megh R. Goyal, PhD, PE
Retired Professor in Agricultural and Biomedical Engineering, University of Puerto Rico—Mayaguez
Campus and Senior Technical Editor-in-Chief in Agriculture Sciences and Biomedical Engineering,
Apple Academic Press Inc., PO Box 86, Rincon, PR 00677, USA. E-mail: goyalmegh@gmail.com

S. K. Gupta, PhD
Independent Researcher, Formerly INAE Distinguished Professor, ICAR-Central Soil Salinity Research
Institute, Karnal 132001, Haryana, India. E-mail: drskg1949@yahoo.com

Mangi Lal Jat, PhD
Senior Cropping System Agronomist, CIMMYT-CCAFS South Asia Leader, International
Maize and Wheat Improvement Centre (CIMMYT), NASC Complex, Pusa 110012, New Delhi, India.
E-mail: jat_ml@yahoo.com

Kamaljit Kaur, PhD
Assistant Biochemist, Department of Biochemistry, College of Basic Sciences and Humanities,
Punjab Agricultural University, Ludhiana 141004, Punjab, India. E-mail: kamaljit_pau@pau.edu

Veda Krishnan, MSc
Scientist, Division of Biochemistry, ICAR-Indian Agricultural Research Institute, New Delhi 110012,
India. E-mail: vedabiochem@gmail.com

Arvind Kumar, PhD
Scientist, ICAR-Central Soil Salinity Research Institute, Karnal 132001, Haryana, India.
E-mail: singh.ak92@gmail.com

Ashwani Kumar, PhD
Scientist (Plant Physiology), ICAR-Central Soil Salinity Research Institute, Karnal 132001, Haryana, India. E-mail: Ashwani.Kumar1@icar.gov.in

Raj Kumar, PhD
Scientist (Fruit Science), ICAR-Central Soil Salinity Research Institute, Karnal 132001, Haryana, India. E-mail: rajhorticulture@gmail.com

Rajesh Kumar, PhD
Senior Scientist, ICAR-National Bureau of Plant Genetic Resources, New Delhi 110012, India. E-mail: kraj.pgr@gmail.com

Sundeep Kumar, PhD
Senior Scientist, ICAR-National Bureau of Plant Genetic Resources, New Delhi 110012, India. E-mail: Sundeep.Kumar@icar.gov.in

Suresh Kumar, PhD
USDA Norman E. Borlaug Fellow, Principal Scientist, Division of Biochemistry, ICAR-Indian Agricultural Research Institute, New Delhi 110012, India. E-mail: sureshkumar3_in@yahoo.co.uk

Athuman Juma Mahinda, MSc
Ph.D. Scholar, Soil Science and Terrestrial Ecosystem Management, Graduate School of Agriculture, Kyoto University, Kyoto 606-8502, Japan. E-mail: mahinda.juma.27u@st.kyoto-u.ac.jp

Anita Mann, PhD
Senior Scientist, ICAR-Central Soil Salinity Research Institute, Karnal 132001, Haryana. E-mail: Anita.Mann@icar.gov.in

Rameshwar Lal Meena, PhD
Principal Scientist (Agronomy), ICAR-Central Soil Salinity Research Institute, Karnal 132001, Haryana, India. E-mail: rlmeena69@gmail.com

Ajay Kumar Mishra, MSc
PhD Scholar, Soil Science and Terrestrial Ecosystem Management Laboratory, Graduate School of Global Environmental Studies, Kyoto University, Kyoto 606-8502, Japan. E-mail: akm8cest@gmail.com

D. K. Sharma, PhD
Emeritus Scientist, ICAR-Central Soil Salinity Research Institute, Regional Research Station, Lucknow, India. E-mail: dk.sharma@icar.gov.in

Prabodh Chander Sharma, PhD
Director, ICAR-Central Soil Salinity Research Institute, Karnal 132001, Haryana, India. E-mail: pcsharma.knl@gmail.com

Hitoshi Shinjo, PhD
Associate Professor, Soil Science and Terrestrial Ecosystem Management Laboratory, Graduate School of Global Environmental Studies, Kyoto University, Kyoto 606-8502, Japan

Amit Kumar Singh, PhD
Scientist, ICAR-National Bureau of Plant Genetic Resources, New Delhi 110012, India. E-mail: amit_singh79@yahoo.com

Anshuman Singh, PhD
Scientist (Fruit Science), ICAR-Central Soil Salinity Research Institute, Karnal 132001, Haryana, India. E-mail: anshumaniari@gmail.com

Anuj Kumar Singh, PhD
Assistant Professor (Plant Physiology), Department of Genetics and Plant Breeding,
C. P. College of Agriculture, S. D. Agricultural University, Sardarkrushinagar 385506, Gujarat, India.
E-mail: anujkumarsinghbhu1@gmail.com

Archana Singh, PhD
Senior Scientist, Division of Biochemistry, ICAR-Indian Agricultural Research Institute,
New Delhi 110012, India. E-mail: sarchana_biochem@iari.res.in

Balraj Singh, PhD
Vice-Chancellor, Agriculture University, Jodhpur 342304, Rajasthan, India.
E-mail: drbsingh2000@yahoo.com

K. P. Singh, PhD
Vice-Chancellor, CCS Haryana Agricultural University, Hisar 125004, Haryana, India.
E-mail: vchauhisar@gmail.com

Rakesh Singh, PhD
Principal Scientist, ICAR-National Bureau of Plant Genetic Resources, New Delhi 110012, India.
E-mail: singhnbpgr@yahoo.com

Vibha Singh, MSc
School of Agriculture and Rural Development, IRDM Faculty Centre, Ramakrishna Mission
Vivekananda University, Ramakrishna Mission Ashrama, Narendrapur, Kolkata 700103, India.
E-mail: svibha90@gmail.com

J. P. Srivastava, PhD
Professor, Department of Plant Physiology, Institute of Agricultural Sciences,
Banaras Hindu University, Varanasi 221005, Uttar Pradesh, India. E-mail: jpsbhu25@ yahoo.co.in

Vinutha Thimmegowda, PhD
Scientist, Division of Biochemistry, ICAR-Indian Agricultural Research Institute, New Delhi 110012,
India.E-mail: vinuthabiochem@gmail.com

Arvind Kumar Verma, PhD
Scientist, ICAR-National Research Centre on Seed Spices, Ajmer 305206, Rajasthan, India.
E-mail: arvindhort@gmail.com

Pandurangam Vijai, PhD
Assistant Professor, Department of Plant Physiology, Institute of Agricultural Sciences,
Banaras Hindu University, Varanasi 221005, Uttar Pradesh, India. E-mail: pvijaivenkat@gmail.com

Dharmmaprakash P. Wankhede, PhD
Scientist, ICAR-National Bureau of Plant Genetic Resources, New Delhi 110012, India.E-mail:wdhammaprakash@gmail.com

R. K. Yadav, PhD
Head and Principal Scientist, ICAR-Central Soil Salinity Research Institute, Karnal 132001,
Haryana, India. E-mail: rk.yadav@icar.gov.in

Sheel Yadav, MSc
Scientist, ICAR-National Bureau of Plant Genetic Resources, New Delhi 110012, India.
E-mail: sheel.y85@gmail.com

Lucina Yeasmin, MSc
School of Agriculture and Rural Development, IRDM Faculty Centre, Ramakrishna Mission
Vivekananda University, Ramakrishna Mission Ashrama, Narendrapur, Kolkata 700103,
West Bengal, India. E-mail: lucina1827@yahoo.com

Vikramjit Kaur Zhawar, PhD
Assistant Biochemist, Department of Biochemistry, College of Basic Sciences and Humanities, Punjab Agricultural University, Ludhiana 141004, Punjab, India. E-mail: vikram97jit@pau.edu

LIST OF ABBREVIATIONS

$^1O_2$	singlet oxygen
ABA	abscisic acid
ABC	ATP binding cassette
ABI	ABA-insensitive locus
ABRE	abscisic acid response element
ACBP	acyl CoA binding protein
ACC	1-aminocyclopropane-1-carboxylic acid
ACO-1	aminocyclopropane-1-carboxylic acid oxidase
ACS	1-aminocyclopropane carboxylic acid synthase
ACS-1	aminocyclopropane-1-carboxylic acid synthase
ADC	arginine decarboxylase
ADH	alcohol dehydrogenase
ADP	adenosine diphosphate
AFLP	amplified fragment length polymorphism
ALDH	aldehyde dehydrogenase
AM	arbuscular mycorrhiza
AM	association mapping
AMF	arbuscular mycorrhizal fungi
AOS	active oxygen species
APRI	alternate partial root-zone irrigation
APX	ascorbate peroxidase
AR	adventitious root
Asl	above sea level
ATE	arginylt-RNA-transferase
ATP	adenosine triphosphate
BADH	betaine aldehyde dehydrogenase
BCE	before Christian era
bgl	below ground level
bHLH	basic helix-loop-helix
BHU	Banaras Hindu University
BIS	Bureau of Indian Standards
BM	brown manuring
bp	base pair
BR	brassinosteroid
C	cytosine

CA	conservation agriculture
Ca^{2+}	calcium divalent cation
CAAS, China	Chinese Academy of Agricultural Sciences
CAM	calmodulin
cAMP	cyclic AMP
CAMTA	calmodulin-binding transcription activator
CAT	catalase
CAX/CHX	Ca^{2+} exchanger or Ca^{2+}/H^+ exchanger
CBL	calceneurin B-like protein
CD	crop diversification
CDR	cold and drought regulatory
CEC	cation exchange capacity
CEL	cellulase
CG	cytosine–guanine
cGMP	cyclic GMP
CHG	cytosine-A, T or C-followed by G
Chl	chlorophyll
Chl *a*	chlorophyll *a*
Chl *b*	chlorophyll *b*
CIMMYT	International Maize and Wheat Improvement Centre
CIPK	CBL-interacting protein kinase
CIPK15	calcineurin B-like interacting protein kinase 15
Cl^-	chloride ion
CMO	choline monoxygenase
CRM	crop residue management
CSSRI	Central Soil Salinity Research Institute
CT	conservation tillage
CYP	cyclophilin
DA-NSCC	depolarization active-non-selective cation channels
DArT	diversity array technology
DDRT-PCR	differential display reverse transcription polymerase chain reaction
DHAR	dehydroascorbate reductase
DI	deficit irrigation
DMS	dimethyl sulfate
DNA	deoxyribonucleic acid
DREB	dehydration-responsive element binding
DSR	direct seeded rice
DTPA	diethylenetriaminepentaacetic acid
E	transpiration

EA	ethanolamine
EC	electrical conductivity
EC_0	salinity of the soil saturation extract at which yield is zero
EC_e	electrical conductivity of soil saturation extract
EC_{IW}	irrigation water salinity
EC_t	threshold electrical conductivity of the soil saturation extract
ein	ethylene insensitive mutant
EL	electrolyte leakage
ERF	ethylene response-factor
ESP	exchangeable sodium percentage
EST	expressed sequenced tags
ET_c	crop evapotranspiration
etr	ethylene receptor mutant
FAD	flavin adenine dinucleotide
FAI	Fertilizer Association of India
FAO	Food and Agriculture Organization
FCI	Fertilizer Corporation of India
FPRI	fixed partial root-zone irrigation
FYM	farm yard manure
G	guanine
G_{50}	germination 50%
G6PDH	glucose-6-phosphate dehydrogenase
GA/GA_3	gibberellic acid
GABA	gamma aminobutyric acid
GalUR	D-galacturonic acid reductase
GBS	genotyping by sequencing
GDH	glutamate dehydrogenase
GEBV	genomic breeding values
GHG	greenhouse gases
GPOX	guaiacol peroxidase
GPX	glutathione peroxidase
GR	glutathione reductase
Gs	stomatal conductance
GSA	glutamate-semialdehyde
GSH	reduced glutathione
GSS	genome survey sequence
GSSG	oxidized glutathione
GST	glutathione-S-transferase

GWA	genome-wide association
GWS	genome-wide selection
H	A, T, or C; except G
H^+-PPases	H^+-pyrophosphatases
H3K9ac	histone 3, lysine 9 is acetylated
H3K9me2	histone 3, lysine 9 is dimethylated
HA-NSCC	hyperpolarization active-nonselective cation channels
HKT	high affinity K^+ transporters or Na^+-influx transporter
HLRDC	Haryana Land Reclamation and Development Corporation
HMG-CoA synthase	hydroxymethylglutaryl-CoA synthase
HNE	4-hydroxy-2-nonenal
HyPRP	hybrid-proline-rich protein
IAA	indole acetic acid
IBA	indole butyric acid
ICAR	Indian Council of Agricultural Research
ICGAR	Institute of Crop Germplasm Resources
ICP/MS	inductively coupled plasma–mass spectrometry
ICP/OES	inductively coupled plasma–optical emission spectrometry
ICRISAT	International Crops Research Institute for the Semi-Arid Tropics
IGP	Indo-Gangetic plain
ISH	index of salt harm
ISSR	inter simple sequence repeat
ITMI	International Triticeae Mapping Initiative
JA	jasmonic acid
KFC	Karnal fodder crop
KK	kaolinite
LEA	late embryogenesis abundant
LOES	low oxygen escape strategy
LOQS	low oxygen quiescent strategy
LP	lipid peroxidation
LSD	Least square difference
MABC	marker-assisted backcrossing
MAPK	mitogen-activated protein kinase
MAS	marker-assisted selection
mbgl	meter below ground level
MDA	malondialdehyde
MDHAR	monodehydroascorbate reductase

miRNA	microRNA
MoA	Ministry of Agriculture
MPK3	mitogen-activated protein kinases
MSAP	*methylation-sensitive arbitrarily primed PCR*
MT	million tons
MYB	myeloblastosis
MYB2	a member of the MYB (myeloblastosis) family of transcription factors
Na^+	sodium ion
NAA	neutron activation analysis
$NADP^+$	nicotinamide adenine dinucleotide phosphate
NADPH	nicotinamide adenine dinucleotide phosphate *(a reduced form of NADP$^+$).*
Nax 1	sodium exclusion locus 1
NBPGR	National Bureau of Plant Genetic Resources
NCA	National Commission on Agriculture
NGS	next-generation sequencing
NHB	National Horticulture Board (India)
NHX	vacuolar Na^+/H^+ antiporter or Na^+/H^+ exchanger
NIL	near isogenic line
NIT1	nitrilase1
NIT2	nitrilase2
NO	nitric oxide
noa/nos	nitric oxide deficient mutant
NSCC	nonselective cation channel
NSGC	national small grains collection
$O_2^{\cdot-}$	superoxide radical
OAT	ornithine-delta-aminotransferase
ODC	ornithine decarboxylase
$^{\cdot}OH$	hydroxyl radical
$OH^{\cdot}$	hydroxyl radical
P5C	pyrroline-5-carboxylate
P5CDH	P5C dehydrogenase
P5CR	P5C reductase
P5CS	Δ1-pyrroline-5-carboxylate synthetase
P-EA	phosphoethanolamine
PA	polyamine
PAH	polycyclic aromatic hydrocarbon
PAL	phenylalanine ammonia-lyase
PAU	Punjab Agricultural University

PBZ	paclobutrazol
PCD	programed cell death
PcMIPS	myoinositol phosphate synthase
PCO_2	partial pressure of carbon dioxide
PCP	pentachlorophenol
PDC	pyruvate decarboxylase
PDH	proline dehydrogenase
PFK	phosphofructokinase
PFP	pyrophosphate: fructose-6-phosphate-1-phophotransferase
PGPR	plant growth promoting rhizobacteria
PGR	plant growth regulator
pH	potency of hydrogen
pH_2	pH of the 1:2 soil water solution
pH_s	pH of the soil saturation paste
PLRDC	Punjab Land Reclamation and Development Corporation
PM	plasma membrane
P_N	net photosynthesis
PO	proline oxidase
POX/POD	peroxidase
ppm	parts per million
PPO	polyphenol oxidase
PRB	permanent raised bed
PRD	partial root-zone drying
PS-I	photosystem I
PS-II	photosystem-II
Ptd-EA	phosphatidyl ethanolamine
Put	putrescine
QTLs	quantitative trait loci
RAPD	random amplification of polymorphic DNA
RBOH	respiratory burst oxidase homolog
rcn1	reduced culm number 1
RdDM	RNA-directed DNA methylation
RDI	regulated deficit irrigation
RES	reactive electrophilic species
RFLP	restriction fragment length polymorphism
RH	residue harvested
RIL	recombinant inbred line
RM	residue management

RNA	ribonucleic acid
RNAi	RNA interference
RNA-seq	RNA sequencing
ROL	radial oxygen loss
ROS	reactive oxygen species
RR	residue retained
RSC	residual sodium carbonate
RSHR	relative salt harm rate
RuBISCO	ribulose-1,5-bisphosphate carboxylase/oxygenase
RuBP	ribulose-1,5-bisphosphate
RUE	resource use efficiency
RW	rice–wheat
RWC	relative water content
S	slope, % decline in yield per unit increase in salinity beyond the threshold
SA	salicylic acid
SAGE	serial amplification of gene amplification
SAM	S-adenosyl methionine
SAMDC	S-adenosylmethionine decarboxylase
SAMS	S-adenosylmethionine synthetase
SAR	sodium absorption ratio
SAS	salt-affected soil
SBI	Spices Board India
SDI	sustained deficit irrigation
SES	salt evaluation score
siRNA	small interfering RNA
SM	smectite
SNP	sodium nitroprusside
SNP	single nucleotide polymorphism
SnRK1	sucrose nonfermenting 1-related kinase 1
SOC	soil organic carbon
SOD	superoxide dismutase
SOM	soil organic matter
SOS	salt overly sensitive
SOS1	salt overly sensitive 1
Spd	spermidine
SPDS	spermidine synthase
Spm	spermine
SPMS	spermine synthase
SSDI	subsurface drip irrigation

SSR	simple sequence repeat
SSRG	salt stress related gene
STC	salt tolerant cultivar
T	thymine
TaSR	*Triticum aestivum* salt responsive
TC	total carbon
TCA	tricarboxylic acid
TGA	total geographical area
TOR	target of rapamycin
TPC	two-pore channel
TSS	total soluble solid
U	uracil
UPBSN	UP Bhumi Sudhar Nigam
USAID	United States Agency for International Development
USDA	United States Department of Agriculture
v/v	volume by volume
V-ATPase	vacuolar type H^+-ATPase
V-PPase	vacuolar pyrophosphatase
VI-NSCC	voltage insensitive-nonselective cation channel
vtc	vitamin C deficient mutant
vte	vitamin E deficient mutant
WSA	water stable aggregate
XET	xyloglucan endotransglycosylase
ZT	zero tillage
ZT-B	zero tillage bed planting
ZT-F	zero tillage flat planting

FOREWORD 1 BY K. P. SINGH

Problems of water logging and soil/water salinity are known to mankind since time immemorial. Nonetheless, salt-induced land and water degradation has attained economic and environmental overtones in many regions around the globe only recently. Severe recuperation of crop productivity in about 20% of the global irrigated lands due to water logging and soil salinity has raised alarm bells. Similarly, a significant proportion of arable lands in rain-fed areas yields poorly due to the intertwined stresses of salinity, water scarcity, and poor soil fertility. Extreme climate variability and its attendant consequences—namely, reduced/increased water flows in perennial rivers, snow melt, intense rainfall, heavy surface run-off, and groundwater depletion—do not auger well for sustainable soil health and are likely to pose huge risks. Forecasts reveal that such climatic aberrations may even lead to upscaling of the salinity problems in many cases. It is projected that under such a scenario, food and nutritional security of an ever-increasing global population may be in jeopardy. Since agricultural expansion into new areas is virtually impossible, increasing global attention is needed to tap the production potential of degraded lands to the highest possible level.

In this scenario, I am of the view that no single prescription to manage these adverse problems in various agro-ecologies will yield desirable results. Integration of technologies based on engineering, hydrological, chemical, and biological interventions will have to be adopted to boost the crop yields in saline and sodic soils.

I am happy to note that the current volume titled *Engineering Practices for Management of Soil Salinity: Agricultural, Physiological, and Adaptive Approaches* includes contributions from experts of various disciplines to give the readers a fair view of integrated approach of reclaiming and managing the salt-affected environment for optimum crop production. While conventional chemical and hydro-technical interventions have withstood the tests and helped to increase food production around the globe, current criticism of these technologies stems from being environmentally unsustainable because of having very high water and environmental footprints. It has resulted in growing interest in salt tolerant crops, which provide several tangible and intangible benefits that include reduced amendment requirements, leaching requirements, consistent and stable yields, and decreased reclamation costs. This has led the plant scientists to unravel the mystery of plant response and adaptations through understanding the morphological, biochemical, physiological, and molecular changes and mechanisms involved in these changes when plants encounter salt stress.

Consistent with this global trend, the editors have devoted a large portion of the publication to describing the recent innovations in screening plant populations, genetic manipulations, and improvement techniques to develop salt tolerant cultivars in a cost-efficient and time-bound manner. Since bioreclamation and saline agriculture is the buzzword and a topical subject in the environment sensitive societies, cultivation of halophytes is also included in one of the chapters. To overcome nutritional security concerns, field practices, including selection of crops and varieties and other technological interventions, are described for the commercial cultivation of high value crops such as fruits and seed spices.

I understand that some of the chapters, besides dealing with firmed up technologies, report the current research trends that are in the initial concept stages. These inclusions provide food for thought to the young researchers pursuing their postgraduation, as undoubtedly these concepts have the potential of being developed into low cost, doable technologies in a foreseeable future.

I congratulate the authors and editors for their commendable efforts in compiling the current innovations and advanced knowledge and techniques on diverse aspects of agricultural salinity management. I am convinced that the technological concepts and breakthroughs brought out in this volume on diverse technologies will help to increase crop productivity and bridge the gap between the growing food and nutritional needs and availability. In addition, people working in the management of saline soils and water will have a fair view of the current knowledge on these subjects, and young researchers will be able to develop their future line of work plans in basic and/or applied lines in diverse subjects of engineering, soil science, agronomy, plant sciences, genetic engineering, and biochemistry.

I am sure that this publication will adorn the shelves of field practitioners, researchers, and policy planners. I am glad to acknowledge Dr. Megh R. Goyal, Senior Editor-in-Chief, who was lecturer/research assistant (Agricultural Engineering) at our university during 1972–1975, and who perfected the acid delinting technology.

K. P. Singh, PhD
Vice-Chancellor
CCS Haryana Agricultural University
Hisar 125004, Haryana, India
Phone: +91-1662-231640
E-mail: vchauhisar@gmail.com

FOREWORD 2 BY S. K. CHAUDHARI

Soil salinity is an incipient problem that tends to exacerbate in irrigated soils lacking adequate drainage. Current catastrophic environmental degradation reported in the Aral Sea basin is a grim reminder of past events where cities and even civilizations vanished because of inappropriate water management and the consequent alarming increase in soil salinization. For example, in 2030 B.C., Mashkan-shapir, a flourishing Mesopotamian city 20 km from the Tigris River, was wiped out as the city failed to support the food needs and livelihood of the people. People were forced to abandon crop production as agricultural fields became essentially useless and very high salinity rendered them uncultivable (https:// www.learner.org/exhibits/collapse/mesopotamia. html). It is a matter of concern that in spite of our current knowledge on the subject that should help to safeguard us from such events, village after village is succumbing to secondary salinity. It is mainly because we are failing in sharing the knowledge, resulting in half-hearted implementation of otherwise potential salinity management technologies.

I am happy to say that we have well-tested technologies that are capable of reversing the land degradation processes. In this category, the hydro-technical interventions (drainage and leaching), chemical technologies (application of amendments notably gypsum), biological reclamation (through salt tolerant grasses, perennial trees and halophytes), and cultivation of salt tolerant crop varieties are worth mentioning. Indian experience has shown that by implementing these interventions, it is possible to obtain higher food grain yields (up to 10 t ha^{-1}) even from the barren salt-affected soils. In addition to tangible economic benefits, such interventions are also helpful in resolving many socioeconomic and environmental problems.

This volume, *Engineering Practices for Management of Soil Salinity: Agricultural, Physiological, and Adaptive Approaches*, provides a glimpse of some of these technologies with emphasis on recent physiological and molecular advances in germplasm screening and modification for sustained crop adaptation to salty soils without any appreciable yield penalty. It seems consistent with the current global trends where large numbers of researchers are trying to develop farmer-friendly biological means of salt stress management through frontier genetic tools. It is my belief that such plant-based solutions will allow us to live with the salts rather than back-breaking efforts required to remove the salts from the system. This approach is gathering momentum as conventional measures to reclaim and cultivate saline and sodic soils are increasingly becoming uneconomical because of resource constraints and are environmentally unsustainable because of the disposal problems.

This publication includes 10 chapters contributed by experts in their field of specialization. Since the chapters cover a wide spectrum of topics, the editors have wisely categorized them into four groups relevant to the subject matter of the publication. I commend the efforts of the authors and editors in preparing this informative and useful publication that comprehensively deals with conventional and the upcoming technologies, including basic research in molecular breeding and genomic approaches. The later studies may prove to be the stepping stone in the development of salt tolerant crop varieties.

Soil salinization as well as its management is dynamic in nature. Many new challenges are emerging among which reclamation of wastewaters and contaminated land and water resources takes the top spot. Impending climate change is likely to alter the whole scenario of land and water salinity. Although some initial guidelines for addressing these challenges emerge from this publication, I would suggest that the editors and publisher may join hands in identifying the current researchers in these areas and bring out a companion publication that addresses these emerging challenges. Undoubtedly, management of saline soils and waters continues and would continue to fascinate the researchers around the globe since it seems to be our best bet to ensure food and nutritional security of the ever-increasing global population and resolve serious socioeconomic and environmental issues, especially in underdeveloped and developing countries.

I once again compliment the authors and editors for preparing this publication that addresses a very vital issue of great significance. I hope that this publication will soon be on the shelves of policy planners, field practitioners, researchers, and students engaged in this field.

S. K. Chaudhari, PhD
Assistant Director General (Soil and Water)
Indian Council of Agricultural Research, KAB II
New Delhi 110012, India
Mobile: +91-9729559063
E-mail: surchaudhari@hotmail.com

PREFACE 1

Abiotic stresses have emerged as one amongst the many stressors that are inflicting a heavy penalty on the agricultural production potential of the soils around the globe. These stresses are categorized as physical (erosion, drought, water logging, surface crusting and sealing, shallow soils and impeded drainage, etc.), chemical (salinity, acidity, toxic contaminants, nutrient deficiencies resulting from nutrient mining, and calcareous or gypsiferous soils), biological (diminished biodiversity and low/high organic matter), and climatic (high or low temperatures, global warming and resulting aberrations in rainfall and extreme climatic regimes). The list, of course, is not exhaustive as many new challenges are noticed under different agro-climatic and ecological conditions. Amongst these, chemical degradations in the form of soil salinity, alkalinity, acidity, and toxic elements, either resulting from natural or anthropogenic reasons, constitute a formidable challenge to increasing production and productivity at the global level in general and in India in particular.

Historically, research to reclaim and manage chemically degraded soils began much before the basic principles of formation of salt-affected soils and ion exchange principles were fully understood. As such, not much headway could be made to reclaim these soils until the end of the 19th century and the beginning of the 20th century. In the later period, many practical solutions and interventions have been developed to manage saline/sodic environments based on a reasonable mix of chemical, engineering, hydrological, agronomic and biological interventions. As such, we have now several technically and economically feasible well-tested technologies.

On the other hand, plant scientists have tried to unravel the mystery of differential plant response to salt stress to understand how plants respond and adapt to salinity stress. Now it is known that it involves complex physiological traits, metabolic pathways, and molecular or gene networks. We have made reasonably good headway in this field, yet being multi-trait mechanisms, we are still far from tying the knots, and end results are sometimes inconsistent. In addition, new challenges in the form of ionic toxicity and plant response to ionic imbalances are still in the realm of theory and perception. A comprehensive approach on this line will not only be able to develop low-cost doable technologies but will also lead to the development of salt-tolerant varieties of plants.

This volume, *Engineering Practices for Management of Soil Salinity: Agricultural, Physiological, and Adaptive Approaches*, is a companion volume to the earlier published book volume titled, *Soil Salinity Management in Agriculture: Technological Advances and Applications*, by Apple Academic Press Inc. As a stand alone volume, it comprehensively deals with the management issues of water logging and soil salinity often termed as "twin problems." The volume contains ten chapters covering reviews on both basic and applied researches in this field. The ten chapters have been grouped under four different subgroups, namely,

i.   Management of Saline/Sodic Stress: Field Practices
ii. Physiological and Molecular Innovations to Enhance Salt Tolerance
iii. Adaptations and Screening of Plants under Water Logging and Salinity Stresses
iv. Non-Conventional and High-Value Crops for Salt-Affected Lands

In the first group, three chapters are included. The two chapters deal with field practices to manage sodic (alkali) stress for optimum cereal and horticultural crops production. The innovative technologies of reclamation encompass use of chemical amendments for field crops and horticultural plants (spot reclamation), crop selection and management, soil management, irrigation water management including irrigation methods, nutrient management, and rainwater management. A combination of these practices helps to blunt the adverse impact of salt stress on plants. The third chapter on conservation agriculture (CA) outlines the practices under CA and salinity management to argue that CA practices aid the post-reclamation salinity management efforts.

The second group of papers provides comprehensive review to understand the morphological, biochemical, physiological, and molecular changes and mechanisms involved in these changes when plants encounter salt stress. While two of the chapters are general in nature reviewing the works for several crops, one of the chapters is crop specific dealing with the cereal crop wheat. Role of these processes and molecular marker including genetic interventions and exogenous application of chemicals for improving salinity tolerance in crop plants is included. These are being flagged as the current hope in fighting the menaces of water logging and soil salinity. Since the salinity problem is expanding globally, role of the genomic approaches to breed high yielding salt tolerant crops to meet the food and nutritional requirement of burgeoning global population is vividly brought out.

Water logging and soil salinity are considered twin problems. With the spread of irrigated areas in arid and semiarid regions, these twin problems are often experienced together compounding the problems and posing management challenges. In many cases, salinity appears to be the main controlling factor in deciding the yield levels. Nonetheless, two papers in this group deal with plant's adaptations and screening under water logging and salinity stresses, giving an insight into as to how the plants adapt to these problems. Major emphasis is on discussing screening protocols under these situations being a prerequisite for any hope of designing successful plant breeding programs.

The last section includes two papers that describe the cropping of halophytic plants and low volume—high value seed spices in salt-affected environment. In the first chapter, the mechanisms of salt tolerance in halophytic plant vis-à-vis glycophytes are included to conclude that halophytes are the future hope to manage salt-affected soils and water. The paper raises the issues of developing agronomy of halophytic plants as well as marketing of the products that need resolution before cultivation of halophytes and saline agriculture become an attractive proposition. The second chapter deals with cultivation of a group of seed spices crops in saline environment. Besides discussing the crop response to stressed environment, nonstructural interventions such as seed priming, nutrient management, use of microbial inoculants, and application of plant growth regulators and bio-stimulants are suggested to blunt the adverse effects of salts on crop productivity.

I believe that impending climate change will have a spiraling effect on water logging and soil salinity scenario around the globe. Sustainability of irrigated agriculture also hinges upon how best we are able to deal with the problems of water logging, salinity of soils and waters, and droughts. Since the book deals with many facets of water logging and salinity management, I hope that the book will be immensely useful to researchers, teachers, and students alike who are involved or are likely to be involved in any manner to manage the crop production in saline environment in the era of climate change. I would appreciate receiving any suggestion(s) from the readers that may help me in improving any forthcoming publication as and when planned.

The chapters included in this publication have been written by experts in their field of specialization. We, the editors, thank all the authors for agreeing to contribute the chapters and for providing manuscripts in time in spite of their preoccupation. We also appreciate these authors for providing unstinted support on priority while the chapters were being edited. It is because of their support that the book could be prepared in a time-bound manner. I also take this opportunity to thank Dr. Megh R. Goyal for meticulous handling of the editing of the chapters and providing suggestions on a day-to-day basis. His learned suggestions helped me to improve the quality of chapters as well as in putting each chapter in proper format. I also thank my coeditor Dr. Anshuman Singh who carried out independent editing of most of the chapters and prepared first draft of many preliminary documents included in the book that considerably reduced my work load. His suggestions were quite forthcoming and useful. I also thank all the members of the editorial staff who have been associated in the editing of this book. In the end, I thank my family especially my wife Sarita for bearing with me in spite of the fact that I broke my promise of sparing whole time to them/her after my superannuation.

—S. K. Gupta, PhD
Lead Editor

PREFACE 2

According to https://en.wikipedia.org/wiki/Soil_salinity, "*The consequences of salinity are: detrimental effects on plant growth and yield; damage to infrastructure (roads, bricks, corrosion of pipes and cables); reduction of water quality for users, sedimentation problems; soil erosion ultimately, when crops are too strongly affected by the amounts of salts. Salinity is an important land degradation problem. Soil salinity can be reduced by leaching soluble salts out of soil with excess irrigation water. Soil salinity control involves water table control and flushing in combination with tile drainage or another form of subsurface drainage. A comprehensive treatment of soil salinity is available from the United Nations Food and Agriculture Organization. High levels of soil salinity can be tolerated if salt-tolerant plants are grown. Sensitive crops lose their vigor already in slightly saline soils, most crops are negatively affected by (moderately) saline soils, and only salinity resistant crops thrive in severely saline soils.*"

One can download free the LeachMod software for simulating leaching of saline irrigated soil from http://www.waterlog.info/leachmod.htm. LeachMod is designed to simulate the depth of the water table and the soil salinity in irrigated areas with a time step as selected by the user (from 1 day to 1 year). The program uses small time steps in its calculations for a better accuracy. In case of a leaching experiment with measured soil salinities, LeachMod can automatically optimize the leaching efficiency by minimizing the sum of the squares of the differences between measured and simulated salinities. The root zone can consist of one, two, or three layers. LeachMod allows the introduction of a subsurface drainage system in a transition zone between root zone and aquifer, and subsequently it determines the drain discharge. When the irrigation/rainfall is scarce and the water table is shallow, LeachMod will calculate the capillary rise and reduce the potential evapotranspiration to an actual evapotranspiration. LeachMod can also take into account upward seepage from the aquifer or downward flow into it. The latter flow is also called natural subsurface drainage. This model is somewhat similar to Salt-Calc. On the one hand, the water management options are fewer (e.g., reuse of drainage or well water for irrigation do not feature here), but the model is more modern in the sense that the variable input for each time step is given in a table so that the calculations over all the time steps are done in one step.

Moreover, by inserting the observed values of soil salinity in the data table, the model optimizes the leaching efficiency of the soil automatically. On July 11, 2015, LeachMod was updated to include more rigorous data checks.

I am a staunch supporter of preserving our natural resources. Importance of wise use of our natural resources has been taken up seriously by universities, institutes/centers, government agencies, and nongovernment agencies. I conclude that the agencies and departments in soil salinity management have contributed to the ocean of knowledge.

Our book also contributes to the ocean of knowledge on soil salinity management. Agricultural and biological engineers (ABEs) with expertise in this area work to better understand the complex mechanics of soil salinity. ABEs are experts in agricultural hydrology principles, such as controlling drainage, and they implement ways to control soil erosion and study the environmental effects of sediment on stream quality.

The mission of this book volume is to serve as a reference manual for graduate and undergraduate students of agricultural, biological, and civil engineering; horticulture, soil science, crop science, and agronomy. I hope that it will be a valuable reference for professionals who work with soil salinity management and for professional training institutes, technical agricultural centers, irrigation centers, agricultural extension service, and other agencies.

My classmate and longtime colleague, Dr. S. K. Gupta, joins me as a lead editor of this volume. Dr. Gupta holds exceptional professional qualities with his expertise in soil salinity management during the last 47 years, in addition to his work as a research scientist and Distinguished Professor at ICAR-Central Soil Salinity Research Institute (Zarifa Farm), Karnal, India. His contribution to the content and quality of this book has been invaluable. I also introduce my coeditor Dr. Anshuman Singh, who is a Scientist (Fruit Science) at the ICAR-Central Soil Salinity Research Institute, Karnal, India. Although I do not know Dr. Singh personally, I can testify to his arduous educational/research efforts in the soil salinity management in fruit science.

I express my deep admiration to my family for understanding and collaboration during the preparation of this book. Our Almighty God, owner of natural resources, must be very happy on the publication of this book. As an educator, there is a piece of advice to one and all in the world: *"Permit that our almighty God, our Creator and excellent Teacher, help us to solve and manage problems in soil salinity with His Grace, because our life must continue trickling on ... and Get married to your profession ...."*

—Megh R. Goyal, PhD, PE
Senior Editor-in-Chief

PREFACE 3

Presence of excess salts in the root zone is a severe constraint on agricultural productivity in several irrigated and rain-fed areas of the world. Severity of the problem is evident by the fact that salinization and the associated problem of water logging diminish crop productivity to varying extents in about 100 countries. Although trace amounts of soluble salts are invariably found in arable lands, certain natural factors such as strong aridity, high salt concentrations in the subsoil horizons and sea water ingress accelerate the process of salt accumulation resulting in reduced soil productivity. Nonetheless, salt-affected soils formed due to such natural causes (primary salinity) can be managed through conventional measures such as chemical amendments and salt leaching with fresh water. In contrast, strenuous efforts are required to deal with the problem of human-induced secondary salinity arising due to pervasive land use and excess irrigation. In many regions such as Indo-Gangetic Plains of India and the South-Western Australia important for the regional and global food security, secondary salinization has attained alarming proportions seriously affecting the land quality, farm productivity, infrastructure and peoples' livelihoods.

Despite the availability of potential reclamation technologies, continued salinization of new areas is worrisome. In addition, the problems of resalinization and resodification of the ameliorated saline and sodic soils are also increasingly becoming noticeable. Rising amendment costs and growing fresh water shortages are adversely impacting soil restoration projects in many parts of the world. Climate change impacts may further reduce fresh water availability with far reaching consequences for irrigated crop production and soil reclamation. In this context, affordable solutions for harnessing the productivity of degraded lands while arresting any further deterioration in land quality assume great significance. Several crops and cultivars possess inherent potential to endure the elevated salt levels in soil and water fueling the interest in developing salt tolerant cultivars in economically important crops. Nonetheless, release of salt tolerant cultivars ascribed to the complex genetic basis of salt tolerance remains rather slow. Availability of molecular and genomic tools can considerably aid to the ongoing efforts to develop such high yielding and salt tolerant genotypes. This coupled with continual refinements in the existing salinity management techniques could slowly pave the way for productivity enhancements in saline areas.

The present volume entitled "Engineering Practices for Management of Soil Salinity: Agricultural, Physiological, and Adaptive Approaches" intends to address such questions. Contributors have made their best efforts to coherently summarize the updated information pertaining to specific areas such that the reader gets a wholesome picture of issues and trends in global salinity research. A total of 10 chapters included in this volume have been grouped into four broad areas, that is, improved field practices, advanced physiological and molecular tools, efficient screening methods, and nonconventional crops for the productive utilization of salt-affected soils.

In the first chapter of Part I, the author has removed the ambiguities in the nomenclature of sodic soils. After a critical scrutiny of different issues and constraints in reclamation using gypsum, plausible guidelines for the sustainable management of degraded sodic soils have been delineated. The second chapter deals with the doable ways and means for profitable fruit growing in salt-affected soils. It has been shown that simple interventions such as selection of salt tolerant scion and rootstock genotypes, improved planting techniques, and innovative irrigation methods could enhance the economic value of salt-affected lands through high value fruit and vine crops. In the third chapter, multifarious advantages accruing from the adoption of resource conservation technologies are discussed. The authors propose to bring more crop acreage under conservation practices in the regions grappling with water scarcity, salinity, and natural resource degradation.

In order to overcome the limitations of conventional morphological descriptors, researchers are increasingly switching over to innovative physiological and molecular tools for the precise elucidation of salt stress regulation in plants. Part II of the book has accordingly been designed to shed light on recent advances in these fields. In addition to innovative physiobiochemical tools for the enhanced understanding of salinity response and adaptation, recent developments in genomic and genetic engineering approaches have been included that could immensely benefit plant breeders in the targeted utilization of genes and gene products for developing the salt tolerant cultivars.

Consistent with the fact that dual stresses of water logging and salinity need even greater efforts for management, two chapters in the third section of the book exclusively deal with the recent advances in screening techniques to expedite the identification of novel genetic stocks carrying the genes responsible for salt and water logging tolerance. Available evidence suggests that integrated application of morphological, physiological, and molecular techniques could accelerate the pace of breeding programs for evolving high yielding cultivars adapted to waterlogged saline lands.

Part IV of the book containing two chapters discusses the importance of nonconventional crops in the productive utilization and phytoremediation of salt-affected lands. A need has long been felt to grow halophytes as economic crops under extreme saline conditions. Their emerging food, fodder, and industrial usage coupled with the inherent salt removal efficiency are the reasons that have propelled interest in halophytes as promising crop resources for the highly deteriorated lands. Over the years, significant strides made in enhancing the salt tolerance of seed spices are now paying dividends. Notwithstanding these achievements, continual refinements in the conventional approaches and novel biological interventions have become absolutely essential for the future viability of seed spice cultivation in traditional areas that are likely to be hit hard by high salinity and climate change impacts in the foreseeable future.

Evidence is mounting that salinization could be a severe threat to global food, economic and environmental security in the coming decades. Climate change impacts, *inter alia*, erratic rainfall, drying and pollution of surface water resources, high evaporative water loss, and fresh water scarcity will create severe obstacles to soil reclamation and management in salinity affected regions. Considering the projected doubling in food demand by 2050, onerous initiatives and programs are needed to raise the productivity of marginal lands while arresting the incipient salinization in the intensively cultivated tracts. Availability of superior cultivars can markedly enhance the crop yields in saline and sodic soils with the aid of other salinity management measures.

I am convinced that advances and breakthroughs reported by the experts in their respective fields will enhance the knowledge and understanding of researchers, academicians, policy makers, and land managers having a stake in the sustainable management of salt-affected lands. I sincerely thank all the contributors for sharing their knowledge and expertise in the form of meticulously drafted manuscripts. The wholehearted support and guidance from Dr. S. K. Gupta, Lead Editor and Prof. Megh R. Goyal, Senior Editor-in-Chief, is duly acknowledged. I am grateful to my wife Ritika Singh for her constant support during the preparation of this manuscript. I also express my gratitude to the Editorial Staff involved in the publication of this volume.

—Anshuman Singh, PhD
Editor

PART I
Management of Saline/Sodic Stress: Field Practices

NOMENCLATURE AND RECLAMATION OF SODIC (ALKALI) SOILS USING GYPSUM: A REVIEW OF HISTORICAL PERSPECTIVE

S. K. GUPTA

ABSTRACT

On a global scale, sodic/alkali lands are spread over an area of 434 M ha, the affected area in India being 3.77 M ha. The current chapter highlights that from reclamation and management points of view there is a need to modify the nomenclature criteria of salt-affected soils. Sodic soils can be grouped under four categories, namely, waterlogged, non-waterlogged, semi-reclaimed and acid-sodic soils. On the other hand, saline soils can be grouped under waterlogged, non-waterlogged, and high sodium absorption ratio saline soils. Until some unanimity emerges, terms sodic and alkali soils may continue to be used synonymously to designate soils having high exchangeable sodium percentage. The use of phosphogypsum (PG) to reclaim sodic lands is likely to increase in future as a result of reduced availability of mineral gypsum. Nonetheless, it may require resolution of several environmental concerns related to widespread use of PG. The chapter includes the salient features of a reclamation and management package of sodic lands. Besides adding appropriate amendment, the package includes integrated operation of management components related to crops selection and agronomic practices, soils, application of irrigation water, changes in plant nutrition strategies, and rainwater management. Socioeconomic issues, stakeholder's participation, and resodification of reclaimed sodic lands are discussed. The chapter also lists a number of management options to avoid or minimize resodification of reclaimed sodic lands.

1.1   INTRODUCTION

Salt-affected soils have been a part of the global land scape. These soils are adversely affecting the productivity of more than 900 million hectare (M ha) in more than 100 countries. Being the hub of irrigation activities, India is no exception. The salinity problem in India has been known since time immemorial as the archaeological evidences from the Indus Valley show that irrigation of agricultural fields was prevalent from the time of the Harappan culture (c. 2400–1750 BCE). However, problems of salt-affected soils came into prominence with recorded evidences only during the middle of the 19th century with the opening of several canal systems by the British. By 1880, some 10,680 km of canals (main and distributaries) existed in the Northwest Province including 6215 km in Punjab.[102]

Although irrigation is not the only cause of development of soil salinity in India or elsewhere as many non-irrigated lands are afflicted with this malice. Nonetheless, expansion of canal irrigation did exacerbate the problems and charted a parallel course with irrigation development. Therefore, soil salinity is often associated with the introduction of canal irrigation. It was in 1855 that a farmer from village Munak (currently in Haryana, India) in the western Yamuna canal command lodged a report of the problems of water logging and soil salinity nearly 15 years after the opening of the canal. By 1864, *usar* (*usar* lands have moderate to severe alkalinity) was well known in Faizabad, Behraich, Lucknow, Sitapur, and Hardoi of United Provinces in India, and more reports were pouring in with the Government of India. In 1876, another alarm was raised about the disastrous spread of *reh* in parts of Aligarh, Meerut, and throughout the Kali *Nadi* Valley of Uttar Pradesh. Through a petition, David Robarts of *pargana* Sikandra Rao, Aligarh district, requested the Board of Revenue for relief due to deterioration of his lands and appearance of saline efflorescence as a result of the introduction of canal irrigation. Similar situation was also reported in the Nira valley canal project in 1889 only 5 years after the opening of canal. The situation was no better in minor irrigation commands as reports of water logging and soil salinity came from tank irrigated areas in Madras Presidency.

Studies were taken up immediately on the nature of *reh* (saline efflorescence) or *usar*, the land affected by saline efflorescence, and causes of *reh* formation. The correspondence on *reh* published in 1864 represents the first-collected documentation of the problem and its scientific analysis. Copies were sent to the Secretary of State in England in 1865, with specimens of *reh* soil collected from village Munak and Asan-Kalan, both in the present district of Karnal, on the Western Yamuna Canal. Center[17] studied sodic soils and saline well water of Punjab and reported that at many places water table has risen from 100 ft (30 m) before irrigation to 8 ft (nearly 3 m) after irrigation. He ascribed occurrence of *reh* in soils to the decomposition of the elements of rocks and soils under the action of air and water. He also discussed nature of *reh* and its varieties, formation of *kankar*, causes of salt accumulation and methods of cure by sluicing and irrigation, manure and cultivation, arboricultural (plantation of trees and grasses), chemical manure, and nitrate of lime. Uses of *reh* for the recovery of Na_2SO_4, etc. have also been mentioned. Records also show that in a report on "Enquiries on *reh* and *oosur* (*usar*)," sent to Government of India by the then Revenue Secretary,[12] several possible ways to reclaim these lands were discussed, which are summarized in Table 1.1. It is also reported that the removal of soil (scraping) or deposit of silt are only temporary remedies, which is now well realized.[12] The experiments on reclamation of the *usar* lands began toward the last quarter of the 19th century. By then, Hilgard[36] has also taken up the land reclamation in the United States.

TABLE 1.1   Possible Methods of Reclaiming *Reh/Usar* in the United Provinces.

Name of the officer	Method: suggested/observed/seen
Mr. Bensen	Manuring and cultivation
Mr. Cadell	Stop irrigation to affected areas
Mr. Dashwood	Manuring
Mr. McConnaughey and Neale	Application of canal silt
Mr. Michael	Manuring with old thatch
Mr. Plowden	Cultivation, irrigation, and rich manuring
Mr. Smeaton	Manuring and application of leaves of *jungly* trees
Mr. Smith	Ditching (drainage), digging, thick manuring, and frequent plowing
Mr. White	Manuring and watering
Mr. Young and Mr. Ahmad	Digging and removing the soil up to 2 ft

Gypsum, a chemical amendment, which has proved to be the most effective antidote in the reclamation of sodic (alkali) soils, was not referred until toward the end of the 19th century in India; although by 1890, it has become quite popular in California, the United States.[19,36]

This chapter looks at history of characterization of sodic (alkali) soils, its extent, and gypsum as an amendment to reclaim sodic lands. It may be mentioned that the purpose is not to give a comprehensive review of more than 100 years of work but refers some important milestones in the journey in India with selected international work. Since the availability of gypsum in India is bound to reduce with restrictions on its mining due to environmental concerns and expanding use in other economic activities, prospects of other chemical gypsum as an amendment is discussed in the Indian context. A complete package of alkali land reclamation is discussed that has been extensively tested and has helped to reclaim more than 2 M ha of barren alkali soils in India.

1.2   TERMINOLOGY OF SALT-AFFECTED SOILS

The salt-affected soil is characterized as a soil having excess soluble salts and/or high exchangeable sodium percentage (ESP) or both on the soil surface and in the root zone to interfere in crop growth. Many terms have been used to describe these soils. In his compilation of works from India and University of California, Hilgard lamented that the term "alkali soils" was being applied almost indiscriminately in the United States and India to all soils containing an unusual amount of soluble mineral salts as most were having pH exceeding 7.0.[36] The term "alkali" in itself is not ambiguous, as it corresponds to "sodic." On the contrary, the term alkaline used to describe a soil with a high pH is a misnomer in the context of sodic soil. According to *the Alberta Environmental Farm Plan,*[103] alkaline soils are soils having pH above the neutral value (generally greater than 7.3). Alkaline soils have sodium absorption ratio (SAR) < 13, ESP < 15, electrical conductivity of soil saturation extract (EC_e) < 4 dS m^{-1}. Hilgard[37] divided salt-affected soils into two classes, namely, "white alkali" and "black alkali." While both these soils have excess of sodium, the former has neutral salts such as NaCl and Na_2SO_4, while the later has Na_2CO_3 and $NaHCO_3$ salts. In the former, soil pH and ESP are not high even though SAR may be quite high. It is expected that SAR will reduce during leaching with or without slight increase in pH. These soils are also termed as high SAR saline soils or simply the saline soils. The black alkali soils are the real sodic soils, the name being derived from the black spots that are often seen on the surface of these soils, being the evaporates of salts and organic matter that imparts black coloration. The nomenclature, however, has become obsolete although many people continue to use these terms.

Gedroiz[33] recognized the salt-affected soils as solonchak, solonetz, and solod. The solonchak is saline soil, while solonetz has a high degree of alkali salts. Solod soils are degraded solonetz soils. Sigmond[89] grouped these soils into three classes, on the basis of sodium and calcium amount, namely, (1) soils rich in alkali and calcium carbonate; (2) soils poor in alkali, but with considerable amount of calcium carbonate; and (3) soils rich in alkali but free of calcium carbonate. In 1932, he revised this classification and categorized the soils as saline, salty alkali, leached alkali, degraded alkali, and regraded alkali. Overstreet et al.[75] visualized alkali soils of saline, sodic, and saline-sodic types. Sodic soils having high amount of exchangeable sodium were characterized as "sodic" alkali soils. The authors justified this nomenclature stating that they have preserved the historical meaning of the term "alkali" as a general class name denoting soils with high sodium contents either in the form of free salts or in the adsorbed state. At the same time, the little-used adjective "sodic" (containing sodium) has been chosen to denote undesirably high amounts of adsorbed sodium.

The Soil Salinity Laboratory of United States[84] classified these soils into nonsaline alkali, saline-alkali, and saline. The preamble of the handbook indicates that in deference to past usage, the term "alkali soil" is employed to refer to soils having a high ESP, and "saline soil" is used in connection with soils having a high value for the electrical conductivity (EC) of the saturation extract. The term sodic soil is not even included in the glossary of terms given in the handbook. Nomenclature Committee appointed by the Board of Collaborators of the US Salinity Laboratory[99] recommended the use of the term "sodic" instead of "alkali" as previously proposed by the Salinity Laboratory. Use of this terminology has been approved by the Soil Science Society of America.[94]

Besides what is stated, a large number of subcategories of salt-affected soils are also recognized in different parts of the world depending on the dominance of a particular chemical constituent (e.g., calcium chloride rich soils or soils containing excessive quantities of exchangeable magnesium–magnesium solonetz, etc.) or a particular morphological character of the soil profile, for example, presence of a structural "B" horizon, etc. Kovda[50] classified these soils on the basis of the amount of total soluble salts and the quantity of chloride, sulfate, and carbonate in the soils. He divided the solonchak into four types: soda, sulfate, chloride, and nitrate. In soda solonchak, the salts found in abundance are sodium carbonate, sodium bicarbonate, and magnesium carbonate. In another classification, alkali soils on the basis of nature and amount of salts present, the thickness of the "A" horizon and the depth of calcium carbonate layers, were classified as solonchak, solonchak–solonetz, meadow solonetz, steppe-like meadow solonetz, solod, and solonetzic meadow.

Clearly, terms sodic and alkali soils have been and are being used interchangeably in the literature, while few authors have qualified by including both these terms together. On the other hand, it has also emerged that sodic soils are not always alkali as these can be acidic/saline as well.[81,82] Degraded sodic soils reported in deltaic regions of West Bengal (India) also have ESP >15 but pH in the range of 5.0–7.0.[18] In several cases reported in India and elsewhere, saline soils have a high sodicity (a high SAR that may be higher than that of even sodic soils), but they are not sodic soils and have usually a good infiltration capacity. Rengasamy[81] elaborately categorized various salt-affected soils as acidic–saline, neutral saline, saline, alkaline–saline, acidic saline–sodic , acidic–sodic, nonalkali sodic–acid, neutral saline–sodic, neutral sodic, nonalkaline sodic and strongly sodic neutral, alkali saline–sodic, and alkali–sodic.

In the Indian context, the most commonly used United States Department of Agriculture (USDA) classification of salt-affected soils has been adopted according to which salt-affected soils are classified into three groups,[84] namely, saline, alkali, and saline-alkali soils, which are illustrated in the next paragraph (Table 1.2).

TABLE 1.2   Classification of Salt-Affected Soils on the Basis of Chemical Characteristics.

Class	EC_e (dS m⁻¹)	ESP[b]	pH_2[c]	SAR (mmol L⁻¹)¹ᐟ²	Indian local name
Alkali	<4	>15	>8.5	>13	*Usar, Rakkar, Bara, Chopan, Kari*
Saline soils	>4[a]	<15	<8.5	<13	*Thur, Uippu, Lona, Shora, Soula, Pokhali, Khar, and Kari*
Saline-alkali soils	>4	>15	Variable	>13	*Usar, Kallar, Karl, Chopan, Bari, Reh, Choudu and Kshar, Shora*

EC_e, electrical conductivity of soil saturation extract; ESP, exchangeable sodium percentage; pH_2, pH of the 1:2 soil water solutions; SAR, sodium absorption ratio.
[a]Present literature sometimes gives this value as 2.0 dS m⁻¹ mainly because many sensitive crops such as gram and vegetables begin suffering at this salinity.
[b]Sometimes SAR (>13) is used instead of ESP. For Indian conditions, ESP is preferable.
[c]If pH_s is measured, the value is taken as 8.2.

Productivity in saline soils (Fig. 1.1A) is reduced due to the presence of excessive salt dominated by sodium chloride, sodium sulfate and chlorides and sulfates of calcium and magnesium. Alkali soils contain excess of salts like bicarbonates, carbonates, and silicate of sodium, capable of alkaline hydrolysis and have sufficient exchangeable sodium to interfere with growth of most crops (Fig. 1.1B). The ESP in many cases may exceed 90% as is the case of *Zarifa viran* Series[14] identified by Central Soil Salinity Research Institute (CSSRI), Karnal (Tables 1.3a and 1.3b). These soils develop cracks after drying (see Fig. 1.3 inset). Saline alkali soils have a combination of harmful quantities of salts and a high content of exchangeable sodium, which interferes with the growth of crop plants. These soils have all the features of a saline soil, and if reclamation procedures are used that do not include gypsum or any other amendment, they become alkali upon leaching. On the basis of ESP and EC_e, the classes of salt-affected soils are shown in Figure 1.2.

(A)   (B)

FIGURE 1.1 **(See color insert.)** A saline (A) and a sodic (B) soil in Haryana (India).

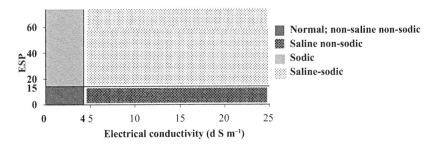

FIGURE 1.2 Categorization of salt-affected soils based on EC_e and ESP of the soils.

Although included in Table 1.2, USDA Handbook does not include SAR to classify the salt-affected soils. It seems that SAR to classify soils was included later as SAR and ESP were found to be correlated in many soils such that SAR 13 approximately equaled to ESP 15 of these soils. In general, this addition created the problem as many saline or high SAR saline soils got characterized as alkali/sodic soils.

TABLE 1.3a Physicochemical Characteristics of *Zarifa Viran* Series (Typic Natrustaf).

Horizon	Depth (cm)	Size class (%)			Coarse fragments (>0.02 mm, % of whole soil)	Organic carbon (%)	Carbonate as CaCO$_3$ <2 mm (%)	pH (1:2.5) soil water suspension	Bulk density (g cc^{-1})	EC (dS m^{-1})	CEC/clay ratio
		Sand (2–0.05 mm)	Silt (0.05–0.002 mm)	Clay (<0.002 mm)							
A11	0–5	43.4	34.6	22.0	–	0.30	0.5	10.3	–	8.2	0.46
A12	5–24	33.4	37.6	29.0	–	0.30	0.9	10.3	1.48	8.0	0.44
B21t	24–56	30.6	36.2	33.2	–	0.20	1.4	9.8	–	1.9	0.45
B22t	56–85	26.0	42.6	31.0	4	0.20	3.3	9.8	1.63	1.4	0.46
B3ca	85–118	33.2	40.0	26.8	10	0.10	12.4	9.6	1.47	1.0	0.41
Cca	118–140	45.0	32.0	23.0	–	0.10	20.5	9.2	–	0.9	0.41

EC, electrical conductivity; CEC, cation exchange capacity.

TABLE 1.3b Physico-chemical Characteristics of *Zarifa Viran* Series (Typic Natrustaf).

Depth (cm)	Exchangeable bases [cmol (p⁺) kg⁻¹]					Water at satu- ration (%)	Water extract from saturation paste (mmol L⁻¹)							DTPA extractable micronutrients (mg L⁻¹)			
	$Ca + Mg$	Na	K	Exchange-able Na (%)	CEC NH_4OAC		$Ca + Mg$	Na	K	CO_3	HCO_3	Cl	SO_4	Zn	Cu	Mn	Fe
0–5	0.2	9.9	0.1	97	10.2	33	0.4	85.3	0.2	30.0	36.5	12.2	5.5	0.36	2.01	12	9
5–24	0.3	12.1	0.4	94	12.8	39	0.4	83.9	0.1	27.0	40.5	14.5	5.5	0.36	2.32	9	16
24–56	0.8	13.4	0.6	90	14.8	32	0.6	18.0	0.1	2.5	8.5	5.5	3.0	0.34	1.38	11	6
56–85	1.8	12.4	0.4	85	14.6	40	0.5	13.5	0.1	2.2	7.6	2.5	2.0	0.21	0.80	8	5
85–118	3.1	7.6	0.5	68	11.2	40	0.8	10.0	0.1	2.2	4.5	2.5	1.0	0.21	0.50	6	4
118–140	5.6	3.8	0.4	39	9.8	44	1.0	9.2	0.1	1.5	5.0	2.5	1.0	–	–	–	–

CEC, cation exchange capacity; DTPA, diethylenetriaminepentaacetic acid.

Note: Clay fraction mineralogy is MI4 KK3 M3; MI = mica, KK = kaolinite, and SM = smectite (4 means half to one-third, 3 means one-third to one-fifth, and 2 means one-fifth to one-twentieth).

From management point of view, soils in India are being categorized only in two categories, that is, saline and alkali/sodic. If ESP is used as a classification criterion, a saline sodic soil with ESP >15 is easily categorized as sodic. If EC_e and SAR are used to categorize soils, some additional calculations are needed to categorize saline-sodic soils either as saline or sodic. If the ratio of the ions $[(CO_3 + HCO_3)/(Cl + SO_4)]$ and/or $[Na/(Cl + SO_4)]$ is less than 1, soil is categorized as saline soil. If these ratios are more than 1, then the soil is categorized as alkali soil.[20]

In the opinion of the author, if management is introduced as a criterion, salt-affected soils in India need to be characterized on the basis of salinity as well as water logging (depth to water table) as both are inseparable being twins in nature. Based on this criterion, salt-affected soils can be classified into six categories, namely, waterlogged saline (shallow water table), non-waterlogged saline, high SAR saline, waterlogged alkali, non-waterlogged alkali, and semi-reclaimed alkali (Table 1.4). The management issues involved in each group of soils is also included in Table 1.4.

TABLE 1.4   Classification of Salt-Affected Soils Based on Management Inputs.

Main class	Subclass	Differences in diagnosis and management
Saline soils	Waterlogged saline soils (water table ≤2.0 m below ground level, bgl)	SAR <15, require subsurface drainage before leaching
	Waterlogged saline soils (water table ≥3.0 m bgl)	SAR <15, do not require subsurface drainage, leaching alone is sufficient
	High SAR saline soils	SAR >15, leaching reduces SAR but some calcium amendment may be required to maintain infiltration rate
Sodic soils	Waterlogged sodic soils (water table ≤2.0 m bgl)	Surface drainage needs strengthening using vertical or bio-drainage, amendment application
	Non-waterlogged sodic soils (water table ≥3.0 m bgl)	Surface drainage and amendment application, special management of semi-reclaimed alkali soils to avoid resodification
	Semi-reclaimed sodic soils	At any one time, these soils constitute a major chunk of salt-affected soils, soils reclaimed as per procedure of sodic soils but referred as such between the intervening period of 1–2 years after reclamation to 10–12 years after reclamation until fit to grow sensitive crops because of self-reclamation, management options required to hasten the rate of reclamation as well as to avoid resodification
	Acid-sodic soils	Lime can be used as an amendment

It is inferred that nomenclature of salt-affected soils has yet not been standardized especially in many developing countries. Even Richards[84] mentioned, "The nomenclature for the problem soils is still in a formative stage." This statement can be easily made even today as not much work has been done since then except some location specific modifications in nomenclature. More work needs to be done to arrive at some universal nomenclature of salt-affected soils.

1.2.1 EXCHANGEABLE SODIUM PERCENTAGE (ESP) LIMIT FOR SODIC SOILS

As per guidelines given by USDA (Table 1.2) and others, a soil is categorized as sodic when the soil ESP exceeds 15. The degree of sodicity increases with increase in ESP being extremely high at ESP 70 (Table 1.5). Even today, ESP >15 is used to classify salt-affected soils as sodic/alkali. On the other hand, it may not be out of place to mention that this threshold for sodic soils is somewhat arbitrary. In some cases, for example, 2 or 3 milli-equivalents of exchangeable sodium per 100 g of soil has equal or even greater usefulness as a critical limit.[84] The change from flocculated (aggregated) to dispersed soil structure occurs gradually with increasing sodium levels. The dispersion of clay particles is strongly dependent on the electrolyte concentration in the soil solution and the relative composition of Ca and Mg on the exchange complex. Clay dispersion is reported to occur at an ESP of five in Ca–Na aggregates but is observed at an ESP of only 3 in Mg–Na samples.[30] An ESP of six was proposed by Northcote and Skene[72] to be the lower limit of soil sodicity. The same value has been reported as the limit to characterize sodicity in Australia.[81] The lower limit is related to low concentrations of soluble salts, low content of calcium, and widespread structural deterioration of Australian soils.[83] Values of five[100] and two[62] have been suggested to cause a deleterious effect on soil structure.

In the Indian context, values ranging between 5 for vertisols to 10 for heavy textured soils have been reported to seriously impact the soil physical properties, and therefore, these values are suggested as the limits to categorize the respective soils as sodic. A Vermaquept has been classified as sodic if it has an ESP 7 (SAR 6) within 1 m from the soil surface, whereas a Haplocalcid has an ESP of 15 (SAR 13) to be designated as sodic. Sodicity problems are more pronounced in clay soils than in loamy, silty, or sandy soils. Within clay soils, soils containing montmorillonite or smectite (swelling clays) are more subject to sodicity problems than illite or kaolinite clay soils. The reason is that the former types of clay have larger specific surface areas and higher cation exchange capacity (CEC). The relative loss in the productivity of fluvisols and vertisols (black soils) at the same ESP reported by Sehgal and Abrol[87] indicates the possibility of different researchers and practitioners to arrive at different conclusions on prescribing the limits to define sodic soils (Table 1.5). The data reported by these authors revealed that the yield loss in fluvisols is nil and in vertisols up to 10% at an ESP of 5. For an ESP >40, yield losses increase to 25–50% in fluvisols as against more than 50% in vertisols. No deleterious effects on crop productivity have been observed in India at ESPs lower than 15, even when some soil physical properties were adversely impacted. From current standpoint therefore, guideline of ESP greater than 15 should continue to be used to characterize sodic soils until more data are made available that may suggest modification in this limit.

TABLE 1.5   ESP and Degree of Sodic Hazard.

Approximate ESP	Sodic hazard
Up to 15	None to slight
15–30	Slight to moderate
30–50	Moderate to high
50–70	High to very high
>70	Extremely high

1.3   EXTENT AND DISTRIBUTION OF SALT-AFFECTED SOILS

The extent of sodic soils is not known precisely. A first attempt to compile information on the extent of salt-affected soils on a worldwide basis was made by Massoud[56] based on the FAO/UNESCO Soil Map of the World. Approximately, 1000 M ha is assessed to be salt-affected.[29] Szabolcs[96] also reported the salt-affected soils of Europe and presented maps showing the distribution of salt-affected soils in most continents. Szabolcs[97] provided the distribution of saline/sodic soils in various continents. Estimates suggest that around 831 M ha area is salt affected, out of which 397 M ha is saline and 434 M ha sodic.[28,31,74] More than 77 M ha of land worldwide has become salt affected due to human causes (secondary salinity); of this, 45.4 M ha are irrigated land (in the world's semiarid and arid regions) and the remaining rain-fed. In India, several estimates ranging from 3.9 to 24.0 M ha have been reported by various agencies, for example:

Abrol and Bhumbla[1]: 7.0 M ha

National Commission on

Massoud[56]: 23.8 M ha

Agriculture[71]: 7.16 M ha

Ministry of Agriculture[63]: 6.3 M ha

NRSA[73]: 3.90 M ha

NBSS & LUP[70]: 6.20 M ha

Singh[91]: 8.11 M ha

If data of Massoud[56] are taken to be true then India has about 23.22 M ha of saline and 0.58 M ha of sodic soils. The current data compiled by CSSRI, Karnal do not support this as only 6.73 M ha area is salt affected with 3.77 M ha sodic and 2.95 M ha saline that include 1.71 M ha inland saline and 1.25 M ha coastal saline.[64] Since the reported extent of coastal saline soils is as high as 3.1 M ha, the mismatch between the figures quoted needs reconciliation. The reported discrepancies also call for a fresh look on the extent of sodic soils at the global level.

1.4   CHARACTERISTICS OF SODIC SOILS

An alkali soil in the Indian context has the following characteristics:

- Characterized by high pH of the soil saturation paste (>8.2) and high ESP (>15).
- Commonly encountered in semiarid and subhumid agro-climatic conditions with annual rainfall of more than 500 mm.
- Low infiltration rate, <5% of the rate in normal lands in the vicinity.
- Low-organic matter content. Black patches may appear at the soil surface.
- Poor internal drainage resulting in dry subsoil and a moist or wet surface layer. Crops fail because of excess surface water or for lack of water even though there may have been adequate rainfall or irrigation.
- Presence of *Kankar* layer ($CaCO_3$ concretions) at depth ranging from 0.5 to 1.8 m depth.
- Underlain by good or sodic ground water.

Several chemical, physical, and microbiological factors get altered because of high pH, ESP or high concentration of salts as in a saline-sodic soil.

The **chemical factors** that adversely affect the plant growth in sodic soils are

- Decreased solubility and availability of micronutrients like Zn, Fe, and Mn due to high pH;
- Deficiency of Ca since nearly all the soluble Ca is precipitated as insoluble $CaCO_3$;
- Excess of Na causing imbalance in the uptake of other ions due to antagonistic effect such as K and Ca nutrition;
- High pH due to high concentrations of HCO_3 and CO_3;
- Increased solubility and accumulation of toxic elements like F, Se, and Mo in plants;
- Presence of $CaCO_3$ and soluble CO_3 and HCO_3 ions;
- Toxicity of Na, HCO_3, and CO_3 ions.

The **physical factors** that affect the plant growth are

- Poor water and air permeability as a result of high dispersion of soil aggregates and clay particles;
- Low water availability due to poor conductance from the lower to the upper soil layers and restricted effective root zone;
- Hard crust on the surface layer which greatly hinders the seedling emergence and reduces the germination percentage and thus plant population;
- Poor internal drainage resulting in water logging and aeration problems;
- Difficulties in preparing seedbed as all parcels of the land do not attain proper moisture conditions at the same time;
- Clod formation and poor tilth of the soils and continuous loss of soil due to air and water erosion; and
- Presence of hard $CaCO_3$ layer acts as a physical barrier for the movement of water, salts, and roots of crops and trees.

The **microbiological factors** are mainly related to low-organic matter content and physical and chemical factors that results in low activity of useful microbes.

1.5   CHEMICAL AMENDMENTS

The chemical amendments are broadly classified into two categories:

- Soluble calcium salts, namely, calcium chloride and gypsum which directly supply soluble Ca for the replacement of exchangeable Na.
- Acidifying materials, namely, sulfuric acid and elemental sulfur which due to their acidulating action and/or increasing the partial pressure of CO_2 supply Ca indirectly by dissolving calcite natively found in sodic soils. The Ca so mobilized is used to replace Na from the exchange complex.

Besides these, many industrial wastes are used as amendments that work on similar principles depending upon their constituents. Amongst them, pyrites, distillery spent wash, dairy waste, phosphogypsum (PG), and press mud have been widely experimented and used.

1.5.1   GYPSUM

Gypsum is one of the earliest forms of fertilizer used in the Europe and the United States. Its fertilizer value was discovered in Europe in the last half of the 18th century with its extensive use in the 18th century. In the United States, it has been applied to agricultural soils for more than 300 years dating back to late 1700s.

Gypsum is also by far the most common amendment used in the reclamation of sodic soils, being in use for more than a century. Gypsum, a dihydrate form of calcium sulfate ($CaSO_4 \cdot 2H_2O$), is available from two major sources in India:

- The natural deposits of gypsum commonly known as mineral gypsum being a secondary mineral.
- Derived as a by-product of industry producing tartaric acid, formic acid, oxalic acid, citric acid, common salt, and phosphoric acid.

Since nearly 90% of it is derived from phosphoric industry, it is commonly referred as Phosphogypsum (PG). Relative composition of two types of gypsum is compared in Table 1.6.[16,85] Maximum limits of various chemical constituents allowed as per Bureau of Indian Standards standard are listed to show that some of the constituents may be in excess of the recommended values.[16] There are other kinds of gypsum as well. Fluoro-gypsum is produced as a by-product during the manufacture of hydrofluoric acid. Since sea water contains 3.5% by weight of $CaSO_4$, marine gypsum is obtained during the process of recovering common salt by solar evaporation of sea water. Salt manufacturers in the coastal regions of Gujarat, Maharashtra, and Tamil Nadu recover marine gypsum as a by-product of salt industry. Currently, a large amount of flue gas desulfurization gypsum commonly referred as FGD gypsum is being used as an amendment in Ohio, the United States. It is produced by removal of sulfur dioxide (SO_2) from flue gas streams when energy sources, generally coal, containing high concentrations of sulfur (S) are burnt.

TABLE 1.6 Chemical Analysis (%) of Mined and By-product (Phospho) Gypsum.

Constituents	Mined gypsum (70–80% purity)[a]	Phosphogypsum[b]	BIS limits for phosphogypsum
$CaSO_4 \cdot 2H_2O$	77.56	85.0	85.0 min
SiO_2	12.37	0.45–4.0	0.40 max
R_2O_3	1.59	0.05–0.3	0.15 max
Magnesia as MgO	0.81	–	0.10 max
Lime as CaO	28.56	32.5–44.0	–
Loss on ignition	20.44	–	–
Chloride as NaCl (% by mass)	0.016	–	0.10 max
Water soluble P_2O_5	–	0.2–1.2	0.40 max
F (% by mass)	–	0.5–1.5	0.40 max
SO_3	–	45–46	0.30 max
pH of 10% aqueous suspension of gypsum			5.0 min

BIS, Bureau of Indian Standards; max, maximum; min, minimum.
[a]Mined by Fertilizer Corporation of India.
[b]Phosphogypsum from Morocco rock phosphate. Range shows the variation due to method to produce phosphoric acid.

The exact quality of PG depends upon the presence of impurities in the rock phosphate. High content of fluorine and presence of some radioactive constituents (uranium and radium) with minor amounts of toxic metals such as arsenic, barium, cadmium, chromium, lead, mercury, selenium, silver, fluoride, and aluminum cause major hindrance in its widespread use. Presence of such impurities besides polluting soil and water resources may impair the product quality. The average composition of PG in the United States has been found to contain fluoride in the range of 5–38 g kg^{-1}, Mo 65 and Cd 0.23 mg kg^{-1}, respectively.[8,9,55,57,76,95] PG produced in North Florida contains roughly 5–10 picocuries per gram (pCi g^{-1}) of radium while PG from Central

Florida contains about 20–35 pCi g^{-1} radium. US Environmental Protection Agency (Federal Register 6/3/92) has permitted the controlled use of PG in agriculture provided radium-226 levels are <10 pCi g^{-1}. This restriction on the maximum radium radioactivity essentially eliminates the use of several kinds of PG such as central and southern Florida PG because its radium-226 levels are commonly in the range of 15–25 pCi g^{-1}.[9]

1.5.1.1 HISTORICAL ACCOUNT: GYPSUM AS AN AMENDMENT

It may be really difficult to pinpoint as to when and by whom gypsum was first applied to reclaim sodic lands. Moreland[65] giving an account of Sleeman visit to *usar* areas in the United Provinces of India around 1850, mentioned that *usar* can be reclaimed by flooding the fields for 2–3 years, cross plowing, manuring, and irrigating. Giving results of other experiments, he did mention the use of fertilizers and gypsum for reclamation, probably those referred by Leather.[51] The report of Mr. Ward, the Chemist in the Royal School of Mines in London, mentioned drainage as a means of reclamation of water logging and soil salinity in Munak village in Karnal but no reference to gypsum or alkali has been made.

Medlicott, then Professor of Geology at Thomson College—Roorkee, was probably the first scientist who made scientific investigation of the problem of *reh* in 1863. His findings are summarized as follows:

> The river water that fed the West and East Jamuna Canals, contained high concentrations of sodium sulphate and sodium chloride, the predominant salt characterizing reh and usar lands. Soils in which salts accumulated through high rates of evaporation and through capillary action were parched with no natural sub-soil drainage. Because of the change in the rate of movement of sub-soil water through percolation, capillary movement and by the poor quality of irrigation water itself, progressive concentration of soluble and partially soluble salts, originally distributed more widely throughout the profile, takes place in the upper strata of the soil to a degree toxic to plant growth. The canal water was an exacerbating agent but not a prime cause and the natural salt lands were made increasingly so by inefficient drainage and insufficient application of water.

By 1877, Hilgard had clarified the composition of alkali in soils and found that as little as 0.08% of sodium carbonate (Na_2CO_3) was sufficient to render the soil practically untillable.[44] Gypsum is the antidote to these, so-called, "true alkali salts." With the cooperation of local farmers, Hilgard carried out field experiments to demonstrate the effectiveness of applying gypsum to soils affected by black alkali. It was an immediate success, and by 1890, gypsum was a widely used treatment for these soils in California. His work was already known in India in 1884 and the government of the North-Western Provinces sent an engineer W.J. Wilson, to California to get familiar with the research studies taking place at Berkeley. The work on land reclamation was simultaneously going on in Australia, Hungary, USSR and the United Kingdom.

History of *Reh* in India, Australia and the United States indicates:

Could planners and policy makers in New South Wales in 1906 have learnt from the past about irrigation salinity? A search for answers to this question starts in India, because British India is known to have been an important source of information, on a great many issues, for the Australian colonies. By the late nineteenth century British engineers had some five decades of experience bringing irrigation to the vast doabs (land between the rivers) of the Indus and Ganges River systems. The British also had first-hand experience of salinity as a direct result of intensive irrigation. Their knowledge of designing, constructing and operating these irrigation systems provided a baseline from which irrigation planners in south-eastern Australia could have proceeded.

The report of the *Reh* Committee was made available in 1886, which focused attention on many important aspects related to canal irrigation, drainage, and spread of sodicity in the state. The committee recommended that

- Indiscriminate use of water through the flush system of irrigation (flood irrigation) should be stopped.
- Provision of deep drainage, straightening and opening of natural drainage ways and preparation of ground water maps were also suggested.
- It was also recommended that experiments in reclaiming *usar* tracts which began in 1874 under the supervision of the newly created Department of Agriculture should be continued.

While touring India in 1889, Voelcker saw tree-planting as a means of arresting soil erosion, and assisting land reclamation. He recommended increasing the water supply to the dry tracts of the north by means of irrigation but was aware of the need for subsoil drainage which is difficult and costly matter in India.[101] Reference to gypsum as an amendment has not been made by him. Leather[51] reported for the first time that the addition of small amounts of gypsum to reduce Na_2CO_3 to 0.1%, in the fine size, applied at the surface of sodic soils, was very beneficial. After that, Leather[52-54] made constant references to gypsum as a soil amendment to reclaim *usar* lands. He mentioned that application of gypsum was made in 1897 at Gursikran in 4×4 yd^2 plots, which were later extended to 0.5 acre (0.2 ha) plots. Gypsum is very useful in improving the water infiltration rate and crop growth but is costly. The gypsum requirement of the soil was calculated as simply the quantity which was equivalent to soluble carbonates present. As per his calculations, gypsum if applied in required quantities would cost somewhere between Rs. 700 and 800 per acre which was obviously prohibitive. Even half the quantity was still expensive. The whole write-up made in the fore-going paragraphs suggests that gypsum might have been used sometimes in the decade between 1888 and 1897. Although it was definitely used in 1897, the period in between needs further review to pinpoint whether gypsum was applied even before 1897.

There are numerous instances thereafter when gypsum was used to reclaim *usar* lands. Nasir[67] and later Barnes and Ali[13] and Nasir[68] applied and reported the use of gypsum to reclaim *usar* lands in Punjab. Henderson[35] mentioned that gypsum application was essential to reclaim some of the soils in Sindh province. Singh and Nijhawan[90] used gypsum to reclaim alkali lands in Punjab. Taylor et al.[98] pointed out that the crop yields declined above pH 8.5. Application of gypsum was suggested to prevent base exchange reaction. Mukerjee et al.[66] reclaimed alkali lands using gypsum as an amendment in the United Provinces of India. Mehta[60] reported that field and laboratory experiments to reclaim the lands began in Punjab in 1918 but failed mainly because land reclamation was much ahead of the researches in the field. The phenomenon of base exchange was not fully understood by then as work on this issue was in progress at that time.[38] The attempts, however, bore fruit by 1939, and the processes of land deterioration and reclamation were clearly understood. Dhawan et al. devoted significant number of years (1950–1958) in developing the gypsum application technology.[26] They concluded that gypsum is comparatively more effective than calcium chloride. Menon and Iyenger[61] and Sanayasi and Iyer[80] used gypsum to reclaim alkali lands in Madras state and Cauvery Mettur area of the state, respectively. Kelley[48] highlighted the role of various amendments and bio-reclamation processes in the reclamation of alkali soils in California, the United States.

The 15 years of post-independence period (1947–1962) saw a great progress in research on saline-sodic soils in Uttar Pradesh (India) and Punjab. Dhar and Mukherji[22] and Dhar[21] tried a number of reclaiming agents particularly molasses and press-mud for the amelioration and reclamation of saline-sodic soils. But the gravity of the problem went on increasing as a consequence of which an *usar* Land Reclamation Committee was constituted in 1938 by the State Government of United Province to advise on all matters related to the characterization, reclamation and prevention of spread of *usar* land. Agarwal et al.[3,5,6] and Khan[49] studied several aspects of soil salinity problem in United Province. In Punjab, scientists like Mackenzie, Taylor, Puri, Asghar, and Mehta devoted themselves to research on the problems of soil salinity and water use. Standards of soil and water quality were established, and the means of reclamation of deteriorated soils were found. The implementation of these research findings in the field started in the late 1930s. Since the problem continued to increase beyond all proportions, "Water Logging Inquiry Committee" was expanded and converted into "Water Logging Board." Commencing from 1927 to 1941 was a period of fruitful research in Punjab on the subject of saline-sodic soils. Puri et al.,[77–79] Singh and Nijhawan,[90] Hoon et al.,[41,43] Mehta,[59] and Taylor et al.[98] in the undivided Punjab made valuable contributions on the subject of sodic soils, their characterization, diagnosis, and reclamation. Singh and Nijhawan[90] reported that calcium salts followed by farmyard manure can improve sodic soils to the status of normal soil. Puri[77] reported the relationship between exchangeable sodium and crop yield in Punjab soils. He observed that treatment of a Montgomery soil with $CaCl_2$ and leaching reduced base saturation from 67% to 20% and increased yields from 32 to 100 kg ha^{-1}. Taylor et al.[98] pointed out that the crop yields declined above pH 8.5. Hoon et al.,[39,40,42] Dhawan et al.,[24–27] Ashgar et al.,[10,11] Kanwar et al.[45–47] contributed greatly in Punjab. Central Board of Irrigation and Power published three research monographs on Land Reclamation.[23,39,61] On the proposal of The Indian Council of Agricultural Research (ICAR), Agarwal and Gupta[4] collected related information from the available sources and wrote a bulletin entitled "Saline-Alkali Soils in India" that was published by ICAR in 1968 and revised in 1979.[7]

1.5.2 PHOSPHOGYPSUM

It is difficult to point out when PG was first applied for land reclamation. Even today gypsum is synonymously used for PG. A number of demonstration on the utility of PG to reclaim sodic lands have been conducted by Hindustan Copper Limited, New Delhi during 1974–1977 in Punjab, Haryana, and Uttar Pradesh in association with research organizations.[32]

Shrotiya and Mishra[88] reported high yields of rice and wheat even in soils with a pH as high as 10 or more when PG @10 t ha^{-1} was applied in each demonstration plot followed by the cultural practices advocated by CSSRI, Karnal (Table 1.7).

TABLE 1.7 Performance of Phosphogypsum in Sodic Land Reclamation.

State	Soil pH before amendment application	Average yield in demonstrations (t ha−1)	
		Rice	Wheat
Haryana	9.5–10.6[a]	4.70	2.84
Punjab	9.0–10.0[b]	4.75	2.98

[a]Range of eight demonstrations.

[b]Range of nine demonstrations.

Mehta and Yadav[58] reported that PG is a promising chemical amendment and can be used effectively for the reclamation of sodic soils. Singh et al.[92] conducted detailed studies on the PG and its effect on the soil properties and crop growth. This study discounted the adverse effects of fluorine in the PG and explained that the adverse effects of fluorine to a great extent are nullified because of the phosphorus present in this amendment.

Current researches in India have proved that PG is a good amendment to reclaim sodic lands and sodic water. The studies also revealed that one time application of PG for land reclamation and/or small doses for water reclamation are unlikely to result in phytotoxic level of fluoride. As a matter of abundant caution, the rice and wheat cultivated on reclaimed lands with PG in fluoride affected areas should be consumed with care for first 2 years of reclamation.

Fears of radioactive elements have also been discounted. The fate of radium-226 in Florida PG was investigated by Mays and Mortvedt.[57] They applied PG containing 25 pCi g^{-1} 226Ra at rates up to 112 Mg ha^{-1} to the surface of a silt loam soil and grew successive crops of corn (*Zea mays* L.), wheat (*Triticum aestivum* L.), and soybean (*Glycine max* L.). Application of PG even @112 Mg ha^{-1} had no effect on the radioactivity levels in grain of corn, wheat, or soybeans. The application @112 Mg ha^{-1} is more than 200 times the normal rate of gypsum used for peanut fertilization. Additionally, they noted no increase in grain Cd levels although at the highest rate they found that corn growth was slow. Nayak et al.[69] also reported that radioactive nuclei present in the PG used in their study has no adverse effect of on crop yield or its quality.

The studies revealed that PG is a better reclaiming agent as compared to mined gypsum. The PG treatment resulted in greater reduction in pH and ESP of both the surface and the lower layers (Table 1.8). It is attributed to higher solubility of PG over mined gypsum, which allows soluble Ca to permeate to lower depths. Application of PG also gave comparatively higher yield over equivalent dose of mined gypsum. PG @10 t ha^{-1} (equivalent to 50% GR) resulted in slightly higher yield of 4.5 t ha^{-1} over the treatment 50% mineral gypsum (4.22 t ha^{-1}) and 25% mineral gypsum + 25% PG (4.32 t ha^{-1}) in the first year of reclamation.

TABLE 1.8 Effect of Phosphogypsum on Soil ESP After First and Third Year of Rice–Wheat Cropping.

Treatments	2006–2007		2008–2009	
	0–15 cm	15–30 cm	0–15 cm	15–30 cm
50 GRm	26.5	31.6	24.6	28.5
50 GRpg	20.4	27.3	18.3	25.7
50 GR$_{PG}$ – KFC^a	18.3	25.2	17.9	23.4
25 GRm + 25 GRpg	17.8	23.5	16.7	21.5
25 GRpg	19.4	25.6	17.9	24.3
12.5 GRm + 12.5 GRpg	25.2	30.8	23.5	27.5
Least significant difference ($p = 0.05$)	1.5	2	1.5	2

[a]KFC means Karnal grass was taken as first fodder crop followed by wheat.

1.5.3 WHAT AILED THE FIELD APPLICATION OF RECLAMATION TECHNOLOGY?

A number of schemes to conduct research on fundamental and applied aspects of the salinity problem were launched by the ICAR/Government of India. These were followed by Government of India sponsored several saline-sodic land-reclamation activities under various projects, yet these could not make a dent on the saline-sodic scene in the country until CSSRI came into picture. "Efforts to reclaim these soils to say the least have been halfhearted," remarked Dr. Bhumbla, founding Director of CSSRI. Amendments, though important, will not be of much use if the other components of technology are not ensured. He listed the following factors due to which gypsum technology failed to pick-up until the beginning of eighth decade of the 20th century.[15]

- Inadequate and inappropriate fertilizer application.
- Inadequate rate of gypsum application, most of which was used up by soluble sodium carbonate and sodium bicarbonate.
- Mixing of gypsum to deeper depths.
- Poor on-farm land development including little attention to land leveling.
- Poor water management.

In nutshell, most attempts were confined to amendment application with little emphasis on the post amendment application management. Training component was completely neglected and socioeconomic constraints were fully ignored. Apparently, any effective and successful sodic land-reclamation program requires both technical and social solutions. While it demands systematically applying technical knowledge in a well-defined sequence, it also calls for coordinating farmers, ensuring water access, and modernization of drainage system. Besides, it requires a coordinated effort of different stakeholders such as farmers (both tenants and land owners), the line departments, the nongovernmental organizations (NGOs), and the commercial banks. Efforts and investments in organizational development and local capacity building are required.

1.6 TECHNOLOGY PACKAGE

Reclamation and management of sodic soils are two distinct activities, though practically both are applied simultaneously to achieve high efficiency or to hasten the process of land reclamation. These technologies have been discussed in details by the authors[34] and are summarized in Table 1.9. Reclamation of sodic soils begins with physical/mechanical processes such as: on-farm land development including *bunding* (dyking) and leveling followed by application of chemical amendment followed by hydraulic processes including flushing especially in heavy textured soils or where excessive salt accumulates at the soil surface and leaching (Table 1.9).

The first and foremost task confronting a developing nation like India was to reduce the cost of reclamation mainly the chemical amendment. The studies established that only 25% of the gypsum dose calculated on the basis of Schoonover[86] method was sufficient to reclaim the highly sodic lands (ESP >90%) of the Indo-Gangetic plains for rice–wheat sequence in

TABLE 1.9 Reclamation and Management Package for Reclamation and Management of Sodic Soils.

Group/subgroup	Technology	Activities/remarks
		Reclamation
	On-farm land development including surface drainage	*Bunding* (dyking), leveling (preferably laser land leveling), farm layout including field drains, tube well for assured source of irrigation, a heavy pre amendment application irrigation
	Application of chemical amendment	Apply adequate dose of amendment, inadequate dose may result in resodification after few years especially with poor management
	Flushing/leaching	Adopt flushing only in heavy textured soils or where salt efflorescence is quite thick to cause problem during tillage, leaching followed by flushing or leaching alone for 7–10 days before transplanting rice crop
		Management
Crop management	Selection of crops and cropping sequences	Select crops and cropping sequences that combine tolerant or medium tolerant crops, e.g., transplanted rice (tolerant) and wheat (medium tolerant) sequence
	Exploitation of varietal differences	Select salt tolerant varieties within the selected tolerant and medium tolerant crops
	Improved agronomic practices	25% higher seed rate or higher plant population in transplanted crops, avoid puddling in initial 2–3 years unless necessary to transplant rice, use rice seedlings of higher age preferably grown in normal lands, avoid prolonged fallowing, practice green manuring
Soil management	Land forming/seeding	Land forming may not play a major role in sodic lands but can be effectively used to manage land drainage; in saline-sodic lands, it can be helpful in managing salts as well as drainage
Irrigation water management	Shallow depth high-frequency irrigation regime	Amount of water being same, follow low-depth high frequency regime
	Switchover to improved surface irrigation techniques	Go for improved surface irrigation techniques in the initial years, delay first irrigation to wheat by about 8–10 days than in normal soils, sprinkler irrigation can be adopted for direct seeded rice–wheat sequence in semireclaimed lands

TABLE 1.9 *(Continued)*

Group/subgroup	Technology	Activities/remarks
Nutrient management	Application of additional nutrients/green manure/farm yard manure	Apply 25% more nitrogen in three splits, can avoid application of phosphorus and potash for few years, zinc application is absolutely necessary, adopt integrated nutrient management
Rainwater management	In situ rainwater conservation; rainwater harvesting and reuse (helps in drainage)	Rainwater management is necessary for effective leaching being the best quality water, for irrigation water conservation, and for effective land drainage

the initial years.[1,2] The current results show that even half of this requirement would suffice if salt tolerant rice and wheat varieties (Fig. 1.3) developed by the CSSRI are cultivated for first 2–3 years before switching over to commonly used high yielding varieties by the farmers.[93]

FIGURE 1.3   (See color insert.) A view of the rice tolerant variety CSR 30 (A) and wheat variety KRL 210 (B).

The efficiency of the reclamation process depends upon the management interventions comprising of crop management, soil management, irrigation water management, chemical management, and rainwater management (Table 1.9). It may be noted that the fine tuning of the package to take care of location specific problems is a continuous process and has been the hallmark of CSSRI's technology development initiatives. Besides, the long-term sustainability of the project hinges upon resolving social issues besides the technical inputs. It calls for cooperation between various stakeholders such as research organization, government departments, NGOs, and of course the farmers.

1.6.1   STAKEHOLDERS' PARTICIPATION: A SUCCESS STORY OF UTTAR PRADESH

The farmers are the kingpin in the success of a reclamation program. To convince the farmers about the success of technology, front line demonstrations were conducted and field visits were arranged to the demonstration sites. The transformation became visible as the technology became a household name in the nearby villages with the nickname of gypsum technology. The next step was to associate Department of Agriculture in CSSRI activities. Government of Haryana was convinced to establish Haryana Land Reclamation and Development Corporation in 1974 on the lines of existing Punjab Land Reclamation and Development Corporation in Punjab. Seeing the performance of land reclamation activities in the two states, Government of Uttar Pradesh also got convinced and established Uttar Pradesh *Bhumi Sudhar Nigam* (UPBSN) in 1995. Unlike Punjab and Haryana, where surface drainage systems were better developed, the drainage component, forming a major input of the program, could be handled much better with this institutional support in Uttar Pradesh as it required regional planning besides the on-farm land development. CSSRI, Regional Research Station, Lucknow provided backstopping not only in fine-tuning the package for local specific needs of Uttar Pradesh but also in extending the land reclamation programs to the farming community in various districts of the state. It is heartening to note that the efforts have brought fruits resulting in reclamation of 0.70 M ha of sodic lands in the state by March 2015.

1.7 SEMIRECLAIMED SODIC LANDS AND RESODIFICATION

Unlike USSR and few other countries, where amendments are applied to reclaim deep profiles, only 15 cm of top soil profile is reclaimed in India that provides good root environment to tolerant shallow rooted crops. The salts remain in the profile at some depth depending upon the water that is passed through the profile. Further reclamation depends upon the management as it takes place only through the self-reclamation processes. Therefore, a new term, semi-reclaimed sodic lands, has emerged under the Indian situation as it is only after 10–12 years after the process of reclamation begins; soils are fit to cultivate most crops including the sensitive ones. A major difficulty with semi-reclaimed lands is that they have the potential of resodification or partially losing the productivity potential requiring repeat application of gypsum. To avoid such a situation, besides the normal management options, additional post reclamation management needs to be adopted to negate or minimize the chances of resodification. Some of these management options are:

- Avoid prolonged fallowing of the land. If water and other resources are available, go for green manuring crop after the harvest of wheat. After 3–4 years of reclamation, a green gram crop can also be taken after wheat harvest to fetch additional income.

- Ensure adequate leaching as changeover to low water requiring crops in subsequent years of reclamation may change the leaching pattern.
- Ensure that no runoff water from adjoining non-reclaimed areas enters the fields.
- If a shallow impermeable layer is present near to the soil surface, or ground water table is shallow, extra care should be taken in applying irrigation water so that upward flux is minimized. Studies from Uttar Pradesh have revealed that in many cases, it is the main cause of resodification especially when inadequate quantity of gypsum has been applied during reclamation.
- If the recommended technological package is used, take account of fertility status especially with regard to phosphorus and potash as their reserves may get exhausted with continuous cropping. If this happens, there is possibility of wrongly attributing productivity loss to resodification. Repeat application of gypsum may not prove to be useful in such cases.
- Monitor land and water resources for their quality. If water has high residual sodium carbonate (RSC), application of gypsum every season or every year may be required to neutralize the RSC to 2.5 meq L^{-1}. Irrigating with high RSC water has been the major cause of resodification in Haryana as farmers sometimes are not aware of the quality of their tube well water.

1.8 CONCLUSION

The problems of soil salinity, sodicity, water logging, and poor quality waters are likely to increase in future due to planned expansion in irrigated area and non-judicious exploitation of natural resources to meet food, fodder, fiber, and timber demand of the burgeoning human and livestock populations. Besides, present rate of land reclamation in India is also not in tune with the target to have a wasteland neutral India by 2030. Amongst the salt-affected soils, alkali soils constitute a major productivity constraint in most productive landscapes around the globe. Indian experiences show that if reclaimed, these soils can contribute 10 t ha^{-1} of food grains. Besides this, activity can provide livelihood to resource poor farming communities in India and other developing countries. Technologies to reclaim and manage these soils are now well understood and successfully implemented on large scale. However in times to come, abiotic stress may be a cocktail of soil salinity, alkalinity, water logging, poor quality waters, and heavy metal toxicity. There will be a need to develop technologies as well as document their impact on soil–water–plant–human continuum. It will require harnessing the potential of frontier sciences such as nanotechnology, biotechnology, cellular, and molecular approaches for developing a holistic approach to land reclamation. Besides technical interventions, research organizations need to work on regulatory, economic (incentives/disincentives), and advisory policy interventions to upstream, upscale, and speed-up the rate of land reclamation to achieve the salt-affected soils neutral world by 2030.

1.9 SUMMARY

On a global scale, productivity of sodic/alkali lands estimated at 434 M ha is adversely affected due to high ESP, the affected area being 3.77 M ha in India. The current chapter looks at the nomenclature of salt-affected soils in general and comments that there is a need to modify the nomenclature criteria from reclamation and management point of view. From these points of view, it is suggested that while saline soils in India can be categorized under three categories, sodic soils can be grouped under four categories including semi-reclaimed alkali lands and acid alkali soils. It has also emerged that terms sodic and alkali soils have been used synonymously to designate soils with high ESP although later has been preferred in India. Gypsum, either mineral or chemical (PG), is one of the major antidote to reclaim alkali lands. Since availability of mineral gypsum is likely to reduce in future because of restrictions on its mining and its increasing use in other economic activities, synthetic gypsum can substitute for mineral gypsum provided environmental concerns are addressed in a scientific manner.

Tracing the history of alkali land reclamation with gypsum, chapter presents the salient features of reclamation and management package being adopted in India. The package is based on reclamation and integrated management comprising of crop, soil, irrigation water, nutrient application, and rainwater management components. Role of socioeconomic issues and stakeholders participation to successfully carry out reclamation programs are highlighted through a brief case study of Uttar Pradesh, India where more than 0.70 M ha land has been reclaimed in about two decades. Resodification of reclaimed alkali lands is emerging as a major issue in land reclamation programs in India, management options are listed to avoid or minimize resodification of reclaimed alkali lands.

KEYWORDS

- alkali soils
- black alkali
- electrical conductivity
- ESP
- gypsum
- resodification
- saline-alkali soils

REFERENCES

1. Abrol, I. P.; Bhumbla, D. R. Saline and Alkali Soils in India: Their Occurrence and Management. In *FAO World Soil Resource Report*; Food and Agricultural Organization: Rome, 1971; Vol. 41, pp 42–51.

2. Abrol, I. P.; Dargan, K. S.; Bhumbla, D. R. *Reclaiming Alkali Soils*; Central Soil Salinity Research Institute: Karnal, 1973;0 p 58; Bull. 2.

3. Agarwal, R. R. Soil Classification for the Alluvium Derived Soils of the Indo-Gangetic Plains in Uttar Pradesh. *J. Ind. Soc. Soil Sci.* **1961,** *9*, 219–232.

4. Agarwal, R. R.; Gupta, R. N. *Saline Alkali Soils in India*; Indian Council of Agricultural Research: New Delhi, 1968, p 221; Tech. Bull. (Agric. Series).

5. Agarwal, R. R.; Mehrotra, C. L.; Gupta, C. P. Spread and Intensity of Soil Alkalinity with Canal Irrigation in Gangetic Alluvium of Uttar Pradesh. *Indian J. Agric. Sci.* **1957,** *27*, 363–373.

6. Agarwal, R. R.; Yadav, J. S. P. Diagnostic Techniques for the Saline and Alkali Soils of the Indian Gangetic Alluvium in Uttar Pradesh. *J. Soil Sci.* **1956,** *7*, 109–121.

7. Agarwal, R. R.; Yadav, J. S. P.; Gupta, R. N. *Saline and Alkali Soils of India*; Indian Council of Agricultural Research: New Delhi, 1979; p 286.

8. Alva, A. K.; Sumner, M. E. Alleviation of Aluminum Toxicity to Soybeans by Phospho-gypsum or Calcium Sulfate in Dilute Nutrient Solutions. *Soil Sci.* **1989,** *147*, 278–285.

9. Alva, A. K.; Sumner, M. E.; Miller, W. P. Reactions of Gypsum or Phospho-gypsum in Highly Weathered Acid Sub-soils. *Soil Sci. Soc. Am. J.* **1990,** *54*, 993–998.

10. Ashgar, A. G.; Dhawan, C. L. Effect of Irrigation and Growing Rice on Saline Soils. I. Manganese, Nitrogen and Phosphate Status of Soil. *Indian J. Agric. Sci.* **1947,** *17*, 199–202.

11. Ashgar, A. G.; Dhawan, C. L. Quality of Irrigation Water. Effect on pH and Exchangeable Bases of Soils. *J. Indian Chem. Soc.* **1948,** *25*, 179–184.

12. Atkins, A. T. *A Note by E.C. Buck and Minutes on `Enquiries on Reh and Oosur'*; Nainital: India, 1874; p 61.

13. Barnes, J. H.; Ali, B. Alkali Soils: Some Bio-chemical Factors in Their Reclamation. *Agric. J. India.* **1917,** *12*, 368–389.

14. Bhargava, G. P.; Sharma, R. C.; Pal, D. K.; Abrol, I. P. In *A Case Study of the Distribution of Salt Affected Soils in Haryana State,* Proc. Int. Symposium Salt Affected Soils; CSSRI: Karnal, **1980;** pp 83–91.

15. Bhumbla, D. R. In *Inaugural Address*, Proc. FCI–FAI Seminar on Use of Gypsum in Reclamation of Alkali Soils; FCI–FAI: New Delhi, 1977; pp 25–34.

16. BIS (Bureau of Indian Standards). *IS:6046.1982—Indian Standard: Specification for Gypsum for Agricultural Use (first revision) UDC 631-821-2 Reaffirmed May 2004: Amendment no. 2 September 2012 to IS 6046:1982: Specification for Gypsum for Agricultural Use (First Revision)*; Indian Standards Institution: Manak Bhavan, New Delhi, 2012; p 12.

17. Center, W. Note on Alkali or Reh Soils and Saline Water. *Rec. Geol. Surv. India.* **1880,** *43*, 253–273.

18. Chakravarti, P.; Chakravarti, S. Soils of West Bengal. *Ann. Indian Nat. Sci. Acad.* **1957,** *23*, 93–101.

19. Chen, L.; Dick, W. A. *Gypsum as Agricultural Amendment: General Use Guidelines*; Ohio State University: Ohio, 2011; p 35; Bulletin No. 945.

20. Chhabra, R. *Chemistry of Salt-affected Soils. Lecture Notes for UPLDC Training Program*; ICAR Research Complex for Eastern Region: Patna, India, 2003, (Unpublished).

21. Dhar, N. R. Molasses and Press Mud in Usar (alkali) Land Reclamation. *J. Indian Chem. Soc.* **1939,** *2*, 105–111.

22. Dhar, N. R.; Mukherji, S. K. In *Reclamation of Alkali Soils by Molasses and Press Mud*, Proc. Sugarcane Technologist Association, India, 1936; Vol. 5, pp 15–24.

23. Dhawan, C. L. *Land Reclamation (Reclamation of Saline and Alkali Soils)*; Central Board of Irrigation and Power: New Delhi, 1964; p 140.

24. Dhawan, C. L.; Dhand, A. D. The Occurrence and Significance of Trace Elements in Relation to Soil Deterioration, II Boron. *Indian J. Agric. Sci.* **1950,** *20*, 479–485.

25. Dhawan, C. L.; Malhotra, M. M. L.; Singh, J. Reclamation of Alkali Soils. Jantar (*Sesbania cannabina*) Root System in Black Alkali Soils Treated with Different Reclaiming Agents. *Bull. Int. Commun. Irrig. Drain.* **1953,** *2*, 76–82.

26. Dhawan, C. L.; Malhotra, M. M. L.; Singh, J. Reclamation of Alkali Soils. *Proc. Nat. Acad. Sci. India.* **1955,** *24V*, 625–639.

27. Dhawan, C. L.; Singh, J. Reclamation of Alkali Soils by Different Crop Rotations. *Proc. Nat. Acad. Sci. India.* **1958,** *27A*, 238–252.

28. Dregne, H.; Kassas, M.; Razanov, B. A New Assessment of the World Status of Desertification. *Desertif. Control Bull.* **1991,** *20*, 6–18.

29. Dudal, R.; Purnell, M. F. Land Resources: Salt Affected Soils. *Reclam. Reveg. Res.* **1986,** *5*, 1–10.

30. Emerson, W. W.; Bakker, A. C. The Comparative Effects of Exchangeable Calcium, Magnesium, and Sodium on Some Physical Properties of Red-brown Earth Sub-soils. II. The Spontaneous Dispersion of Aggregates in Water. *Aust. J. Soil Res.* **1973,** *11*, 151–157.

31. FAO. *Production Yearbook*; Food and Agricultural Organization: Rome, Italy, 1997, Vol. 51, pp 218–222.

32. FCI–FAI. Proc. FCI–FAI Seminar on Use of Gypsum in Reclamation of Alkali Soils; FCI–FAI: New Delhi, 1977; p 190.

33. Gedroiz, K. K. *Alkali Soils, Their Formation, Properties and Reclamation*; Reinhold Publishing Corporation: New York, NY, 1912; p 176.

34. Gupta, S. K.; Gupta, I. C. *Genesis, Reclamation and Management of Sodic (Alkali) Soils*; Scientific Publishers Ltd.: Jodhpur, 2016; p 322.

35. Henderson, G. S. *Alkali or Kallar Experiments and Completion Report of the Daulatpur Reclamation Station*; Sindh. Deptt. of Agri. Bombay, Govt. Central Press: Bombay, 1915; Bull. 64.

36. Hilgard, E. W. *Alkali Lands, Irrigation and Drainage in Their Mutual Relations*; College of Agriculture Annual Report 1986, University of California, Sacramento, CA,1986; p 56.

37. Hilgard, E. W. *Soils, Their Formation, Properties, Composition, and Relations to Climate and Plant Growth*; The Macmillan Co.: New York, 1906; pp 422–484.

38. Hissink, D. J. Base Exchange in Soils. *Trans. Faraday Soc.* **1924,** *20*, 551–566.

39. Hoon, R. C. *Land Reclamation*; Central Board of Irrigation and Power: New Delhi, 1955; pp 160–178; Publication 2.

40. Hoon, R. C.; Ahluwalia, G. S. The Physico-chemical Changes in Deteriorated Soils as a Result of Soil Reclamation Operation, II. *J. Irrig. Power.* **1953,** *10*, 38–42.

41. Hoon, R. C.; Dewan, R. The Physico-chemical Changes in Deteriorated Soils as a Result of Soil Reclamation Operations, I. *J. Irrig. Power.* **1949,** *6*, 265–272.

42. Hoon, R. C.; Jain, L. C.; Bhatnagar, B. B. The Physico-chemical Changes in Deteriorated Soils as a Result of Soil Reclamation Operation, IV. The Occurrence of Gypsum in Soils. *J. Irrig. Power.* **1950,** *7*, 60–68.

43. Hoon, R. C.; Mehta, M. L. *A Study of the Soil Profile of the Punjab Plains with Reference to Their Natural Flora*; Res. Pub. Punjab Irrigation Research Institute: Lahore, 1937; Vol. 3, p 3.

44. Jenny, H. E. W. *Hilgard and the Birth of Modern Soil Science*; Istituto di Chimica Agraria, Universitàndi Pisa, Pisa, Italy, 1961; p 10.

45. Kanwar, J. S. Reclaiming Deteriorated Soils. *Fert. News.* **1961,** *6*, 19–21.

46. Kanwar, J. S. In *Reclamation of Saline-alkali Soils*, Proc. Symposium Salinity and Alkalinity; Indian Agricultural Research Institute: New Delhi, 1962.

47. Kanwar, J. S.; Bhumbla, D. R. *Reclamation of Alkaline and Saline Soils in the Punjab*; Indian Council of Agricultural Research: New Delhi, 1960; Report.

48. Kelley, W. P. *The Reclamation of Alkali Soils*; Bull. 617; University of California, Agricultural Experiment Station, Berkeley, California, 1937; p 40.

49. Khan, A. D. *Usar Soils: Their Identification and Reclamation*; Report No. I; Agric. Animal Husbandry: Uttar Pradesh, India, 1950; pp 29–30.

50. Kovda, V. A. Alkaline Soda-saline Soils. *Agrokem. Talajtan. Suppl.* **1965,** *14*, 15–48.

51. Leather, J. W. Reclamation of *Reh* and *Usar* Lands. *Agric. Ledger,* **1897,** *7*, 1–13.

52. Leather, J. W. *Annual Report. Imp. Dep. Agric. 1904–5*; Government Central Press: Calcutta, India; 1906; p 67.

53. Leather, J. W. *Usar Investigations at Aligarh*. In *Investigations on Usar Lands in United Provinces*; Government Press: Allahabad, India; 1910; p 96.

54. Leather, J. W. *Investigations on Usar Lands in United Provinces*; Government Press: Allahabad, 1914; p 96.

55. Lin, Z.; Myhre, D. L.; Martin, H. W. In *Effects of Lime and Phospho-gypsum on Fibrous Citrus-root Growth and Properties of Sodic Horizon Soil*, Proceedings of the Soil and Crop Science Society of Florida, 1988, Vol. 47; pp 67–72.

56. Massoud, F. I. *Salinity and Alkalinity as Soil Degradation Hazard. FAO/UNDP Expert Consultation on Soil Degradation, AGL: 50/74/10*; FAO: Rome, 1974; p 21.

57. Mays, D. A.; Mortvedt, J. J. Crop Response to Soil Applications of Phospho-gypsum. *J. Environ. Qual.* **1986,** *15,* 78–81.

58. Mehta, K. K.; Yadav, J. S. P. Phospho-gypsum for Reclamation of Sodic Soils. *Indian Farming,* **1977,** *27,* 6–7.

59. Mehta, M. L. *The Formation and the Reclamation of Thur Lands in Punjab*; Punjab Irrigation Research Institute, Amritsar, India, 1940, p 64; Report 3.

60. Mehta, M. L. *Land Reclamation*; Central Board Irrigation Power: New Delhi, 1951; p 117.

61. Menon, P. K. R.; Iyenger, R. T. Arid, Alkali and Saline Soils. *Proc. Nat. Acad. Sci. India,* **1955,** *24A,* 637–646.

62. Mitchell, J. K. *Fundamentals of Soil Behavior*; John Wiley and Sons Inc.: New York, 1976; p 422.

63. MoA (Ministry of Agriculture). *Draft Report on Status of Land Degradation in India*; Ministry of Agriculture: New Delhi, 1994; p 102.

64. Mondal, A. K., Obi Reddy, G. P.; Ravisankar, T. Digital database of Salt Affected Soils in India Using Geographic Information System. *J. Soil Salinity Water Qual.* **2011,** *3,* 16–29.

65. Moreland, W. H. Reh. *Agric. Ledger.* **1901,** *13,* 416–463.

66. Mukerjee, B. K.; Aggarwal, B. R.; Mukerjee, P. Studies in Gangetic Alluvium of United Provinces I Cultivated Soils of Unnao District. *Indian J. Agric. Sci.* **1946,** *16,* 263–279.

67. Nasir, S. M. *Some Observations on the Barren Soils of the Bari Doab Colony in the Punjab*; Imp. Agric. Research Institute: Lahore, 1913; Bulletin 14.

68. Nasir, S. M. *Some Observations on the Barren Soils of the Lower Bari Doab Colony in the Punjab*; Imp. Agriculture Research Institute, Pusa, India, 1923; p 11; Bulletin 145.

69. Nayak, A. K.; Sharma, D. K.; Mishra, V. K.; Jha, S. K.; Singh, C. S.; Shahabuddin, Md.; Srivastava, M.; Kale, S. P.; Singh, G. B. *Phospho-gypsum: An Alternate Chemical Amendment for Sodic Lands*; CSSRI, RRS: Lucknow, 2011; p 42; Tech. Bull. 1/2011.

70. NBSS & LUP. *Soils of India*; National Bureau of Soil Survey and Land use Planning: Nagpur, 2002; p 130; Pub. No. 94.

71. NCA (National Commission on Agriculture). *Reports of the Commission*; Ministry of Agriculture and Irrigation, Government of India: New Delhi, 1976; p 373; Part V.

72. Northcote, K. H.; Skene, J. K. M. *Australian Soils with Saline and Sodic Properties*; CSIRO Australia, Division of Soils, Canberra, 1972; p 62; Soil Publication No. 2.

73. NRSA (National Remote Sensing Agency). *Mapping Salt-affected Soils of India on 1:250,000 Scale*; NRSA: Hyderabad, India, 1996.

74. Oldeman, L. R.; Van Englen, V. W. P.; Pulles, J. H. M. *The Extent of Human-induced Soil Degradation.* In *World Map of Status of Human-Induced Soil Degradation: An Explanatory Note*; Oldeman, et al., Eds.; International Soil Reference and Information Centre (ISRIC): Wageningen, 1991; pp 27–33.

75. Overstreet, R.; Martin, J. C.; King, H. M. Gypsum, Sulfur, and Sulfuric Acid for Reclaiming an Alkali Soil of the Fresno Series. *Hilgardia.* **1951,** *21,* 113–141.

76. Pavan, M. A.; Bingham, F. T.; Peryea, F. J. Influence of Calcium and Magnesium Salts on Acid Soil Chemistry and Calcium Nutrition of Apple. *Soil Sci. Soc. Am. J.* **1987,** *51,* 1526–1530.

77. Puri, A. N. *The Relation between Exchangeable Sodium and Crop Yield in Punjab Soils and a New Method to Characterize Alkali Soils*; Research Publication, Punjab Irrigation Research Institute: Amritsar, India, 1933; Vol. 5, p 4.

78. Puri, A. N.; Taylor, E. M.; Mckenzie; Ashgar, A. G. *Soil Deterioration in the Canal Irrigated Areas of Punjab. III. Formation and Characteristics of the Soil Profiles in the Alkaline Alluvium of the Punjab*. Research Publication, Punjab Irrigation Research Institute: Amritsar, India, 1937; Vol. 4, p 7.

79. Puri, G. S. *Indian Forest Ecology*; Oxford Book and Stationary Co.: New Delhi, 1960.

80. Sanayasi, R. M.; Iyer, G. T. A. Reclamation of Alkaline Lands in the Cauvery Metur Project Area. *Proc. Nat. Acad. Sci. India.* **1955**, *24A*, 606–610.

81. Rengasamy, P. World Salinization with Emphasis on Australia. *J. Exp. Bot.* **2006**, *57*, 1017–1023.

82. Rengasamy, P. Soil Processes Affecting Crop Production in Salt-affected Soils. *Funct. Plant Biol.* **2010**, 37, 613–620.

83. Rengasamy, P.; Olsson, K. A. Sodicity and Soil Structure. *Aus. J. Soil Res.* **1991**, *29*, 935–952.

84. Richards, L. A., Ed. *Diagnosis and Improvement of Saline and Alkali Soils*; U.S. Dept. of Agriculture, Riverside, CA, 1954; p 160; Handbook No. 60.

85. Rohtagi, B. B.; Saxena, S. K.; Singh, R. In *Availability of Quality Gypsum from Indigenous Sources as a Soil Amendment*, Proc. FCI–FAI Seminar on Use of Gypsum in Reclamation of Alkali Soils; FCI–FAI, New Delhi, 1977; pp 37–55.

86. Schoonover, W. R. *Examination of Soils for Alkali*; Univ. of California. Extension Service: Berkeley, California, 1952.

87. Sehgal, J.; Abrol, I. P. *Soil Degradation in India—Status and Impact. Transactions. 15th International Society of Soil Science*; International Society of Soil Sciences: Acapulco, Mexico, 1994; Vol. 7a, p 212.

88. Shrotiya, G. C.; Mishra, U. N. Utilization of Phospho-gypsum for Agricultural Purposes. *Fert. News.* **1976**, *21*, 37–38.

89. Sigmond, A. A. J. D. *Hungarian Alkali Soils and Methods of Their Reclamation*. Calif. Agr. Expr. Sta. Special Publication, University of California: USA, 1922; p 156; English Translation Published in 1927.

90. Singh, D.; Nijhawan, S. D. A Study of the Physico-chemical Changes Accompanying the Process of Reclamation in Alkali Soils. *Indian J. Agric. Sci.* **1932**, *2*, 1–18.

91. Singh, N. T. *Salt Affected Soils in India*. In Land and Soils; Khoshoo, T. N., Deekshatulu, B. L., Eds.; Indian National Science Academy: New Delhi, 1992; pp 65–100.

92. Singh, N. T.; Hira, G. S.; Bajwa, M. S. Use of Amendments in Reclamation of Sodic Soils in India. *Agro. Es. Talaj.* **1981**, *30*, 158–177.

93. Singh, Y. P. *Sustainable Reclamation and Management of Sodic Soils: Farmers' Participatory Approaches* (Chapter 9). In Soil Salinity Management in Agriculture: Technological Advances and Applications; Gupta, S. K., Goyal, M. R., Eds.; Apple Academic Press: USA, 2016; pp 289–315.

94. Soil Science Society of America, Committee on Terminology. Supplemental Report of Definitions Approved by the Committee on Terminology. *Soil Sci. Soc. Amer. Proc.* **1960**, *24*, 235–236.

95. Sumner, M. E. *Gypsum as an Amendment for the Subsoil Acidity Syndrome*; Final Report, Florida Institute of Phosphate Research, Bartow, FL, 1990, Project 83-01-024R.

96. Szabolcs, I. *Salt Affected Soils in Europe*; Martinus Nijhoff: The Hague, 1974; p 63.

97. Szabolcs, I (with Bibliography Compiled by Varallyay,G.). *Review of Research on Salt Affected Soils*; UNESCO: Paris, 1979; p 137.

98. Taylor, E. M.; Puri, A. N.; Asghar, A. G. Soil Deterioration in the Canal-irrigated Area of the Punjab. I. Equilibrium Between Ca and Na Ions in Base Exchange Reactions. *Indian J. Agric. Sci.* **1934,** *4,* 8.

99. US Salinity Laboratory Staff. Report of the Nomenclature Committee Appointed by the Board of Collaborators of the U.S. Salinity Laboratory. *Soil Sci. Am. Proc.* **1958,** *22,* 270.

100. van Beekom, C. W. C.; van den Berg, C.; de Boer, Th. A.; van der Molen, W. H.; Verhoeven, B.; Westerhof, J. J.; Zuur, A. J. Reclaiming Land Flooded with Salt Water. *Neth. J. Agric. Sci.* **1953,** *1,* 153–163.

101. Voelcker, J. A. *Report on the Improvement of Indian Agriculture*; Eyre and Spottiswoode: London, 1897; p 507.

102. Whitcombe, E. *Agrarian Conditions in Northern India. United Provinces Under British Rule. 1860–1900*; University of California Press: Berkley, USA, 1972; Vol. 1, p 330.

103. www.albertaefp.com. *The Alberta Environmental Farm Plan.* Agricultural Research and Extension Council of Alberta, Alberta, Canada (accessed March 10, 2015).

CHAPTER 2

SOIL SALINITY MANAGEMENT IN FRUIT CROPS: A REVIEW OF OPTIONS AND CHALLENGES

ANSHUMAN SINGH, D. K. SHARMA, RAJ KUMAR,
ASHWANI KUMAR, R. K. YADAV, and S. K. GUPTA

ABSTRACT

Salinity is a severe impediment to the sustainable management of over 800 M ha arable land area globally; especially in irrigated arid and semiarid regions. Unsound on-farm irrigation management is the major driver of secondary salinization in irrigated lands where even proven salinity management technologies fail to deliver the expected dividends. Salinity and associated problems like waterlogging alter cell physiology and metabolism in ways that greatly reduce plant growth and economic yields. About two-thirds of the fruit crops grown commercially in world are categorized as salt sensitive. Despite their high sensitivity to salt-induced soil perturbations, fruit crops perform well in salt-affected soils when grown with the aid of improved management practices. The information presented in this chapter leads to the conclusion that agronomic interventions such as selection of salt tolerant scion and rootstock cultivars, refined planting techniques, balanced nutrition, and drip irrigation can be of considerable help in commercial fruit cultivation in salt-affected lands otherwise considered to be unsuitable for high-value crops.

2.1 INTRODUCTION

Salinity has been a part of the global landscape since time immemorial and is a severe threat to agricultural production in many parts of the world. Although, soil salinity related stresses are common in most agro-ecological regions, yet the adverse impacts of salinity are often more pronounced in arid and semiarid climates as a result of restricted salt leaching due to scanty rainfall and high evaporation.[1,153] Presence of native salts and the existence of shallow saline groundwater exacerbate salinity build-up in arid and semiarid irrigated lands.[38] In humid coastal regions, rainfall far exceeds evapotranspiration that has caused water losses resulting in net downward movement of soluble salts, but the temporal dynamics of rainfall results in development of salinity especially during the post monsoon periods. Estimates by Dregne et al.,[41] Oldeman et al.,[133] and FAO[50] suggest that around 831 M ha area is salt affected out of which 397 M ha is saline and 434 M ha is sodic. More than 77 M ha of land worldwide has become salt affected due to anthropogenic causes (secondary salinity); of this, 45.4 M ha are irrigated land (in the world's semiarid and arid regions) and the remaining rain-fed. Clearing of perennial trees and pastures for growing the annual field crops is the major cause of secondary salinization in dry lands. Replacement of perennial vegetation with food crops induces deep percolation of water[100,114] causing gradual rise in water table (Fig. 2.1) and the consequent accumulation of salts in crop root zone.[66,72,179] Faulty irrigation practices and neglect of drainage component in irrigation projects is the major cause of secondary salinization in irrigated lands.[185] Irrigation-induced salinity, attributed to deep percolation of water and rise in the groundwater table, accelerates salt movement from the lower to upper soil horizons (root zone) adversely *affects* soil properties and plant growth. In India, 6.73 M ha area is salt affected, 3.77 M ha being sodic and 2.96 *being* saline. Central Soil Salinity Research Institute (CSSRI)[31] projected that M ha total area under salt-affected soils would increase to 16.20 M ha by 2050; an increase of about 9.5 M ha in the extent of the twin problems of water logging and soil salinity.

Over the years, the problem of secondary salinization has substantially increased in both irrigated[151] and rain fed lands[54] resulting in heavy crop losses and in extreme cases abandonment of the arable lands. A relevant example in the case is massive water logging and salinity build-up in large tracts of north-western India—an agriculturally important region crucial to the food security of the country and Aral Sea Basin in former USSR. In extreme cases, secondary salinization and prolonged submergence render the affected lands entirely unsuitable for crop production and cause environmental degradation. Farm incomes and rural employment drastically decrease increasing the distress amongst the farming and nonfarming communities.[165] What adds to worry is the fact that sometimes even proven technologies fail to alleviate water logging and salt hazard due to diverse environmental and socioeconomic limitations.[70] As the success of soil reclamation projects as well as sustained yields from the reclaimed lands are inextricably linked to the availability of fresh water, emerging issues in agricultural water management must be addressed to draw the future course of action for salinity management.[118]

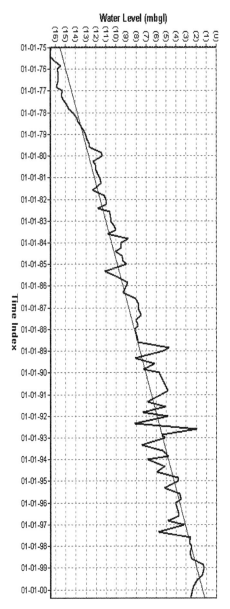

FIGURE 2.1   (See color insert.) Rise in water table (0.6 m per annum) upon introduction of irrigation in Bhakra Canal Command in Haryana, India. *Note*: mbgl, meter below ground level.

2.1.1   SOIL SALINITY AND PLANT GROWTH

Although plant life evolved under extremely saline conditions in the seas and oceans, yet most of the domesticated crops do not exhibit tolerance to salts. In most of the halophytic (salt tolerant) species, concentrations of both organic and inorganic solutes far exceed than those of glycophytes (salt sensitive) which suggest the existence of a "metabolic and regulatory network" for salt stress adaptation in halophytes for survival and growth in saline environments.[57]

It seems that in the process of evolution, most of the characters conferring the salt tolerance have been lost. In this adaptation process, plants in contrast to nutrient-rich (eutrophic) environment of marine water have to establish themselves in fresh water and land environments low in nutrients (oligotrophic). It is, therefore, possible that a transition from marine to fresh water and ultimately to land environment would have necessitated adaptive changes in root system for nutrient acquisition leading to changes in osmotic adjustment capacity in land plants.[59] Besides, adaptation to the terrestrial environment characterized by alternate wetting-drying conditions, availability of good quality rain water, soil nutrients, and acclimatization to harsh conditions necessitated evolutionary changes in shoot and root system such that salt tolerance gradually diminished with plant evolution.[164]

This chapter focuses on the overview of salt tolerance in fruit crops and the measures available to sustain fruit cultivation in saline environments. The constraints to grow fruit crops in salt-affected lands, with special reference to climate change induced stressors and water scarcity are critically examined. The physiological bases of salinity tolerance are briefly reviewed to understand the diverse mechanisms underlying salt tolerance in fruit species. Taking into account the loopholes in conventional measures of salinity mitigation, the future research priorities to enhance salt tolerance through genetic improvement, agronomic manipulations, and irrigation management are thoroughly explored.

2.2   SALINITY AND SODICITY

Salinity as a generic term refers to "saline" and "sodic" soils and the marginal quality "saline" and "sodic" groundwater unsuitable for human consumption and crop irrigation. Saline soils contain excess of soluble salts, mainly chlorides and sulfates of Na^+, Ca^{2+}, and Mg^{2+}, raising the soil saturation extract electrical conductivity ($EC_e \geq 4$ dS m^{-1}) leading to "osmotic stress" and "specific ion effects" in the plants. These soils have ESP < 15 and pH_s < 8.2. The sodic soils, on the contrary, have high exchangeable sodium percentage (ESP > 15), which adversely affects water and air flux, water holding capacity, root penetration, and seedling emergence.[126,165] These soils have variable EC_e and pH_s > 8.2. Sodic soils having EC_e > 4.0 dS m^{-1} are sometimes referred as saline–sodic although reclamation and management procedures are similar to sodic soils.

In sodic soils, excess Na^+ replaces other cations such as Ca^{2+}, Mg^{2+}, and K^+ bound to the clay particle leading to structural problems and poor aggregate stability. Clay particles become less tightly bound to each other with consequent dispersion of soil aggregates. Sodic soils show surface crusting and hard setting and suffer from water inundation and oxygen deficiency affecting the crop emergence and root growth. Clay dispersion also increases the compaction and strength of subsoil restricting the water and oxygen flux to plant roots.[34] In sodic soils, besides the salt tolerance of the plants, their tolerance to water stagnation assumes significance because the plants are exposed simultaneously to both the stresses.

Although widespread salinization is a severe limitation to global agriculture,[54,126,140,150] some optimistic accounts unravel the scope for technology-driven improvements in agricultural production from marginally to moderately salt-affected lands.[184,201] One recent finding suggests that about 85% of the saline/sodic soils have only minor to moderate constraints for crop production[201] and thus can be put to productive use by capitalizing the potential of available technologies with suitable modifications. Emphasis on appropriate solutions for enhancing the productivity of degraded lands in tandem with arresting the losses caused by secondary salinization holds the promise for sustainable gains. While embarking on soil reclamation projects, ecosystem characteristics must be fully considered so as to fine-tune the existing practices for site specific requirements. Efforts should be made to integrate appropriate plant species, soil erosion control measures, rainwater harvesting, and efficient irrigation methods with the routine practices to obtain sustainable results.[97] The failure of soil reclamation and management projects in most cases is attributed to high costs, lack of technical know-how, and practical difficulties in implementation, necessitating the emphasis on low-cost, durable technologies to provide long-term solution to this environmental menace in low-income countries facing insurmountable constraints such as fresh water shortages, massive land degradation, poverty, and feeble policy support.[98,115,165]

2.3   EFFECTS OF SALT STRESS IN FRUIT CROPS

Salinity adversely affects cell physiology and metabolism in a number of ways. The cell-specific events, which affect key metabolic pathways and cause injury in salt stressed plants, include cell membrane damage due to electrolyte leakage (EL) and lipid peroxidation (LP), oxidative stress caused by the free oxygen radicals, impaired water relations, altered gas exchange characteristics, and ion toxicities. Depending on factors such as salt concentration, crop species and growth stage, these metabolic impairments ultimately lead to the appearance of damage symptoms, stunted growth and yield reduction in plants exposed to salinity stress.

2.3.1   CELL MEMBRANE DAMAGE

Abiotic stresses including salinity trigger EL, mainly K^+ efflux, from plant cells. As the most abundant macronutrient, K^+ plays a crucial role in plant metabolism. Evidence is growing that K^+ efflux mediated by selective and nonselective cation channels is also vital for adaptation to stress conditions[40] as it facilitates a switch over from "normal" to "hibernated" state during initial hours of stress exposure and ensures growth resumption when stress is released.[163] Under saline conditions, excessive Na^+ uptake causes membrane depolarization leading to K^+ efflux via depolarization-activated outward-rectifying K^+ channels.[162] Under moderate stress, K^+ efflux saves metabolic energy for adaptation. Severe stress, on the contrary, accentuates K^+ loss leading to programmed cell death (PCD),[39] which is an ordered series of events displacing the damaged cells for cellular differentiation and tissue homeostasis in stressed plants.[162]

Salt-induced PCD-mediated replacement of damaged root cells facilitates the production of new roots with better adaptive capacity to grow in the saline substrate.[84] As salinity triggered leakage of solutes is common in higher plants, EL measurements could provide a reliable estimate of salt-induced cell injury. EL increased linearly in olive as NaCl concentration increased from 33 to 166 mmol L^{-1}.[139] In *kinnow* (*Citrus nobilis* × *C. deliciosa*) budded on *Citrus jambhiri*, salt treatment (50 mmol L^{-1} NaCl) enhanced EL by about 85% compared to control.[89] Malondialdehyde (MDA) level, an end product of LP, is also a reliable indicator of cell membrane damage in salt-treated plants.[131]

In citrus rootstock *Carrizo citrange* seedlings, NaCl application increased MDA production by 1.3–1.5-fold than the control. The fact that salt-induced transient increase in leaf MDA levels was moderate and even slightly decreased with time seems to be due to an effective antioxidant protection.[10] Plants of strawberry cultivars exposed to 80 mmol L^{-1} of NaCl differed in MDA production with Korona and Elsanta cultivars showing 1.35 and 1.5-fold increase in fruit MDA concentration, respectively, than the control.[91] Cashew genotype CCP06 showed marked increase (~56%) in MDA content in root plasma membranes under NaCl-induced salinity (8 $dS\ m^{-1}$) as compared to BRS 189, which recorded only marginal difference with control plants and thus succeeded in avoiding oxidative damage to the membrane lipids.[7] These examples tend to show the correlation between low EL and MDA levels, and better membrane integrity in salt stressed plants. The genotypes showing smaller EL and MDA values following salinity treatments are likely to be more salt tolerant than others. In saline substrates, Na^+ displaces Ca^{2+} ions involved in pectin-associated crosslinking and plasma membrane binding in cell membranes.[49] A variety of membrane proteins, lipids, and osmo-protectant compounds such as glycinebetaine and proline protect cell membranes in salt tolerant genotypes.[110]

2.3.2 OXIDATIVE STRESS

Although normally produced during plant metabolism, stress conditions induce rapid generation of harmful active oxygen species (AOS), also referred to as reactive oxygen species (ROS), such as superoxide radicals (O_2^-), singlet oxygen $(^1O_2)$, hydrogen peroxide (H_2O_2), and hydroxyl radical (OH) in plant cells.[119] Under stress conditions, the ability of plants to scavenge AOS is greatly reduced causing excessive accumulation of free radicals. Given their highly reactive nature, the AOS disrupt cellular function by causing oxidative damage to cell membranes and organelles, vital enzymes, photosynthetic pigments, and bio-molecules such as lipids, proteins and nucleic acids. To overcome the potential damage, plants synthesize diverse antioxidant compounds for the detoxification and removal of the deleterious free radicals with the degree of protection depending on factors such as species/cultivar, growth stage, and the type and duration of stress.[119] In fruit plants, protective antioxidant systems consist of both enzymatic and nonenzymatic components and are discussed under the Section 2.5.

2.3.3 ALTERATIONS IN LEAF WATER RELATIONS

Due to presence of soluble salts, osmotic potential of soil water increases (becomes more negative) making it difficult for roots to absorb the water.

In order to maintain turgor and water uptake for growth, plants tend to keep internal water potential below that of the soil by accumulating inorganic ions and metabolically compatible solutes. Although uptake of solutes (Na^+ and Cl^-) does not require energy use, yet their accumulation often proves toxic to cell cytosol. In contrast, synthesis of compatible solutes such as proline requires energy use and thus reduces plant growth.[184] As with most of the woody crops, fruit species show decrease in leaf water potential (Ψ_w) and leaf osmotic potential (Ψ_s) with consequent changes in leaf turgor potential (Ψ_p) under saline conditions.[24] Salinity (50 mmol L^{-1} of NaCl) decreased Ψ_w and Ψ_s in *Citrus xlimonia* seedlings.[117] High salinity in irrigation water (8.5 dS m^{-1}) reduced Ψ_w and Ψ_s in both younger and older leaves of strawberry cultivars Rapella, Ostara, Fern, and Selva, but the consequent reduction in Ψ_p was more in older than in younger leaves.[13] Accumulation of Na^+ and Cl^- in leaves caused a reduction in Ψ_w and Ψ_s resulting in Ψ_p values equal to or above the control level in citrus under NaCl (50 mmol L^{-1}) induced salinity.[106] Predawn leaf water potential (Ψ_{pd}) significantly decreased (~60%) in loquat plants treated with 50 mmol L^{-1} NaCl.[62] There were only slight differences in Ψ_w in control and salinized pear trees.[130] In cherry cvs. "*Bigarreau Burlat*" (BB) and "*Tragana Edessis*" (TE) grafted on "Mazzard" rootstock, saline irrigation (25 or 50 mmol L^{-1} NaCl) decreased Ψ_w and Ψ_s. Although Ψ_p in NaCl-treated BB plants was considerably higher than control plants, salinity stress did not affect Ψ_p of TE plants. The fact that Ψ_p did not change in TE and increased in BB with increasing salinity appears to be contradictory to less tolerance in BB.[136]

2.3.4 LOSS OF LEAF PIGMENTS

Salinity in root zone decreases the concentration of leaf pigments with sensitive genotypes showing greater reduction than tolerant ones. Occasionally, salt tolerant cultivars may exhibit high leaf chlorophyll (Chl) levels particularly under moderate salinity.[182] Both Chls (Chl-*a* and Chl-*b*) and carotenoids are affected in salinized plants albeit at different rates.[2] Data presented in Table 2.1 illustrate that crop genotypes vary in their response to salinity induced loss of leaf Chl. Nutritional deficiencies in leaves may partly account for the degeneration of pigment molecules which can be alleviated to some extent by the application of amendments and fertilizers. Under certain conditions, salt tolerant genotypes exhibit marginal increase in leaf Chl relative to control plants. Salt stressed (soil EC_e 6.5 dS m^{-1}) *bael* (*Aegle marmelos* Correa cv. NB-5) plants had higher leaf Chl than control plants.[173] Both Chl and carotenoid concentrations were increased up to 68 mmol L[-1] NaCl level but declined with further increase in salinity in pineapple under in vitro conditions. Salinized plants of cv. "N36" had higher pigment concentration than cultivars "Morris" and "Sarawak."[14] Chl is a membrane bound pigment and its integrity depends on membrane stability. As cell membranes are damaged under saline conditions, Chl seldom remains intact. Again, salt-induced increase in chlorophyllase activity and accumulation of Na[+] and Cl[-] ions in the leaves accentuates the rate of Chl degradation.[5,173]

TABLE 2.1 Loss of Leaf Chlorophyll Under Salinity Stress.

Crop	Effects	References
Aonla	Total leaf chlorophyll decreased with increasing salinity (5–15 dS m[-1]); maximum and minimum reductions were noted in cultivars NA-18 and Chakaiya, respectively	[148]
Date palm	Cultivar Lulu showed significantly lower total leaf chlorophyll values of 1.34 and 1.07 mg g[-1] under 15 dS m[-1] and 30 dS m[-1] seawater treatments, respectively. Application of Pentakeep-v™ (5-aminolevulinic acid-based fertilizer; 300 mL/pot), however, significantly enhanced the total chlorophyll in salinized plants	[206]
Guava	Salinity induced (30 mmol L[-1] NaCl) reduction in leaf chlorophyll was attributed to decrease in chlorophyll synthesis caused by N and Mg deficiencies	[5]
Mango	Increasing salinity (15–45 mmol L[-1] NaCl) decreased chlorophyll concentration; rate of chlorophyll degradation was rapid in cv. "Alphonso," intermediate in cvs. "Taimour," "Ewaise," and "Hindy Bisinnara" and slow in "Zebda"	[123]
Mulberry	Chlorophyll and total carotenoids significantly decreased under salinity (12 dS m[-1])	[2]
Olive	Chlorophyll *a*, *b*, and (*a* + *b*) levels decreased in salt (40–160 mmol L[-1] NaCl) treated plants; reduction in chlorophyll *a* was very high in cv. "Roghani" than in "Zard"	[125]

2.3.5 GAS EXCHANGE CHARACTERISTICS

Both stomatal and nonstomatal factors affect photosynthetic assimilation in salt-treated plants. While restricted stomatal conductance limits CO_2 supply to the chloroplast cells, limited CO_2 transport in mesophyll cells may be caused by cell membrane leakage, leaf shrinkage induced alterations in the structure of intercellular spaces and biochemical regulations.[127] Diffusive resistance, that is, decrease in CO_2 conductance in stomata and mesophyll cells is seen at mild to moderate salt levels. Disruption of biochemical pathways under high salinity, especially under high irradiance, is a pointer to severe oxidative stresses in salt stressed plants.[56]

Nonstomatal inhibition of photosynthesis is attributed to salt-induced impairments in photosynthetic apparatus independent of stomatal closure. It is seen that salinity either decreases the efficiency of ribulose-1,5-bisphosphate (RuBP) carboxylase or limits RuBP regeneration capacity.[20] Poor stomatal conductivity and the consequent decline in CO_2 availability in chloroplast decrease the carboxylation of RuBP leading to low carbon fixation in salt stressed plants.[92] Salt-induced structural defects in thylakoid—the site of light-dependent reactions of photosynthesis—such as swelling and expansion of thylakoid membrane[65] and deformities in grana arrangement[145] appear to be hastened by light as dark growing plants, despite excessive accumulation of Na^+ and Cl^- in leaves, tend to maintain thylakoid integrity[120] suggesting the involvement of light induced oxidative stress in destabilization process.[121] Under extreme salinity and high light intensity conditions, complete loss of thylakoid structure is likely.[121] Under saline conditions, excess toxic ions (Na^+ and Cl^-) can also despair carbon assimilation in leaf tissue.[127] It is seen that plants exposed to mild saline stress show rapid recovery after stress is relieved. Under high salinity, however, only 40–60% of the maximum photosynthetic rate is attained subsequent to stress release.[26]

Salinity (EC_{iw}: 5–20 dS m^{-1}) adversely affected stomatal conductance and net photosynthesis in *Ziziphus rotundifolia* and *Z. nummularia* seedlings, but reductions were relatively higher in *Z. nummularia*. Higher CO_2 uptake and better photosynthesis in *Z. rotundifolia* appeared to be due to osmotic adjustment, high Chl retention and restricted translocation of Na^+ from root to shoot.[116] At 60 mmol L^{-1} of NaCl salinity, mango rootstock *Mangifera zeylanica* showed about 2.6-fold higher CO_2 assimilation as compared to *M. indica* "13-1." Significantly lower Na^+ contents were found in roots and young leaves of *M. zeylanica* as compared to "13-1."[159] Supplemental nutrition with 10 mmol L^{-1} Ca(NO$_3$)$_2$ alleviated salt stress (60 mmol L^{-1} NaCl) in guava seedlings as was evident from over 2-fold increase in leaf photosynthesis over salinized plants growing in absence of Ca(NO$_3$)$_2$.[47] Irrigation with seawater (15 and 30 dS m^{-1}) not only significantly decreased the gas exchange but also accelerated the respiratory carbon loss resulting in reduced photosynthetic assimilation and increased CO_2 compensation point in seedlings of date cultivar Lulu. Treatment with 0.08% ALA(5-aminolevulinic acid)-based fertilizer Pentakeep-v™(300 mL pot[-1]), however, alleviated salt stress as evident by significant improvements in the biochemical efficiency of carbon fixation and the rate of electron transport required for RuBP regeneration at 15 dS m[-1] salinity.[205]

Rooted cuttings of grape cultivars Rish-Baba and Sahebi showed decrease in photosynthesis, stomatal conductance, and transpiration with increasing salinity and photosynthesis virtually ceased in both the cultivars at the highest salinity levels (10.1 and 11.4 dS m[-1]). Internal CO_2 level initially decreased but thereafter steadily increased indicating the role of nonstomatal factors such as Chl content play as major determinants of leaf photosynthesis in salt stressed grapes.[76] Addition of 200 mmol L[-1] NaCl in irrigation water caused the highest photosynthetic reductions in olive cultivars having inherently high photosynthesis such as Kerkiras than in those with low photosynthesis (e.g., Chalkidikis). No relationship was observed between Na^+ accumulation and photosynthesis reduction in either young or old leaves. Strong correlation between low photosynthesis and low mesophyll conductance in both type of cultivars revealed that low chloroplast CO_2 concentration was the main limitation of photosynthesis in olive leaves.[108] Moderate salinity (up to 50 mmol L[-1] NaCl) did not have any significant effect on CO_2 assimilation and stomatal conductance in olive cultivars except in Megaritiki in which appreciable decrease was noticed.[25]

2.3.6   OSMOTIC STRESS AND ION TOXICITY

Salinity-induced growth suppression may either be due to osmotic effect or ionic injury. The reduction in growth corresponds to the decline in soil water potential, osmotic stress being the dominant factor. It is possible to distinguish osmotic stress from specific ion effects in a given crop by using isosmotic (i.e., having equal osmotic pressure) concentrations of different salts.[199] Ideally, isosmotic solutions of inert osmoticum such as polyethylene glycol that do not penetrate the cell wall are used along with salt solutions.[208] If the adverse effects of a particular salt outweigh than those of other isosmotic solutions, then injury is attributed to Na^+ and Cl^- ions.[199] In fast-growing species such as peach, high transpiration rate decreases salt exclusion by roots and thus excess salts accumulate in leaves and shoots. In salt tolerant crops such as olive, the absorbed salts are partitioned and retained in basal stems and older leaves minimizing damage to the actively growing young leaves.[112] High salt concentrations cause "physiological drying" of soil—plant available water becomes unusable due to trapping by salt ions—hindering water absorption by the roots. Under osmotic stress, fruit plants such as citrus tend to accumulate compatible osmo-protectants and inorganic ions to lower the leaf water potential for resuming the water uptake. Although helpful in improving the water relations, this strategy also proves detrimental due to continual accumulation of toxic Cl^- ions in the foliage.[10]

Of the total number of fruit species commercially grown in different parts of the world for which salinity tolerance threshold has been worked out by different researchers, about 75% are sensitive to salinity. In comparison, salt sensitive genotypes in other crops range from 5% to 15% of the species listed.[16] The two-phase inhibition of plant growth in saline soils, which involves an initial osmotic shock followed by ion injuries, differs with the crop. In annual plants, salt-induced injury symptoms generally appear within few days but in perennial crops salt injury may become noticeable after months or even years. While osmotic stress hampers growth in both tolerant and sensitive genotypes, specific salt effects often prove deleterious to sensitive ones.[126] In contrast to annual crops, which tend to become more salt tolerant with increase in age, salt tolerance generally declines in fruit crops as they grow older. Translocation of salts stored in roots to the growing leaves coupled with relatively slower growth in trees may partly explain this tendency. It is generally seen that salt sensitive species such as citrus and stone fruits often absorb toxic amounts of Na^+ in otherwise normal soils. Under certain conditions, Na^+ and Cl^- may not be the predominant ions in saline soils and the use of rootstocks that restrict the uptake of these toxic ions may render specific salt effects relatively unimportant; and osmotic inhibition will thus virtually cause most of the deleterious effects in salinized fruit plants.[16]

Fruit species vary in their sensitivity to the harmful effects of Na^+ and Cl^- ions. In citrus fruits, for example, the adverse effects of salinity are mainly due to accumulation of Cl^- ions in leaf and shoot tissues. When leaf Cl^- concentration reaches around 1.5% on dry weight basis, ethylene production is triggered resulting in leaf abscission. Therefore, salt tolerance in many citrus species depends on ability of roots to prevent or reduce Cl^- uptake from the saline soils.[107] In salt-sensitive fruit species, visible injury symptoms are attributed to Na^+ or Cl^- only if affected leaves contain more than 0.2% Na^+ or 0.5% Cl^-.[16,174] Salinity adversely affects mineral nutrition in fruit crops. Reduced N uptake by salinized plants is attributed higher Cl^- uptake which decreases the leaf and shoot-NO_3^- concentration. Decreased P availability in saline soils is due to ionic strength effects that reduce P activity and the low solubility of Ca–P minerals. Sodium-induced K^+ deficiency also adversely affects growth and yield in different crops. Reduction in K^+ uptake in plants by Na^+ is a competitive process and occurs regardless of whether the solution is dominated by Na^+ salts of Cl^- or SO_4^{2-}. The presence of adequate Ca^{2+} in the substrate influences the K^+/Na^+ selectivity by shifting the uptake ratio in favor of K^+ at the expense of Na^+. Improvement in Ca^{2+}-mediated membrane integrity invariably leads to reduction of K^+ leakage from root cells and a more favorable root-K^+ status.[69]

2.4 TREE GROWTH AND DEVELOPMENT UNDER SALT STRESS

Salinity adversely affects seedling establishment, vegetative growth, and fruit yield in fruit crops. These adverse effects are briefly discussed under this section.

2.4.1 SEED GERMINATION AND SEEDLING ESTABLISHMENT

Available evidence suggests that salinity delays and decreases the seed germination as well as adversely affects seedling establishment in fruit crops (Table 2.2). The extent of damage varies with genotype and the type of salt. Some genotypes exhibit distinctly high salt tolerance as compared to others which even fail to germinate. Decrease in osmotic potential of the substrate retards water imbibition by seeds resulting in embryo injury[207] and delayed radicle protrusion.[194] Accumulation of Na^+ and Cl^- ions to toxic levels and poor reserve mobilization in seed cotyledons also hamper seed metabolism.[194] Certain salts are more harmful than others, for example, NaCl being more harmful to seed germination in guava than $CaCl_2$ and Na_2SO_4.[90] In certain cases, salt-induced inhibition of germination may partially be overcome by seed treatment with growth regulators, but they fail to enhance germination at high salt concentrations.[96] Screening of diverse germplasm using seeds can be of paramount significance in polyembryonic species such as mango and citrus where elite, true breeding polyembryonic genotypes are used as clonal rootstocks. Evaluation of salt tolerance in seeds may also significantly enhance the understanding on biochemical regulation of salt stress adaptation in seed tissues.[172]

TABLE 2.2 Effects of Salinity on Seed Germination and Seedling Growth.

Crop	Effects	References
Citrus	Based on relative delay in seed germination and reductions in seedling height, stem girth, fresh and dry aerial and root biomass; Rangpur lime was adjudged tolerant, rough lemon moderately tolerant and trifoliate orange susceptible rootstock to salinity	[172]
Guava	Saline irrigation decreased seed germination with about 25% germination recorded at EC_{iw} 1.5–3.0 dS m^{-1}. Irrigation with 4.5 dS m^{-1} saline water gave an intermediate response but germination was below 10% when salinity reached 6 dS m^{-1}. High salinity also considerably reduced seedling survival	[22]
Mulberry	Among 43 genotypes, in vitro screening indicated considerable salt tolerance in English Black, Rotundiloba, KPG-3, Ko-litha-3, Mysore Local, and Sultanpur which showed survival up to 25.4 dS m^{-1} salinity	[193]
Peelu	Stem and root elongation was retarded by increasing salinity, but seedlings survived and grew up to 16.5 dS m^{-1} salinity	[147]
Strawberry	In vitro NaCl salinity (100 mmol L^{-1}) had negligible effect in "Silver Jubilee" and "Purpuratka." "Dukat" showed an intermediate response (4.6% reduction), but "Clone no. 1386" and "Selva" were most salt sensitive as they recorded severe reduction (~30%) in seed germination	[45]

2.4.2 ROOT AND SHOOT GROWTH INHIBITION

Root zone salinity above 2 dS m^{-1} restricts the vegetative growth and even short term exposure to 5 dS m^{-1} salinity causes substantial growth reduction in many fruit crops.[46] The adverse salt effects on vegetative growth, *inter alia*, are seen as reductions in plant height, stem girth, leaf area, and root length. The examples presented in Table 2.3 show that increasing salinity almost invariably impairs the plant growth with complete cessation and plant mortality occurring above a threshold value depending on the genotype[206]. Salinity reduces shoot growth by suppressing leaf initiation and expansion as well as internode growth and by accelerating the leaf abscission.[207] Distinctly higher tolerance in a genotype as compared to others may be due to differences in genetic make-up and growth habit. For instance, polyembryonic mango rootstocks exhibit greater salt tolerance than mono-embryonic types.[160] Salinity significantly decreased plant height, stem diameter, leaf area and dry matter production in leaves, shoots, and roots of guava plants. Salt effects were pronounced when EC_{iw} exceeded 3 dS m^{-1}.[22]

TABLE 2.3   Adverse Effects of Salinity on Plant Growth.

Crop	Effects of salinity	References
Aonla	Plant establishment and survival were affected at 10 dS m^{-1} and above salinity	[149]
Avocado	The maximum reduction in growth was seen at 60 mmol L^{-1} NaCl	[129]
Ber	While pruning weight and tree canopy area significantly decreased at 5 dS m^{-1} salinity, there was no survival at 20 dS m^{-1}	[79]
Date palm	Vegetative growth significantly decreased at 12 dS m^{-1}	[189]
Loquat	50 mmol L^{-1} NaCl decreased plant growth by 40%	[77]
Mango	Polyembryonic rootstocks Kurukkan and Olour survived up to 4.2 dS m^{-1} salinity while Kerala-1 and Kerala-2 failed to grow above 1.3 dS m^{-1}	[43]
Pomegranate	Cultivar Dholka did not exhibit adverse effects up to 8.6 dS m^{-1}, but heavy reduction in plant growth occurred at 16.4 dS m^{-1} and above salinity levels	[12]

Grape rootstocks (1103 P, 41 B, 140 Ru, and 5 BB) exposed to different levels of NaCl (0, 5000, 10,000, 15,000, and 20,000 ppm) for 50 days showed reductions in fresh and dry weights of shoots and roots, plant dry mass, shoot length, and leaf number. Genotype 41 B showed the maximum tolerance index on the basis of root dry weight, while 5 BB was found to be the most susceptible.[33] Shoot length in olive cultivars Koroneiki, Kalamata, and Megaritiki was adversely affected above 25 mmol L^{-1} NaCl, while in others (Amphissis, Kothreiki, and Mastoidis) significant reduction was noted above 50 mmol L^{-1} NaCl.[25] In mango rootstock "13-1," reduction in vegetative growth was stronger under NaCl-induced salinity as Na$_2$SO$_4$-treated plants recorded significantly less decrease and delayed appearance of damage symptoms.[160] NaCl-induced salinity (EC$_{iw}$ of 8 and 16 dS m^{-1}) markedly reduced plant growth in cashew genotypes BRS 189 and CCP06. However, reductions in leaf area and plant dry mass were significantly lower in BRS 189 as it excluded Na$^+$ and Cl$^-$ ions from the shoots.[7]

2.4.3   VISIBLE INJURY SYMPTOMS

Salt-treated plants show an array of injury symptoms in leaves, shoots, and roots. Initially, leaves show marginal chlorosis and leaf tip burn. With increase in the level and duration of salinity, symptoms gradually affect the entire leaf and even shoot. Under severe stress conditions, appearance of necrotic lesions and heavy leaf defoliation are observed. Salt injury symptoms in guava cultivar Paluma appeared as chlorosis and subsequent necrosis of the affected leaf tissues. Appearance of these symptoms even with the use of marginally saline water (2 dS m^{-1}), due to accumulation of Na$^+$ and Cl$^-$ ions to toxic levels, revealed the extreme salt sensitivity of cv. Paluma.[170] Salt-induced leaf tip burning appeared in olive cultivars Koroneiki, Amphissis, and Mastoidis at 100 mmol L^{-1} of NaCl. Severe toxicity symptoms such as dead leaf edge, stem tip necrosis, and eventual leaf abscission occurred at 200 mmol L^{-1} of NaCl.[25] Salt application (400 mmol L^{-1} NaCl) caused necrosis in fully expanding leaves in date palm cultivar Madjol. The necrotic symptoms started with leaflet tip burning and progressed along the leaflet margins.[4]

Cherry (*Prunus cerasus*) rootstocks CAB 6P and Gisela 5 exposed to salinity (4 dS m^{-1}) and boron (0.2 mmol L^{-1}) became progressively wilted and suffered from leaf burn. In CAB 6P plants, leaves remained attached to the plants till the end of the experiment, whereas Gisela 5 plants showed higher leaf shedding.[176] Marginal leaf scorch followed by necrosis and leaf drop was noted in kiwifruit cultivar Hayward with potassium chloride application. The symptoms developed when Cl$^-$ concentrations reached 1.5% of leaf dry weight.[144] NaCl (50 and 70 mmol L^{-1})-treated *loquat* plants showed significant defoliation and foliar necrosis.[62] Root zone salinity above 10.1 dS m^{-1} caused severe yellowing of younger leaves and appearance of puckered lesions between leaf veins in mulberry cultivar "Tollygunj."[192]

2.4.4 FRUIT YIELD AND QUALITY

Salinity adversely affects flowering and fruit set which in turn reduces the final yield and fruit quality. The reproductive growth of citrus is particularly sensitive to saline flooding.[78] Citrus trees irrigated with 20 mmol L^{-1} NaCl water produced more vegetative flushes at the expense of flowering, as numbers of reproductive and mixed flushes were reduced by salinity.[83] Salinity stress reduced flowering intensity and fruit retention in Japanese plum (*Prunus salicina* Lindl).[78] Development of fewer flowers might account for yield reduction in salt stressed avocado trees.[40]

Osmotic stress and impaired water relations adversely affect flowering and the subsequent fruit set leading to lower yields in salinized *ber* trees.[79] Salinity-induced yield reduction in Washington Navel orange trees was due to decrease in the number of fruits per tree rather than reduction in average fruit weight.[73] Mango cv. "Osteen" grafted onto Gomera-3 rootstock produced smaller fruits (and thus low fruit yield) when irrigated even with marginally saline water (EC$_{iw}$ 1.02–2.50 dS m^{-1}).[44]

Six-year-old "Fino 49" lemon trees grafted on *Citrus macrophylla* rootstock and drip irrigated with saline (15 and 30 mmol L^{-1} NaCl) water exhibited significant reduction in fruit yield with increasing salinity. Salt stress decreased the number of fruits per tree due to heavy fruitlet drop. A significant increase in fruit yield was observed in the last year by increasing the quantity of water.[63] Salinity stress reduced fruit yield by up to 27% and 64% in strawberry cvs. Korona and Elsanta, respectively. Fruit quality parameters such as taste, aroma, and texture were decreased by more than 24% in Elsanta, but Korona showed nonsignificant differences with control plants. Total soluble solids (TSS: Brix) and TSS: acidity ratio in fruits were decreased by 20% in Korona and 35% in Elsanta. The ability of Korona to retain Cl^{-} ions in roots appeared to alleviate salt stress resulting in better growth and relatively higher fruit yield.[156] Salinity reduced the fruit set and yield in *ber* trees but had no adverse effect on fruit quality. About 50% yield reduction was recorded at soil EC$_e$ value of 11.3 dS m^{-1}.[79] Saline irrigation (5.5–6.5 dS m^{-1}) caused occasional decrease in fruit size but increased the percent dry weight, percent oil, and oil yield per unit fruit weight in olive cultivars "Manzanillo" and "Uovo de Piccione."[95]

2.5 ENZMATIC AND NONENZYMATIC RESPONSES

Accumulation of free radicals in salt stressed plants enhances the activity of antioxidant molecules in fruit crops (Table 2.4). The main antioxidant enzyme is superoxide dismutase (SOD), several SOD isozymes being Mn–SOD, Cu/Zn–SOD, and Fe–SOD. SOD is a metalloprotein that catalyzes the conversion of superoxide radical into hydrogen peroxide. To avoid hydrogen peroxide accumulation, a compound even more damaging than superoxide radical, enzymes catalase (CAT) and ascorbate peroxidase (APX) are activated.[10,81] Salinized trees also synthesize proline to overcome the osmotic stress and cellular dehydration (Table 2.5). The stress protection activities of proline are attributed to its involvement in osmotic adjustment, stabilization of subcellular structures, and the elimination of free radicals.[74] In halophytes, proline is the major component of amino acid pool under salt stress. While proline levels remain low under nonsaline conditions, salinized plants show manifold increase in proline concentration.[178] The increase in proline content is mostly positively correlated with the level of salt tolerance and salt tolerant genotypes generally show elevated proline concentrations as compared to salt sensitive ones.

TABLE 2.4   Antioxidant Defense System Under Salinity.

Crop	Effects	References
Cashew nut	NaCl-induced salinity (EC$_e$ 0.7–20.6 dS m^{-1}) increased the activities of guaiacol peroxidase and ascorbate peroxidase in roots leading to significant decrease in lipid peroxidation in root tissues	[36]
Citrus	Activities of antioxidant enzymes (SOD, CAT, APX, and GR) increased under salinity (30, 60, or 90 mmol L^{-1} NaCl) in leaves of salt-sensitive *Carrizo citrange*	[10]
Grape	NaCl (12 mmol L^{-1}) salinity significantly increased the activities of CAT and SOD in leaves of rootstock 1616C and cv. Razaki	[204]
Mango	Salinized "Olour" seedlings treated with 1500 ppm paclobutrazol had higher SOD (24%), CAT (46%) and POD (163%) activities indicating that paclobutrazol enhanced free radical scavenging capacity in salt-treated plants	[177]
Mulberry	Salinity (EC$_e$ 1.7–6.9 dS m^{-1}) up-regulated SOD, POD, and CAT levels	[181]
Olive	In 1-year-old self-rooted plants, exposure to 66 or 166 mmol L^{-1} NaCl for 10 days resulted in increased activity of SOD in cv. Leccino, but almost no changes in cv. Oblica	[68]

APX, ascorbate peroxidase; CAT, catalase; SOD, superoxide dismutase; GR, glutathione reductase

TABLE 2.5   Proline Accumulation for Osmo-protection.

Crop	Effects	References
Aonla	Free proline and total free amino acid contents increased in tested cultivars (NA-6, NA-7, NA-10, NA-18, Chakaiya, and Anand 1) with increasing salinity (0–15 dS m^{-1})	[148]
Cashew	The total free amino acids and free proline increased in leaves but remained unchanged in roots of salinized plants	[190]
Citrus	Free proline increased in the leaves of "Verna" and "Fino" lemon scions budded on Sour orange but not in those budded on Macrophylla under 40 or 80 mmolL^{-1} NaCl	[132]
Mulberry	Proline accumulation increased at low (1–4 dS m^{-1}) but decreased at high (8–12 dS m^{-1}) salinity; accumulation was higher in the genotype "BC2-59" as compared to "S-30" and "M-5"	[2]
Walnut	Kaman 1 and Bilecik cultivars showed higher accumulation of proline than Kaman-5 in response to saline irrigation (EC$_{iw}$ of 1.5, 3 and 5 dS m^{-1})	[3]

2.6 SALINITY MANAGEMENT IN FRUIT CROPS: LIMITATIONS, SCOPE, AND CHALLENGES

Over the years, many promising technologies have been developed for reclaiming salt-affected lands. The huge success of many simple and cost-effective practices such as gypsum-based sodic land reclamation package and salt tolerant varieties in harnessing the productivity of saline and sodic lands is beyond doubt.[86] Notwithstanding the potential of these and other technologies, a range of hurdles including socioeconomic constraints have greatly diminished the appeal of otherwise doable technologies among the farming communities."[165]

Traditional reclamation of both saline and sodic soils demands substantial quantities of good quality water for salt leaching; a practice that has become somewhat irrelevant with increasing fresh water scarcity.[86] Gypsum, the preferred amendment for alleviating sodicity stress, is available in finite amount. In recent decades, constraints such as decreased availability of mined gypsum, deterioration in product quality, and high market prices have increasingly made gypsum use a costly and less efficient proposition in sodic land reclamation generating interest in other easily available soil amendments such as farm yard manure (FYM), distillery spent wash, and press-mud.[86,165] Emphasis on tapping the potential of nanoscale materials as soil ameliorants and conditioners has also increased.[105] Although subsurface drainage is a proven technology to restore the productivity of waterlogged saline lands, yet difficulties in the disposal of saline drainage water and poor community participation have hindered its widespread adoption.[70]

Given the fact that cost effective and environmentally safe disposal of saline drainage effluents is a cumbersome task, efforts to harness the potential of waterlogged lands through saline aquaculture have gathered momentum.[6] There is evidence to prove that conjunctive use of marginal quality drainage water and fresh water in irrigation may give good results when tried in combination with measures such as salt tolerant varieties.[118]

The growing realization that many currently used technologies suffer from formidable limitations and that newer, more robust techniques in offing might be out of reach of the resource poor farmers has necessitated refinements in the existing technologies so as to make them practically and economically viable in addressing the emerging constraints. In recent decades, greater emphasis has been placed on plant-based solutions for salinity management. Salt tolerant trees and crop cultivars are increasingly being advocated as a lasting solution to the salinity problem, especially in moderately affected soils, while eliminating the costs incurred in the repeated use of chemical amendments.[86] Conventional as well as emerging technologies and approaches for enhancing the salt tolerance in fruit crops, their practical utility and limitations, and the future research agenda have been critically examined in this chapter under subsections, such as

- Genetic improvement for salt tolerance,
- Agronomic practices for salinity alleviation, and
- Irrigation management in salt-affected soils.

2.6.1   GENETIC IMPROVEMENT FOR SALT TOLERANCE

Despite availability of a large pool of data on physiological and biochemical bases of salinity adaptation in plants, meager success in developing salt tolerant genotypes is ascribed to three factors:

- Salt tolerance is a complex quantitative trait governed by many genes greatly influenced by environmental factors.
- Physiological research and plant breeding efforts continue to be largely disconnected domains hindering the progress in genetic improvement.[126]
- Breeding for salt tolerance has remained a low priority area in crop breeding projects in majority of the countries.[58]

Throughout the 20th century, breeding for abiotic stress tolerance got little attention of plant breeders who almost exclusively focused on improving crop yield and quality. Nonetheless, efforts to exploit the genetic variability culminated into the release of promising salt tolerant cultivars in crops like rice and wheat.[11,86] Limited success in developing tolerant genotypes through conventional approaches is also partly due to low genetic variation in available germplasm and constraints in transferring the trait of interest from wild relatives to commercial varieties.[11,203]

Breeding for salt tolerance is even more tedious in woody fruit crops having long juvenile phase, gigantic tree size and complex floral biology. Prolonged juvenility, usually 5 years or more in majority of the tree fruits, characterized by the absence of flowers and fruits, restricts the scope for genetic improvement through backcrossing, inbreeding, and hybridization. During this phase, selection is done on the basis of morphological traits only. With increase in age, most of the trees attain huge size which makes crossing a cumbersome task. Numerous floral induction techniques such as canopy management, grafting on dwarfing rootstocks, and identification of early flowering mutants and use of precocious parents in crossing programs have been tried with limited success. Of late, genetic transformation with floral induction genes has proved effective in accelerating flower initiation in many fruit species. Overexpression of the *FRUITFULL(FUL)*-homolog *BpMADS4* induced early flowering in apple. Similarly, *Arabidopsis LFY* gene has been successfully tested for reducing the juvenile phase in citrus. In pear, overexpression *citrus FT homolog* induced in vitro flowering from transgenic shoots.[55]

Further improvements in transgenic development with new floral regulatory genes are expected to significantly reduce the breeding cycle in fruit crops. As previously mentioned, genotypic differences in fruit crops can be genetically manipulated to improve salt tolerance. However, in light of the fact that conventional approaches have largely failed to deliver, emphasis should be on emerging molecular and nanotechniques for identifying new genes and sources of salt tolerance and their subsequent integration in established cultivars either through genetic transformation or hybridization.

2.6.2   DEVELOPMENT OF ROOTSTOCKS

In spite of concerted efforts over the decades, only a few commercial salt tolerant cultivars have been released due to complex inheritance of salt tolerance and the related problem of linkage drag in conventional salinity improvement programs. This situation has gradually shifted the emphasis on alternative strategies such as development of salt tolerant rootstocks[29] even in vegetable crops,[29,101,209] where self-rooted plants are commonly used. Grafting protects the shoot system from salt shock as tolerant rootstocks either exclude Na^+ and Cl^- ions or restrict their translocation to the leaves and shoots.[30,33,61,200] Besides limiting salt transport to the foliage, grafting also offers opportunity to study the effects of the genes transcribed in the roots toward the performance of shoot system.[28]

A number of studies have been conducted to identify fruit rootstocks with better Na^+ and Cl^- exclusion capacity. Among 21 polyembryonic mango rootstocks—Orange, Golden Tropic, Banana, Red Haruman, and Pico—showed good Na^+ and Cl^- exclusion capacity when irrigated with marginally saline water containing 480 ppm of NaCl.[82] Mango cv. Amrapali scions grafted on nondescript seedling showed lesser leaf Na^+ accumulation while those grafted on Olour had lower leaf Cl^- when irrigated with 50 mmol L^{-1} of NaCl water.[35]

Peach cultivar Armking grafted on GF_{677} retained most of the Na^+ and Cl^- ions in basal (old) leaves and stems. In contrast, Armking grafted on

Mr.S.$_{2/5}$ recorded higher translocation of Na$^+$ and Cl$^-$ compared to the young leaves.[112] Grape cv. Sultana grafted on different rootstocks (e.g., Ramsey, 1103 Paulsen and J17–69) was drip-irrigated with normal and saline (1.75 and 3.50 dS m^{-1}) water over a 5-year period. Rootstock 1103 Paulsen was adjudged to be the best chloride excluder as revealed by the lowest concentrations of accumulated Cl$^-$ in petioles, laminae, and grape juice at high salinity.[196] Cashew nut grafts having BRS 226 as rootstocks exhibited higher transpiration and more accumulations of Na$^+$ and Cl$^-$ in leaves compared to grafts having CCP 09 as rootstocks under 200 mmol L^{-1} NaCl salinity.[53] West Indian avocado rootstocks VC 207 and VC 256 exhibited high salinity tolerance as compared to Duke 7 which is attributed to partitioning of Cl$^-$ ions in roots.[30] These observations point to differential response of root system with regard to salt exclusion and uptake, which should be taken into account while selecting the appropriate rootstock(s) for saline soils.

2.6.3 MARKER-ASSISTED SELECTION AND QUANTITATIVE TRAIT LOCI MAPPING

Quantitative trait loci (QTL) mapping refers to the identification of a single gene or a few genes involved in the expression of complex phenotypic traits. Rapid strides in molecular marker and statistical techniques have significantly improved the prospects of locating such QTL—essentially a particular stretch of chromosome—responsible for allelic variation in a complex quantitative trait.[113] As a basic requirement, two contrasting parents are crossed to generate the mapping population[113] for establishing the linkage relationships between molecular markers and the "trait of interest."[197] Of late, QTL mapping is increasingly being applied in fruit crops to enhance the understanding with regard to the effect(s) of marker loci on important tree characteristics so as to expedite the genetic improvement by culling the undesirable types during progeny evaluation[197] resulting in considerable savings in time, money and labor required in monitoring and advancing the breeding population. The practical utility of QTL mapping in locating the genes involved in salt tolerance is increasingly becoming evident. A range of traits such as reduced salt entry and transport inside the plants and alleviation of drought (osmotic) stress have been considered to be mediated by the QTL linked to salinity stress.[126]

Molecular markers can be used to identify QTLs which could improve the efficiency of selection. QTL mapping in particular could facilitate the selection of traits controlled by several genes and highly influenced by environmental factors. The development of high-density linkage maps containing markers (such as simple sequence repeats, restriction fragment length polymorphisms, and amplified fragment length polymorphisms) will facilitate pyramiding traits of interest to attain substantial improvement in salt tolerance. QTL mapping in a salinized (40 mmol L^{-1} NaCl) backcross progeny population of citrus [*Citrus grandis* × (*Poncirus trifoliate* × *Citrus grandis*)] indicated that many regions of the citrus genome could have a major impact on growth[187] and Na$^+$/Cl$^-$ accumulation[188] under salt stress. QTLs linked to vegetative stage salinity tolerance were identified in a backcross tomato population (salt sensitive *Lycopersicon esculentum* line NC84173 × and salt tolerant *Lycopersicon pimpinellifolium* line LA722) indicating the need to exploit interspecific variation for enhancing salt tolerance by marker-assisted selection.[60]

2.6.4   GENETIC TRANSFORMATION

Development of transgenic lines in superior scion and rootstock cultivars may provide a lasting solution to different biotic and abiotic stresses including salinity. Some notable examples of genetic transformation for enhanced salt tolerance in fruit crops are summarized in Table 2.6.

TABLE 2.6   Genetic Transformation for Developing Salt Tolerant Genotypes.

Crop	Genetic transformation	References
Apple	Vacuolar Na$^+$/H$^+$ antiporter *MdNHX1* from the salt tolerant rootstock Luo-2 was introduced into rootstock M.26 by *Agrobacterium*-mediated transformation	[103]
Banana	Transgenic plants overexpressing *MusaDHN1* exhibited improved salt tolerance due to enhanced accumulation of proline and reduced malondialdehyde levels	[167]
Carrizo citrange	Yeast gene *HAL2* implicated in salt tolerance was integrated through *Agrobacterium*-mediated transformation	[23]
Mulberry	Stable transformation with barley *hva1* gene resulted in better physiological relations of transgenic lines under salinity	[99]
Pear	*MdSPDS1*-over-expressing transgenic plants were obtained by *Agrobacterium*-mediated transformation; selected transgenic lines produced spermidine synthase and exhibited tolerance to 150 mmol L^{-1} NaCl	[198]
Strawberry	Stable expression of osmotin gene in cv. Chandler through co-cultivation with *Agrobacterium tumefeciens*	[85]

The stable expression of transgenes increases the ability of transformed lines to tolerate excess salts in the growing medium. An array of physiological and biochemical events account for increased salt tolerance in transgenic lines. For example, stable expression of transgene *hva1* imparted salt tolerance in transgenic mulberry line ST30 exhibited on account of better cellular membrane stability, high photosynthetic yield, less photooxidative damage and better water use efficiency.[99] Although immense potential of genetic engineering in solving present and emerging problems in agriculture is well documented, yet there are biological, regulatory, and ideological constraints in the commercialization of genetically transformed products necessitating the search for appropriate alternative solutions.[171]

2.6.5   TRANSGRAFTING

Transgrafting refers to the grafting of a transgenic rootstock with a conventional wild-type scion variety. Given the fact that wild-type scion will flower and fruit and not the genetically engineered rootstock, there are ample possibilities for addressing the regulatory and consumer concerns over the flow of genetically engineered pollen. Advent of transgrafting means that fruit breeders can exclusively focus on developing genetically engineered rootstocks to benefit the fruit growers as well as to allay the consumer concerns related to genetic transformation of scion cultivars.[75]

2.6.6   IN VITRO SCREENING

In comparison to field trials and pot studies, characterized by soil heterogeneity and salinity interactions with other environmental factors, in vitro screening for identifying the underlying physiological mechanisms of salinity tolerance is a relatively fast technique with short generation time. Rapid in vitro screening for salt tolerance has been reported in almond,[168] apple,[169] citrus,[138] cherry,[48] mulberry,[191] and pistachio.[27] In vitro screening may unravel potentially novel insights into the physiological and nutritional bases of salt tolerance leading to further investigations to establish the correlation between in vitro and in vivo behavior of genotypes under saline conditions. Repeated observations and comparative studies with field grown plants can establish in vitro evaluation as an effective preselection criterion for salt tolerance.

2.7 AGRONOMIC PRACTICES FOR SALINITY ALLEVIATION

A number of agronomic interventions are available to enhance the salt toler-ance in fruit crops. In combination with appropriate salt tolerant cultivars and irrigation management, these techniques can significantly alleviate salinity stress.

2.7.1 SELECTION OF SALT TOLERANT CULTIVARS

As elsewhere indicated in this chapter, considerable intergeneric and intervarietal genotypic differences exist in different fruit crops under saline conditions. These differences should be taken into account while selecting a scion and/or rootstock for cultivation in salt-affected soils. Research efforts have led to the identification of a number of fruit crops for sustain-able land use in salt-affected environments. Singh et al.[175] reported that fruit species—such as Indian jujube (*Zizyphus mauritiana*) guava (*Psydium guajava*), pomegranate (*Punica granatum*), Indian gooseberry (*Emblica officinalis* Gaertn.), Christ's thorn (*Carrisa congesta* Wight), tamarind (*Tamarindus indica*), and Black plum (*Syzygium cuminii* L. Skeels)—can be successfully grown in sodic soils (~ESP 30–45). Presence of water in root zone can cause substantial damage in tropical and sub-tropical fruits.[157,158] For fruit species sensitive to water logging and frost, proper drainage in the rainy season and protection of the young tender saplings during winter is necessary for the overall success of the plantations. It has been shown that pomegranate and *bael* require good drainage. They showed 75% and 42.4% sapling mortality respectively, due to prolonged water stagnation in sodic soils. Besides salinity, other prevailing constraints—such as drought, waterlogging and mineral deficiencies—should be considered to select the appropriate genotypes. For example in areas having perpetual water short-ages, water use efficient cultivars are likely to give better results.

2.7.2 PLANTING TECHNIQUES

Different planting techniques have been described for raising tree plantations in salt-affected lands lying barren and unproductive. Depending upon the sodicity levels, topography, and resources, one of the following techniques can be successfully used for tree plantation:

- Pit method (refilled with reclaimed alkali soil or normal soil: Fig. 2.2a);
- Auger hole method (Fig. 2.2b);
- Combination of pit and auger hole method (Fig. 2.2b);
- Ridge plantation;
- Trench plantation.

The depth of the pits in shallow depth of *kankar* (CaCO$_3$ concretions) layer situations and of auger whole in medium depth *kankar* layer situations is adjusted so that the same is broken to facilitate the free movement of tree roots beyond the *kankar* layer.

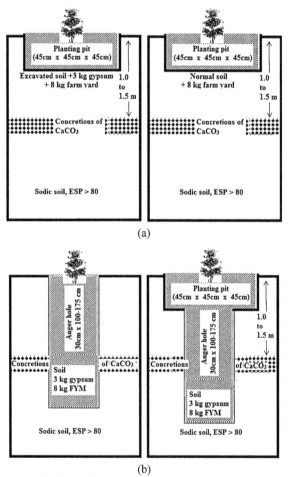

FIGURE 2.2 **(See color insert.)** Planting methods in sodic soils: (a) planting methods in sodic soils: planting pit technique and (b) planting methods in sodic soils: auger hole technique (Courtesy: Authors[70]).

In sodic soils suffering from the problems such as high pH, high ESP, low water infiltration, and hard calcareous pan, planting in gypsum amended auger holes is suggested to obtain good results.[32] CSSRI perfected such a technology using tractor powered auger, which is widely used to plant forest plantations both in normal and sodic soils (Fig. 2.3). For saline lands, subsurface planting and furrow irrigation method have been found appropriate.[186] In this method, saplings are planted in the auger holes in furrows during the rainy season. The V-shaped furrows (20 cm deep and 60 cm wide at the top) are made using a tractor-driven furrow maker. Saplings are irrigated with saline water during the early establishment stage. Besides reducing the water application costs, the furrows help in creating a favorable low salinity environment by permitting the downward and lateral water flux and salt leaching below the root zone.

FIGURE 2.3 Tractor operated auger used for pit preparation in sodic soils (Courtesy: Authors).

2.7.3 FRUIT-BASED AGRO-FORESTRY SYSTEMS

Adoption of fruit-based agro-forestry systems is an attractive option to harness the productivity of degraded lands. Moreover, the productivity of such systems can be further enhanced by growing locally adapted food crops, vegetables, and medicinal plants in interspaces, at least during the initial few years of tree growth.[17] In a semiarid climate, a fruit-based agro-forestry system was developed by integrating cluster bean/pearl millet and barley in Indian gooseberry (*Emblica officinalis* Gaertn), Christ's thorn (*Carrisa congesta* Wight), and Bengal quince (*Aegle marmelos* Correa) orchards.

Irrigation with saline water (4–10.5 dS m^{-1}) once in each dry month ensured very good survival (80–90%) of trees.[165]

2.7.4   NUTRIENT MANAGEMENT

A number of simple strategies have been suggested to counteract the adverse effects of excess salts on plant mineral nutrition. These include applications of chemical amendments and manures,[88] avoiding the use of fertilizers having higher Na$^+$ and Cl$^-$ concentrations,[19] use of controlled release fertilizers, fertigation and use of fertilizers having low osmotic effect,[183] and supplemental nutrition.[47]

Five-year-old guava plants in a saline–sodic soil (pH$_s$ 8.8 and EC$_e$ 4.70 dS m^{-1}) gave the maximum fruit yield (28.13 kg plant^{-1}) when treated with 100% of gypsum requirement, 40 kg FYM and 2 kg urea per plant as compared to control.[88] Supplemental calcium nutrition [10 mmol L^{-1} Ca(NO$_3$)$_2$] not only enhanced K$^+$ and Ca^{2+} translocation to leaves but also restricted Na$^+$ uptake by roots in salt stressed (30–60 mmol L^{-1} NaCl) guava plants thus giving a high K$^+$/Na$^+$ ratio in leaves.[47] Calcium nutrition often has ameliorative effects on plant growth in saline soils.[19] Supplementary CaSO$_4$ at 2.5 and 5 mM concentrations ameliorated the negative effects of salinity (40 mmol L^{-1} of NaCl) on plant dry matter and Chl content in plum rootstocks Marianna GF 8-1, Myrobolan B and Pixy. The addition of CaSO$_4$ maintained membrane stability and significantly lowered the leaf and root Na$^+$ concentrations.[18] Integrated nutrient management is often more effective in sustaining growth of the plants than chemical nutrients alone, especially in sodic soils.

2.7.5   USE OF MYCORRHIZAL INOCULANTS

Arbuscular mycorrhizal (AM) fungi mitigate the detrimental effects of salinity by regulating key physiological functions such as accumulation of compatible solutes to avoid cell dehydration, regulation of ion and water uptake by roots, reduction of oxidative stress by enhancing the antioxidant capacity and stabilizing photosynthesis for sustained growth.[141] Red tangerine (*Citrus tangerine* Hort. ex Tanaka) seedlings inoculated with AM fungi (*Glomus mosseae* and *Paraglomus occultum*) had better shoot and root growth and produced significantly higher biomass under 100 mmol L^{-1} NaCl salinity as compared to nonmycorrhizal controls. Inoculation with AM fungi significantly increased root length and root surface area, improved photosynthesis and reduced leaf Na^+ concentrations resulting in favorable ionic balance in terms of high K^+/Na^+ ratio.[202] AMF-inoculated (*Glomus mosseae*, *Glomus intraradices,* and *Glomus claroideum*) olive plants exhibited higher survival, better plant growth, and greater ability for N, P, and K uptake from saline substrate having 6 g of NaCl kg^{-1}. The *G. mosseae* was the most efficient strain in reducing the detrimental effects of salinity; it increased shoot and root growth by 239% and by 468%, respectively, in salt stressed plants. Mycorrhizal plants showed the lowest biomass reduction under salinity (34%), while growth was reduced by 78% in control plants.[142]

2.7.6   APPLICATION OF PLANT GROWTH SUBSTANCES

Depletion of endogenous hormones such as auxins, gibberellins, and cytokinins in salt-stressed plants[128] is a pointer to their protective role under stress conditions. Accordingly, the exogenous application of these and other plant growth substances has been tried to alleviate salinity stress in a range of crops. Auxin indole acetic acid economizes water use in stressed plants by increasing the stomatal resistance. Gibberellins are implicated in ion homeostasis and in modulating the endogenous ABA levels in salinized plants.[143] Similarly, cytokinins such as kinetin alleviate salinity stress by reducing toxicity symptoms, promoting growth and up-regulating endogenous polyamine levels which have a protective role in salt stress.[111] The protective functions of polyamines, the ubiquitous organic compounds in higher plants, under stress conditions are widely recognized.[109] The most common polyamines in higher plants are putrescine, spermidine and spermine. Use of 50 ppm putrescine reduced membrane injury index and increased relative water content, photosynthetic rate, pigment content, and endogenous proline and antioxidant levels in NaCl-stressed salt-sensitive citrus rootstock *Karna khatta*.[166] There may be differential accumulation of polyamines under stress. For instance, there was increased accumulation of putrescine but negligible changes in spermidine and spermine contents in apple callus under salt stress.[104]

Paclobutrazol (PBZ)—an antigibberellin growth retardant—alleviates salt stress in pomegranate,[155] citrus,[166] strawberry,[87] and mango.[94] The ameliorating effects of PBZ are ascribed to upregulation of antioxidant enzymes, reduced uptake of Na^+ and Cl^- ions, improved nutrient acquisition and improved water balance under saline conditions. PBZ interrupts GA biosynthesis, and there is an increase in ABA and cytokinin contents, which ultimately help in maintaining better water balance in the plants.[166] It is also suggested that PBZ promotes salt tolerance by increasing the levels of photosynthetic pigments and by preventing damage to the cell membranes.[94]

Salicylic acid (SA) has long been recognized as a key molecule in the signal transduction pathway of different biotic and abiotic stresses. Available data indicate that exogenous applications of low concentrations of SA mostly improve plant growth under stress conditions.[80] Foliar spray (0.5 or 1 mmol L^{-1}) of SA resulted in better growth in salt stressed (30, 60, or 90 mmol L^{-1} NaCl) pistachio seedlings by preventing cell EL and maintaining favorable Chl and water relations and photosynthetic capacity in leaves.[15] Application of 0.50 and 1.00 mmol L^{-1} SA solutions alleviated salt stress (25, 50, and 75 mmol L^{-1} NaCl) in Valencia orange seedlings caused significant improvements in leaf Chl, leaf relative water content, net photosynthesis, proline accumulation and reduction in EL giving better cellular membrane integrity.[93]

2.8   IRRIGATION MANAGEMENT IN SALT-AFFECTED SOILS

Dwindling fresh water supplies have necessitated the best productive use of available water in agricultural production. In addition, the menace of poor quality water is bound to increase in future implying that fruit production will have to be sustained with lesser fresh water and profitable use of poor quality water. It is suggested that improvements in orchard design and adoption of efficient irrigation practices may significantly lower water use in fruit crops. Currently available techniques for utilizing saline water in fruit crops are briefly discussed in this section:

2.8.1   DRIP IRRIGATION

Due to significant reduction in water consumption, application of nutrients with water (fertigation), and ease in the use of marginal quality water, drip irrigation is increasingly becoming popular in fruit crops. Use of poor-quality water through drip, however, requires some changes from standard irrigation practices such as selection of appropriately salt tolerant crops, adhering to a maintenance schedule to avoid clogging, and leaching at appropriate intervals, if required. When using poor quality water, drip irrigation offers many advantages over traditional methods. The fact that plant canopy does not come into direct contact with saline water, the chances of salt injury to produce are greatly minimized. Again, owing to high-application frequency, salt concentrations in the root zone remain manageable resulting in alleviating adverse effects of osmotic stress.[122] Besides considerable reduction in water and energy use costs coupled with the significant increase in yield, direct water application into root zone means virtually no surface run-off and soil displacement.[146] As irrigation channels and bunds are not required, additional lands can be put under crops.

Addition of soluble nutrients and pesticides in irrigation water (called chemigation) means efficient use of these resources resulting in reduced cost of cultivation and control of environmental pollution as agro-chemical loads into soil and groundwater are minimized. Saline water of 8 dS m^{-1} was used successfully to grow Indian Jujube under drip irrigation in sandy soils of Rajasthan, India.[8] In pitcher irrigation, an indigenous alternative of drip irrigation, grape yield was at the same level with EC_{iw} of 4 dS m^{-1} than with 2 dS m^{-1} in surface irrigation.[71] Similarly, alkali water application to sapota (family Sapotaceae) and Indian jujube was possible using distillery spent wash as an amendment using pipe irrigation, another indigenous technology to use poor quality waters.

2.8.2   SUBSURFACE DRIP IRRIGATION

Subsurface drip irrigation (SSDI) refers to the application of water below the soil surface through emitters with discharge rates mostly equal to the surface drip.[21] They are similar in design but vary with surface drip in that tubes are buried in soil. In addition to the advantages of surface systems, SSDI curtails the water loss due to evaporation and deep percolation and virtually eliminates the surface runoff. It also permits precise application of water and nutrients in the root zone.[152] Long-term use of saline water through surface drip may result in gradual downward movement of salts to the root zone increasing osmotic stress and salt toxicity to the crops. The assumption that salt accumulation can be overcome by adopting the subsurface drip irrigation (SSDI) is based on the premise that salt front is partially driven down into the deeper soil bulk and to the periphery of the root zone under SSDI and thus minimizing the risk of damaging the main roots of the plants. Moreover, the improved moisture conditions in the vicinity of the emitter offset the inhibiting effects of the presence of the salts in the saline water.[134] Utility of SSDI in using domestic sewage water, likely to be a major resource under the climatic change scenario, has been established to grow papaya and guava.[135]

2.8.3 DEFICIT IRRIGATION

Deficit irrigation (DI) refers to the application of irrigation water below the crop evapo-transpiration (ET_c) requirement resulting in enhanced water uptake from the soil. Two approaches for DI include sustained DI (SDI) and regulated DI (RDI). In SDI, a fixed amount of water corresponding to a fraction of ET_c is given at regular intervals. In RDI, less irrigation is provided at specific growth stage(s) least sensitive to water stress so as to minimize the yield loss. During rest of the crop growth, full irrigation is done.[52] In contrast to supplemental irrigation in humid and subhumid regions to stabilize crop yields by alleviating water stress, DI aims to minimize irrigation water use while preventing the yield loss.[51] DI is likely to be more successful in regions where rainfall restores the soil moisture extracted by plants under deficit water supplies.[64] Satisfactory results also depend on understanding of crop response to drought[64] and salinity[51]; especially when saline or reclaimed waste water is used. Although significant water savings are possible with DI, this practice might accelerate salt accumulation with prolonged use of saline water as observed in mandarin (*Citrus clementina* cv. "Orogrande")[124] and peach (*Prunus persica* L. Batsch cv. Calrico) orchards.[9] In certain cases, as observed in grapefruit (*Citrus Paradisi* Macf. cv. Star Ruby), salts accumulating in summer months were washed away in rainy season[137] implying the need to practice RDI in regions where rainfall is adequate to leach the salts from root zone.

2.8.4 PARTIAL ROOT-ZONE DRYING (PRD)

In partial root-zone drying (PRD), only half of the root system is irrigated while the rest is kept in a dry state.[161] Considering the difficulties involved in prudent monitoring of soil water status in DI, efforts were made to find solutions for more controlled water applications leading to the development of PRD technique.[180] Two variants of conventional PRD—alternate partial root-zone irrigation (APRI) and fixed partial root-zone irrigation (FPRI)—are known. While APRI involves alternate watering on both sides of the root zone, FPRI consists of fixed watering on one side of the root zone.[102]

Trials conducted in peach,[67] olive,[195] and grapes[37,42] reveal the potential of PRD in curtailing the irrigation water use. Drying roots produce ABA which regulates stomatal conductance in such a way that transpiration induced water loss is minimized while photosynthetic assimilation is not affected as wet roots ensure proper turgidity in leaves.[154,180] Irrigation is usually alternated in about 15 days as extended drying may hamper plant vigor and productivity.[180]

2.9 CONCLUSION

Further expansion in arable area is virtually impossible; therefore, growing food demand and nutritional security concerns in India need to be addressed mainly by bringing chemically degraded lands under crops and tree cover. Although majority of the fruit crops are sensitive to salinity, yet their commercial cultivation is possible in saline and sodic soils, especially those suffering from slight to moderate constraints. In this chapter, a comprehensive review is presented that looks at salinity/sodicity impacts on seed germination and seedling establishment, shoot and root growth inhibition, visible injury symptoms, and fruit yield and quality. Several doable techniques such as selecting salt tolerant scion cultivars and rootstocks, planting techniques, spot soil amelioration, and novel irrigation techniques are described that can help to successfully raise fruit plantations in salt-affected soils. Besides direct benefit in terms of higher incomes to the orchardists, fruit trees owing to their large canopy and deep root system should eventually improve the soil and environmental quality and ensure long-term economic security to resource poor farmers owning such marginal lands.

2.10 SUMMARY

Relentless land degradation, fresh water scarcity, and climate change induced changes in crop growing conditions have posed formidable risks to global food security, especially in resource poor developing countries having dismal adaptive capacity to cope up with these challenges. Continued shrinkage of agricultural lands necessitates the best productive use of degraded, salt-affected lands for sustaining the growing food requirements while preserving the environmental balance.

In the beginning of this chapter, an overview of global salinity menace and its implications for food security, nutritional security and environmental sustainability were briefly discussed. Then, effects of salinity on fruit crop physiology were critically examined under various heads, namely, cell membrane damage, oxidative stress, alterations in leaf water relations, loss of leaf pigments, gas exchange characteristics, and osmotic stress and ion toxicity. Authors then discussed the impacts of these changes on seed germination and seedling establishment, shoot and root growth inhibition, visible injury symptoms and fruit yield and quality. After presenting a brief account of available technologies for salinity management in agriculture, authors discussed the current and emerging techniques such as genetic improvement for salt tolerance, development of root stocks, marker-assisted selection and QTL, mapping genetic transformation, transgrafting and in vitro screening for enhancing the salt tolerance in fruit crops.

Many cost-effective and doable technologies have been described which have the potential to harness the productivity of salt-affected soils and poor quality water through fruit crops. Irrigation management was specially included to emphasize on climate resilience agriculture where water scarcity would be one of the major challenges and shift to unconventional resource, the poor quality ground water may be imminent.

KEYWORDS

- electrolyte leakage
- fruit crops
- gas exchange
- genetic improvement
- ionic balance
- salt stress
- salt tolerance

REFERENCES

1. Abrol, I. P.; Yadav, J. S. P.; Massoud, F. I. *Salt-affected Soils and Their Management*; FAO Soils Bulletin 39: Rome, Italy, 1988; p 143.
2. Agastian, P.; Kingsley, S. J.; Vivekanandan, M. Effect of Salinity on Photosynthesis and Biochemical Characteristics in Mulberry Genotypes. *Photosynthetica* **2000**, *38*, 287–290.
3. Akça, Y.; Samsunlu, E. The Effect of Salt Stress on Growth, Chlorophyll Content, Proline and Nutrient Accumulation, and K/Na Ratio in Walnut. *Pak. J. Bot.* **2012**, *44*, 1513–1520.
4. Al-Abdoulhadi, I. A.; Ali-Dinar, H.; Ebert, G.; Büttner, C. Effect of salinity on leaf growth, leaf injury and biomass production in date palm (*Phoenix dactylifera* L.) cultivars. *Indian J. Sci. Technol.* **2011**, *4*, 1542–1546.

5. Ali-Dinar, H. M.; Ebert, G.; Ludders, P. Growth, Chlorophyll Content, Photo-synthesis and Water Relations in Guava (*Psidium guajava* L) Under Salinity and Different Nitrogen Supply. *Gartenbauwissenschaft* **1999**, *64*, 54–59.

6. Allan, G. L.; Fielder, D. S.; Fitzsimmons, K. M.; Applebaum, S. L.; Raizada, S.; Burnell, G.; Allan, G. Inland Saline Aquaculture. In *New Technologies in Aquaculture: Improving Production Efficiency, Quality and Environmental Management*; Woodhead Publishing (Elsevier Ltd.), 2009; pp 1119–1147.

7. Alvarez-Pizarro, J. C.; Gomes-Filho, E.; de Lacerda, C. F.; Alencar, N. L. M.; Prisco, J. T. Salt-induced Changes on H^+-ATPase Activity, Sterol and Phospholipid Content and Lipid Peroxidation of Root Plasma Membrane from Dwarf-cashew (*Anacardium occidentale* L.) Seedlings. *Plant Growth Regul.* **2009**, *59*, 125–135.

8. Anonymous. *Biennial Report-2006–08 Management of salt Affected Soils and Use of Saline Water in Agriculture*; Central Soil Salinity Research Institute: Karnal. 2009, pp 142–143.

9. Aragüés, R.; Medina, E. T.; Martínez-Cob, A.; Faci, J. Effects of Deficit Irrigation Strategies on Soil Salinization and Sodification in a Semiarid Drip-irrigated Peach Orchard. *Agric. Water Manage.* **2014**, *142*, 1–9.

10. Arbona, V.; Flors, V.; Jacas, J.; García-Agustín, P.; Gómez-Cadenas, A. Enzymatic and Non-enzymatic Antioxidant Responses of Carrizo Citrange, a Salt-Sensitive Citrus Rootstock, to Different Levels of Salinity. *Plant Cell Physiol.* **2003**, *44*, 388–394.

11. Ashraf, M.; Akram, N. A. Improving Salinity Tolerance of Plants Through Conventional Breeding and Genetic Engineering: An Analytical Comparison. *Biotechnol. Adv.* **2009**, *27*, 744–752.

12. Asrey, R.; Shukla, H. S. Salt Stress and Correlation Studies in Pomegranate (*Punicagranatum* L.). *Indian J. Hortic.* **2003**, *60*, 330–334.

13. Awang, Y. B.; Atherton, J. G.; Taylor, A. J. Salinity Effects on Strawberry Plants Grown in Rockwool. I: Growth and Leaf Water Relations. *J. Hortic. Sci.* 1993, *68*, 783–790.

14. Aziz, B. A.; Suraya, A. N.; Zain, H. S. M. The Effect of NaCl on the Mineral Nutrient and Photosynthesis Pigments Content in Pineapple (*Ananascomosus*) In Vitro Plantlets. *Acta Hortic.* **2011**, *902*, 245–252.

15. Bastam, N.; Baninasab, B.; Ghobadi, C. Improving Salt Tolerance by Exogenous Application of Salicylic Acid in Seedlings of Pistachio. *Plant Growth Regul.* **2013**, *69*, 275–284.

16. Bernstein L. *Salt Tolerance of Fruit Crops*. USDA Agricultural Information Bulletin 292; United States Department of Agriculture, 1980; pp 8.

17. Bhadauria, S.; Sengar, R. M. S.; Mohan, D.; Singh, C.; Kushwah, B. S. Sustainable Land use Planning Through Utilization of Alkaline Wasteland by Biotechnological Intervention. *Am. Eurasian J. Agric. Environ. Sci.* **2010**, *9*, 325–337.

18. Bolat, I.; Kaya, C.; Almaca, A.; Timucin, S. Calcium Sulfate Improves Salinity Tolerance in Rootstocks of Plum. *J. Plant Nutr.* **2006**, *29*, 553–564.

19. Boman, B. J.; Zekri, M.; Stover, E. Managing Salinity in Citrus. *HortTechnology* **2005**, *15*, 108–113.

20. Brugnoli, E.; Lauteri, M. Effects of Salinity on Stomatal Conductance, Photosynthetic Capacity, and Carbon Isotope Discrimination of Salt-tolerant (*Gossypiumhirsutum* L.) and Salt-sensitive (*Phaseolus vulgaris* L.) C_3 Non-halophytes. *Plant Physiol.* **1991**, *95*, 628–635.

21. Camp, C. R. Subsurface Drip Irrigation: A Review. *Trans. ASAE* **1998**, *41*, 1353–1367.

22. Cavalcante, L. F.; Cavalcante, I. H. L.; Pereira, K. S. N.; De Oliveira, F. A.; Gondim, S. C.; De Araújo, F. A. R. Germination and Initial Growth of Guava Plants Irrigated with Saline Water. *Rev. Bras. Engenharia Agrícola e Ambiental* **2005,** *9*, 515–519.

23. Cervera, M.; Ortega, C.; Navarro, A.; Navarro, L.; Pena, L. Generation of Transgenic Citrus Plants with the Tolerance-to-salinity Gene HAL2 from Yeast. *J. Hortic. Sci. Biotechnol.* **2000,** *75*, 26–30.

24. Chartzoulakis, K. S. Salinity and Olive: Growth, Salt Tolerance, Photosynthesis and Yield. *Agric. Water Manage.* **2005,** *78*, 108–121.

25. Chartzoulakis, K.; Loupassaki, M.; Bertaki, M.; Androulakis, I. Effects of NaCl Salinity on Growth, Ion Content and CO_2 Assimilation Rate of Six Olive Cultivars. *Sci. Hortic.* **2002,** *96*, 235–247.

26. Chaves, M. M.; Flexas, J.; Pinheiro, C. Photosynthesis Under Drought and Salt Stress: Regulation Mechanisms from Whole Plant to Cell. *Ann. Bot.* **2009,** *103*, 551–560.

27. Chelli-Chaabouni, A.; Mosbah, A. B.; Maalej, M.; Gargouri, K.; Gargouri-Bouzid, R.; Drira, N. In Vitro Salinity Tolerance of Two Pistachio Rootstocks: *Pistaciavera* L. and *P. atlantica* Desf. *Environ. Exp. Bot.* **2010,** *69*, 302–312.

28. Colla, G.; Roupahel, Y.; Cardarelli, M.; Rea, E. Effect of Salinity on Yield, Fruit Quality, Leaf Gas Exchange, and Mineral Composition of Grafted Watermelon Plants. *HortScience* **2006,** *41*, 622–627.

29. Colla, G.; Rouphael, Y.; Rea, E.; Cardarelli, M. Grafting Cucumber Plants Enhance Tolerance to Sodium Chloride and Sulfate Salinization. *Sci. Hortic.* **2012,** *135*, 177–185.

30. Crowley, D.; Arpaia, M. L.; Smith, W.; Clark, P.; Bender, G.; Witney, G. In *Rootstock Selections for Improved Salinity Tolerance of Avocado*, Proceedings of California Avocado Research Symposium, California, USA, 2003; Vol. 1, pp 116–119.

31. CSSRI. *CSSRI Vision 2050*. ICAR-Central Soil Salinity Research Institute, Karnal, India, 2015; p 31.

32. Dagar, J. C.; Singh, G.; Singh, N. T. Evaluation of Forest and Fruit Trees Used for Rehabilitation of Semiarid Alkali–Sodic Soils in India. *Arid Land Res. Manage.* **2001,** *15*, 115–133.

33. Dardeniz, A.; Muftuoglu, N. M.; Altay, H. Determination of Salt Tolerance of Some American Grape Rootstocks. *Bang. J. Bot.* **2006,** *35*, 143–150.

34. Davies, S.; Lacey, A. *Identifying Dispersive Soils*. Farmnote No. 386; Department of Agriculture and Food, Government of Western Australia, 2009; p 2.

35. Dayal, V.; Dubey, A. K.; Awasthi, O. P.; Pandey, R.; Dahuja, A. Growth, Lipid Peroxidation, Antioxidant Enzymes and Nutrient Accumulation in Amrapali Mango (*Mangifera indica* L) Grafted on Different Rootstocks Under NaCl Stress. *Plant Knowledg. J.* **2014,** *3*, 15–22.

36. De Abreu, C. E.; Prisco, J. T.; Nogueira, A. R.; Bezerra, M. A.; Lacerda, C. F. D.; Gomes-Filho, E. Physiological and Biochemical Changes Occurring in Dwarf-cashew Seedlings Subjected to Salt Stress. *Braz. J. Plant Physiol.* **2008,** *20*, 105–118.

37. De la Hera, M. L.; Romero, P.; Gomez-Plaza, E.; Martinez, A. Is Partial Root-zone Drying an Effective Irrigation Technique to Improve Water use Efficiency and Fruit Quality in Field-grown Wine Grapes Under Semiarid Conditions? *Agric. Water Manage.* **2007,** *87*, 261–274.

38. De Oliveira, A. B.; Gomes-Filho, E.; Alencar, N. L. M. *Comparison Between the Water and Salt Stress Effects on Plant Growth and Development*; S. Akinci, Ed.; INTECH (Open Access Publisher): London, United Kingdom, 2013; pp 67–94.

39. Demidchik, V. Mechanisms and Physiological Roles of K$^+$ Efflux from Root Cells. *J. Plant Physiol.* **2014,** *171*, 696–707.

40. Downton, W. J. S. Growth and Flowering in Salt-stressed Avocado Trees. *Aust. J. Agric. Res.* **1978,** *29*, 523–534.

41. Dregne, H.; Kassas, M.; Razanov, B. A New Assessment of the World Status of Desertification. In *Desertification Control Bulletin*; United Nations Environment Program: Nairobi, 1991; Vol. 20, pp 6–18.

42. Du, T.; Kang, S.; Zhang, J.; Li, F.; Yan, B. Water Use Efficiency and Fruit Quality of Table Grape Under Alternate Partial Root-zone Drip Irrigation. *Agric. Water Manage.* **2008,** *95*, 659–668.

43. Dubey, A.; Srivastav, M.; Singh, R.; Pandey, R.; Deshmukh, P. Response of Mango (*Mangifera indica*) Genotypes to Graded Level of Salt Stress. *Indian J. Agric. Sci.* **2006,** *76*, 670–672.

44. Durán-Zuazo, V. H.; Martínez Raya, A.; Aguilar Ruiz, J. Impact of Salinity on the Fruit Yield of Mango (*Mangifera indica* L. cv. Osteen). *Eur. J. Agron.* **2004,** *21*, 323–333.

45. Dziadczyk, P.; Dyrda, D. E.; Hortyński, J. A. NaCl Influence on the In Vitro Germination of *Fragaria* × *ananassa* (Duch.) and *F. vesca* (L.) Seeds. *Small Fruits Rev.* **2004,** *4*, 11–19.

46. Ebert, G. Salinity Problems in (Sub-)tropical Fruit Production. *Acta Hortic.* **2000,** *531*, 99–106.

47. Ebert, G.; Eberle, J.; Ali-Dinar, H.; Ludders, P. Ameliorating Effects of Ca(NO$_3$)$_2$ on Growth, Mineral Uptake and Photosynthesis of NaCl-stressed Guava Seedlings (*Psidium guajava* L.). *Sci. Hortic.* **2002,** *93*, 125–135.

48. Erturk, U.; Sivritepe, N.; Yerlikaya, C.; Bor, M.; Ozdemir, F.; Turkan, I. Response of the Cherry Rootstock to Salinity In Vitro. *Biol. Plant.* **2007,** *51*, 597–600.

49. Essah, P. A.; Davenport, R.; Tester, M. Sodium Influx and Accumulation in Arabidopsis. *Plant Physiol.* **2003,** *133*, 307–318.

50. FAO. *Production Yearbook Volume 48*; Food and Agricultural Organization: Rome, 1997.

51. Fereres, E.; Soriano, M. A. Deficit Irrigation for Reducing Agricultural Water Use. *J. Exp. Bot.* **2007,** *58*, 147–159.

52. Fereres, E.; Goldhamer, D. A.; Sadras, V. O. Yield Response to Water of Fruit Trees and Vines: Guidelines. In *Crop Yield Response to Water*; Steduto, P., Hsiao, T. C., Fereres, E., Raes, D. Eds.; FAO Irrigation and Drainage Paper Vol. 66; Food and Agriculture Organization of the United Nations: Rome, 2012; pp 246–296.

53. Ferreira-Silva, S. L.; Silva, E. N.; Carvalho, F. E. L.; De Lima, C. S.; Alves, F. A. L.; Silveira, J. D. Physiological Alterations Modulated by Rootstock and Scion Combination in Cashew Under Salinity. *Sci. Hortic.* **2010,** *127*, 39–45.

54. Fitzpatrick, R. W. Land Degradation Processes. In *Regional Water and Soil Assessment for Managing Sustainable Agriculture in China and Australia*; McVicar, T. R., Rui, L., Walker, J., Fitzpatrick, R. W., Changming, L. Eds.; ACIAR Monograph, 2010; Vol. 84, pp 119–129.

55. Flachowsky, H.; Hanke, M. V.; Peil, A.; Strauss, S. H.; Fladung, M. A Review on Transgenic Approaches to Accelerate Breeding of Woody Plants. *Plant Breeding* **2009,** *128*, 217–226.

56. Flexas, J.; Bota, J.; Loreto, F.; Cornic, G.; Sharkey, T. D. Diffusive and Metabolic Limitations to Photosynthesis Under Drought and Salinity in C$_3$ Plants. *Plant Biol.* **2004,** *6*, 269–279.

57. Flowers, T. J.; Colmer, T. D. Salinity Tolerance in Halophytes. *New Phytologist* **2008,** *179,* 945–963.

58. Flowers, T. J.; Yeo, A. R. Breeding for Salinity Resistance in Crop Plants: Where Next? *Funct. Plant Biol.* **1995,** *22,* 875–884.

59. Flowers, T. J.; Galal, H. K.; Bromham, L. Evolution of Halophytes: Multiple Origins of Salt Tolerance in Land Plants. *Funct. Plant Biol.* **2010,** *37,* 604–612.

60. Foolad, M. R.; Chen, F. Q. RFLP Mapping of QTLs Conferring Salt Tolerance During the Vegetative Stage in Tomato. *Theor. Appl. Genet.* **1999,** *99,* 235–243.

61. Fort, K.; Walker, A. Breeding Salt Tolerant Rootstocks. *FPS Grape Program Newsletter* **2011,** *10,* 9–11.

62. García-Legaz, M. F.; López-Gómez, E.; Beneyto, J. M.; Navarro, A.; Sánchez-Blanco, M. J. Physiological Behavior of Loquat and Anger Rootstocks in Relation to Salinity and Calcium Addition. *J. Plant Physiol.* **2008,** *165,* 1049–1060.

63. García-Sánchez, F.; Carvajal, M.; Porras, I.; Botıa, P.; Martınez, V. Effects of Salinity and Rate of Irrigation on Yield, Fruit Quality and Mineral Composition of 'Fino 49' Lemon. *Eur. J. Agron.* **2003,** *19,* 427–437.

64. Geerts, S.; Raes, D. Deficit Irrigation as an On-farm Strategy to Maximize Crop Water Productivity in Dry Areas. *Agric. Water Manage.* **2009,** *96,* 1275–1284.

65. Geissler, N.; Hussin, S.; Koyro, H. W. Elevated Atmospheric CO_2 Concentration Ameliorates Effects of NaCl Salinity on Photosynthesis and Leaf Structure of *Aster tripolium* L. *J. Exp. Bot.* **2009,** *60,* 137–151.

66. Ghoulam, C.; Foursy, A.; Fares, K. Effects of Salt Stress on Growth, Inorganic Ions and Proline Accumulation in Relation to Osmotic Adjustment in Five Sugar Beet Cultivars. *Environ. Exp. Bot.* **2002,** *47,* 39–50.

67. Goldhammer, D. A.; Salinas, M.; Crisosto, C.; Day, K. R.; Soler, M.; Moriana, A. In *Effects of Regulated Deficit Irrigation and Partial Root Zone Drying on Late Harvest Peach Tree Performance*, Proceedings of the 5th International Peach Symposium, 2001; Vol. 592, pp 343–350.

68. Goreta, S.; Bučević-Popović, V.; Pavela-Vrančić, M.; Perica, S. Salinity-induced Changes in Growth, Superoxide Dismutase Activity, and Ion Content of Two Olive Cultivars. *Z. Pflanzenernähr Bodenk* **2007,** *170,* 398–403.

69. Grattan, S. R.; Grieve, C. M. Salinity-mineral Nutrient Relations in Horticultural Crops. *Sci. Hortic.* **1999,** *78,* 127–157.

70. Gupta, S. K. *Subsurface Drainage for Waterlogged Saline Soils.* Unpublished Report Submitted to Agricultural Education Division, ICAR, New Delhi; Central Soil Salinity Research Institute: Karnal, 2013; p 353.

71. Gupta, S. K.; Dubey, S. K. Pitcher Irrigation for Water Conservation and use of Saline Water in Vegetable Production. *Indian Farming* **2001,** *51,* 33–34.

72. Gupta, S. K.; Gupta, I. C. *Genesis and Management of Sodic (Alkali) Soils.* Scientific Publishers Pvt. Ltd.: Jodhpur, India, 2017; p. 320.

73. Haggag, L. F. Response of Washington Navel Orange Trees to Salinity of Irrigation Water. *Egypt. J. Hortic.* **1997,** *24,* 67–74.

74. Hare, P. D.; Cress, W. A. Metabolic Implications of Stress-induced Proline Accumulation in Plants. *Plant Growth Regul.* **1997,** *21,* 79–102.

75. Haroldsen, V. M.; Chi-Ham, C. L.; Bennett, A. B. Transgene Mobilization and Regulatory Uncertainty for Non-GE Fruit Products of Transgenic Rootstocks. *J. Biotechnol.* **2012,** *161,* 349–353.

76. Hatami, E.; Esna-Ashari, M.; Javadi, T. Effect of Salinity on Some Gas Exchange Characteristics of Grape (*Vitis vinifera*) Cultivars. *Int. J. Agric. Biol.* **2010**, *12*, 308–310.

77. Hernández, J. A.; Aguilar, A. B.; Portillo, B.; López-Gómez, E.; Beneyto, J. M.; García-Legaz, M. F. The Effect of Calcium on the Antioxidant Enzymes from Salt-treated Loquat and Anger Plants. *Funct. Plant Biol.* **2003**, *30*, 1127–1137.

78. Hoffman, G. J.; Catlin, P. B.; Mead, R. M.; Johnson, R. S.; Francois, L. E.; Goldhamer, D. Yield and Foliar Injury Responses of Mature Plum Trees. *Irrig. Sci.* **1989**, *10*, 215–229.

79. Hooda, P. S.; Sindhu, S. S.; Mehta, P. K.; Ahlawat, V. P. Growth, Yield and Quality of Ber (*Zizyphus mauritiana* Lamk.) as Affected by Soil Salinity. *J. Hortic. Sci. Biotechnol.* **1990**, *65*, 589–594.

80. Horváth, E.; Szalai, G.; Janda, T. Induction of Abiotic Stress Tolerance by Salicylic Acid Signaling. *J. Plant Growth Regul.* **2007**, *26*, 290–300.

81. Hossain, Z.; López-Climent, M. F.; Arbona, V.; Pérez-Clemente, R. M.; Gómez-Cadenas, A. Modulation of the Antioxidant System in Citrus Under Waterlogging and Subsequent Drainage. *J. Plant Physiol.* **2009**, *166*, 1391–1404.

82. Hoult, M. D.; Donnelly, M. M.; Smith, M. W. Salt Exclusion Varies amongst Polyembryonic Mango Cultivar Seedlings. *Acta Hortic.* **1997**, *455*, 455–458.

83. Howie, H.; Lloyd, J. Response of orchard 'Washington Navel' Orange, *Citrus sinensis* (L.) Osbeck to Saline Irrigation Water. 2. Flowering, Fruit Set and Fruit Growth. *Aust. J. Agric. Res.* **1989**, *40*, 371–380.

84. Huh, G. H.; Damsz, B.; Matsumoto, T. K.; Reddy, M. P.; Rus, A. M.; Ibeas, J. I.; Narasimhan, M. L.; Bressan, R. A.; Hasegawa, P. M. Salt Causes Ion Disequilibrium-induced Programmed Cell Death in Yeast and Plants. *Plant J.* **2002**, *29*, 649–659.

85. Husaini, A. M.; Abdin, M. Z. Development of Transgenic Strawberry (*Fragaria × ananassa* Duch.) Plants Tolerant to Salt Stress. *Plant Sci.* **2008**, *174*, 446–455.

86. ICAR-CSSRI. *ICAR-Central Soil Salinity Research Institute Vision 2050*; Indian Council of Agricultural Research: New Delhi, 2015; p 31.

87. Jamalian, S.; Tehranifar, A.; Tafazoli, E.; Eshghi, S.; Davarynejad, G. H. Paclobutrazol Application Ameliorates the Negative Effect of Salt Stress on Reproductive Growth, Yield, and Fruit Quality of Strawberry Plants. *Hortic. Environ. Biotechnol.* **2008**, *49*, 1–6.

88. Jamil, M.; Sadiq, M.; Mehdi, S. M.; Hussain, S. S. Fruit Yield Improvement of Deteriorated Guava Plants in Salt Affected Soil. *Soil Environ.* **2011**, *30*, 166–170.

89. Kakade, V.; Dubey, A.; Awasthi, O.; Dahuja, A. Responses of Physiochemical Attributes of Kinnow Budded on Jatti Khatti to Triazole Treatment Under Salinity. *Acta Agron. Hung.* **2012**, *60*, 433–447.

90. Kaul, M. K.; Mehta, P. K.; Bakshi, R. K. Note on Effect of Different Salts on Seed Germination of *Psidium guajava* L. Cv. L-49 (Sadar). *Curr. Agric.* **1988**, *12*, 83–85.

91. Keutgen, A. J.; Pawelzik, E. Quality and Nutritional Value of Strawberry Fruit Under Long Term Salt Stress. *Food Chem.* **2008**, *107*, 1413–1420.

92. Khavari-Nejad, R. A.; Mostofi, Y. Effects of NaCl on Photosynthetic Pigments, Saccharides, and Chloroplast Ultrastructure in Leaves of Tomato Cultivars. *Photosynthetica* **1998**, *35*, 151–154.

93. Khoshbakht, D.; Asgharei, M. R. Influence of Foliar-applied Salicylic Acid on Growth, Gas-exchange Characteristics, and Chlorophyll Fluorescence in Citrus Under Saline Conditions. *Photosynthetica* **2015**, *53*, 410–418.

94. Kishor, A.; Srivastav, M.; Dubey, A. K.; Singh, A. K.; Sairam, R. K.; Pandey, R. N.; Dahuja, A.; Sharma, R. R. *Paclobutrazol Minimizes the Effects of Salt Stress in Mango (Mangifera indica L.). J. Hortic. Sci. Biotechnol.* **2009**, *84*, 459–465.

95. Klein, I.; Ben-Tal, Y.; Lavee, S.; De Malach, Y.; David, I. Saline Irrigation of cv. Manzanillo and Uovo Di Piccione Trees. *Acta Hortic.* **1994**, *356*, 176–180.

96. Kozlowski, T. T. Response of Woody Plants to Flooding and Salinity. *Tree Physiol.* **1997**, *1*, 1–17.

97. Lal, R. Carbon Sequestration in Drylands. *Ann. Arid Zone (India)* **2000**, *39*, 1–10.

98. Lal, R. Soil Management in the Developing Countries. *Soil Sci.* **2000**, *165*, 57–72.

99. Lal, S.; Gulyani, V.; Khurana, P. Overexpression of HVA1 Gene from Barley Generates Tolerance to Salinity and Water Stress in Transgenic Mulberry (*Morus indica*). *Transgen. Res.* **2008**, *17*, 651–663.

100. Lambers, H. Dryland Salinity: A Key Environmental Issue in Southern Australia. *Plant Soil* **2003**, *257*, 5–7.

101. Lee, J.; Kubota, C.; Tsao, S. J.; Bie, Z.; Hoyos-Echevarria, P.; Morra, L.; Oda, M. Current Status of Vegetable Grafting: Diffusion, Grafting Techniques, Automation. *Sci. Hortic.* **2010**, *127*, 93–105.

102. Li, F.; Liang, J.; Kang, S.; Zhang, J. Benefits of Alternate Partial Root-zone Irrigation on Growth, Water and Nitrogen use Efficiencies Modified by Fertilization and Soil Water Status in Maize. *Plant Soil* **2007**, *295*, 279–291.

103. Li, Y.; Zhang, Y.; Feng, F.; Liang, D.; Cheng, L.; Ma, F.; Shi, S. Overexpression of a Malus Vacuolar Na+/H+ Antiporter Gene (MdNHX1) in Apple Rootstock M. 26 and Its Influence on Salt Tolerance. *Plant Cell Tissue Organ Cult.* **2010**, *102*, 337–345.

104. Liu, J.; Nada, K.; Honda, C.; Kitashiba, H.; Wen, X.; Pang, X.; Moriguchi, T. Polyamine Biosynthesis of Apple Callus Under Salt Stress: Importance of the Arginine Decarboxylase Pathway in Stress Response. *J. Exp. Bot.* **2006**, *57*, 2589–2599.

105. Liu, R.; Lal, R. Nano-enhanced Materials for Reclamation of Mine Lands and Other Degraded Soils: A Review. *J. Nanotechnol.* **2012**, 461–468.

106. Lloyd, J.; Kriedemann, P. E.; Aspinall, D. Contrasts between *Citrus* Species in Response to Salinisation: An Analysis of Photosynthesis and Water Relations for Different Rootstock-scion Combinations. *Physiol. Plant.* **1990**, *78*, 236–246.

107. López-Climent, M. F.; Arbona, V.; Pérez-Clemente, R. M.; Gómez-Cadenas, A. Relationship Between Salt Tolerance and Photosynthetic Machinery Performance in Citrus. *Environ. Exp. Bot.* **2008**, *62*, 176–184.

108. Loreto, F.; Centritto, M.; Chartzoulakis, K. Photosynthetic Limitations in Olive Cultivars with Different Sensitivity to Salt Stress. *Plant Cell Environ.* **2003**, *26*, 595–601.

109. Maiale, S.; Sánchez, D. H.; Guirado, A.; Vidal, A.; Ruiz, O. A. Spermine Accumulation Under Salt Stress. *J. Plant Physiol.* **2004**, *161*, 35–42.

110. Mansour, M. M. F. Plasma Membrane Permeability as an Indicator of Salt Tolerance in Plants. *Biol. Plant.* **2013**, *57*, 1–10.

111. Mansour, M. M. F. Nitrogen Containing Compounds and Adaptation of Plants to Salinity Stress. *Biol. Plant.* **2000**, *43*, 491–500.

112. Massai, R.; Remorini, D.; Tattini, M. Gas Exchange, Water Relations and Osmotic Adjustment in Two Scion/Rootstock Combinations of *Prunus* Under Various Salinity Concentrations. *Plant Soil* **2004**, *259*, 153–162.

113. Mauricio, R. Mapping Quantitative Trait Loci in Plants: Uses and Caveats for Evolutionary Biology. *Nat. Genet.* **2001**, *2*, 370–381.

114. McFarlane, D. J.; Williamson, D. R. An Overview of Water Logging and Salinity in Southwestern Australia as Related to the 'Ucarro' Experimental Catchment. *Agric. Water Manage.* **2002,** *53*, 5–29.

115. McWilliam, J. R. The National and International Importance of Drought and Salinity Effects on Agricultural Production. *Funct. Plant Biol.* **1986,** *13*, 1–13.

116. Meena, S. K.; Gupta, N. K.; Gupta, S.; Khandelwal, S. K.; Sastry, E. V. D. Effect of Sodium Chloride on the Growth and Gas Exchange of Young *Ziziphus* Seedling Rootstocks. *J. Hortic. Sci. Biotechnol.* **2003,** *78*, 454–457.

117. Melgar, J. C.; Syvertsen, J. P.; Martínez, V.; García-Sánchez, F. Leaf Gas Exchange, Water Relations, Nutrient Content and Growth in Citrus and Olive Seedlings Under Salinity. *Biol. Plant.* **2008,** *52*, 385–390.

118. Minhas, P. S. Saline Water Management for Irrigation in India. *Agric. Water Manage.* **1996,** *30*, 1–24.

119. Misra, N.; Gupta, A. K. Effect of Salinity and Different Nitrogen Sources on the Activity of Antioxidant Enzymes and Indole Alkaloid Content in *Catharanthus roseus* seedlings. *J. Plant Physiol.* **2006,** *163*, 11–18.

120. Mitsuya, S.; Kawasaki, M.; Taniguchi, M.; Miyake, H. Light Dependency of Salinity-induced Chloroplast Degradation. *Plant Prod. Sci.* **2003,** *6*, 219–223.

121. Mitsuya, S.; Takeoka, Y.; Miyake, H. Effects of Sodium Chloride on Foliar Ultrastructure of Sweet Potato (*Ipomoea batatas* Lam.) Plantlets Grown Under Light and Dark Conditions In Vitro. *J. Plant Physiol.* **2000,** *157*, 661–667.

122. Mmolawa, K.; Or, D. Root Zone Solute Dynamics Under Drip Irrigation: A Review. *Plant Soil* **2000,** *222*, 163–190.

123. Morsy, M. H. Growth Ability of Mango Cultivars Irrigated with Saline Water. *Acta Hortic.* **2003,** *609*, 475–482.

124. Mounzer, O.; Pedrero-Salcedo, F.; Nortes, P. A.; Bayona, J. M.; Nicolás-Nicolás, E.; Alarcón, J. J. Transient Soil Salinity Under the Combined Effect of Reclaimed Water and Regulated Deficit Drip Irrigation of Mandarin Trees. *Agric. Water Manage.* **2013,** *120*, 23–29.

125. Mousavi, A.; Lessani, H.; Babalar, M.; Talaei, A. R.; Fallahi, E. Influence of Salinity on Chlorophyll, Leaf Water Potential, Total Soluble Sugars, and Mineral Nutrients in Two Young Olive Cultivars. *J. Plant Nutr.* 2008, *31*, 1906–1916.

126. Munns, R. Genes and Salt Tolerance: Bringing Them Together. *New Phytologist* **2005,** *167*, 645–663.

127. Munns, R.; James, R. A.; Läuchli, A. Approaches to Increasing the Salt Tolerance of Wheat and Other Cereals. *J. Exp. Bot.* **2006,** *57*, 1025–1043.

128. Murkute, A. A.; Sharma, S.; Singh, S. K. Citrus in Terms of Soil and Water Salinity: A Review. *J. Sci. Ind. Res.* **2005,** *64*, 393.

129. Musyimi, D. M.; Netondo, G. W.; Ouma, G. Effects of Salinity on Gas Exchange and Nutrients Uptake in Avocados. *J. Biol. Sci.* **2007,** *7*, 496–505.

130. Myers, B. A.; Dennis, W. W.; Callian, L.; Hunter C. C. Long Term Effects of Saline Irrigation on the Yield and Growth of Mature Williams Pear Trees. *Irrig. Sci.* **1995,** *16*, 35–46.

131. Najafian, S.; Rahemi, M.; Tavallali, V. Effect of Salinity on Tolerance of Two Bitter Almond Rootstocks. *Am.-Eurasian J. Agric. Environ. Sci.* **2008,** *3*, 264–268.

132. Nieves, M.; Cerda, A.; Botella, M. Salt Tolerance of 2 Lemon Scions Measured by Leaf Chloride and Sodium Accumulation. *J. Plant Nutr.* **1991,** *14*, 623–636.

133. Oldeman, L. R.; Van Englen, V. W. P.; Pulles, J. H. M. The Extent of Human-induced Soil Degradation. In *World Map of Status of Human-induced Soil Degradation: An explanatory Note*; Oldeman, et al., Eds.; International Soil Reference and Information Centre: Wageningen, 1991; pp 27–33.

134. Oron, G.; DeMalach, Y.; Gillerman, L.; David, I.; Rao, V. P. Improved Saline-water Use Under Subsurface Drip Irrigation. *Agric. Water Manage.* **1999,** *39,* 19–33.

135. Pandey, R. S.; Singh, A. Study on Backpressure in Subsurface Drip Irrigation Under Field Irrigation Utilizing Sewage Water. In: *Annual Report 2014–15, CSSRI, Karnal*; Central Soil salinity Research Institute: Karnal, 2015; p. 39.

136. Papadakis, I. E.; Veneti, G.; Chatzissavvidis, C.; Sotiropoulos, T. E.; Dimassi, K. N.; Therios, I. N. Growth, Mineral Composition, Leaf Chlorophyll and Water Relationships of Two Cherry Varieties Under NaCl-induced Salinity Stress. *Soil Sci. Plant Nutr.* **2007,** *53,* 252–258.

137. Pedrero, F.; Maestre-Valero, J. F.; Mounzer, O.; Nortes, P. A.; Alcobendas, R.; Romero-Trigueros, C.; Bayona, J. M.; Alarcón, J. J.; Nicolás, E. Response of Young 'Star Ruby' Grapefruit Trees to Regulated Deficit Irrigation with Saline Reclaimed Water. *Agric. Water Manage.* **2015,** *158,* 51–60.

138. Pérez-Tornero, O.; Tallón, C. I.; Porras, I.; Navarro, J. M. Physiological and Growth Changes in Micropropagated *Citrus macrophylla* Explants due to Salinity. *J. Plant Physiol.* **2009,** *166,* 1923–1933.

139. Perica, S.; Goreta, S.; Selak, G. V. Growth, Biomass Allocation and Leaf Ion Concentration of Seven Olive (*Olea europaea* L.) Cultivars Under Increased Salinity. *Sci. Hortic.* **2008,** *117,* 123–129.

140. Podmore, C. Irrigation Salinity—Causes and Impacts. *Primefact* **2009,** *937,* 1–4.

141. Porcel, R.; Aroca, R.; Ruiz-Lozano, J. M. Salinity Stress Alleviation Using Arbuscular Mycorrhizal Fungi. A Review. *Agron. Sustainable Dev.* **2012,** *32,* 181–200.

142. Porras-Soriano, A.; Soriano-Martín, M. L.; Porras-Piedra, A.; Azcón, R. Arbuscular-Mycorrhizal Fungi Increased Growth, Nutrient Uptake and Tolerance to Salinity in Olive Trees Under Nursery Conditions. *J. Plant Physiol.* **2009,** *166,* 1350–1359.

143. Prakash, L.; Parthapasenan, G. Interactive Effect of NaCl Salinity and Gibberellic Acid on Shoot Growth, Content of Abscisic Acid and Gibberellin like Substances and Yield of Rice (*Oryza sativa*). *Plant Sci.* **1990,** *100,* 173–181.

144. Prasad, M.; Burge, G. K.; Spiers, T. M.; Fietje, G. Chloride-induced Leaf Breakdown in Kiwifruit. *J. Plant Nutr.* **1993,** *16,* 999–1012.

145. Rahman, S.; Matsumuro, T.; Miyake, H.; Takeoka, Y. Salinity-induced Ultrastructural Alterations in Leaf Cells of Rice (*Oryza sativa* L.). *Plant Prod. Sci.* **2000,** *3,* 422–429.

146. Raman, S. In *Status of Research on Micro-irrigation for Improving Water Use Efficiency in Some Horticultural Crops.* Proceedings of the National Seminar on Problems and Prospects of Micro-irrigation: A Critical Appraisal, Bangalore, Nov 19–20, 1999; Institution of Engineers: India, 1999; pp 31–45.

147. Ramoliya, P. J.; Pandey, A. N. Effect of Salinization of Soil on Emergence, Growth and Survival of Seedlings of *Cordia rothii. For. Ecol. Manage.* **2003,** *176,* 185–194.

148. Rao, V. K.; Rathore, A. C.; Singh, H. K. Screening of Aonla (*Emblica officinalis* Gaertn.) Cultivars for Leaf Chlorophyll and Amino Acid Under Different Sodicity and Salinity Levels. *Indian J. Soil Conserv.* **2009,** *37,* 193–196.

149. Rao, V. K.; Singh, H. K. Response of Sodicity and Salinity Levels on Vegetative Growth and Nutrient Uptake of Aonla Genotypes. *Indian J. Hortic.* **2006,** *63,* 359–364.

150. Rengasamy, P. World Salinization with Emphasis on Australia. *J. Exp. Bot.* **2006,** *57,* 1017–1023.

151. Ritzema, H. P.; Satyanarayana, T. V.; Raman, S.; Boonstra, J. Subsurface Drainage to Combat Waterlogging and Salinity in Irrigated Lands in India: Lessons Learned in Farmers' Fields. *Agric. Water Manage.* **2008,** *95,* 179–189.

152. Roberts, T.; Lazarovitch, N.; Warrick, A. W.; Thompson, T. L. Modeling salt Accumulation with Subsurface Drip Irrigation Using HYDRUS-2D. *Soil Sci. Soc. Am. J.* **2009,** *73,* 233–240.

153. Rozema, J.; Flowers, T. Crops for a Salinized World. *Science* 2008, *322,* 1478–1480.

154. Sadras, V. O. Does Partial Root-Zone Drying Improve Irrigation Water Productivity in the Field? A Meta-analysis. *Irrig. Sci.* **2009,** *27,* 183–190.

155. Saeed, W. T. Pomegranate Cultivars as Affected by PBZ, Salt Stress and Change in Fingerprints. *Bull. Faculty Sci. Cairo Univ.* **2005,** *56,* 581–615.

156. Saied, A. S.; Keutgen, A. J.; Noga, G. The Influence of NaCl Salinity on Growth, Yield and Fruit Quality of Strawberry cvs. 'Elsanta' and 'Korona'. *Sci. Hortic.* **2005,** *103,* 289–303.

157. Schaffer, B.; Andersen, P. C.; Ploetz, R. C. Responses of Fruit Crops to Flooding. *Horticult. Rev.* 1992, *13,* 257–313.

158. Schaffer, B.; Davies, F. S.; Crane, J. H. Responses of Subtropical and Tropical Fruit Trees to Flooding in Calcareous Soil. *HortScience* **2006,** *41,* 549–555.

159. Schmutz, U. Effect of Salt Stress (NaCl) on Whole Plant CO_2-gas Exchange in Mango. *Acta Hortic.* **2000,** *509,* 269–276.

160. Schmutz, U.; Ludders, P. Physiology of Saline Stress in One Mango (*Mangifera indica* L.) Rootstock. *Acta Hortic.* **1993,** *341,* 160–167.

161. Sepaskhah, A. R.; Ahmadi, S. H. A Review on Partial Root-zone Drying Irrigation. *Int. J. Plant Prod.* **2010,** *4,* 241–258.

162. Shabala, S. Salinity and Programmed Cell Death: Unraveling Mechanisms for Ion Specific Signaling. *J. Exp. Bot.* **2009,** *60,* 709–712.

163. Shabala, S.; Pottosin, I. Regulation of Potassium Transport in Plants Under Hostile Conditions: Implications for Abiotic and Biotic Stress Tolerance. *Physiol. Plant.* **2014,** *151,* 257–279.

164. Shannon, M. C. Adaptation of Plants to Salinity. *Adv. Agron.* **1997,** *60,* 75–120.

165. Sharma, D. K.; Singh, A. Salinity Research in India: Achievements, Challenges and Future Prospects. *Water Energy Int.* **2015,** *58,* 35–45.

166. Sharma, D. K.; Dubey, A. K.; Srivastav, M.; Singh, A. K.; Sairam, R. K.; Pandey, R. N.; Dahuja, A.; Kaur, C. Effect of Putrescine and Paclobutrazol on Growth, Physiochemical Parameters, and Nutrient Acquisition of Salt-sensitive Citrus Rootstock Karnakhatta (*Citrus karna* Raf.) Under NaCl Stress. *J. Plant Growth Regul.* **2011,** *30,* 301–311.

167. Shekhawat, U. K. S.; Srinivas, L.; Ganapathi, T. R. MusaDHN-1, A Novel Multiple Stress-inducible SK3-type Dehydrin Gene, Contributes Affirmatively to Drought- and Salt-stress Tolerance in Banana. *Planta* 2011, *234,* 915–932.

168. Shibli, R. A.; Shatnawi, M. A.; Swaidat, I. Q. Growth, Osmotic Adjustment, and Nutrient Acquisition of Bitter Almond Under Induced Sodium Chloride Salinity In Vitro. *Commun. Soil Sci. Plant Anal.* **2003,** *34,* 1969–1979.

169. Shibli, R.; Mohammad, M.; Abu-Ein, A.; Shatnawi, M. Growth and Micronutrient Acquisition of Some Apple Varieties in Response to Gradual In Vitro Induced Salinity. *J. Plant Nutr.* **2000,** *23,* 1209–1215.

170. Silva, A. B. F.; Fernandes, P. D.; Gheyi, H. R.; Blanco, F. F. Growth and Yield of Guava Irrigated with Saline Water and Addition of Farmyard Manure. *Rev. Bras. de Ciências Agrárias* **2008,** *3,* 354–359.

171. Singh, A.; Singh, A. K. The Future Prospects of Fruit Biotechnology—An Indian Perspective. In *Horticulture for Food and Environment Security*; Chadha K. L., Singh, A. K., Singh, S. K., Dhillon, W. S. Eds.; The Horticultural Society of India: New Delhi, India, 2012; pp 67–71.

172. Singh, A.; Saini, M. L.; Behl, R. K. Seed Germination and Seedling Growth of Citrus (Citrus Species) Rootstocks Under Different Salinity Regimes. *Indian J. Agric. Sci.* **2004,** *74,* 246–248.

173. Singh, A.; Sharma, P. C.; Kumar, A.; Meena, M. D.; Sharma, D. K. Salinity Induced Changes in Chlorophyll Pigments and Ionic Relations in Bael (*Aegle marmelos* Correa) Cultivars. *J. Soil Salin. Water Qual.* **2015,** *7,* 40–44.

174. Singh, G.; Singh, N. T.; Abrol, I. P. Agroforestry Techniques for the Rehabilitation of Degraded Salt-affected Lands in India. *Land Degrad. Dev.* **1994,** *5,* 223–242.

175. Singh, I. S.; Singh, R. K.; Khanna, S. S. *Management of Sodic Soils Through Plantation.* Bulletin of the N. D. University of Agriculture and Technology, Faizabad, India, 1994; p 51.

176. Sotiropoulos, T. E.; Therios, I. N.; Almaliotis, D.; Papadakis, I.; Dimassi, K. N. Response of Cherry Rootstocks to Boron and Salinity. *J. Plant Nutr.* **2006,** *29,* 1691–1698.

177. Srivastav, M.; Kishor, A.; Dahuja, A.; Sharma, R. R. Effect of Paclobutrazol and Salinity on Ion Leakage, Proline Content and Activities of Antioxidant Enzymes in Mango (*Mangifera indica* L.). *Sci. Hortic.* **2010,** *125,* 785–788.

178. Stewart, G. R.; Lee, J. A. The Role of Proline Accumulation in Halophytes. *Planta* **1974,** *120,* 279–289.

179. Stirzaker, R. J.; Cook, F. J.; Knight, J. H. Where to Plant Trees on Cropping Land for Control of Dryland Salinity: Some Approximate Solutions. *Agric. Water Manage.* **1999,** *39,* 115–133.

180. Stoll, M.; Loveys, B.; Dry, P. Hormonal Changes Induced by Partial Root Zone Drying of Irrigated Grapevine. *J. Exp. Bot.* **2000,** *51,* 1627–1634.

181. Sudhakar, C.; Lakshmi, A.; Giridarakumar, S. Changes in the Antioxidant Enzyme Efficacy in Two High Yielding Genotypes of Mulberry (*Morus alba* L.) Under NaCl Salinity. *Plant Sci.* **2001,** *161,* 613–619.

182. Sudhir, P.; Murthy, S. D. S. Effects of Salt Stress on Basic Processes of Photosynthesis. *Photosynthetica* **2004,** *42,* 481–486.

183. Syvertsen, J. P.; Boman, B.; Tucker, D. P. H. Salinity in Florida Citrus Production. *Proc. Florida State Hortic. Soc.* **1989,** *102,* 61–64.

184. Tester, M.; Davenport, R. Na^+ Tolerance and Na^+ Transport in Higher Plants. *Ann. Bot.* **2003,** *91,* 503–527.

185. Thomas, R. P.; Morini, S. *Management of Irrigation-induced Salt-affected Soils.* Jointly Published by CISEAU, IPTRID and AGLL, FAO: Rome, Italy; 2005; pp 4.

186. Tomar, O. S.; Gupta, R. K. Performance of Some Tree Species in Saline Soils Under Shallow and Saline Water Table Conditions. *Plant and Soil* **1985,** *27,* 329–335.

187. Tozlu, I.; Guy, C. L.; Moore, G. A. QTL Analysis of Morphological Traits in an Intergeneric BC1 Progeny of *Citrus* and *Poncirus* Under Saline and Non-saline Environments. *Genome* **1999,** *42,* 1020–1029.

188. Tozlu, I.; Guy, C. L.; Moore, G. A. QTL Analysis of Na+ and Cl-accumulation Related Traits in an Intergeneric BC1 Progeny of Citrus and Poncirus Under Saline and Non-saline Environments. *Genome* **1999**, *42*, 692–705.

189. Tripler, E.; Ben-Gal, A.; Shani, U. Consequence of Salinity and Excess Boron on Growth, Evapotranspiration and Ion Uptake in Date Palm (*Phoenix dactylifera* L., cv. Medjool). *Plant Soil.* **2007**, *297*, 147–155.

190. Viégas, R. A.; da Silveira, J. A. G. Ammonia Assimilation and Proline Accumulation in Young Cashew Plants during Long Term Exposure to NaCl-salinity. *Rev. Bras. de Engenharia Agrícola e Ambiental* **1999**, *11*, 153–159.

191. Vijayan, K.; Chakraborti, S. P.; Ghosh, P. D. In Vitro Screening of Mulberry (*Morus* spp.) for Salinity Tolerance. *Plant Cell Rep.* **2003**, *22*, 350–357.

192. Vijayan, K.; Chakraborti, S. P.; Ercisli, S.; Ghosh, P. D. NaCl Induced Morpho-biochemical and Anatomical Changes in Mulberry (*Morus* spp.). *Plant Growth Regul.* **2008**, *56*, 61–69.

193. Vijayan, K.; Chakraborti, S. P.; Ghosh, P. D. Screening of Mulberry (*Morus* spp.) for Salinity Tolerance through In Vitro Seed Germination. *Indian J. Biotech.* **2004**, *3*, 47–51.

194. Voigt, E. L.; Almeida, T. D.; Chagas, R. M.; Ponte, L. F. A.; Viégas, R. A.; Silveira, J. A. G. Source-sink Regulation of Cotyledonary Reserve Mobilization During Cashew (*Anacardium occidentale*) Seedling Establishment Under NaCl Salinity. *J. Plant Physiol.* **2009**, *166*, 80–89.

195. Wahbi, S.; Wakrim, R.; Aganchich, B.; Tahi, H.; Serraj, R. Effects of Partial Root Zone Drying (PRD) on Adult Olive Tree (*Olea europaea*) in Field Conditions Under Arid Climate: I. Physiological and Agronomic Responses. *Agric. Ecosyst. Environ.* **2005**, *106*, 289–301.

196. Walker, R. R.; Blackmore, D. H.; Clingeleffer, P. R.; Correll, R. L. Rootstock Effects on Salt Tolerance of Irrigated Field-Grown Grapevines (*Vitis vinifera* L. cv. Sultana) 2. Ion Concentrations in Leaves and Juice. *Aust. J. Grape Wine Res.* **2004**, *10*, 90–99.

197. Wang, D.; Karle, R.; Iezzoni, A. F. QTL Analysis of Flower and Fruit Traits in Sour Cherry. *Theor. Appl. Genet.* **2000**, *100*, 535–544.

198. Wen, X. P.; Pang, X. M.; Matsuda, N.; Kita, M.; Inoue, H.; Hao, Y. J.; Honda, C.; Moriguchi, T. Over-expression of the Apple Spermidine Synthase Gene in Pear Confers Multiple Abiotic Stress Tolerance by Altering Polyamine Titers. *Transgen. Res.* **2008**, *17*, 251–263.

199. West, D. W. In *Stress Physiology in Trees-salinity*. Proceedings of the Symposium on Physiology of Productivity of Subtropical and Tropical Tree Fruits, 1985; Vol. 175, pp 321–332.

200. Whiting, J. Rootstock Breeding and Associated R&D in the Viticulture and Wine Industry. *Wine Vitic. J.* **2012**, *6*, 52–54.

201. Wicke, B.; Smeets, E.; Dornburg, V.; Vashev, B.; Gaiser, T.; Turkenburg, W.; Faaij, A. The Global Technical and Economic Potential of Bioenergy from Salt-affected Soils. *Energy Environ. Sci.* **2011**, *4*, 2669–2681.

202. Williams, W. D. Anthropogenic Salinization of Inland Waters. *Hydrobiologia* **2001**, *466*, 329–337.

203. Wu, Q. S.; Zou, Y. N.; He, X. H. Contributions of Arbuscular Mycorrhizal Fungi to Growth, Photosynthesis, Root Morphology and Ionic Balance of Citrus Seedlings Under Salt Stress. *Acta Physiol. Plant.* **2010**, *32*, 297–304.

204. Yamaguchi, T.; Blumwald, E. Developing Salt-tolerant Crop Plants: Challenges and Opportunities. *Trends Plant Sci.* **2005**, *10*, 615–620.

205. Yildirim, O.; Aras, S.; Ergül, A. Response of Antioxidant Systems to Short-term NaCl Stress in Grapevine Rootstock-1616 C and *Vitis vinifera* L. cv. Razaki. *Acta Biol. Cracoviensia* **2004,** *46*, 151–158.

206. Youssef, T.; Awad, M. A. Mechanisms of Enhancing Photosynthetic Gas Exchange in Date Palm Seedlings (*Phoenix dactylifera* L.) Under Salinity Stress by a 5-Aminolevulinic Acid-based Fertilizer. *J. Plant Growth Regul.* **2008,** *27*, 1–9.

207. Zekri, M. Effects of NaCl on Growth and Physiology of Sour Orange and Cleopatra Mandarin Seedlings. *Sci. Hortic.* **1991,** *47*, 305–315.

208. Zhang, H.; Irving, L. J.; McGill, C.; Matthew, C.; Zhou, D.; Kemp, P. The Effects of Salinity and Osmotic Stress on Barley Germination Rate: Sodium as an Osmotic Regulator. *Ann. Bot.* **2010,** *106*, 1027–1035.

209. Zhu, J.; Bie, Z.; Huang, Y.; Han, X. Effect of Grafting on the Growth and Ion Concentrations of Cucumber Seedlings Under NaCl Stress. *Soil Sci. Plant Nutr.* **2008,** *54*, 895–902.

APPENDIX A

LIST OF ENGLISH, LOCAL, AND BOTANICAL NAMES OF FRUIT PLANTS REFERRED IN THIS CHAPTER

English name	(Local name)	Botanical name
Almond	(Badam)	*Prunus dulcis* (Mill.) DA Webb
Apple	(Seb)	*Malus × domestica* Borkh.
Avocado	(Avocado)	*Perseaamericana* Mill.
Banana	(Kela)	*Musa paradisiaca* L.
Bengal quince	(bael)	*Aegle marmelos* Correa
Black plum	(Jamun)	*Syzygium cuminii* L. Skeels.
Cashew nut	(Kaju)	*Annacardium occidentale* L.
Christ's thorn	(Karonda)	*Carrisa congesta* Wight
Date palm	(Khajoor)	*Phoenix dactylifera* L.
Duchesne	(Loquat; Japanese medlar)	*Eriobotryajaponica* (Thunb.) Lindl.
Grape	(Angoor)	*Vitis vinifera* L.
Guava	(Amrud)	*Psidium guajava* L.
Indian gooseberry	(Aonla)	*Emblica officinalis* Gaertn.
Indian jujube	(Ber)	*Zizyphus mauritiana* Lam.
Japanese plum	(Aloo bukhara)	*Prunus salicina* Lindl.
Kiwifruit	(Kiwifruit)	*Actinidia chinesis* Planch.
Mandarin lime	(Rangpur lime)	*Citrus × limonia* Osbeck
Mango	(Aam)	*Mangifera indica* L.
Mulberry	(Shahtoot)	*Morus* spp.
Olive	(Jaitoon)	*Oleaeuropaea* L.
Peach	(Aadoo)	*Prunus persica*(L) Batsch
Pear	(Nashpati)	*Pyrus communis* L.
Pineapple	(Ananas)	*Ananascomosus* (L.) Merr.
Pistachio	(Pista)	*Pistacia vera* L.
Pomegranate	(Anar)	*Punica granatum* L.
Salvadora	(Peelu)	*Salvadora oleoides* Decne.
Strawberry	(Strawberry)	*Fragaria × ananassa*
Sweet cherry	(Sweet cherry)	*Prunus cerasus* L.
Tamarind	(Imli)	*Tamarindus indica* L.
Walnut	(Akhrot)	*Juglans regia* L.

CHAPTER 3

ROLE OF CONSERVATION AGRICULTURE IN MITIGATING SOIL SALINITY IN INDO-GANGETIC PLAINS OF INDIA

AJAY KUMAR MISHRA, ATHUMAN JUMA MAHINDA,
HITOSHI SHINJO, MANGI LAL JAT, ANSHUMAN SINGH, and
SHINYA FUNAKAWA

ABSTRACT

Land degradation mainly due to soil salinity poses a major challenge to food security in the Indo-Gangetic Plain (IGP). Since IGP is in a transition phase from extensive to intensive agriculture, it needs to be addressed with utmost care to ensure resource conservation and environmental sustainability. Conservation agriculture (CA) based on the proven principles of minimal tillage, residue recycling, and crop diversification has emerged as a viable option for a paradigm shift toward economically and ecologically sustainable agriculture. The implication of CA for soil salinity reduction in IGP is of core significance that might build resilience to climate change. CA has shown great potential in ameliorating soil salinity and alleviation of secondary salinization in IGP region. Management practices of CA, such as minimum tillage, zero tillage, permanent raised beds, use of cover crop, crop rotation, and direct seeding; have proved to increase substantial amount of SOC, which plays a central role in reducing soil salinity, by improving soil physical condition that enhances infiltration rates, which leaches the salts down beyond the root zone of the crops. When CA is supplemented with some of the salinity management practices described in this chapter, it can help halt the land degradation due to salinity leading to sustainable food security regime.

3.1   INTRODUCTION

Over a third of the global arable land is under highly intensive, continuous cropping systems relying upon excessive agrochemicals and conventional tillage (CT) that significantly contributes to physical, chemical, and biological land degradation.[30] All these forms of land-degradation pose severe challenges in achieving the goals of food security, as they have adverse impacts on agronomic productivity, soil health and the environment. The magnitude and the forms of the land degradation depend on the climatic conditions and topographic features of the area.[30] For instance, chemical land degradation, particularly arising due to salinization and sodication (or sodification), is likely to be the prominent threat to crop productivity in arid and semiarid regions besides coastal ecology.[11,77]

Worldwide, salt-affected soils cover approximately 3% of the major food producing arable land, of which 402 and 434 M ha are classified as saline and sodic soils, respectively.[88] In India, salt-affected soils cover an area of 6.73 M ha, of which 44% is saline and 56% is sodic. Furthermore, intensive agricultural practices with poor irrigation and drainage practices are expected to aggravate the problems especially in the irrigation commands and coastal regions. It is projected that there will be a twofold increase in salt-affected area in 2025, in which the Indo-Gangetic Plains (IGP), India's breadbasket, is most likely to be affected.[88] Besides, intense pressure from uncontrolled population growth that will fiercely compete for good quality land and water resources leaving only marginally and poor-quality resources for agriculture will adversely impact the sustainable agricultural growth.[29,31] Topographically, the IGP is divided into four regions, namely,

a) Trans IGP comprising Punjab in Pakistan, and Punjab and Haryana in India;
b) Upper and middle IGP comprising western, central, and eastern Uttar Pradesh and Bihar;
c) Terai (an extension of the IGP) in Uttarakhand in India and parts of Nepal; and
d) Lower IGP comprising West Bengal in India and parts of Bangladesh.

In Pakistan, the rice–wheat (RW) cropping system occupies about 10% of the total cultivated area (2.1 M ha), mainly in two zones: the Punjab RW zone comprising about 1.2 M ha, and the Sindh RW zone occupying the remaining area.[4] The Indian portion of the IGP depicted in Figure 3.1, with a brief description provided in Table 3.1 occupies 15% of the total land mass that produce more than 50% of the grains in the country and supports more than 2 million people.[8,63,77]

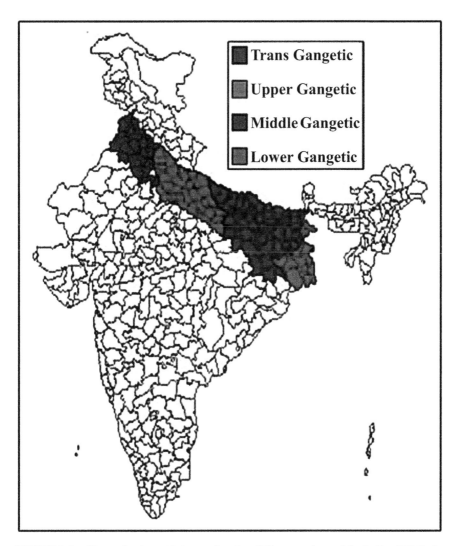

FIGURE 3.1 **(See color insert.)** A map showing different regions of the Indian IGP (Indo-Gangetic plains).

Indian IGP is already witnessing stagnation in food and fiber production. It is likely that production might decrease in future, unless salt-affected areas

TABLE 3.1　Description of the States Under Indian IGP.

Attributes	Punjab	Haryana	Delhi	Uttar Pradesh	Bihar	West Bengal	India
Geographical area (M ha)	5.04	4.42	0.148	24.1	9.42	8.88	328.7
% of TGA	1.53	1.35	0.045	7.33	2.86	2.70	15.8
Climatic zone	Semiarid, subtropical	Semiarid, subtropical	Humid, subtropical	Humid, subtropical	Subtropical	Tropical savanna—humid subtropical	Tropical wet, tropical savanna—humid dry, subtropical humid, and montane
Population (million)	27.7	25.4	0.250	199.6	103.8	91.34	1.21[a]
Population density (persons km^{-2})	550	573	5854.7	820	1102	1000	383.3
Annual min–max. temp. (°C)	17–30	23–30	31–19	19–32	20–31	22–32	10–40
Annual rainfall (mm)	733	700	790	1023	1116	1800	1083

TGA, total geographical area.

[a]Billion.

Data collected from various sources.

are reclaimed, and barren land is converted into the productive agricultural land.[2,11] Some interventions to reduce the magnitude of salinity problem in IGP are being implemented notable amongst them are

- Application of chemical amendments;
- Hydro-technical interventions such as land drainage and leaching;
- Biological interventions, such as the use of salt-tolerant varieties including agroforestry and improved cultural and irrigation practices together with the application of manures.

However, improvement in agricultural productivity is still below the mark as against the overgrowing population in the IGP region.[43,54,60] It appears that conservation agriculture (CA) needs to be effectively adopted and out-scaled into the existing agricultural system, to achieve the potential productivity of the IGP in a sustainable manner.[54]

3.1.1 CONCEPTUAL FRAMEWORK OF CONSERVATION AGRICULTURE

CA refers to the resource-conserving set of management practices that strives to achieve acceptable profits together with high and sustained production levels while concurrently improving soil quality and protecting the environment.[41] FAO defined CA based on following four principles[11,32]:

a) Minimum soil disturbance;
b) Permanent soil cover;
c) Crop diversification; and
d) Minimal soil compaction.

To achieve these general goals, one or more of the listed practices can be adopted (Table 3.2). The listed components protect the soil quality, conserve water and nutrients, promote soil biological activity and contribute to integrated pest management. Although there are many direct and indirect advantages of CA, yet the key benefits are improved soil organic carbon (SOC) content and hence soil health enhanced input use efficiency and finally increased the potential to reduce greenhouse gas (GHG) emissions.[12]

CA has expanded over the years, from a meager 2.8 M ha worldwide in 1973–1974, to about 157 M ha around the globe.[61] Major CA practicing countries are the USA (35.6 M ha), Brazil (31.8 M ha), Argentina (29.2 M ha), Canada (18.3 M ha), and Australia (17.7 M ha). Other countries where the area under CA exceeds 1.0 M ha are China, Russia, Paraguay, Kazakhstan, India, and Uruguay.[61,62] Technically, the concept of CA is relatively new in the Asian continent but is gaining momentum in many Asian countries including India. In the Indian IGP alone, its extent is approximated to be 1.5 M ha.[83] The potential of CA has already emerged to increase yield, reduce runoff, increase the SOC, and reduce negative impacts of the impending climate change. Additionally, it has ensured short and long-term economic and environmental benefits to the farming communities.[61,83]

TABLE 3.2 Principles and Practices for Conservation Agriculture.[a]

Principle	Practices
Diversified crop rotation	Including cover crops, legumes in crop rotation, resource-conserving crops
Minimal compaction	Avoiding use of heavy machinery, GPS-controlled machinery movement
Minimum soil disturbance	No-tillage, minimal tillage
Permanent soil cover	Cultivating cover crops, residue management including organic mulching, more crops in rotation

[a]Adapted from Ref. [33].

Besides what has been stated above, limited information on the role of CA in mitigating the detrimental effect of soil salinity has been generated.

This chapter reviews and explicitly explores the scope and role of CA practices in countering the adverse effects of soil salinity in Indian IGP.

3.2 TRANSFORMATION OF AGRICULTURE SYSTEMS IN IGP AND CONSERVATION AGRICULTURE

Like any other part of the world, agriculture in IGP had shifted from hunting, nomad, and root digging, in search of food or to secure livelihood to a stage of stabilized civilization of growing selected crops in a well-planned environ-ment. Archaeological reports from several important sites in the Southern, Central, and Western parts of the IGP suggest that a major shift has occurred from incipient agricultural activities to well-developed agricultural practices over a span of last 10,000 years.[64,73,85,90,95,96] RW crops are being grown in South-Asian countries like India, Pakistan, Nepal, Bangladesh, and Bhutan, as well as China for more than 1000 years.

The RW cropping system (growing RW on the same piece of the land in a year) has been developed through the introduction of rice in the traditional wheat-growing areas and vice versa.[78] This cropping system is one of the world's largest agricultural production systems, covering an area of 26 M ha and spread over the IGP in South Asia and China.[66] However, either of the two crops can also be grown sequentially, with other crops such as maize, pigeon pea, sugarcane, and lentil depending on the season and rainfall characteristics.[42,49]

In the mid-1960s, Indian IGP witnessed green revolution due to its favorable climatic conditions, fertile land that favored crop diversification and thus facilitated shifting to RW cultivation.[8,50] In the RW cropping system, rice is grown in *Kharif* season (June–October) and wheat in *Rabi* season (November–April). Although three major cropping systems of the IGP are RW, rice–fallow, and rice–mustard, yet RW alone occupies about 72% of the total cultivated land area.[55] The success of RW system is highly attributed to its adaptability, availability of high-yielding varieties of wheat and rice along with their mechanization compatibility.[54,59] Diversification attempts have resulted in the adoption of other farming systems such as maize–wheat, sugarcane–wheat, cotton–wheat, rice–mustard–jute, rice–potato, rice–vegetables–jute, etc. in the IGP.[63] The maize–wheat sequence is now the fifth dominant cropping system of India occupying nearly 2 M ha in IGP.[55]

In the early 1980s, the art and science of CA started to emerge with zero tillage (ZT) whereby dry seeding was the major practice being introduced by the International Crops Research Institute for the Semiarid Tropics. The aim was to expand cropping intensities that would increase crop yields in the arid and semiarid areas of the tropics.

Dry seeding practice was partly adopted and only practiced by few farmers due to various practical challenges and climate conditions.[84] By 1990, ZT was well tested, perfected and widely adopted in irrigated areas of IGP.[54] As a result of this, the area under ZT was recorded as 1.90 M ha in 2005 which further increased by 32% in the next 2 years.[58] The data of progressive adoption of CA at the global scale, the whole of IGP and Indian IGP as reported by Jat et al.[57] and Kassam et al.[61] is illustrated in Figure 3.2. The documented benefits of CA based on various studies are listed in Table 3.3.

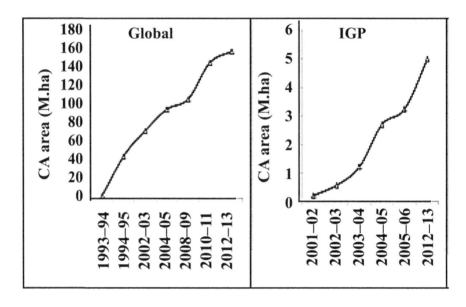

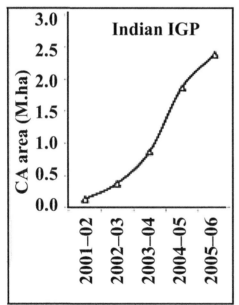

FIGURE 3.2   Progressive Global, IGP, and Indian IGP uptake of CA.

The focus of CA practices, so far, has been to achieve agricultural sustainability by means of resources conservation and judicious use of external inputs while maintaining high yields and profits. Despite these agronomic and environmental benefits, CA has not been widely adopted in all the regions of IGP.[83] The potential caveats for wide-scale adoption of CA include the high initial cost of machinery, the inability of simple ZT machine to handle loose straw left on the surface after combine harvesting of previous crop and lack of scale-appropriate and locally adapted machinery in some locations.[71] Some key points are highlighted in Table 3.4.

TABLE 3.3   Potential Benefits of CA for Higher Adoption and Out-scaling in Indian IGP.

Potential of CA in Indian IGP	References
1. Advancement in sowing date	[67,84]
2. Avoiding crop residue burning to reduce the loss of nutrients and minimize environmental pollution	[40,70,94]
3. Carbon sequestration and build-up of SOM, a practical strategy to mitigate GHG emissions and climate resilient and smart agriculture	[82]
4. Crop diversification and intensification	[9,83]
5. Enhancement of production and productivity (4%–10%)	[35,59]
6. Enhancement of soil quality (physical, chemical, and biological)	[17,37]
7. Enhancement of water and nutrient use efficiency	[10,25,82]
8. Improvement in resource use efficiency (RUE) through residue decomposition, soil structural improvement, increased plant nutrients availability	[26,83,94]
9. Reduced cost of production	[3,19]
10. Reduction in GHG emissions and improved environmental sustainability	[76,79,83]
11. Reduction in weeds incidence, such as *Phalaris minor* in wheat	[18,68]
12. Use of cover crops, surface residues as mulch	[72,94]

TABLE 3.4   CA Prospects for Higher Economics and Environmental Conservation.

Key points	Prospects	References
Crop diversification	Cropping sequences/rotations and agroforestry systems, when adopted in appropriate spatial and temporal patterns, can further enhance natural ecological processes. Future investigations require compatibility of crops in CA	[92,93]
Environmental benefits reduced crop residue burning	The burning of crop residues contributes to loss of plant nutrients. Improvement of soil, air, and water quality should be highlighted and quantified	[53,54,65, 82,83]
Increased yields	Inconsistent results showing varying yields; however, CA is knowledge and skill intensive practice and needs to be adopted as a package	[50,54,59, 65,82]

TABLE 3.4   *(Continued)*

Key points	Prospects	References
Reduced incidence of weeds	Reduction of incidence of *Phalaris minor* (a major weed in wheat); management of weed particularly in DSR as a practice of CA need to be investigated as solution-based research	[15,18,68]
Reduction in the cost of production	Economic-based quantification of ecosystem services provided by CA and direct reduced cost as savings on account of diesel, labor and input costs, particular herbicides would be the future thrust area of research	[3,19]
Resource improvement	Improved mechanization, better technology for residue recycling, and region-specific modification and trials required for adoption and out scaling of CA	[44,48,82, 84,91]
Saving in water and nutrients	Experimental results and farmers experience indicate that considerable saving in water (up to 20–30%) and nutrients are achieved with zero-till planting particularly in laser leveled, and bed planted crops. However, scattered and limited pieces of literature are available on the amount of macro and micronutrient required in CA	[55,56, 58,84]
Soil carbon sequestration	Soil carbon sequestration offers multiple benefits toward improving soil quality and environmental safeguard. However, to gain a significant amount of carbon in soil needs a long-term trial of CA in Indian IGP	[13,14,37, 39,79]
Soil quality enhancement	CA improves soil quality almost under all the climatic condition. Furthermore, drivers and indicators of soil quality in CA need to be investigated in Indian IGP	[17,36,42, 55,61]

3.3   NATURE AND PROPERTIES OF SOILS OF INDIAN IGP

The nature and properties of the soils of Indian IGP are highly influenced by its surroundings and climatic conditions. Toward the north of Himalaya, this great plain is bound by the basins of three distinct rivers systems known as Indus, Ganga, and Brahmaputra[63] and the tributaries of Ganges and Indus, namely, Beas, Yamuna, Ravi, Chambal, Sutlej, and Chenab.[87] The plain comprises one of the world's greatest stretches of flat and deep alluvium. The nature and properties of the alluvium vary in texture from sandy to clayey, calcareous to non-calcareous and acidic to alkaline (sodic). Though the overall topographic situation remains fairly uniform with elevations of 150 m above sea level (asl) in the Bengal basin, to 300 m asl in the Punjab plain, the local geomorphic variations are significantly appealing.[77]

In terms of climate, the temperature regime in the Indian IGP is hyper-thermic. The region comprises four major agroecological zones of arid, semiarid, humid, and per-humid. Soils under arid and semiarid parts of the IGP lack organic carbon due to the high rate of decomposition. The adverse climatic conditions induce precipitation of $CaCO_3$. It deprives the soils of Ca^{2+} ions on the soil exchange complex with a concomitant development of sodicity in the subsoils, which impairs the hydraulic conductivity of soils.[77] Several hypotheses have emerged to explain the formation of sodic soils in the IGP, but probably the consensus on this issue is yet to develop. The differences in climate, especially precipitation in agro-ecological zones, have contributed to the formation of a variety of soils that represent mainly three soil orders namely, Entisols, Inceptisols, and Alfisols.[86] Recent studies reveal that Mollisols and Aridisols also form the part of IGP soils. Various soil-forming processes in the IGP have been in operation, such as calcification, leaching, lessivage, salinization and alkalinization, gleization, and homogenization.

3.3.1   SALINITY AND WATER LOGGING PROBLEMS IN INDIAN IGP

The major causes for the development of salt-affected soil in the Indian IGP are climate, poor internal drainage, shallow water table, saline nature of irrigation water, salt blown by the wind, saline nature of parent rock material, excessive use of basic fertilizers and climate regime.[24]

The problem of water logging is acute and is caused by both water stagnation and shallow water table (>2 m below the ground). There are considerable variations in waterlogged areas of the first kind mainly depending upon the climatic conditions. With anticipated climatic variations, the waterlogged area in the IGP is likely to go a sea change wherein both the extent and degree of waterlogging will increase. Waterlogged area for 2 years and the net change during this period is shown in Figure 3.3.[52] In Haryana, Bihar, and West Bengal, the area under waterlogging increased drastically in less than a decade. The increased area was more than 35% from 2003–2005 to 2011–2013.

The topographic features and groundwater hydrology of arid and semi-arid climate are often responsible for in situ accumulation or transport and deposit of salts. The area affected by soil salinity in different states of IGP (Table 3.5) reveal that salt-affected soils in the Indian IGP cover an area of 2.35 M ha, which is more than one-third of the total salt-affected soils of India. Poor quality water and improper irrigation management further exuberate the problems of soil salinity.[74] The introduction of irrigation without making proper provisions of drainage is the major cause for the development of salinity in canal commands.[47] Various estimates made by Central Soil Salinity Research Institute (CSSRI), Karnal, Haryana, India revealed that the problems of soil salinity/sodicity will extend to about 2.5 times more than the current affected area (Table 3.6).[47]

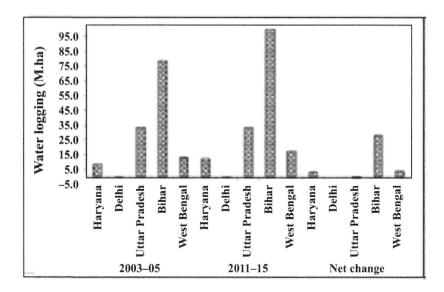

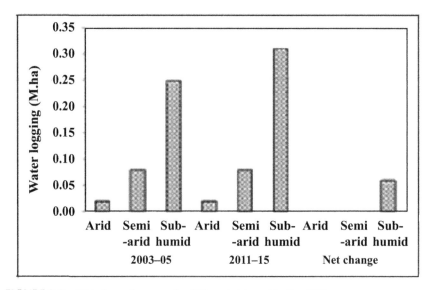

FIGURE 3.3   Waterlogged area under different states of Indian IGP.

TABLE 3.5 Salt-affected Soils in Various States of Indian IGP.

State in India	Area under salt-affected soils (ha)		
	Saline	Sodic	Total
Bihar	47,301	105,852	153,153
Haryana	49,157	183,399	232,556
Punjab	–	151,717	151,717
Uttar Pradesh	21,989	1,346,971	1,368,960
West Bengal	441,272	–	442,172
Total	**559,719**	**1,787,939**	**2,347,658**

TABLE 3.6 CSSRI Projections of the Extent of Salt-affected Soils in India.

Year	Projected salt-affected area (M ha)	References
Current	6.73	[74]
2020	$11.7^a = (9 + 2.7)$	[20]
2025	13.0	[21]
2030	15.5	[22]
2050	16.2	[23]

[a]Salt-affected soils and water logging in irrigation commands.

3.4 RECLAMATION/MANAGEMENT OF SALT-AFFECTED SOILS

Salt-affected soils slowly turn unproductive when the accumulation of soluble salts or exchangeable sodium percentage (ESP) or both exceeds the minimum threshold level for the crops to survive. From reclamation and management point of view, the salt-affected soils in India have been broadly placed into two categories:

a) **Sodic soils**, in general, are characterized by high soil pH going up to 10.8 in extreme cases, high ESP going as high as more than 90, low organic carbon, poor infiltration and poor fertility status. These soils are dominated by sodium carbonate and sodium bicarbonate salts.

b) **Saline soils** have high electrical conductivity (>4 dS m^{-1}), low ESP (<15) and low pH$_2$ (<8.5). The dominant salts in saline soils include Cl and SO_4^{2-} of Na, Ca, and Mg.[1]

The reclamation and management practices of the two groups of soils prepared from the text versions of the two kinds of soils are listed in

Table 3.7.[45–47] Flushing as mentioned in this table is adopted only in heavy textured soils having excessive salt accumulation on the soil surface.[70,75,98] It may be noted that basically, most management practices are similar in nature for the two kinds of soils, although in practice, these may vary widely. For example, the crops and cropping sequences for the two groups of soils may be different because the crop tolerance to salinity and sodicity stresses could be vastly different for various crops and their varieties.[45] Nonetheless, the list provides a basic framework of discussions to highlight the role of CA in salinity management.

TABLE 3.7 Reclamation and Management Technologies for Saline and Sodic Soils.

Group/Subgroup	Technology	
	Saline soils	**Sodic soils**
Reclamation	Subsurface drainage for waterlogged saline lands	On-farm land development including surface and vertical drainage
	On-farm land development including surface drainage	Application of chemical amendment
	Flushing/leaching.	Flushing/leaching
	Management	
Crop management	Select crops and cropping sequences as per salt tolerance	Selection of crops and cropping sequences
	Within crops, exploit varietal differences in salt tolerance	Exploitation of varietal differences
	Improved cultural practices	Improved agronomic practices including cultivation of green manure crop after wheat crop
Soil management	Land forming/seeding and irrigation	Land forming/seeding
Irrigation water management	Shallow depth-high frequency irrigation	Shallow depth-high frequency irrigation regime
	Pre/post-sowing irrigation Switch-over to improved irrigation techniques Leaching for salt balance	A switchover to improved irrigation techniques
Chemical management	Application of additional nutrients, organic manures Application of chemical amendments (only if required)	Application of additional nutrients/green manure/FYM
Rainwater management	In situ rainwater conservation; rainwater harvesting and reuse; fallowing	In situ *rainwater conservation;* rainwater harvesting and reuse (helps in drainage)

3.5 ROLE OF CONSERVATION AGRICULTURE IN MITIGATING SOIL SALINITY IN INDIAN IGP

Although practicing CA, in the long run, is known to reduce both the menace of soil salinity as well as provide the necessary protection against secondary salinization, yet authors have looked at various practices that are used to mitigate salinity in IGP (Table 3.6) and those used in CA. The three principles namely disturb the soil as little as possible:

- Keep the soil covered as much as possible;
- Mix crops are nearly sacrosanct to the concept of CA but can be applied in a wide range of conditions by varying practices depending on location-specific factors that come into play; and
- Rotate crops.

For example, CA can be practiced on different types and sizes of farms with different combinations of crops and sources of power.

The basic practices of application of amendments to reclaim sodic soils and subsurface drainage to reclaim waterlogged saline lands are not part of the CA but can be easily used as an add-on within the general principles of CA. Once the land reclamation process is initiated through these practices, management practices required to minimize the salt stress on the plants or to hasten the process of reclamation especially in sodic soils of the IGP and that of CA can complement each other to be mutually beneficial. Practices of precision land leveling and bed/furrow farming in CA can help minimize the drainage problem commonly encountered in saline/sodic lands of IGP during the monsoon season.

CA stands as the best-adapting strategy to build up organic matter in soils and create a healthy soil ecosystem. In Indian IGP, the SOC and soil inorganic carbon stocks from 0 to 150 cm depth are approximated to be 2.0 and 4.58 Pg, respectively.[7,13,77] It is projected that a change to no-tillage across the RW region of Indian IGP would increase SOC levels at 0–30 cm depth by an average of 5 Mg C ha[-1] over a period of 20 years.[39] It is likely that, in places where there is high clay content such as West Bengal, the SOC content could be higher than those with coarse-textured soil as shown in Figure 3.4.[39] It means a shift to CA practices will help accumulate SOC, which will ultimately hasten amelioration of soil salinity/sodicity as discussed in the following sections. To sustain crop productivity in sodic soils, application of farmyard manure, green manure or residue management are equally effective, later involving no cost to the farmers.[80,97]

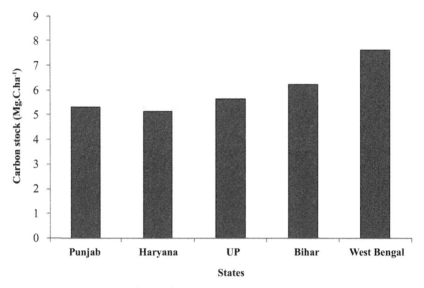

FIGURE 3.4   Soil carbon stock up to 30 cm of soil depth in different states of Indian IGP (note the increasing trend from trans to lower IGP).

3.5.1   SOIL ORGANIC CARBON (SOC) BUILD-UP THROUGH RESIDUE MANAGEMENT

The SOC build-up improves physical properties notably resulting in higher infiltration rates and hydraulic conductivity of the soils. This positive feature helps to improve greater intake of water and its redistribution in the root zone to help leach the salts. CA practices sufficiently increase the soil water-holding capacity due to substantial increment in soil organic matter that acts like a sponge sucking water. The appreciable level of soil moisture in CA helps in diluting the soluble salts, decrease their accumulation on the soil surface and leaching out soluble salt out of the rhizosphere.

Permanent crop cover in CA helps bind soil particles and thus improve soil structure with appreciable increase in mesopores and macropores for rapid percolation and infiltration of the water. These pores play a significant role in facilitating leaching of the soluble salts in saline soils. During rainfall period or irrigation time the water effectively washes away the soluble salt beyond the root zone of the crop. The surface soil is, therefore, left free from salinity problem. Permanent vegetation also provides crop cover that reduces the evaporation loss of water from the soil. It may be noted that because of the above, improving SOM is an essential pre-requisite to sustain a high productivity from saline/sodic lands of the IGP (Table 3.7), which is an outcome of CA practices.

SOC build-up in the soil helps to create microclimatic conditions as the substrate for soil microorganisms. Higher organic matter means rapid growth and diversity of microorganism that fastens the decomposition rate, decomposed material releases proton (H^+) in the soil and buffers the soil pH. Soil microbial activities lead to acid secretion for degradation of the residues which also helps in reducing sodicity of the calcareous alkali soils of the IGP. In fact, it hastens the process of land reclamation in such soils because of the enhanced self-reclamation process, which is usually slow.

Furthermore, the addition of organic materials to the soil surface prevents direct contact of sunlight to the soil, thus preventing soil moisture loss by evaporation. Prevention of evaporation means prevention of salt movement from lower layers to the root zone of the crop. The stored soil moisture due to the presence of soil cover encourages decomposition and degradation of organic matter which in turn provides complex functional groups that holds the cations of the soluble salts such as Ca and Mg.

3.5.2 COMPATIBILITY OF SALT-TOLERANT VARIETIES IN CONSERVATION AGRICULTURE

Selection of salt-tolerant crop varieties offers tremendous possibilities of utilizing saline soils and water. It has been observed that several salt-tolerant varieties also perform well under CA involving no-till and direct seeding of rice.[89] As such, selection of crops and varieties for CA can be complementary to the salinity management in IGP.

3.5.3 EXPERIMENTAL EVIDENCE

CA can play a pivotal role in combating soil salinity problems in IGP; although there is limited scientific evidence, which confirms the role of the CA in reducing soil salinity in the region. In Haryana (a Trans IGP region of India), ZT, residue management, direct seeding, and brown manure are being practiced as part of CA and these practices have provided the potential ability to ameliorate soil salinity by increased soil carbon fractions and water-stable aggregates, which play important roles in soil moisture retention and diffusivity.[16,17,34] Many other studies have indicated the contribution of CA in mitigating the soil salinity in different regions of Indian IGP. The indices of reduction in soil salinity under different cropping system in Indian IGP are shown in Table 3.8.

The permanent raised beds (PRBs) with enough crop residues in CA of Indian IGP have great potential to reduce soil salinity in salt-affected areas.[28] Under permanent raised-bed planting with residue retention, sodicity was reduced by 2.64 and 1.80 times in 0–5 and 5–20 cm layers, respectively.[38] A study under a cotton–wheat–maize sequence with two tillage methods ("CA"—PRBs and CT) combined with two residue levels [residue harvested; residue retained (RR)], compared to preexperiment levels, revealed that salinity in the top 30 cm soil was increased significantly during cotton (May–October), while a negligible change occurred during wheat (October–June) and maize (July–September) seasons.

In the absence of crop residues, soil salinity on top of the beds was increased compared to CT without crop residue retention.[27] When retaining crop residues, the soil salinity under PB was reduced by 32% in the top 10 cm and by 22% over the top 90 cm soil profile compared to CT without crop residue retention. Thus, PB + RR seem a promising option to slow down on-going soil salinization in salt-affected agro-ecologies such as those in the irrigated arid lands of Central Asia.[28]

In comparison to CT, values of ESP, and dispersion index were much lower in an irrigated Vertisol after 9 years of minimum tillage.[51] Plant residue reduced the evaporation of moisture from the soil, thus retaining more soil moisture by 3.5% and reduced the seasonal accumulation of salts by 1.6–4 times compared to the site where conventional agriculture was practiced. Retention of more moisture in the soil and reduction of salinization both eventually lead to saving of water consumption for irrigation and flushing of salts. It can play a major role in regions where the acute shortage of water resources is perceived.[6]

Raised bed and furrow practices in CA along with planting practices can help manage salts using proper watering sequence.[28] With the permanent skip furrow irrigation method, salts are eliminated across the bed from the irrigated side of the furrow, where crops are grown, to the dry sides where there are no plants.[69] With enough mulch residues applied, reduced evaporation of water during the drying periods will keep the flux downward reducing the impact of salts on plants.[81] Such and other management practices of root zone salinity can improve emergence, stand establishment and finally crop yields in saline fields.[28]

It is now emerging that an array of these and other management practices and technologies interlinked with CA may improve salt-affected soils as well as mitigate negative consequences of climate change. The key challenge is to assure that such practices do not reduce yields and follow all principles

TABLE 3.8 CA-based Management Practices in Indian IGP and Its Indices of Soil Salinity Alleviation.

CA components	Region in IGP	Cropping system	Climate	Indices of salinity alleviation	References
ZT, RM, DSR, and BM	Trans (Haryana)	Rice–wheat	Semiarid, Tropical	Increased carbon fractions, TC and WSA	[16]
ZT	Trans (Punjab)	Soybean–wheat	Semiarid, tropical	Soil moisture retention and soil temperature	[5]
ZT-B, ZT-F, RM, and CD	Upper (Delhi)	Wheat–cotton and wheat–green gram–maize	Semiarid, subtropical	Carbon stock, carbon retention efficiency and soil bulk density	[25]
ZT, DSR, and RM	Trans (Haryana)	Rice–wheat	Semiarid, tropical	Soil water retention, unsaturated hydraulic conductivity and diffusivity	[17]
ZT, DSR, and RM	Upper (Uttar Pradesh)	Rice–wheat	Subtropical	Soil carbon and WSA	[72]
ZT and RM	Upper (Uttarakhand)	Maize–wheat	Subtemperate	Higher moisture in the soil profile	[37]
ZT, RM, and PRB	Upper (Uttar Pradesh)	Rice–wheat	Semiarid, subtropical	Steady-state infiltration rate and soil aggregation	[55]
ZT and DSR	Upper (Uttar Pradesh)	Rice–wheat	Semiarid, subtropical	Infiltration, WSA, and penetration resistance	[36]

ZT, zero tillage; ZT-B, zero tillage bed planting; ZT-F, zero tillage flat planting; RM, residue management; CD, crop diversification; BM, brown manuring; PRB, permanent-raised beds; DSR, direct seeded rice; TC, total carbon; WSA, water stable aggregate.

of CA-based management practices. It will pave the way to harness multiple benefits of CA practice.

3.6 CONCLUSION

Conventional agriculture coupled with intensive cropping and excessive application of agrochemicals is significantly contributing to land degradation and impacting the sustainability of agriculture. The physical, chemical and biological degradations have adversely impacted the agronomic productivity, soil health and the environment. CA has emerged as a low-cost alternative for not only to halt the land degradation processes but also reverse the processes to improve soil health. While exploring the contributions of CA interventions such as minimum tillage, ZT, PRB, use of cover crop, crop rotation, direct seeding, and residue management, it has emerged that these substantially improve the SOC, enhance infiltration rates, and reduce evaporation rates. Although the role of CA in the reclamation of salt-affected soils may be limited, yet the positive role in leaching of the salts beyond the root zone depths of the crops is well documented. Similarly, reduced evaporation of soil moisture curtails the upward movement of the water to minimize salt accumulation at the soil surface.

It is concluded that CA practices supplemented with some of the salinity management practices described in this chapter can help to reclaim salt-affected soils that may ultimately lead to sustainable food security regime. More intensive research combining the salinity management and CA practices may help to address the key challenges of CA in mitigating salinity problems at low cost around the globe in general and in Indian IGP in particular.

3.7 SUMMARY

Over a third of the global arable land is under intensive, continuous cropping systems. It uses excessive agrochemicals and conventional agriculture that significantly contributes to land degradation. Just like physical and biological degradations, chemical land degradation poses severe challenges in achieving food security, due to its adverse impacts on agronomic productivity, soil health and the environment. In India, 6.73 M ha is salt-affected, of which 44% is saline, and 56% is sodic. Furthermore, extensive agricultural practices with poor irrigation and drainage management are expected to aggravate the problems. It is projected that by 2050, there will be a two-and-a-half-fold increase in salt-affected area, in which the IGP is most likely to be affected. Already Indian IGP has witnessed the stagnation in food and fiber production, which is likely to decrease further unless salt-affected agricultural areas are reclaimed, and marginally barren lands are converted into productive agricultural land. To get the potential productivity of IGP sustainably, CA has to be effectively adopted and out scaled into the existing salt mitigating interventions.

CA has shown great potential in ameliorating soil salinity and alleviation of secondary salinization in IGP region. Management practices of CA, such as minimum tillage, ZT, PRB, use of cover crop, crop rotation, direct seeding have proved to increase substantial amount of SOC, which plays a central role in reducing soil salinity, by improving soil physical condition that enhances infiltration rates, which leaches the salts down beyond the root zone of the crops. Coupled with practices of permanent crop cover and crop residues on the surface evaporation of soil moisture is curtailed that limits the upward movement of the salts minimizing accumulation of soluble salts at the soil surface. When CA is supplemented with some of the salinity management practices described in this chapter, it can help halt the land degradation due to salinity leading to sustainable food security regime.

Although several pieces of evidence have been presented whereby CA has been reported to have a significant effect in managing soil salinity in terms of enhancing soil quality, yet there is no clearly defined guidelines to identify and spearhead the set of practices for mitigating soil salinity. For example, information on the types, amount and allowance period of cover crops and residues, which should be applied to alleviate soil salinity, are not clearly brought out. Therefore, intensive research is needed in addressing the key challenges of CA in mitigating salinity problems in Indian IGP.

KEYWORDS

- agricultural productivity
- carbon sequestration
- crop residue management
- Indo-Gangetic plain
- salinity management
- salt-tolerant varieties
- water management

REFERENCES

1. Abrol, I. P.; Yadav, J. S. P.; Masood, F. I. Salt-Affected Soils and Their Management. Food and Agriculture Organization, Rome; *Soils Bull* **1988,** *39,* 119.

2. Aggarwal, P. K.; Joshi, P. K.; Ingram, J. S. I.; Gupta, R. K. Adapting Food Systems of the Indo-Gangetic Plains to Global Environmental Change: Key Information Needs to Improve Policy Formulation. *Environ. Sci. Policy* **2004,** *7,* 487–498.

3. Aryal, J. P.; Sapkota, T. B.; Jat, M. L.; Bishnoi, D. K. On-farm Economic and Environmental Impact of Zero-tillage Wheat: A Case of North-West India. *Exp. Agric.* **2015,** *91,* 1–16.

4. Aslam, M. *Improved Water Management Practices for the Rice–Wheat Cropping Systems in Sindh Province, Pakistan.* Report No. R-70-1a; International Irrigation Management Institute; Lahore, Pakistan, 1998; p 96.

5. Aulakh, M. S.; Manchanda, J. S.; Garg, A. K.; Kumar, S.; Dercon, G.; Nguyen, M. L. Crop Production and Nutrient Use Efficiency of Conservation Agriculture for Soybean–Wheat Rotation in the Indo-Gangetic Plains of Northwestern India. *Soil Tillage Res.* **2012,** *120,* 50–60.

6. Aybergenov, B. A.; Asamatdinov, K. I. *Feasibility of Conservation Agriculture for Saline Irrigated Lands on the Southern Aral Sea.* Uzbekistan, 2010; p 5. http://aciar.gov.au/files/node/14068/feasibility_of_conservation_agriculture_for_saline_44561.pdf (accessed Dec 20, 2016).

7. Bakker, D. M.; Hamilton, G. J.; Hetherington, R.; Spann, C. Salinity Dynamics and the Potential for Improvement of Water Logged and Saline Land in a Mediterranean Climate Using Permanent Raised Beds. *Soil Tillage Res.* **2010,** *110,* 8–24.

8. Balasubramanian, V.; Adhya, T. K.; Ladha, J. K. Enhancing Eco-Efficiency in the Intensive Cereal-based Systems of the Indo-Gangetic Plains. In: *Issues in Tropical Agriculture Eco-Efficiency: From Vision to Reality*; CIAT Publication Cali: Colombia, 2012; pp 1–7.

9. Bhan, S.; Behera, U. K. Conservation Agriculture in India Problems, Prospects and Policy Issues. *Int. Soil Water Cons. Res.* **2014,** *2,* 1–12.

10. Bhatt, R.; Kukal, S. S. Soil Moisture Dynamics During Intervening Period in Rice–Wheat Sequence as Affected by Different Tillage Methods at Ludhiana, Punjab, India. *Soil Environ.* **2015,** *34,* 82–88.

11. Bhatt, R.; Kukal, S. S.; Busarid, M. A.; Arora, S.; Yadav, M. Sustainability Issues on Rice–Wheat Cropping System. *Int. Soil Water Cons. Res.* **2016,** *4,* 64–74.

12. Bhattacharyya, R.; Tuti. M. D.; Bisht, J. K.; Bhatt, J. C., Gupta, H. S. Conservation Tillage and Fertilization Impacts on Soil Aggregation and Carbon Pools in the Indian Himalayas Under an Irrigated Rice–Wheat Rotation. *Soil Sci.* **2012,** *177,* 218–228.

13. Bhattacharyya, T.; Pal, D. K.; Easter, M.; Batjes, N. H.; Milne, E.; Gajbhiye, K. S.; Chandran, P.; Ray, S. K.; Mandal, C.; Paustian, K.; Williams, S.; Killian, K.; Coleman, K.; Falloon, P.; Powlson, D. S. Modelled Soil Organic Carbon Stocks and Changes in the Indo-Gangetic Plains, India from 1980 to 2030. *Agric. Ecosyst. Environ.* **2007,** *122,* 84–94.

14. Bhattacharyya, T.; Prakash, V.; Kundu, S.; Srivastva, A. K.; Gupta, Z. S. Soil Aggregation and Organic Matter in Sandy Clay Loam Soil of the Indian Himalayas Under Different Tillage and Crop Regimes. *Agric. Ecosyst. Environ.* **2009,** *132,* 126–134.

15. Bhullar, M. S.; Pandey, M.; Kumar, S.; Gill, G. Weed Management in Conservation Agriculture in India. *Indian J. Weed Sci.* **2016,** *48,* 1–12.

16. Choudhury, S. G.; Srivastava, S.; Singh, R.; Chaudhari, S. K.; Sharma, D. K.; Singh, S. K.; Sarkar, D. Tillage and Residue Management Effects on Soil Aggregation, Organic Carbon Dynamics and Yield Attribute In Rice–Wheat Cropping System Under Reclaimed Sodic Soil. *Soil Tillage Res.* **2014,** *136,* 76–83.

17. Chaudhari, S. K.; Bardhan, G.; Kumar, P.; Singh, R.; Mishra, A. K.; Rai, P.; Singh, K.; Sharma, D. K. Short-term Tillage and Residue Management Impact on Physical Properties of a Reclaimed Sodic Soil. *J. Indian Soc. Soil Sci.* **2015,** *63,* 30–38.

18. Chauhan, B. S.; Singh, B. G.; Mahajan, G. Ecology and Management of Weeds Under Conservation Agriculture: A Review. *Crop Protect.* **2012,** *38,* 57–65.

19. Chhetri, A. K.; Aryal, J. P.; Sapkota, T. B.; Khurana, R. Economic Benefits of Climate-smart Agricultural Practices to Smallholder Farmers in the Indo-Gangetic Plains of India. *Curr. Sci.* **2016,** *110,* 1251–1256.

20. CSSRI. *Vision 2020*: *CSSRI Perspective Plan.* Central Soil Salinity Research Institute, Karnal, 1997; p 32.

21. CSSRI. *CSSRI-Perspective Plan: Vision 2025.* Central Soil Salinity Research Institute, Karnal, 2007; p 127.

22. CSSRI. *CSSRI Vision 2030.* ICAR-Central Soil Salinity Research Institute, Karnal, 2011; p 26.

23. CSSRI. *CSSRI Vision 2050.* ICAR-Central Soil Salinity Research Institute, Karnal, 2015; p 31.

24. Dagar, J. C. *Salinity Research in India: An Overview. Bull. Natl. Inst. Ecol.* **2005,** *15,* 69–80.

25. Das, T. K.; Bhattacharyya, R.; Sudhishri, S.; Sharma, A. R.; Saharawat, Y. S.; Bandyopadhyay, K. K.; Sepat, S.; Bana, R. S.; Aggarwal, P.; Sharma, R. K.; Bhatia, A.; Singh, G.; Datta, S. P.; Kar, A.; Singh, B.; Singh, P.; Pathak, H.; Vyas, A. K.; Jat, M. L. Conservation Agriculture in an Irrigated Cotton–Wheat System of the Western Indo-Gangetic Plains: Crop and Water Productivity and Economic Profitability. *Field Crops Res.* **2014,** *158,* 24–33.

26. Department of Agriculture Cooperation & Farmers Welfare. *Integrated Nutrient Management.* Government of India: New Delhi. http://agricoop.nic.in/dacdivision/ mmsoil280311.pdf (accessed July 17, 2015).

27. Devkota, M.; Gupta, R. K.; Martius, C.; Lamers, J. P. A.; Devkota, K. P.; Sayref, K. D.; Vlek, P. L. G. Soil Salinity Management on Raised Beds with Different Furrow Irrigation Modes in Salt-affected Lands. *Agric. Water Manage.* **2015,** *152,* 243–250.

28. Devkota, M.; Martius, C.; Gupta, R. K.; Devkota, K. P.; McDonald, A. J.; Lamers, J. P. A. Managing Soil Salinity with Permanent Bed Planting in Irrigated Production Systems in Central Asia. *Agric. Ecosyst. Environ.* **2015,** *202,* 90–97.

29. Erenstein, O.; Sayre, K.; Wall, P.; Hellin, J.; Dixon, J. Conservation Agriculture in Maize- and Wheat-based Systems in the (Sub) Tropics: Lessons from Adaptation Initiatives in South Asia, Mexico, and Southern Africa. *J. Sustainable Agric.* **2012,** *36,* 180–206.

30. FAO. *Global Agriculture Towards 2050: High level Expert Forum*; Food and Agriculture Organization: Rome, 2009. http://www.fao.org/fileadmin/templates/wsfs/ docs/Issues_papers/ HLEF2050_Global_Agriculture.pdf (accessed Oct 20, 2016).

31. FAO. *Conservation Agriculture: Principles, Sustainable Land Management and Ecosystem Services*; Food and Agriculture Organization: Rome, 2011; pp 1–4. http:// www.fao.org/ag/ca/CA Publications/CA_Teramo_Kassam_Friedrich.pdf.

32. FAO. *Food and Agriculture Organization of the United Nations*; 2014. http://www.fao. org/ag/ca/6c.html (accessed Jan 17, 2015).

33. FAO. Agriculture and Consumer Protection Group. *What is Conservation Agriculture?* 2015. http://www.fao.org/ag/ca/1a.html (accessed Dec 22, 2016).

34. Gabriel, J. L.; Almendros, P.; Hontoria, C.; Quemada, M. The Role of Cover Crops in Irrigated Systems: Soil Salinity and Salt Leaching. *Agric. Ecosyst. Environ.* **2012,** *158,* 200–207.

35. Gathala, M. K.; Ladha, J. K.; Kumar, V.; Saharawat, Y. S.; Kumar, V.; Sharma, P. K.; Sharma, S.; Pathak, H. Tillage and Crop Establishment Affects Sustainability of South Asian Rice–Wheat System. *Agronomy J.* **2011a,** *103,* 961–971.

36. Gathala, M. K.; Ladha, J. K.; Saharawat, Y. S.; Kumar, V.; Sharma, P. K. Effect of Tillage and Crop Establishment Methods on Physical Properties of a Medium-textured Soil Under a Seven-year Rice–Wheat Rotation. *Soil Sci. Am. J.* **2011b,** *75,* 1851–1862.

37. Ghosh, B. N.; Dogra, P.; Sharma, N. K.; Alam, N. M.; Singh, R. J.; Mishra, P. K. Effects of Resource Conservation Practices on Productivity, Profitability and Energy Budgeting in Maize–Wheat Cropping System of Indian Sub-himalayas. *Proc. Indian Nat. Sci. Acad. India Sect. B: Biol. Sci.* **2016,** *86,* 595.

38. Govaerts, B.; Sayre, K. D.; Lichter, K.; Dendooven, L.; Deckers, J. Influence of Permanent Raised Bed Planting and Residue Management on Physical and Chemical Soil Quality in Rainfed Maize/Wheat Systems. *Plant Soil* **2007,** *291,* 39–54.

39. Grace, P. R.; Antle, J.; Aggarwal, P. K.; Ogle, S.; Paustian, K.; Basso, B. Soil Carbon Sequestration and Associated Economic Costs for Farming Systems of the Indo-Gangetic Plain: A Meta-analysis. *Agric. Ecosyst. Environ.* **2012,** *146,* 137–146.

40. Gupta, P. K.; Sahai, S.; Singh, N.; Dixit, C. K.; Singh, D. P.; Sharma, C.; Tiwari, M. K.; Gupta, R. K.; Garg, S. C. Residue Burning in Rice–Wheat Cropping System: Causes and Implications. *Curr. Sci.* **2004,** *87,* 1713–1717.

41. Gupta, R.; Seth, A. A Review of Resource Conserving Technologies for Sustainable Management of the Rice–Wheat Cropping Systems of the Indo-Gangetic Plains. *Crop Prot.* **2007,** *26,* 436–447.

42. Gupta, R. K.; Naresh, R. K.; Hobbs, P. R.; Jiaguo, Z.; Ladha, J. K. Sustainability of Post-Green Revolution Agriculture: The Rice–Wheat Cropping Systems of the Indo-Gangetic Plains and China. In *Improving the Productivity and Sustainability of Rice–Wheat Systems: Issues and Impact*; Ladha, J. K., et al. Eds.; ASA Special Publication 65; ASA, Madison, 2003; pp 1–25.

43. Gupta, R. K.; Naresh, R. K.; Hobbs, P. R.; Ladha, J. K. Adopting Conservation Agriculture in the Rice–Wheat System of the Indo-Gangetic Plains: New Opportunities For Saving Water. In *Water Wise Rice Production. Proceedings of the International Workshop on Water Wise Rice Production*; Los Banos, Philippines; April 8–11, 2002; pp 207–222.

44. Gupta, R. K.; Singh, Y.; Ladha, J. K.; Singh, B.; Singh, J.; Singh, G.; Pathak, H. Yield and Phosphorus Transformations in a Rice–Wheat System with Crop Residue and Phosphorus Management. *Soil Sci. Soc. Am. J.* **2007,** *71,* 1500–1507.

45. Gupta, S. K. Saline and Sodic Soil: Reclamation and Management Challenges. In *Souvenir 5th National Seminar Climate Resilient Saline Agriculture: Sustaining Livelihood Security*; Bikaner; 21–23 January 2017; pp 9–17.

46. Gupta, S. K.; Gupta, I. C. *Salt Affected Soils: Reclamation and Management*; Scientific Publishers Ltd.: Jodhpur, 2014; p 321.

47. Gupta, S. K.; Gupta, I. C. *Genesis, Reclamation and Management of Sodic (Alkali) Soils*. Scientific Publishers Ltd.: Jodhpur, 2016; p 322.

48. Hobbs, P. R.; Gupta, R. Resource Conserving Technologies for Wheat in the Rice–Wheat System. In *Improving the Productivity and Sustainability of Rice–Wheat Systems: Issues and Impacts*; Ladha, J. K., Eds.; ASA: Madison, WI, USA; ASA Special Publication *65*, 2003; pp 149–171.

49. Hobbs, P. R.; Morris, M. *Meeting South Asia's Food Requirements from Rice–Wheat Cropping Systems: Priority Issues Facing Researchers in the Post-green Revolution Era*. Natural Resource Group Paper 96-01; CIMMYT, Mexico, D. F., 1996; p 46.

50. Hobbs, P. R.; Sayre, K.; Gupta, R. The Role of Conservation Agriculture in Sustainable Agriculture. *Phil. Trans. R. Soc. B.* **2008**, *363*, 543–555.

51. Hulugalle, N. R.; Entwistle, P. Soil Properties, Nutrient Uptake and Crop Growth in an Irrigated Vertisol after Nine Years of Minimum Tillage. *Soil Tillage Res.* **1997**, *42*, 15–32.

52. Indian Space Research Organization. *Desertification and Land Degradation Atlas of India* (*Based on IRS AWiFS data of 2011–2013 and 2003–2005*). Space Applications Centre, ISRO: Ahmedabad, India, 2016; p 219.

53. Jain, N.; Bhatia, A.; Pathak, H. Emission of Air Pollutants from Crop Residue Burning in India. *Aerosol Air Qual. Res.* **2014**, *14*, 422–430.

54. Jat, M. L.; Dagar, J. C.; Sapkota, T. B.; Singh, Y.; Govaerts, B.; Ridaura, S. L.; Saharawat, Y. S.; Sharma, R. K.; Tetarwal, J. P.; Jat, R. K.; Hobbs, H.; Stirling, C. Climate Change and Agriculture: Adaptation Strategies and Mitigation Opportunities for Food Security in South Asia and Latin America. *Adv. Agron.* **2016**, *137*, 127–236.

55. Jat, M. L.; Gathala, M. K.; Saharawat, Y. S.; Tetarwal, J. P.; Gupta, R.; Singh, Y. Double No-till and Permanent Raised Beds in Maize–Wheat Rotation of North-Western Indo-Gangetic Plains of India: Effects on Crop Yields, Water Productivity, Profitability and Soil Physical Properties. *Field Crops Res.* **2013**, *149*, 291–299.

56. Jat, M. L., Saharawat, Y. S., Gupta, R. Conservation Agriculture in Cereal Systems of South Asia: Nutrient Management Perspectives. *Karnataka J. Agric. Sci.* **2011**, *24*, 100–105.

57. Jat, M. L.; Singh, G.; Ravi, M. K.; Saharawat, Y. S.; Kumar, V.; Gathala, M. K.; Sidhu, H. S.; Gupta, R. K. Innovations through Conservation Agriculture: Progress and Prospects of Participatory Approach in Indo-Gangetic Plain. In *Proceedings of 4th World Congress on Conservation Agriculture*; National Academy of Agricultural Sciences: New Delhi; 4–7 February 2009; pp 60–64.

58. Jat, M. L.; Gathala, M. K.; Ladha, J. K.; Saharawat, Y. S.; Jat, A. S.; Kumar, V.; Sharma, S. K.; Kumar, V.; Gupta, R. Evaluation of Precision Land Leveling and Double Zero-till Systems in Rice–Wheat Rotation: Water Use, Productivity, Profitability and Soil Physical Properties. *Soil Tillage Res.* **2009**, *105*, 112–121.

59. Jat, R. K.; Sapkota, T. B.; Singh, R. G.; Jat, M. L.; Kumar, M.; Gupta, R. Seven Years of Conservation Agriculture in a Rice–Wheat Rotation of Eastern Gangetic Plains of South Asia: Yield Trends and Economic Profitability. *Field Crops Res.* **2014**, *164*, 199–210.

60. Joshi, P. K.; Tyagi, N. K. Salt Affected and Waterlogged Soils in India: A Review. In *Strategies for Changing Indian Irrigation;* ICAR, New Delhi, India and IFPRI: Washington D. C., USA, 1994; pp 237–252.

61. Kassam, A. H.; Friedrich, T.; Derpsch, R.; Kienzle, J. *Overview of the Worldwide Spread of Conservation Agriculture. Field Actions Sci* Report 8, 2015. Revues.org/3720 (accessed Jan 15, 2016).

62. Kassam, A. H.; Friedrich, T.; Shaxson, F.; Bartz, H.; Mello, I.; Kienzle, J.; Pretty, J. *The Spread of Conservation Agriculture: Policy and Institutional Support for Adoption and*

Uptake. Field Actions Sci. Report 7, 2014. http://factsreports.revues.org/3720 (accessed Jan 15, 2016).

63. Koshal, A. K. Changing Current Scenario of Rice–Wheat System in Indo-Gangetic Plain Region of India. *Int. J. Sci. Res.* **2014,** *4,* 1–14.

64. Kumar, S.; Parkash, B.; Manchanda, M. L.; Singhvi, A. K.; Srivastava, P. Holocene Landform and Soil Evolution of the Western Gangetic Plains: Implications of Neotectonics and Climate. *Z. Geomorphol NF Supplement* **1996,** *103,* 283–312.

65. Kumar, V.; Saharawat, Y. S.; Gathala, M. K.; Jat, A. S.; Singh, S. K.; Chaudhary, N.; Jat, M. L. Effect of Different Tillage and Seeding Methods on Energy Use Efficiency and Productivity of Wheat in the Indo-Gangetic Plains. *Field Crops Res.* **2013,** *142,* 1–8.

66. Ladha, J. K.; Fischer, K. S.; Hossain, M.; Hobbs, P. R.; Hardy, B. *Improving the Productivity and Sustainability of Rice–Wheat Systems of the Indo-Gangetic Plains: A Synthesis of NARS-IRRI Partnership Research.* Discussion Paper No. 40; International Rice Research Institute: Los Baños, Philippines, 2000; p 31.

67. Malik, R. K.; Gupta, R. K.; Singh, C. M.; Yadav, A.; Brar, S. S.; Thakur, T. C.; Singh, S. S.; Singh, A. K.; Singh, R.; Sinha, R. K. Accelerating the Adoption of Resource Conservation Technologies in Rice Wheat System of the Indo-Gangetic Plains. In *Proceedings of Project Workshop*; Directorate of Extension Education, Chaudhary Charan Singh Haryana Agricultural University: Hisar, India; June 1–2, 2005.

68. Malik, R. K.; Kumar, V.; Yadav, A.; McDonald, A. Conservation Agriculture and Weed Management in South Asia: Perspective and Development. *Indian J. Weed Sci.* **2014,** *46,* 31–35.

69. Meiri, A.; Plaut, Z, Crop production and management under saline conditions. *Plant Soil* **1985,** *89,* 253–271.

70. Mishra, A. K.; Chaudhari, S. K.; Kumar, P.; Singh, K.; Rai, P.; Sharma, D. K. Consequences of Straw Burning on Different Carbon Fractions and Nutrient Dynamics. *Indian Farm.* **2014,** *64,* 11–12, 15.

71. Mishra, A. K.; Shinjo, H.; Jat, H. S.; Jat, M. L.; Jat, R. K.; Funakawa, S. Farmers Perspective on Adaptation and Up-scaling of Conservation Agriculture Based Management Practices in Indo-Gangetic Plains of India. In *Proceedings of Int. Conf. on Conservation Agriculture and Sustainable Land Use*; 2016; p 57.

72. Mishra, V. K.; Srivastava, S.; Bhardwaj, A. K.; Sharma, D. K.; Singh, Y. P.; Nayak, A. K. Resource Conservation Strategies for Rice–Wheat Cropping Systems on Partially Reclaimed Sodic Soils of the Indo-Gangetic Region, and Their Effects on Soil Carbon. *Nat. Resour. Forum* **2015,** *39,* 110–122.

73. Misra, V. N. Pre-historic Human Colonization of India. *J. Biosci.* **2001,** *26,* 491–531.

74. Mondal, A. K.; Sharam, R. C., Singh, Gurbachan and Dagar, J. C. Computerized Database on Salt Affected Soils in India. Central Soil Salinity Research Institute, Karnal; Technical Bulletin *CSSRI/Karnal1/2/2010*; 2010; p 28.

75. Digital Database of Salt Affected Soils in India using Geographic Information System. *J. Soil Salini. Water Qual.* **2011,** *3,* 16–29.

76. Nikita. *Alkali Soils: Factors, Effects and Reclamation.* (accessed July 17, 2015).

77. Padre, A. T.; Rai, M.; Kumar, V.; Gathala, M.; Sharma, P. C.; Sharma, S.; Nagar, R. K.; Deshwal, S.; Singh, L. K.; Jat, H. S.; Sharma, D. K.; Wassmann, R.; Ladha, J. Quantifying Changes to the Global Warming Potential of Rice Wheat Systems with the Adoption of Conservation Agriculture in Northwestern India. *Agric. Ecosyst. Environ.* **2016,** *219,* 125–137.

78. Pal, D. K.; Bhattacharyya, T.; Srivastava, P.; Chandran, P.; Ray, S. K. Soils of the Indo-Gangetic Plains: Their Historical Perspective and Management. *Curr. Sci.* **2009**, *96*, 1193–1202.

79. Paroda, R. S. Global Conventions and Partnerships and Their Relevance to Conservation Agriculture. In *Proc. 4th World Congress on Conservation Agriculture*; National Academy of Agricultural Sciences: New Delhi, 4–7 February 2009; pp 30–35.

80. Pathak, H.; Byjesh, K.; Chakrabarti, B.; Aggarwal, P. K. Potential and Cost of Carbon Sequestration in Indian Agriculture: Estimates from Long-term Field Experiments. *Field Crops Res.* **2011**, *120*, 102–111.

81. Paul, D. Osmotic Stress Adaptations in Rhizobacteria. *J. Basic Microbiol.* **2012**, *52*, 1–10.

82. Richards L. A. (Ed.) *Diagnosis and Improvement of Saline and Alkali Soils.* In USDA Agriculture Handbook: Washington, DC, USA; No. 60, 1954; p 160.

83. Saharawat, Y. S.; Ladha, J. K.; Pathak, H.; Gathala, M.; Chaudhary, N.; Jat, M. L. Simulation of Resource-conserving Technologies on Productivity, Income and Greenhouse Gas Emission in Rice–Wheat System. *J. Soil Sci. Environ. Manage.* **2012**, *3*, 9–22.

84. Sapkota, T. B.; Jat, M. L.; Aryal, J. P.; Jat, R. K.; Chhetri, A. K. Climate Change Adaptation, Greenhouse Gas Mitigation and Economic Profitability of Conservation Agriculture: Some Examples from Cereal Systems of Indo-Gangetic Plains. *J. Integr. Agric.* **2015**, *14*, 1524–1533.

85. Sapkota, T. B.; Majumdar, K.; Jat, M. L.; Kumar, A.; Bishnoi, D. K.; Mcdonald, A. J.; Pampolino, M. Precision Nutrient Management in Conservation Agriculture Based Wheat Production of Northwest India: Profitability, Nutrient Use Efficiency and Environmental Footprint. *Field Crops Res.* **2014**, *155*, 233–244.

86. Saxena, A.; Prasad, V.; Singh, I. B.; Chauhan, M. S.; Hasan, R. On the Holocene Record of Phytoliths of Wild and Cultivated Rice from Ganga Plain: Evidence for Rice-based Agriculture. *Curr. Sci.* **2006**, *90*, 1547–1552.

87. Shankarnarayana, H. S., Sarma, V. A. K. Soils of India. In *Benchmark Soils of India*; Murthy, R. S. et al., Eds.; ICAR-NBSS&LUP: Nagpur, 1982; pp 41–70.

88. Sharma, B. R.; Upali, A. A.; Sikka, A. *Indo-Gangetic River Basins: Summary Situation Analysis.* Report by IWMI and NRAAI: New Delhi, India, 2008; p 47.

89. Sharma, D. K. *Sustainable Technologies for Crop Production Under Salt-affected Soil in India.* 2015. https://www.jircas.affrc.go.jp/program/proD/english/files/ 2015/03/2014-session-21.pdf (accessed Dec 15, 2016).

90. Sharma, D. K.; Singh, A.; Sharma, P. C. Role of ICAR-CSSRI in Sustainable Management of Salt-affected Soils: Achievement, Current Trends and Future Perspective. In *Proceedings of 4th International Agronomy Congress;* New Delhi, Nov 22–26, 2016; pp 91–103.

91. Sharma, G. R.; Misra, V. D.; Mandal, D.; Misra, B. B.; Pal, J. N. (Eds.). *Beginning of Agriculture.* Abinash Prakashan, Allahabad, 1980; p 320.

92. Sidhu, H. S.; Singh, M.; Singh, Y.; Blackwell, J.; Lohan, S. K.; Humphreys, E.; Jat, M. L.; Singh, V.; Singh, S. Development and Evaluation of the Turbo Happy Seeder for Sowing Wheat into Heavy Rice Residues in NW India. *Field Crops Res.* **2015**, *184*, 201–212.

93. Singh, R. K.; Singh, A.; Pandey, C. B. Agro-biodiversity in Rice–Wheat-based Agro Ecosystems of Eastern Uttar Pradesh, India: Implications for Conservation and Sustainable Management. *Int. J. Sustainable Dev. World* **2014**, *21*, 46–59.

94. Singh, S.; Ladha, J. K.; Gupta, R. K.; Bhusan, L.; Rao, A. N.; Sivaprasad, B.; Singh, P. P. Evaluation of Mulching, Intercropping with Sesbania and Herbicide Use for Weed Management in Dry-seeded Rice (*Oryza sativa L.*). *Crop Prot.* **2007,** *26*, 518–524.

95. Singh, Y.; Sidhu, H. S. Management of Cereal Crop Residues for Sustainable Rice–Wheat Production System in the Indo Gangetic Plains of India. *Proc. Indian Natl. Sci. Acad.* **2014,** *80*, 95–114.

96. Williams, M. A. J.; Clarke, M. F. Quaternary Geology and Prehistoric Environments in the Son and Belan Valleys, North Central India. *Mem. Geol. Soc. India* **1995,** *32*, 282–308.

97. Williams, M. A. J.; Pal, J. N.; Jaiswal, M.; Singhvi, A. K. River Response to Quaternary Climatic Fluctuations: Evidence from the Son and Belan Valleys, North-Central India. *Quat. Sci. Rev.* **2006,** *25*, 2619–2631.

98. Yaduvanshi, N. P. S. Nutrient Management for Sustained Crop Productivity in Sodic Soils: A Review. Chapter 12, In *Soil Salinity Management in Agriculture: Technological Advances and Applications;* Gupta, S. K. and Goyal, M. R. Eds.; Apple Academic Press: USA, 2016; pp 365–394.

PART II

Physiological and Molecular Innovations to Enhance Salt Tolerance

CHAPTER 4

PHYSIOLOGICAL AND BIOCHEMICAL CHANGES IN PLANTS UNDER SOIL SALINITY STRESS: A REVIEW

VIKRAMJIT KAUR ZHAWAR and KAMALJIT KAUR

ABSTRACT

This chapter highlights five major mechanisms that operate under salt stress to improve salinity tolerance in plants. These are osmolyte metabolism, antioxidant-detoxification metabolism, phenolic/cell wall metabolism, ion homeostasis, and signal transduction. The content of proline and glycine betaine increases under salt stress as a result of higher synthesis and decreased catabolism. Betaine aldehyde dehydrogenase, the biosynthetic enzyme for glycine betaine, serves as detoxification enzyme under salinity stress. Proline and glycine betaine improve salt tolerance through activating antioxidant metabolism and decreasing oxidative toxicity under salt stress. Although glycine betaine and proline may result in high level of antioxidant metabolism under salt stress but recent researches have indicated that all constituents of antioxidant metabolism may not be required to be upregulated for salt tolerance. Plants could use phenolics, ascorbate, and glutathione either nonenzymatically or enzymatically to detoxify deleterious reactive oxygen species (ROS). Lignin present in cell wall has been found to increase under abiotic and biotic stresses. Ion homeostasis regulation can occur in root of plant, in vacuoles to keep ion concentration low in cytosol, and may help to retrieve ions from xylem stream. Salt tolerance has been related to higher ion exclusion, sequestration of ions in vacuoles and higher ratios of K^+/Na^+, K^+/Ca^{+2}, Ca^{+2}/Na^+ in different plant species. Chapter discusses the role of several signaling mechanisms, namely, calcium signaling, phytohormonal signaling, ROS signaling, ROS/NO signaling, ethylene signaling, and refers to role of polyamines in signaling. While the calcium signaling is involved in SOS1/SO2/SOS3 pathway for ion homeostasis and proline accumulation, phytohormonal signaling and ROS/NO-signaling are found to regulate salt responses in plants. Polyamines, ethylene, and glycine betaine share common precursor of *S*-adenosyl methionine for their synthesis, hence could interact with one another under salinity. The review provides the readers a good insight into the salt tolerance mechanisms that come into play when plants are subjected to salt stress.

4.1   INTRODUCTION

With the likely addition of 2.3 billion mouths by 2050, world agriculture faces a daunting task of ensuring food and nutritional security to the new guests on this planet.[48] Salinity is one of the major abiotic stresses affecting agriculture and is a major hurdle in more than 100 countries spread over all the continents. The extent and distribution of saline environment is likely to expand with increasing soil salinization of irrigated lands, use of poor quality water including sea water for irrigation and impending climate change. Excessive salt content in the root zone results in osmotic stress that causes physiological water deficit, ion toxicity and imbalance in the uptake of ions leading to nutrient deficiencies. High salt content also induces many metabolic changes that lead to molecular damage to adversely impact plant growth. On practical front, many technologies have been developed to manage saline/sodic environment based on chemical, hydrological, engineering, agronomic, and biological interventions. A separate line of studies has been to understand how plants respond to salinity stress at different levels to unravel the basis of plant adaptation or tolerance to salinity stress. The plant adaptation to abiotic salt stress involves complex physiological traits, metabolic pathways, and molecular or gene networks. A comprehensive approach on this line is imperative for the development of salt-tolerant varieties of plants.

Salinity is a major abiotic stress limiting growth and productivity of plants in many areas of the world. Recent research studies have identified various adaptive responses to salinity stress at molecular, cellular, metabolic, and physiological levels, although mechanisms underlying salinity tolerance are far from being completely understood. To cope with salinity stress, plants employ various mechanisms, at the cellular level and/or at the whole plant level, which are controlled by a variety of genes and signaling pathways. These are expressed and activated at different times during the life of a plant. Transcriptomic, proteomics and metabolomics are important "–omic" techniques in the post genomic era.[50]

- The *transcriptome* is the set of all messenger RNA molecules in one cell or a population of cells, and it is a term applied to the total set of transcripts in a given organism [https://en.wikipedia.org/wiki/Transcriptome] or to the specific subset of transcripts present in a particular cell type and is also referred to as expression profiling.
- Proteomics is the study of large-scale proteins in an organism encoded by its genome.[188]
- *Metabolomics* focuses on a global profile of the low molecular weight (1000 Da) metabolites which are the end products of metabolisms in bio fluids, tissues, and even whole organism.[23]

Plants proteome, metabolome, and ionic profiles in both halophytes and glycophytes are found to be affected by salinity stress, degree of stress affecting the complexity of the plant's response to them. Root and leaf proteome of creeping bent grass,[180] root proteome of tomato,[109] leaf proteome of barley,[49,135] and leaf transcriptomes of poplar[23,38] and leaf proteome of mangrove *Kandelia candel*[174] under salt stress indicated the major genes/proteins contributing to salt tolerance were related to ion homeostasis, antioxidant-detoxification, osmo-protectant synthesis, signal transduction, cell wall reinforcement, and energy relation.

Transcriptomic and proteomic studies have indicated that a wide variety of gene networks are affected under salt stress and therefore, tolerance is contributed by complex cross talks among various cellular components and signaling networks.[31]

This chapter discusses five major mechanisms that can be operative under salt stress to provide salinity tolerance in plants. These include osmolyte metabolism, antioxidant-detoxification metabolism, phenolic/cell wall metabolism, ion homeostasis, and signal transduction. Exogenous application of various chemicals and their response in imparting salt tolerance are also reviewed.

4.2 OSMOLYTE METABOLISM MECHANISM

Proteomic and transcriptomic studies have shown the important contribution of osmolyte metabolism in salt tolerance. As such, various osmolytes like proline, glycine betaine, trehalose, mannitol, and sugars like sucrose have been widely studied under salt stress. Recent studies indicate that some osmolytes like proline and glycine betaine play signaling role also besides their role in water retention. Amount of osmolytes increased under salt stress in different plant species due to their increased synthesis or decreased catabolism. Proline and glycine betaine are well studied and are briefly discussed in this section. In exogenous application studies, it has been observed that such applications induced osmolyte accumulation and improved plant adaptation to salt stress.

4.2.1 PROLINE

Proline synthesis in plants occurs mainly from glutamate, which is reduced to glutamate-semialdehyde (GSA) by P5C synthase (P5CS) enzyme and spontaneously converted to pyrroline-5-carboxylate (P5C). P5C reductase (P5CR) reduces P5C intermediate to proline. The catabolism of proline occurs in mitochondria via sequential action of proline dehydrogenase or proline oxidase [proline dehydrogenase (PDH) or PO] producing P5C from proline, and P5C dehydrogenase (P5CDH), which regenerates glutamate from P5C. Alternatively, proline can also be synthesized from ornithine which is transaminated first by ornithine-delta-aminotransferase (OAT) producing GSA and P5C and then converted to proline.[78] Proline plays a key role of compatible osmolyte besides acting as an enzyme protectant, free radical scavenger, cell redox balancer, cytosolic pH buffer, and stabilizer of subcellular structures to bring about salinity tolerance in plants. Among various compatible osmolytes, proline is the only molecule that has been shown to protect plants against singlet oxygen and free radicals induced damage. Moreover, proline synthesis uses NADPH and increases $NADP^+$ availability to photosystem I, PSI, and prevents its over reduction during drought and salt stress.[78,99]

Proline accumulation is related to salt tolerance[88,143,181] and also regarded as an indicator of stress severity.[5,99,107] The accumulation of proline under salinity stress has been observed in different plant species. Proline content was increased with increase in salt concentration as well as duration of salt stress in *Pennisetum glaucum*.[156] It is attributed either to increase of its synthesis or decrease of its oxidation. Proline accumulation caused by NaCl was related to an increase in OAT, P5CS, P5CR activity and a decrease in PDH activity in *Oryza sativa*.[9] In cactus pear, salt stress increased the expression of P5CS gene and induced proline accumulation.[155] Salt stress decreased PO and increased γ-glutamyl kinase activity and proline content in *Phyllanthus amarus*.[80] Transgenic Arabidopsis expressing P5CS gene from common bean showed salt tolerance due to higher accumulation of proline.[28] Transgenic sorghum overexpressing P5CS gene over accumulated proline, which helped to protect photosynthetic and antioxidant activity under salt stress.[159] Acclimation to salt stress upregulated proline synthesis enzyme activity and inhibited proline oxidation activity to accumulate proline in cucumber cell cultures.[120] P5CS gene expression has been related to proline content under salt stress in canary grass, where both were increased in leaves but decreased in roots.[33] Proline export or transport is increased under salt stress.[89]

Regulatory pathways of proline accumulation under stresses are not so clear. Exogenous treatment with abscisic acid (ABA), a stress-inducing hormone, when applied alone did not increase proline but under water stress, its accumulation was related to endogenous ABA level.[36,168] Many genes that respond to water stress or salt stress do not respond to exogenous supply of ABA, indicating the existence of both ABA-dependent and -independent pathway such that proline accumulation could be contributed by both pathways.[89] Reactive oxygen species (ROS) (H_2O_2) produced by activated NADPH oxidases under salt stress as well as osmotic stress, increased proline accumulation in *Arabidopsis thaliana*.[18] ABA increase of proline may be mediated through ROS-signal.[170] Calcium signaling has been suggested to regulate proline under salinity stress. In soybean[186] under salt stress, CAM (calmodulin) activates MYB2 TF which activates MYB2-dependent genes where P5CS1 (for proline synthesis) is one of such genes. In rice suspension cells under salinity, calcium participated in the regulation of proline.[148] Nitrogen being a constituent of proline, studies have shown a link between the production of proline and N-assimilation. Enzymes of proline metabolism are also altered under N treatment.[78] A regulatory interaction exists between ethylene, proline and N for salt tolerance where nitrogen differentially regulates proline production and ethylene formation to alleviate the adverse effect of salinity on photosynthesis in mustard.[77]

4.2.2 GLYCINE BETAINE

Quaternary ammonium compounds especially betaine serve as an organic osmolyte under stresses. Different forms are glycine betaine, proline betaine, β-alanine betaine, hydroxyproline betaine, etc. For their synthesis, betaine aldehyde is synthesized from choline by enzyme choline mono-oxygenase (CMO) and then betaine aldehyde dehydrogenase (BADH) converts betaine aldehyde to betaine. Among these compounds, glycine betaine occurs abundantly in plants exposed to dehydration due to salinity.[13,105,191] The accumulation of glycine betaine has been well documented in many halophytic species.[52] It helps to maintain osmotic pressure in a cell; it protects cells by stabilizing quaternary structures of antioxidant enzymes, bio-membranes, and oxygen-evolving photosystem II (PSII) complex.[99,139]

Synthesis of glycine betaine has been shown to have S-adenosyl methionine (SAM) as methyl group donor to ethanolamine (EA) or to phosphor-EA or phosphatidyl EA (Ptd-EA) to form choline, where SAM is common intermediate between ethylene synthesis and choline synthesis. Ethylene-insensitive mutants *etr1-1* and *ein2-2* of *A. thaliana* were found more sensitive to salt stress, indicating the importance of ethylene signaling in salt tolerance, where importance of ethylene has been linked to glycine betaine synthesis.[24] Greater salt tolerance has been linked to higher amount of glycine betaine and ethylene under salt stress in wheat.[91] Two populations of xerohalophyte *Atriplex halimus* L., Monastir (originating from salt-affected coastal site) and Sbikha (from non-saline semiarid area) showed differential tolerance to salt and water stress as well as their ability to accumulate proline and glycine betaine. Monastir was more resistant to higher salinity and exhibited a greater ability to produce glycine betaine in response to salt stress, while Sbikha was more resistant to water stress and displayed higher rate of proline accumulation under water stress.[17]

4.2.2.1 TRANSGENICS FOR HIGHER ACCUMULATION OF GLYCINE BETAINE

Different transgenics have been developed with higher accumulation of glycine betaine accompanied with improved salt and other abiotic stress tolerance. Transgenic tobacco plants expressing plastid-expressed CMO gene from beet showed improved salt and drought tolerance due to higher accumulation of glycine betaine in leaves, roots and seeds, higher net photosynthetic rate, apparent quantum yield of photosynthesis under salt stress.[193] Transgenic tobacco plants expressing BADH gene from spinach showed higher salt tolerance mainly due to improved CO_2 assimilation.[184] Dual gene transformed tobacco plants expressing BADH gene and NHX1 (vacuolar Na^+/H^+ antiporter) gene showed higher accumulation of betaine as well as Na^+, produced higher salt tolerance than single gene transformed plants.[197]

Transgenic potato plants expressing chloroplastic BADH gene from spinach showed improved tolerance to different abiotic stresses (salt, oxidative stress and low temperature) due to stronger photosynthetic activity, reduced ROS production and increased antioxidant defense.[47] Transgenic potato plants with ability to synthesize glycine betaine in chloroplasts due to transfer of bacterial choline oxidase gene showed higher salt/drought toler-ance due to lesser oxidative stress, higher photosynthesis, water content and accumulated higher levels of vegetative biomass.[4]

BADH besides its role in glycine betaine synthesis plays additional roles in salt tolerance probably through catalyzing other aldehydes in addition to betaine aldehyde, to alleviate the toxic effects.[161] *Oryza sativa* transgenic with RNAi-mediated suppressed OsBADH1 gene showed no alteration in glycine betaine content under salt stress but showed reduced salt tolerance due to increased oxidative stress.[161] Similarly, two genes of *A. thaliana*, ALDH10A8 and ALDH10A9 that code for BADH, were found weakly induced by ABA, salt stress, chilling stress, methyl viologen, and dehydra-tion. Their role through creating T-DNA insertion mutant for these genes in *A. thaliana* indicated that enzymes encoded by these genes oxidize many other aldehydes like 4-aminobutyratealdehyde, 3-aminopropionealdehyde besides betaine aldehyde; hence may serve as detoxification enzymes controlling the level of amino aldehydes under stress conditions.[114]

4.2.3 EXOGENOUS APPLICATION TO INDUCE SALT TOLERANCE

Exogenous application of different osmolytes provides salt tolerance in different plant systems through activating antioxidant potential and improving ion homeostasis. In salt-stressed tobacco bright yellow-2,[12,13,67,68] rice,[63,123] and mung bean,[69] application of proline or glycine betaine increased antioxidant defense [like ascorbate peroxidase (APX), dehydroascorbate reductase (DHAR), glutathione reductase (GR)], and detoxification system [glutathione peroxidase (GPX), glutathione-*S*-transferase (GST), glyoxyl-ases I and II] improved redox state of ascorbate and glutathione, decreased oxidative damage to lipids and protein and decreased H_2O_2. Both proline and glycine betaine though decreased H_2O_2 but did not alter superoxides and nitric oxides in salt-stressed tobacco bright yellow-2 cells.[12] Due to their action of improved antioxidant and membrane integrity and decreased ROS, exogenous supply of both proline or glycine betaine suppressed cell death in salt-stressed tobacco bright yellow-2 suspension cells.[13]

In another study using glycine–betaine-pretreated wild seedlings and BADH—transgenic seedlings of tobacco, it was found that glycine betaine can act as a cofactor in salt stress induction of calcium signaling pathway prior to induction of CAM and heat shock transcription factor genes.[101] Proline or glycine betaine could also regulate ion homeostasis as application of proline/glycine betaine suppressed Na(+)-enhanced apoplastic flow so to reduce Na(+)-uptake by rice seedlings under salt stress.[157] Exogenous supply of trehalose has also been found to upregulate antioxidant enzymes genes and promotes recovery of rice seedlings under salt stress.[123] Besides, its role in antioxidant defense, exogenous proline improved growth, water status, and photosynthetic activity also in young olive tree under saline conditions.[15] Pretreated wheat seedlings with exogenous mannitol improved salt tolerance though improved activities of antioxidant enzymes like super-oxide dismutase (SOD), POX, catalase (CAT), APX, and GR and decreased oxidative toxicity measured as malondialdehyde (MDA) under salt stress.[145] Exogenous sucrose (osmolyte as well as signaling molecule) enhanced salt tolerance in *A. thaliana* seedlings though increasing antioxidant enzymes like SOD, POX, CAT, increasing protein content, chlorophyll-*a* (Chl-*a*), Chl-*b*, anthocyanin, and transcription of genes involved in anthocyanin biosynthesis.[131] Proline was found to be better protectant than glycine betaine under salinity in some studies.[63,68] Wheat grain socked in ascorbic acid or thiamine or sodium salicylate reduced salt stress induced proline accumulation in shoots and roots of wheat seedlings, where vitamins or sodium salicylate could ameliorate salinity stress.[5]

4.3   ROS SCAVENGING AND ANTIOXIDANT MACHINERY

4.3.1   ANTIOXIDANT ENZYMES

Antioxidant enzymes are found to increase under salt stress to counteract the increase in ROS. For example, salt stress increased gene expression of antioxidant enzymes or increased activities of enzymes like SOD, CAT, APX, DHAR, peroxidase (POX), GR, GPX, GST in perennial rye grass,[73] tomato,[55,109] barley,[128] chickpea.[45]

Different proteomics and transcriptomics studies[38,49,109,135,180] have also indicated the major contribution of antioxidant defense related genes or proteins for salt tolerance. Higher antioxidant enzymes have been linked to tolerance in many studies that compared species/genotypes of varying level of salt tolerance. Increased level of SOD, CAT, GR, APX has been related to salt tolerance in chickpea genotypes,[134] in nodal cuttings of potato,[3] species of *Plantago*,[146] halophytes like *Centaurea tuzgoluensis*[185] and *Cakile mari-time*,[43] rice cultivars,[44] barley,[128] in alfalfa varieties.[175] However, the results are not uniform in some cases so that sometimes some antioxidant enzymes were related, while others are not related to salt tolerance. For example, higher CAT was related to tolerance, while higher APX was related to susceptibility in lettuce,[25] higher POX, APX, GR but not CAT was related to tolerance in barnyard grass,[2] higher CAT with tolerance in safflower,[154] higher SOD and CAT but not DHAR with tolerance in barley,[128] and higher CAT but not SOD and POX with tolerance in pea.[122] Quantitative trait locus (QTL)-study related to antioxidants for salt tolerance in cultivated tomato versus wild tomato indicated a very complex antioxidant response of plant toward salt stress. Though higher levels of antioxidants (enzymes and compounds) were related to tolerance, but direct correlation between them was difficult to prove with some under all conditions, while others were related only under saline or non-saline conditions.[54]

Salt tolerance has been related to rapid induction and sustained expression of antioxidant enzyme genes/activities in perennial rye grass,[71] *Medicago truncatula*,[111] *Plantago* species,[146] indica rice,[113] and leaf transcriptomic study in poplar.[38] However, when halophyte (*Cakile maritima*) and glycophyte (*A. thaliana*) were compared, it was found that antioxidant enzymes were increased at early stage of stress and then decreased in halophyte but remained high throughout the stress period in glycophyte.[43] Another halophyte *Centaurea tuzgolueensis* showed varying antioxidant response for short and long term exposure to salt stress but tolerance for both was related to increased ability of plant to scavenge ROS.[185]

Root proteomics of four contrasting genotypes of tomato[109] showed that more variations in proteomes were due to genotype effect than salt effect. Some other studies[119] have also indicated that different genotypes/species could use different antioxidant tolerance under salinity. In different tissues of plants exposed to salinity, antioxidant response could vary. In several reports,[19,20,26,133,180] different antioxidants were found in leaf, root, and shoot of plants under salt stress where roots generally showed less induction of antioxidants but not accompanied with high damage. Similarly in two maize hybrids[138] and in barley cultivars,[177] salt-related metabolic changes were mainly found in shoots and being negligible in roots. Compared to osmotic stress, roots contribute more than shoots in salt tolerance during ionic stress, for example, in microarray, shoots showed more number of genes alteration than roots under osmotic stress while opposite was true for ionic salt stress.[112] Similarly in in vitro intact plants and micropropagated shoots of *Carrizo citrange* plants, shoots were independent of roots for induction of proline, ABA and MDA under osmotic stress, whereas in ionic stress of salinity, response of roots was higher.[127] Antioxidant response was also different in young and mature leaves of maize under salt stress, where induction of CAT was higher in mature leaf while of APX and SOD was higher in young leaf.[19] Nodules of common bean were affected more than the roots under salt stress.[81]

Within the cell, differential antioxidant response was found in apoplast and symplast of pea leaves under salt stress where SOD, DHAR, ascorbate, and glutathione were present but APX, MDHAR, and GR were absent in apoplast, while in symplast, all enzymes were found.[66] Chloroplasts of two types of cells, bundle sheath cell and mesophyll cell, in C4-plant maize[124] under salt stress produced different antioxidant response, where bundle sheath cell chloroplast was less damaged than of mesophyll cell chloroplast which was due to higher H_2O_2-scavenging due to higher APX and ascorbate in bundle sheath chloroplast. Chloroplast may be the main source of ROS under salt stress as salt sensitivity is found to increase under high light intensity.[2] Leaf proteome of creeping bent grass[180] also showed the association of salt tolerance with maintenance of functionality and integrity of thylakoid membranes.

There may be varieties of modes of antioxidant mechanism to detoxify ROS/oxidative stress; plants may vary in using such mode, but ultimate aim of these modes would be the removal of deleterious ROS and deleterious effects of ROS. Hence, it would be more appropriate to relate plasticity and redundancy of antioxidant metabolism to tolerance rather than just increase of an antioxidant.[119,150–152]

Various transgenic crops have also been developed with different antioxidant enzymes for salt tolerance. Transgenic Arabidopsis expressing APX gene isolated from *Jatropha curcas* showed enhanced salt tolerance due to higher root growth, Chl content, total APX activity and less H_2O_2 content.[30] Transgenic tobacco overexpressing cytosolic APX gene isolated from *Lycium chinense* Mill. showed salt tolerance due to less H_2O_2, higher APX activity, proline content, and net photosynthetic rate under salt stress.[178] Transgenic tobacco—expressing thylakoid APX gene isolated from *J. curcas* carrying signal peptide to be localized in chloroplast—showed enhanced salt tolerance due to less H_2O_2, higher Chl content, and APX activity.[104] Transgenic cotton overexpressing SOD1, APX1 and CAT1 genes showed higher salt tolerance due to Chl content, photochemical yield, biomass accumulation under salt stress. Moreover, transgenic plants overexpressing these genes in chloroplast had higher tolerance than expressing these genes in cytoplasm.[108]

4.3.2   DETOXIFICATION ENZYMES

ROS accumulation in plants (either due to higher stress levels or weak antioxidant mechanism of plant) damages important bio-molecules like fatty acids/lipids, nucleic acid, proteins, carbohydrates and leads to formation of their oxidized products like MDA, 4-hydroxy-2-nonenal, reactive electrophilic species, and protein carbonyls.[41,116] Plant resistant to abiotic stress enhances activity of antioxidant mechanism (antioxidant enzymes APX, CAT, GR, SOD, and nonenzymatic antioxidants like ascorbate, glutathione that are involved in directly detoxifying the ROS).[58] Detoxification mechanism of the plant acts on damages, either reversing the damage or repairing the damage in the cell. It acts through enzymes like GPXs, GST, proteases and total glutathione content along with glutaredoxins/peroxiredoxins/thioredoxins, etc.[40,42,53,116,160] GSTs are heterogeneous family of enzymes that catalyze the conjugation of reduced glutathione to electrophilic sites of variety of compounds. These compounds can be thiobarbituric acid reactive substances (TBARS), toxic derivatives of lipid peroxides, or nucleic acids resulting from oxidative damage, which on glutathionylation are either converted to nontoxic compounds or targeted to plant vacuoles for their removal form cell.[41,55] GPXs are family of isozymes, which catalyze the reduction of H_2O_2 or organic/lipid peroxides by using reduced glutathione as reducing agent. GPX detoxifying phospholipid hydroperoxides formed during oxidative stress play important role in detoxification mechanism of plants.[37,55]

Various studies have established the importance of these mechanisms. In early response of soybean plants to salinity stress, it was found that exogenous nitric oxide activated some isoforms of GST while decreased other forms when applied individually; however, under salt stress, nitric oxide as well as salt stress signal act as potent regulators of GST gene and enzyme expression through both ABA-dependent and -independent pathways.[39] Transgenic tobacco with over expression of GPX/GST gene conferred salt and chilling tolerance due to enhanced ROS scavenging and less oxidative damage.[140] Transgenic arabidopsis with over expression of each of two wheat GPX genes showed improved salt and H_2O_2 tolerance.[188] This report also indicated that besides GPX playing protective role in oxidative stress, it may also be the signaling component in ROS, ABA, and salt-responsive pathways. Proteomic studies in leaves of creeping bent grass indicated the contribution of detoxification enzymes like UPD-sulfoquinovose synthase, methionine synthase, and glucan exohydrolase, GST along with antioxidant enzyme CAT in salt tolerance.[180] Methylglyoxal detoxification system (glyoxylases I and II) contributed to salt tolerance in rice.[44]

Glucose-6-phosphate dehydrogenase (G6PDH) plays central role in controlling NADPH. It reduced glutathione (GSH) levels through increased activity of GR and GPX under salt stress in reed callus, where H_2O_2 produced through plasma-membrane NADPH oxidase was the positive regulator of G6PDH at activity as well as its protein amount level.[176] In the same study, swamp reed (SR) ecotype growing in water ponds produced higher increase for G6PDH, GSH, GPX, GR than the dune reed (DR) ecotype growing in desert and sand area under mild salt stress; but at higher salt concentrations, response was opposite, that is, lesser increase in SR than DR.

4.3.3   NONENZYMATIC ANTIOXIDANTS

4.3.3.1   ASCORBATE

Ascorbate is an important antioxidant that serves as electron donor to many important enzymatic as well as nonenzymatic reactions. During salinity stress, ascorbate-deficient mutant vtc-1 (having 30–60% of ascorbate content of wild type plant) showed the decrease in ascorbate to total ascorbate ratio, decreased activity of enzymes of ascorbate–glutathione cycle, accumulated high amount of H_2O_2, followed by decrease of photosynthetic efficiency in *A. thaliana*.[75,112] Higher redox state of ascorbate along with decreased oxidative toxicity has been related to tolerance in wheat.[20,86] Transgenic potato expressing Gal-UR gene (D-galactouronic acid reductase) overproduced ascorbate accompanied with high reduced state of glutathione, high activity of glutathione-dependent glyoxylase enzyme and reduced accumulation of cytotoxic methylglyoxal under salt stress.[166] Exogenously supplied ascorbic acid also improved salt tolerance in seedlings of *Citrus aurantium*.[96] and in wheat[144] through increased endogenous antioxidants (like total POX, APX, SOD, CAT, ascorbate, glutathione, total phenols, etc.), improved membrane integrity and decreased lipid peroxidation, toxicity symptoms. When combined with melatonin, higher improvement was achieved in salinity tolerance in *C. aurantium*.[95,96]

4.3.3.2   GLUTATHIONE

Glutathione pool [GSH and GSSG (oxidized glutathione)] plays an important role in antioxidant and detoxification.[160] Similar to ascorbate, GSH serves as antioxidant to various enzymatic and nonenzymatic reactions. GSH plays an important role in ascorbate–glutathione cycle as well as many detoxification reactions catalyzed by GPX, GST, and other redox proteins like glutaredoxins/thioredoxins, etc.[53,160] Maintaining high ratio of GSH/ GSSG functions as a redox couple that plays important role in salt tolerance in different plant species.[112] Exogenously supplied GSH also improved salt tolerance in wheat[144] in nodal explants of chickpea[149] through increased antioxidant defense and reduced oxidative damage.

4.3.3.3 TOCOPHEROLS

The α-Tocopherol is a lipid antioxidant that can scavenge ROS and protect lipids from oxidation. Exogenous application of α-tocopherol enhanced salt tolerance in *Vicia faba* plants[147] and in wheat[144] under saline conditions through improved growth, yield, water content, membrane stability index, and increased endogenous antioxidants. Besides playing a role of antioxidant under salt stress, it may also affect signaling and gene expression. *A. thaliana* mutant impaired in tocopherol biosynthesis accumulated γ-tocopherols instead of α-tocopherols and showed down-regulated ethylene signaling pathway during salt stress and jasmonic acid (JA) pathway during age-maturation of leaf.[27] Interplay between resveratrol (a phytoalexin found in red wine) and α-tocopherol was found to play role in salinity adaptation of citrus seedlings, where α-tocopherol activation of antioxidant enzymes under salt stress was increased when it was combined with resveratrol supply.[95]

4.3.3.4 CAROTENOIDS

Salt stress increased carotenoids and anthocyanins so extensively in diverse tomato genotypes that their presence can visibly be assessed in tomato products,[21] and it was therefore suggested that saline soils can be exploited to obtain tomatoes with higher levels of such secondary metabolites. Transgenic euhalophyte *Salicornia europaea* with inhibition of beta-lycopene cyclase gene showed decreased salt tolerance. Transgenic arabidopsis expressing beta-lycopene cyclase gene showed improved salt tolerance due to increased carotenoids and less ROS, higher photosynthetic activity.[29] Transgenic sweet-potato with RNA interference for beta-carotene hydroxylase gene showed enhanced salt tolerance due to higher levels of beta-carotene, carotenoids, and decreased levels of ROS along with increased gene expression of phytoene synthase, lycopene beta-cyclase, and high amount of ABA.[92]

4.3.3.5 POLYAMINES

Polyamines are new class of growth substances well known for their anti-senescence and antistress effects due to their acid neutralizing, antioxidant properties as well as their membrane and cell-wall stabilizing abilities.[125,167] Putrescine (diamine) is the precursor of polyamines like spermine (triamine) and spermidine (tetramine); it is synthesized by two carboxylases in plants ornithine decarboxylase and arginine decarboxylase. Spermine (triamine) and spermidine (tetramine) are formed from putrescine by the subsequent addition of an aminopropyl moiety from decarboxylated SAM.[126] In salt-stressed spinach leaves, polyamine putrescine increased POX, CAT, and proline content but decreased polyphenol oxidase (PPO) while its antagonist ethephon showed opposite effects on these enzymes and proline.[125] In three varieties of indica rice, exogenously applied polyamines (spermidine or spermine) have ameliorated salinity stress by improving proline, anthocyanin, reducing sugar, POX, protease levels, and decreasing protein carbonylation.[141] In salt-stressed seedlings of medicinal plant *Panax ginseng*, application of spemidine increased plant growth, prevented Chl degradation, increased antioxidant enzymes such as CAT, APX, GPX, decreased H_2O_2, and superoxides. Exogenous spermine to salt-stressed *Cucumis sativus*[153] improved antioxidant systems of chloroplast and improved photochemical efficiency of PSII. It alleviated the harmful effects of soil salt stress through increased endogenous antioxidants in wheat.[144]

Polyamine and ethylene shared common precursor of SAM, where ethylene promotes senescence while polyamine promotes antisenescence. Growth-promoting hormones like auxins/gibberellins increased polyamine while decrease of ABA biosynthesis (in ABA biosynthetic mutant) increased polyamine synthesis.[10,125]

4.4 PHENOLIC METABOLISM AND CELL WALL CHANGES

Phenolic compounds include soluble phenolics and wall phenolics (lignin) that impart stress tolerance.[60,97,118] These are used for cell-wall hardening due to their deposition in form of lignin or their use to make crosslinks between other cell-wall polysaccharides/proteins.[102,103,118] These reactions provide mechanical barrier against penetration of stressor including metal chelating compounds and help plant to reduce water loss[60,118,169] under stress. POXs localized in cell wall helps in catalyzing such crosslinks and lignification.[102,103,163,194] Various studies[118] have correlated POXs with lignin and lignin with growth suppression under different stresses. Some studies have given importance to POX-catalyzed other crosslinks in growth suppression than lignification.[102,103] Phenolics also play role of ROS scavengers where these can be used directly or enzymatically through POX to detoxify ROS. Studies have related total phenolics, POX activities with decreased ROS and increased stress tolerance.[198] Hormonal (ABA) regulation has also been suggested in some studies[61,84,183] for their regulation under stresses but pathways leading to their deposition are not resolved.

Under salt stress, phenolics was increased in pea,[122] in roots of *Morinda citrifolia*,[1] sugarcane,[171] and *Cistus heterophyllus*.[106] The increase in phenolics was related to salt tolerance in sugarcane[171] and to antioxidant activity in *C. heterophyllus*,[106] but not related to salt tolerance in pea cultivars.[122] Phenolics was increased but lignin did not change in roots of *Matricaria chamomilla* under salt stress, thus indicating the role of phenolics as antioxidants.[97] In salt-stressed adventitious roots of *M. citrifolia*,[1] phenolics synthesis was increased due to increased phenylalanine ammonia-lyase (PAL). Increase of a major phenolic called avicularin in shoot of halophyte (*Mesembryanthemum edule*) of arid provenance was related to its salt tolerance.[46] *A. thaliana* transgenic overexpressing cytosolic Cu/Zn SOD gene showed significant thickening of inter-vesicular cambium of inflorescence stem due to increased lignin and upregulation of lignin biosynthetic genes like PAL-1 and POX (PRXR9GE). Such changes were related to its higher salt tolerance.[59] Salinity source like NaCl, KCl and $CaCl_2$ induced phenolic acids, flavonoids, polyphenols, antioxidant activity in leaves of artichoke and cardoon.[22]

Relationship of salt tolerance with lignin has not been well established. Some studies positively relate it while other with no effect or negatively relate it to salt tolerance. In rice seedlings, salt stress decreased growth and increased cell wall POXs, where increased POXs were related to decreased growth due its involvement in cell wall crosslinks through ferulates.[102,103] Decreased lignin with increased root extension was related to drought and salt tolerance in wheat and such response was improved on ABA-supply in sensitive cultivar.[85] Studies on halophytes species of Chenopodiaceae family[62] showed less importance of lignin than extensin for salt tolerance, as lignin localized in the xylem is reduced, while extensin localized in phloem is increased under salt stress for translocating organic solutes to maintain turgor. Salinity may increase lignin in roots to prevent penetration of salt water or to retain salt in cells but salt stress decreases lignin in stem. Moreover, this is also accepted that lignification may provide resistance in supporting and counteracting high osmotic pressure, which halophyte has to face at the level of rhizosphere.[62] Salt stress reduced lignin and cellulose in wheat seedlings; however, presoaking of wheat grains in ascorbic acid, thiamine, or sodium salicylate partially alleviated this adverse effect of salt stress.[5]

In most plants, root growth was inhibited under salt stress. However, for halophytes and few tolerant plant species, root growth was maintained or even promoted under salinity.[72,129] Transcriptome analysis in root tips of two contrasting Bermuda grass genotypes for salt tolerance indicated that there is differential regulation of transcription factors/genes regulating cell wall loosening/stiffening between these two genotypes where promoting cell-wall loosening and root growth may be the part of salt tolerance.[70]

4.5 ION HOMOESTASIS

Salt enters in plant through roots and then is pulled upward through transpirational pull. Various strategies by plant can be adopted to get rid of salt entry, such as effluxing ions from root cells to soil, by closing stomata to reduce transpiration and to avoid salt transport to aerial parts of plant, to reduce entry of ions into xylem, by directing the salt accumulation to older and mature leaves which have to dehisce.[64,119]

4.5.1 HALOPHYTES VERSUS GLYCOPHYTES FOR ION HOMEOSTASIS

Major differences between halophytes and glycophytes for ion homeostasis is that halophytes often accumulate large amount of ions and use them as osmolyte to prevent water loss and turgor loss, while glycophytes tend to get rid of ions to avoid toxicity, but become short of osmolytes, often losing turgor. Halophytes have developed efficient compartmentation of salt ions into vacuoles and hence keep cytosolic and organeller compartments free of higher concentration of ions for proper functioning of enzymes/proteins. Moreover, halophytes often direct ion flux into xylem so as to increase salt concentration in aerial parts, which lower water potential to facilitate water uptake and transport and lowers metabolic cost of production of osmolytes.[64,100]

Glycophytes, to which most crop species belong, vary in response from very salt sensitive to moderately salt resistant. Of cereals, rice is the most sensitive and barley is the most tolerant crop, while bread wheat is moderately resistant. Tall wheatgrass is a halophytic relative of wheat and is one of the most tolerant monocots. Among dicots, some legumes are very sensitive but some like alfalfa very tolerant. Similarly, *A. thaliana* is salt sensitive but its tolerant relative is *Thellungiella halophile*.[119]

4.5.2   TRANSPORT OF IONS FROM SOIL TO PLANT AND REGULATION FOR ION HOMEOSTASIS

In wheat root growing in 150 mM NaCl, it has been shown by quantitative and calibrated X-ray microanalysis[119] on root tissue 10 cm from tip that as water stream moves across cortex, Na^+ concentration decreases, that is, highest in epidermis and lowest in endodermis. This indicates the removal of Na^+ ions from water stream by root cells and storing them in vacuoles. In stele, vacuolar Na^+ concentration was highest in pericycle and minimum in xylem vessel. These results indicate that major part of regulation seems to be for minimizing the entry of ions into transpiration stream and into aerial parts of plant.

4.5.2.1   INFLUX AND EFFLUX OF IONS BETWEEN ROOT CELLS AND SOIL ENVIRONMENT

Net influx of ions from outside to root cells is determined by its influx into cells and its efflux out of cells. Influx of ions can be due to passive movement of ions through voltage independent or weakly voltage dependent nonselective cation channels (NSCCs) or other sodium transporters like some members of high affinity K^+ transporters (HKT) family.[79,119] First entry of ions occur through NSCCs and then later through HKT. NSCCs have been classified according to their voltage dependence or their responsiveness to certain ligands or physical stimuli. Depolarization activated (DA)-NSCCs and voltage insensitive (VI)-NSCCs are more efficient, give rapid influx of ions while hyperpolarization-activated (HA)-NSCCs are activated later and weakly selective for monovalent cations. Rapid influx of Na^+ through DA-NSCCs and VI-NSCCs contribute to tolerance by producing a rapid and efficient signal for adaptation and these channels are the predominant type of channels in tolerant species.[79] In contrast, sensitive species have more (HA)-NSCCs number, produce weak signal for adaptation. Late entry of sodium ions through HKT does not contribute to tolerance as *Arabidopsis* mutant with defective hkt1-1 showed enhanced salt tolerance.[119]

Efflux of ions has been best studied in SOS1/SOS2/SOS3-pathway[76,79,119,179] of root cells where SOS1 is Na^+/H^+ antiporter located in plasma-membrane of the cell, is activated by SOS2-SOS3 complex through calcium signaling. Though not known yet, but it is likely that other genes encoding Na^+/H^+ antiporters also exist. Moreover, members of other gene families like family of CHX genes or mechanism of pumping out Na^+ through Na^+-translocating ATPases may also exist.

Chloride ions can be more toxic than Na^+ for some species like soybean, woody perennials such as avocado, grapevines and citrus.[119] These species actually exclude Na^+ very well but increase Cl^- ions in leaf blade. Thermodynamically in most situations, Cl^- influx requires energy and is probably catalyzed by a $Cl^-/2H^+$ symporter, but passive diffusion could occur if membrane potential is depolarized and cytosolic Cl^- ion concentration becomes low. However, under saline conditions, cytosolic Cl^- concentration may become higher then Cl^- permeable anion channels that could be useful in reducing net influx of Cl^- ions.[64] Many studies have been undertaken to find whether Na^+ ion or Cl^- ion is more toxic. In different species of wheat, salinity tolerance has been related to low sodium accumulation of leaf but not of Cl^- accumulation. But opposite is reported in citrus. For most species, as high as 400 mM concentration of Cl^- ions were tolerated, even sensitive species like citrus can tolerate up to 250 mM of concentration. Unfortunately, the data about differences between tolerant and sensitive species for Cl^- tolerance is largely unknown.[119]

4.5.2.2   COMPARTMENTATION OF IONS TO VACUOLES

Besides influx-efflux, ions are sequestered in vacuoles. This is achieved by tonoplast Na^+/H^+ antiporter belonging to Na^+/H^+ exchanger (NHX) family in *Arabidopsis*.[79,119] About six members of NHX gene family have been identified in *A. thaliana* and where NHX1–4 are vacuolar and NHX5–6 are endosomal. Activities of Na^+/H^+ antiporter (SOS1) on plasma membrane and Na^+/H^+ antiporter (NHX) on tonoplast depend on the maintenance of H^+ gradients across membrane to provide driving force for Na^+ transport. For this, H^+-pumping transporters are located on plasma membrane and tonoplast. These are P-H^+-ATPase on plasma membrane and V-H^+-ATPase or tonoplast H^+-pyrophosphatase (AVP1) on tonoplast. These transporters energize the pumping of Na^+ efflux from cell and influx into vacuoles. Different studies have shown that V-ATPase together with H^+-pyrophosphatase (AVP1) is the major enzyme system responsible for maintaining the high concentration of H^+ inside the vacuole relative to cytoplasm. Gene expression encoding P-H^+-ATPase is upregulated during salt stress. Constitutive expression of NHX1 or gene encoding vacuolar H^+-pyrophosphatase (AVP1) improved the tolerance in *Arabidopsis*[119] and finger millet[8] indicating the importance of efficient sequestration of ions in salt tolerance. Besides sequestering toxic sodium ions in vacuole, this strategy lowers the water potential of the entire protoplast and so additional water loss is prevented. This way, it solves both ionic and osmotic components of salinity stress. In one study, it has been found that leaf concentration of Na^+ ions over 200 mM (on tissue basis) has not affected the leaf function but the same concentration when applied in vitro, enzymes of the same leaf were inhibited completely. Similarly, enzymes of halophytes were found more susceptible than enzymes of glycophytes to this concentration of Na^+ in the same laboratory experiment. This indicates that compartmentation of Na^+ is more evolved than evolution of enzymes for salt tolerance.[79,119] Data of Cl^- transport at tonoplast (Cl^- sequestration) is mostly lacking though it is proposed to be through H^+/anion antiporter or anion permeable channel on tonoplast.

4.5.2.3   INFLUX AND EFFLUX OF IONS BETWEEN STELAR CELLS AND XYLEM

Net influx of ions into transpiration stream is determined by efflux of ions from stellar cells and influx back into these cells from xylem apoplast. Na^+/H^+ antiporter, SOS1 is expressed in stellar cells, so could be involved in efflux of Na^+ ions from stellar cells. However, this statement needs clarification as knockout of this gene increased shoot sodium ion concentration rather than reducing it. Efflux of ions back from xylem has also been evidenced as some members of HKT gene family have been found to be involved in this process, for example, AtHKT1;1 in *Arabidopsis* root, HKT1;5 in rice and wheat and TmHKT1;4-A2 (Nax1) in durum wheat. These genes have been found to be associated with salinity tolerance by giving higher K^+/Na^+ ratio in leaf.

Cl^- loading to xylem is most likely passive mediated by anion channels. These channels are down-regulated by ABA, hence indicating the existence of tolerance mechanism for Cl^- ions also. The control of Cl^- transport to aerial parts may be due to reduced loading of Cl^- via anion channels or it may also be due to increased active retrieval of Cl^- from xylem stream through anion/H^+ symporter.

4.6 SIGNAL TRANSDUCTION

4.6.1 CALCIUM SIGNALING

Sodium concentration sensing or turgor sensing could be the perception. Either extracellular sodium concentration sensed at plasma-membrane level or sensed intracellular is not well known. Similarly turgor sensing is also not understood. But first recorded response in roots is increase in cytosolic free calcium ion concentration. Recently, this increase is suggested to be preceded by rapid increase of [Na$^+$] through NSCCs. This increase of [Na$^+$] is very short-lived and is dissipated through deactivation of VI-NSCCs by cAMP or cGMP proposed to be generated by cyclases which are turgor-sensitive membrane proteins. Appropriate target of sodium signal is actually concomitant influx of calcium which then carries on signaling after dissipation of sodium signal.[79,119]

Under salt stress, rapid increase of free cytosolic Ca^{2+} occurs within 1–5 s via influx through either NSCCs or a mechano-selective calcium channel (MSC) in the plasma-membrane, which can be amplified through release from internal stores especially from vacuole like through TPC1 (two-pore channel 1). Increased calcium activates Ca^{2+}/CaM-dependent protein kinases (CDPK) which activates ATPases (P-H$^+$-ATPase) of plasma membrane to maintain membrane integrity and ion homeostasis.

The best characterized signaling pathway specific to salt stress is SOS1/SOS2/SOS3 pathways.[76,79,119,179] The SOS3 is calceneurin B-like protein (CBL4) which senses increased concentration of calcium and gets dimerized and then interacts with SOS2. The SOS2 is CBL-interacting protine kinase (CIPK24) and is serine/threonine protein kinase with an SNF1/AMPK-like catalytic domain and a unique regulatory domain. Regulatory domain has positive modulator (SOS3) binding site and negative modulator (ABI2)-binding site. The catalytic and regulatory domains interact with one another and repress kinase activity, presumably by blocking substrate access to the catalytic site. But when SOS3 binds to its site on regulatory domain, it makes catalytic site accessible to substrate hence relieve repression of kinase activity. ABI2 when binds to its site on regulatory domain, it dephosphory-lates and deactivates SOS2 and SOS1. SOS2–SOS3 complex is then targeted to plasma membrane where it phosphorylates and activates SOS1 (Na$^+$/H$^+$ antiporter). Besides SOS1, SOS2–SOS3 may regulate other salt tolerance effectors like it seem to activate V-Na$^+$/H$^+$ antiporter (NHX), V-H$^+$-ATPase, calcium exchanger (CHX), AVP1 in tonoplast while inhibit Na$^+$ transporter AtHKT1 (Na$^+$-influx transporter) in plasma membrane. Several other salt stress related genes have also been identified to be regulated through this network. Moreover, SOS3 homolog, that is, SOS3-like calceneurin B-like protein 8 (SCABP8) has been identified as probable calcium sensor in shoot as SOS3 acts primarily in root. Further analyses have shown that SOS3 and SCABP8 are only partially redundant in their functions and each plays additional and unique roles in the plant salt stress response.

Additionally, it has been indicated that some transcription factors (CAMTA1–4 and CAMTA6) bind to CAM and get activated. MYB2 transcription factor interacts directly to a specific CAM in soybean and activates MYB2-dependent genes including P5CS1 (for proline synthesis).[79] AtNIG1 is a basic helix–loop–helix (bHLH)-type transcription factor in *A. thaliana* which binds to calcium under salt stress. Calcium can also stimulate NADPH oxidase (a primary source of stress related oxidative burst in plants and thus generates a further important stress signal).

Additionally, sustained high amount of calcium in the cytosol is also deleterious as it can cause degradative processes or cell death due to aggregation of proteins, nucleic acids and impairment of integrity of lipid membrane. Vacuole represents storage of calcium in plant cells. Ca^{2+} exchangers (CAXs) are located in tonoplast for calcium transport and these are auto-inhibited. Upon calcium increase, these are activated by Ca^{2+}/CAM and hence help to restore the level of cytosolic $[Ca^{2+}]$. The six members of CAX (CAX1–6) were identified in *A. thaliana*, CAX1 also activated through SOS-pathway.

Different studies have related salt tolerance to higher K^+/Na^+, K^+/Ca^{2+}, Ca^{2+}/Na^+ ratios like in root and leaf in safflower.[154] Higher expression of genes encoding Na^+/H^+ antiporters, H^+-pumps, K^+-uptake has been related to salt tolerance in poplar.[38] Other study related tolerance due to detoxification of ROS over ion toxicity in barnyard grass.[2] Proteomic studies in roots of creeping bent grass indicated higher levels of vacoular H^+-ATPase in salt tolerant cultivar.[180]

4.6.2 ROS AND NO SIGNALING

ROS (especially H_2O_2) and nitric oxide (NO) are known signaling molecules with wide variety of functions including germination, growth, programmed cell death and senescence. Phyto-hormones as well as stresses including salt stress activate membrane located in NADPH oxidase to produce ROS as signal.[112] Exogenous supply of ROS/NO has been reported to enhance salt tolerance in different plants. ABA induction of antioxidant response under water and salt stresses[112] has been shown to be mediated through

ROS production, where ROS may activate antioxidant potential through mitogen-activated protein kinase (MAPK)-signaling. In wheat seedlings, ROS (H_2O_2) improved salt tolerance due to improved photosynthetic capacity, leaf gas exchange, leaf water relation, and ion homeostasis and membrane stability.[172] In maize, H_2O_2-pretreatment before salt stress improved salt tolerance through increased antioxidant enzymes [SOD, APX, CAT, guaiacol POX (GPOX), GR] and decreased lipid peroxidation.[35] In rice, pretreatment with H_2O_2 or NO (SNP as NO donor) before salt stress, gave better growth, higher antioxidant enzymes, higher PS-II activity, higher gene expression of stress-related genes like SPS, HSP26, and P5CS. Moreover, pretreatment with NaCl before salt stress also gave better growth but this was due to increased H_2O_2 contents.[165] However in another report[164] in roots of rice seedlings, exogenous H_2O_2 when supplied alone enhanced activities of APX, GR, and gene expression of OsAPX and OsGR, but under salt stress, it was not involved in such regulations. Similarly in microalga *Ulva fasciata* under hyper-salinity, only upregulation of FeSOD was found mediated through H_2O_2, while other antioxidant enzymes like APX, CAT, GR, MnSOD seem to be modulated by some other factors.[158] Strong evidence has been provided in citrus plants that H_2O_2 and NO elicit long-lasting systemic primer-like antioxidant activity under physiological and salt stress conditions.[162]

Endogenous NO is increased in root tips under salt stress in sunflower seedlings, and it was followed by relative reduction of Na^+/K^+ ratio which may be due to regulation of transporter protein in root.[34] Seedlings of yellow seeds of *Atriplex centralasiatica* were more salt tolerant than seedlings derived from brown seeds. Through studying physiology to transcriptomic microarray analysis, it was found that superior salt tolerance of seedlings derived from yellow seeds was related to superior ability of plant to achieve ion homeostasis, redox equilibrium and signal transduction, a rapid response to salt stress and ultimately better growth where nitric oxide was the main regulator and the main difference between these two types of seeds.[182] Nitric oxide deficient mutants (Atnoa1 or Atnos1) in Arabidopsis displayed poor salt tolerance mainly due to poor ion homeostasis, higher oxidative toxicity and poor antioxidant response.[195,196] Such NO-deficient mutant of Arabidopsis (Atnoa1) restores salt tolerance when OsNOA1 gene was over expressed, and this tolerance was accompanied with reduced Na^+/K^+ ratio.[130] NO seems to interact with H_2O_2, implying H_2O_2 might be downstream of NO to regulate the activity of PM H^+-ATPase.[110] There is emerging evidence that NO may play systemic signaling role also for establishing salt tolerance in plants. Moreover, plants possess prime-like mechanism that allows them to memorize previous NO-exposure events and generates defense responses following salt stress.[115] Deciphering the protective role of nitric oxide against salt stress at physiological and proteomic levels in maize indicated that G-protein SIGNALING could the early event that works upstream to NO biogenesis.[11]

4.6.3 HORMONAL SIGNALING

4.6.3.1 SALICYLIC ACID (SA)

Exogenous supply of salicylic acid (SA) alone or combined with calcium improved plant growth, photosynthetic pigments, proline, activities of antioxidant enzymes POD, CAT, SOD, GR, APX, and activity of carbonic anhydrase in wheat under salt stress where combined application was more effective than individual application.[7] In *M. chamomilla* under salt stress, exogenous SA reduced Na^+ concentration in rosettes, superoxides, MDA in both rosettes and roots, increased flavonoids, PPO, and GPOX in roots.[97] Spraying SA to nickel or salt exposed plants of Indian mustard (*Brassica juncea*) improved the activity of nitrate reductase, carbonic anhydrase, antioxidant enzymes, and proline level.[187] Exogenous SA alleviated salt stress in *Brassica juncea* (mustard) through improved photosynthesis, growth, ascorbate–glutathione metabolism and sulfur assimilation. Its positive effects on photosynthesis and ascorbate–glutathione metabolism overlapped to similar effects observed by exogenous reduced glutathione in same plant material.[121] Positive effects of SA on photosynthesis and growth were suggested due to increase of reduced glutathione and decrease of oxidative stress.[90,121] In pea plants, SA negatively affected the response to salt stress by decreasing antioxidant potential and increasing damage; however, it could enhance resistance of salt-stressed plants to possible pathogen attack as suggested by increased PR1b (pathogen related gene) expression.[14]

An interaction between SA, glycine betaine, and ethylene has been suggested in salt-stressed *mung* bean where exogenous SA increased methionine and glycine betaine and decreased ethylene formation due to its inhibition of 1-aminocyclopropane carboxylic acid synthase (ACS) activity.[90] Cross talk of SA, NO, and ROS under salt stress[56] has been indicated in tomato where SA or salt stress increased ROS and NO individually but when SA was applied under salt stress decreased both ROS and NO and increased cell viability at whole plant level as well as at protoplast level.

4.6.3.2   ABSCISIC ACID

Abscisic Acid (ABA) has been shown to increase expression and activity of antioxidant enzymes and amount of nonenzymatic antioxidants ascorbate and glutathione under water stress and salt stress.[27,82,83,107,112] For this, ABA uses ROS signaling as it produced ROS through activation of membrane bound NADPH oxidase.[74,189,190] ABA-induced stomatal closure in guard cells under water or salt stress is partially dependent on NADPH oxidase activity.[32] ABA also increased LEA proteins/genes under abiotic stresses (like water stress, salt stress, cold stress, etc.) in different plant species.[6,20,87,93,94,142] Being involved in osmotic adjustment, protection of cellular structures and stomatal regulation, ABA often appears an efficient strategy to reduce water deficit caused either by salinity or by drought.[16,17,136]

In addition to its positive impact on stressed plant responses, ABA is also considered a senescing hormone and can thus assume dual functions in plant behavior.[57] In xero-halophyte *Atriplex halimus*[16] under salinity, increased level of ABA (made through exogenous application of ABA) increased ethylene and produced leaf senescence. Ionic stress could also produce higher amounts of ABA than osmotic stress of iso-osmotic concentration as found in *Phaseolus vulgaris*[117] and in salt-bush.[16] In some studies, lower concentrations of ABA under salt stress has been related to salt tolerance like in salt tolerant *Brassics napus* versus salt sensitive *Brassica carinata*[65,192] and in wheat cultivars.[181] However, it is only growth reduction and not decreased photosynthesis of *B. carinata* under salinity that was related to higher concentrations of ABA.[65] Similarly in wheat cultivars, low content of ABA of salt tolerant cultivar under salinity was suggested due to low salinity stress intensity threshold of the cultivar.[181] Different studies have also shown that mutations in positive regulators of ABA-signaling such as abi3, abi4 and abi5.[51,132] and another QTL (RAS1)[137] confer NaCl-resistant and ABA-insensitive germination capacity to Arabidopsis; however, this capacity may not be related to salt tolerance as abi4 mutant when transferred to nonsaline medium, grew poorer and gained less weight compared to its wild type.[132,169]

4.6.3.3   ETHYLENE

Increase in ethylene evolution has been found under salinity in chickpea roots where chickpea is considered sensitive to salinity.[98] Ethylene has been indicated as downstream signal to nitric oxide under salt stress in *A. thaliana* to regulate PM-H[+]-ATPase activity, ion homeostasis (improved Na[+]/K[+] ratio).[173] Due to down regulated JA/ethylene pathway in vte4 not in vte1 and wild type, vte4 performed better under salt stress than vte4 and WT in *A. thaliana*.[27] Ethylene-insensitive mutant s*etr1-1* and *ein2-2* of *A. thaliana* were more sensitive to salt stress,[24] hence indicating the importance of ethylene signaling in salt tolerance, where importance of ethylene has been linked to glycine betaine synthesis in *A. thaliana*[24] and in wheat.[91] It is because glycine betaine and ethylene share common precursor SAM for their synthesis. SAM also serves a common precursor in ethylene and polyamine synthesis, where ethylene has been more related to senescence while polyamine with antisenescence property.[10,125] A regulatory interaction exists between ethylene, proline and N for salt tolerance where nitrogen differentially regulates proline production and ethylene formation to alleviate the adverse effect of salinity on photosynthesis in mustard.[77]

4.7 CONCLUSION

Plants respond to salinity stress at molecular, cellular, metabolic, and physiological levels, being further controlled by a variety of genes and signaling pathways. Transcriptomic, proteomics, and metabolomics techniques have confirmed that under salt stress a wide variety of gene networks are affected and salt tolerance is contributed by complex cross talks amongst various cellular components and signaling networks. A critical comprehensive review of five major mechanisms, operative under salt stress that allow plants to adapt or tolerate salt stress, reveals that plants under salt stress respond by accumulating higher amounts of osmolytes (amino acid, sugar alcohols, sugars, etc.), mainly attributed to increased synthesis and decreased catabolism. This response enhances water retention capacity of the plants and helps to prevent water loss under matric/osmotic stresses. Similarly, high level of antioxidant metabolism is related to salinity tolerance, although recent researches have indicated that all constituents of antioxidant metabolism may not be required to be upregulated for salt tolerance. Since soluble phenolics like ascorbate and glutathione are used by the plants to detoxify deleterious ROS, it has also been found related to salinity tolerance in different plants. Lignin, a polyphenolic compound present in cell wall, causes cell-wall hardening-cum-growth suppression. Its accumulation especially in root of the plant may prevent entry of stressors like biotic stressors and metal ions. Mechanism of ion homeostasis in glycophytes imparts salt tolerance to the plants through higher ion exclusion, sequestration of ions in vacuoles and higher ratios of K^+/Na^+, K^+/Ca^{2+}, Ca^{2+}/Na^+ in different plant species. In the mechanism of signal transduction, calcium signaling helps in ion homeostasis and in proline accumulation under salt stress. Among other signaling networks, phytohormonal SIGNALING like ABA, SA, ethylene and ROS/NO-signaling are found to regulate salt responses in plants.

Practical applications of these findings are also reviewed wherein response of exogenous application of various chemicals to impart salt tolerance is collated. Exogenous application of different osmolytes provides salt tolerance in different plant systems through activating antioxidant potential and improving ion homeostasis. Application of proline or glycine betaine is reported to increase antioxidant defense in salt-stressed tobacco bright yellow-2, rice, and mung bean. Pretreated wheat seedlings with exogenous mannitol showed higher salt tolerance through improved activities of antioxidant enzymes like SOD, POX, CAT, APX, and GR and decreased oxidative toxicity measured as MDA. Salt stress induced proline accumulation in shoots and roots of wheat seedlings socked in ascorbic acid or thiamine or sodium salicylate was lower while vitamins or sodium salicylate helped to ameliorate salinity stress. It has been found that exogenous nitric oxide activated some isoforms of GST while decreased other forms when applied individually. Under salt stress, nitric oxide as well as salt stress signal act as potent regulators of GST gene and enzyme expression. Exogenously supplied GSH improved salt tolerance in some plant species such as wheat. An interaction between SA, glycine betaine and ethylene has been suggested in salt-stressed *mung* bean where exogenous SA increased methionine and glycine betaine and decreased ethylene formation due to inhibition of ACS activity. These and other similar studies points to a way forward for technology development to impart salt tolerance and choose breeding lines for varietal development programs.

4.8 SUMMARY

Five major mechanisms can be operative under salt stress to enhance salinity tolerance in plants. These include osmolyte metabolism, antioxidant-detoxification metabolism, phenolic/cell wall metabolism, ion homeostasis, and signal transduction.

Osmolytes (amino acid, sugar alcohols, sugars, etc.) enhance water retention capacity and help to prevent water loss under osmotic stresses like water stress and salt stress. Proline and glycine betaine, widely studied for their role in salt stress amelioration, are found to increase under salt stress mainly due to increased synthesis and decreased catabolism although their regulation is not well understood. Proline has been found upregulated through ABA-dependent as well as ABA-independent pathways. While ROS signaling could upregulate it, ABA could also use ROS-signaling to upregulate it. Its accumulation may also involve calcium signaling. Glycine betaine shares its synthetic pathway with ethylene from common precursor of SAM. Hence its regulation under salinity may involve ethylene signaling. BADH is the biosynthetic enzyme for glycine betaine and catalyzes the conversion of betaine aldehyde to betaine. It could also serve the role of detoxification enzyme under salinity through acting on other toxic aldehydes, hence removing their toxicity. Both glycine betaine and proline may also play role of signal molecules as their exogenous supply in different plant species improved salt tolerance through activating antioxidant metabolism and decreasing oxidative toxicity under salt stress.

Antioxidant metabolism constitutes antioxidant and detoxification enzymes (APX, CAT, GR, POX, SOD, GST, GPX, and glyoxylases I and II) and nonenzymatic antioxidants like ascorbate, glutathione, α-tocopherol, carotenoids, etc. High level of antioxidant metabolism under salt stress is related to salinity tolerance. But recent researches have indicated that all constituents of antioxidant metabolism may not be required to be upregulated for salt tolerance. Plasticity and redundancy of antioxidant metabolism in a plant should be related to tolerance rather an increase of an antioxidant factor.

Phenolics are synthesized through phenyl propanoid pathway. Plants use soluble phenolics as ROS scavengers under stresses. Phenolics like ascorbate and glutathione can be used directly (nonenzymatically) or enzymatically (through POXs) to detoxify deleterious ROS. Increased level of phenolics and POXs has been related to salinity tolerance in different plants. Lignin is present in cell wall and is a polyphenolic compound. Its deposition in cell wall causes cell-wall hardening-cum-growth suppression. Its content increases under stresses including abiotic and biotic stresses. Its accumulation especially in root of the plant may prevent entry of stressors like biotic stressors and metal ions. It could also prevent entry of ions under salt stress. But in other studies, cell-wall loosening-cum-growth promotion is related to salinity tolerance.

Mechanism of ion homeostasis may differ between halophytes and glycophytes. In glycophytes, there are three major steps where regulation can occur in root of plant: (1) at the entry of salt into root cell from soil environment where Na^+ exclusion through ion transporters like SOS1 could be used by plant; (2) to store the ions into vacuoles so to keep ion concentration low in cytosol for proper working of enzymes; and (3) to retrieve ions from xylem stream, so as to avoid entry of ions in aerial parts of plant. Ion homeostasis as regulated through calcium signaling is well studied. Salt tolerance has been related to higher ion exclusion, sequestration of ions in vacuoles and higher ratios of K^+/Na^+, K^+/Ca^{2+}, Ca^{2+}/Na^+ in different plant species.

In signal transduction, calcium signaling is well studied specially for SOS1/SO2/SOS3 pathway for ion homeostasis. Calcium signaling may also be involved in proline accumulation under salt stress. Among other signaling networks, phytohormonal signaling like ABA, SA, ethylene, and ROS/NO-signaling are found to regulate salt responses in plants. Phytohormones like ABA and SA could also interact with ROS/NO for regulations under salinity. ABA has been found to increase ROS production through its activation of NADPH oxidases, involve ROS/NO signals downstream to activate antioxidant response in plants under osmotic stresses. ROS signaling could upregulate antioxidant metabolism under stresses through MAPK-pathway. ROS/NO could also play systemic signaling role for establishing salt tolerance. Ethylene signaling has also been found important for salt tolerance. Polyamines as growth promoting substances in plants may also have signaling role similar to hormones. Polyamines, ethylene, and glycine betaine share common precursor of SAM for their synthesis, hence could interact with one another under salinity.

Practical applications of these findings are also reviewed wherein role of exogenous application of various chemicals to impart salt tolerance in plants is collated. Exogenous application of different osmolytes, proline or glycine betaine, mannitol, ascorbic acid or thiamine or sodium salicylate has been reviewed along with the mechanisms that become operative upon application of these chemicals. For example, an interaction between SA, glycine betaine, and ethylene has been suggested in salt-stressed *mung* bean where exogenous SA increased methionine and glycine betaine and decreased ethylene formation due to inhibition of ACS activity. It seems that there is need to carry forward these studies to arrive at doable technologies to impart salt tolerance and hasten the varietal development programs.

KEYWORDS

- **antioxidant metabolism**
- **compartmentation**
- **exogenous applications**
- **ion homeostasis**
- **photosynthesis**
- **reactive oxygen species**
- **salt stress**
- **turgor**

REFERENCES

1. Abdullahi Baque, M.; Lee, E. J.; Paek, K. Y. Medium Salt Strength Induced Changes in Growth, Physiology and Secondary Metabolite Content in Adventitious Roots of *Morinda citrifolia*: The Role of Antioxidant Enzymes and Phenylalanine Ammonia Lyase. *Plant Cell Rep.* **2010**, *29*, 685–694.
2. Abogadallah, G. M.; Serag, M. M.; Quick, W. P. Fine and Coarse Regulation of Reactive Oxygen Species in the Salt Tolerant Mutants of Barnyard Grass and Their Wild Type Parents under Salt Stress. *Physiol. Plant.* **2010**, *138*, 60–73.
3. Aghaei, K.; Ehsanpour, A. A.; Komatsu, S. Potato Responds to Salt Stress by Increased Activity of Antioxidant Enzymes. *J. Integr. Plant Biol.* **2009**, *51*, 1095–1103.
4. Ahmad, R., Kim, M. D., Back, K. H., Kim, H. S., Lee, H. S., Kwon, S. Y., Murata, N.; Chung, W. I.; Kwak S. S. Stress Induced Expression of Choline Oxidase in Potato Plant Chloroplasts Confers Enhanced Tolerance to Oxidative, Salt and Drought Stresses. *Plant Cell Rep.* **2008**, *27*, 687–698.
5. Al-Hakimi, A. M. A.; Hamada, A. M. Counteraction of Salinity Stress on Wheat Plants by Grain Soaking in Ascorbic Acid, Thiamine or Sodium Salicylate. *Biol. Plant* **2001**, *44*, 253–261.
6. Ali-Benali, M.; Alary, R.; Joudrier, P.; Gautier, M. F. Comparative Expression of Five Lea Genes During Wheat Seed Development and in Response to Abiotic Stresses by Real-time Quantitative RT-PCR. *Biochim. Biophys. Acta* **2005**, *1730*, 56–65.
7. Al-Whaibi, M. H.; Siddiqui, M. H.; Basalah, M. O. Salicylic Acid and Calcium Induced Protection of Wheat against Salinity. *Protoplasma* **2012**, *249*, 769–778.
8. Anjaneyulu, E.; Reddy, P. S.; Sunita, M. S.; Kishore, P. B.; Meriga, B. Salt Tolerance and Activity of Antioxidative Enzymes of Transgenic Finger Millet Overexpressing a Vacuolar H(+)-pyrophosphatase Gene (SbVPPase) from Sorghum Bicolor. *J. Plant Physiol.* **2014**, *171*, 789–798.
9. Bagdi, D. L.; Shaw, B. P. Analysis of Proline Metabolic Enzymes in *Oryza sativa* Under NaCl Stress. *J. Environ. Biol.* **2013**, *34*, 677–681.

10. Bagni, N.; Bongiovanni, B.; Franceshetti, M.; Tassoni, A. Abscisic Acid Mutants Affect Polyamine Metabolism. *Plant Biosyst.* **1998,** *131,* 181–187.

11. Bai, X.; Yang, L.; Yang, Y.; Ahmad, P.; Yang, Y.; Hu, X. Deciphering the Protective Role of Nitric Oxide Against Salt Stress at the Physiological and Proteomic Levels in Maize. *J. Proteome Res.* **2011,** *10,* 4349–4364.

12. Banu, M. N.; Hoque, M. A.; Watanabe-Sugimoto, M.; Islam, M. M.; Uraji, M.; Matsuoka, K.; Nakamura, Y.; Murata, Y. Proline and Glycinebetaine Ameliorated NaCl Stress Via Scavenging of Hydrogen Peroxide and Methylglyoxal but Not Superoxide or Nitric Oxide in Tobacco Cultured Cells. *Biosci. Biotechnol. Biochem.* **2010,** *74,* 2043–2049.

13. Banu, N. A.; Hoque, A.; Watanabe-Sugimoto, M.; Matsuoka, K.; Nakamura, Y.; Shimoishi, Y.; Murata, Y. Proline and Glycinebetaine Induce Antioxidant Defense Gene Expression and Suppress Cell Death in Cultured Tobacco Cells Under Salt Stress. *J. Plant Physiol.* **2009,** *166,* 146–156.

14. Barba-Espin, G.; Clemente-Moreno, M. J.; Alvarez, S.; Garcia-Legaz, M. F.; Hernandez, J. A.; Diaz-Vivancos, P. Salicylic Acid Negatively Affects the Response to Salt Stress in Pea Plants. *Plant Biol.* **2011,** *13,* 909–917.

15. Ben Ahmed, C.; Ben Rouina, B.; Sensoy, S.; Boukhriss, M.; Ben Abdullah, F. Exogenous Proline Effects on Photosynthetic Performance and Antioxidant Defense System of Young Olive Tree. *J. Agric. Food Chem.* **2010,** *58,* 4216–4222.

16. Ben Hassine, A.; Lutts, S. Differential Responses of Saltbush *Atriplex halimus* L. Exposed to Salinity and Water Stress in Relation to Senescing Hormones Abscisic Acid and Ethylene. *J. Plant Physiol.* **2010,** *167,* 1448–1456.

17. Ben Hassine, A.; Ghanem, M. E.; Bouzid, S.; Lutts, S. An Inland and a Coastal Population of the Mediterranean Xero-halophyte Species *Atriplex halimus* L. in Their Ability to Accumulate Proline and Glycinebetaine in Response to Salinity and Water Stress. *J. Exp. Bot.* **2008,** *59,* 1315–1326.

18. Ben-Rejeb, K.; Lefebvre-De Vos, D.; Le Disquet, I.; Leprince, A. S.; Bordenave, M.; Maldiney, R.; Jdey, A.; Abdelly, C.; Savoure. A. Hydrogen Peroxide Produced by NADPH Oxidases Increases Proline Accumulation during salt or Mannitol Stress in *Arabidopsis Thaliana. New Phytol.* **2015,** *208,* 1138–1148.

19. Bernstein, N.; Shoresh, M.; Xu, Y.; Huang, B. Involvement of the Plant Antioxidative Response in the Differential Growth Sensitivity to Salinity of Leaves vs Roots during Cell Development. *Free Radical Biol. Med.* **2010,** *49,* 1161–1171.

20. Bhagi, P.; Zhawar, V. K.; Gupta, A. K. Antioxidant Response and Lea Genes Expression under Salt Stress and Combined Salt Plus Water Stress in Two Wheat Cultivars Contrasting in Drought Tolerance. *Indian J. Exp. Biol.* **2013,** *51,* 746–757.

21. Borghesi, E.; Gonzalez-Miret, M. L.; Escudero-Gilete, M. L.; Malorgio, F.; Heredia, F. J.; Melendez-Martinez, M. J. Effects of Salinity Stress on Carotenoids, Anthocyanins and Color of Diverse Tomato Genotypes. *J. Agric. Food Chem.* **2011,** *59,* 11676–11682.

22. Borgognone, D.; Cardarelli, M.; Rea, E.; Lucini, L.; Colla, G. Salinity Source Induced Changes in Yield, Mineral Composition, Phenolic Acids and Flavonoids in Leaves of Artichoke and Cardoon Grown in Floating System. *J. Sci. Food Agric.* **2014,** *94,* 1231–1237.

23. Brosche, M.; Vinocur, B.; Alatalo, E. R.; Lamminmäki, A.; Teichmann, T.; Ottow, E. A.; Djilianov, D.; Afif, D.; Bogeat-Triboulot, M. B.; Altman, A.; Polle, A.; Dreyer, E.; Rudd, S.; Paulin, L.; Auvinen, P.; Kangasjärvi, J. Gene Expression and Metabolite Profiling of *Populus euphratica* Growing in the Negev Desert. *Genome Biol.* **2005,** *6*(R101), 2–17.

24. Cao, W. H.; Liu, J.; He, X. J.; Mu, R. L.; Zhou, H.; Chen, S. Y.; Zhang, J. S. Modulation of Ethylene Responses Affects Plant Salt Stress Responses. *Plant Physiol.* **2007,** *143,* 707–719.

25. Carassay, L. R.; Bustos, D. A.; Golberg, A. D.; Taleisnik, E. Tipburn in Salt-affected Lettuce (*Lactuca sativa* L.) Plants Results from Local Oxidative Stress. *J. Plant Physiol.* **2012,** *169,* 285–293.

26. Cavalcanti, F. R.; Lima, J. P. M. S.; Ferreira-Silva, S. L.; Viegas, R. A.; Silveira, J. A. G. Roots and Leaves Display Contrasting Oxidative Response during Salt Stress and Recovery in Cowpea. *J. Plant Physiol.* **2007,** *164,* 591–600.

27. Cela, J.; Chang, C.; Munne-Bosch, S. Accumulation of γ-Rather Than α-Tocopherol Alters Ethylene Signaling Gene Expression in the vte4 Mutant of *Arabidopsis thaliana.* *Plant Cell Physiol.* **2011,** *52,* 1389–1400.

28. Chen, J. B.; Yang, J. W.; Zhang, Z. Y.; Feng, X. F.; Wang, S. M. Two P5CS Genes from Common Bean Exhibiting Different Tolerance to Salt Stress in Transgenic Arabidopsis. *J. Genet.* **2013,** *92,* 461–469.

29. Chen, X.; Han H.; Jiang, P.; Nie, L.; Bao, H.; Fan, P.; Lv, S.; Feng, J.; Li, Y. Transformation of Beta-lycopene Cyclase Genes from *Salicornia Europaea* and Arabidopsis Conferred Salt Tolerance in Arabidopsis and Tobacco. *Plant Cell Physiol.* **2011,** *52,* 909–921.

30. Chen, Y.; Cai, J.; Yang, F. X.; Zhou, B.; Zhou, L. R. Ascorbate peroxidase from *Jatropha Curcas* Enhances Salt Tolerance in Transgenic Arabidopsis. *Genet. Mol. Res.* **2015,** *14,* 4879–4889.

31. Chinnusamy, V.; Zhu, J.; Zhu, J. K. Salt Stress Signaling and Mechanisms of Plant Salt Tolerance. *Genet. Eng. (N Y)* **2006,** *27,* 141–177.

32. Cho, D.; Shin, D.; Jeon, B. W.; Kwak, J. M. ROS-mediated ABA Signaling. *J. Plant Biol.* **2009,** *52,* 102–113.

33. Cong, L. L.; Zhang, X. Q.; Yang, F. Y.; Liu, S. J.; Zhang, Y. W. Isolation of the P5CS Gene from Red Canary Grass and Its Expression under Salt Stress. *Genet. Mol. Res.* **2014,** *13,* 9122–9133.

34. David, A.; Yadav, S.; Bhatla, S. C. Sodium Chloride Stress Induces Nitric Oxide Accumulation in Root tips and Oil Body Surface Accompanying Slower Oleosin Degradation in Sunflower Seedlings. *Physiol. Plant.* **2010,** *140,* 342–354.

35. De Azevedo Neto, A. D.; Prisco, J. T.; Enéas-Filho, J.; Medeiros, J. V.; Gomes-Filho, E. Hydrogen Peroxide Pre-treatment Induces Salt-stress Acclimation in Maize Plants. *J. Plant Physiol.* **2005,** *162,* 1114–1122.

36. De Carvallo, M. H. C. Drought Stress and Reactive Oxygen Species. *Plant Signal. Behav.* **2008,** *3,* 156–165.

37. Del Buono, D.; Ioli, G.; Scarponi, L. Glutathione Peroxidases in *Lolium multiflorum* and *Festuca arundinacea*: Activity, Susceptibility to Herbicides and Characteristics. *J. Environ. Sci. Health. Part B* **2011,** *46,* 715–722.

38. Ding, M.; Hou, P.; Shen, X.; Wang, M.; Deng, S.; Sun, J.; Xiao, F.; Wang, R.; Zhou, X.; Lu, C.; Zhang, D.; Zheng, X.; Hu, Z.; Chen, S. Salt Induced Expression of Genes Related to Na (+)/ K (+) and ROS Homeostasis in Leaves of Salt Resistant and Salt Sensitive Poplar Species. *Plant Mol. Biol.* **2010,** *73,* 251–269.

39. Dinler, B. S.; Antoniou, C.; Fotopoulos, V. Interplay between GST and Nitric Oxide in the Early Response of Soybean (*Glycine max* L.) Plants to Salinity Stress. *J. Plant Physiol.* **2014,** *171,* 1740–1747.

40. Dixon, D. P.; Edwards, R. Glutathione Transferases. *Arabidopsis Book* **2010**, *8*, E0131, DOI: 10.1199/Tab.0.131. https://www.ncbi.nlm.nih.gov/pmc/articles/PMC3244946/ (accessed May 18, 2017).

41. Dixon, D. P.; Davis, B. G.; Edwards, R. Functional Divergence in the Glutathione Transferase Super Family in Plants: Identification of Two Classes with Putative Functions in Redox Homeostasis in Arabidopsis Thaliana. *J. Biol. Chem.* **2002**, *277*, 30859–30869.

42. Dixon, D. P.; Skipsey, M.; Edwards, R. Roles of Glutathione Transferases in Plant Secondary Metabolism. *Phytochemistry* **2010**, *71*, 338–350.

43. Ellouzi, H.; Hamed, K. B.; Cela, J.; Munne-Bosch, S.; Abdelly, C. Early Effects of Salt Stress on the Physiological and Oxidative Status of *Cakile aritima* (Halophyte) and *Arabidopsis thaliana* (Glycophyte). *Physiol. Plant.* **2011**, *142*, 128–143.

44. El-Shabrawi, H.; Kumar, B.; Kaul, T.; Reddy, M. K.; Singla Pareek, S. L.; Sopory, S. K. Redox Homeostasis, Antioxidant Defense and Methylglyoxal Detoxification as Markers for Salt Tolerance in Pokkali Rice. *Protoplasma* **2010**, *245*, 85–96.

45. Eyidogan, F.; Oz, M. T. Effect of Salinity on Antioxidant Responses of Chickpea Seedlings. *Acta Physiol. Plant.* **2007**, *29*, 485–493.

46. Falleh, H.; Jalleli, I.; Ksouri, R.; Boulaaba, M.; Guyot, S.; Magne, C.; Abdelly, C. Effect of Salt Treatment on Phenolic Compounds and Antioxidant Activity of Two *Mesembryanthemum edule* provenances. *Plant Physiol. Biochem.* **2012**, *52*, 1–8.

47. Fan, W.; Zhang, M.; Zhang, H.; Zhang, P. Improved Tolerance to Various Abiotic Stresses in Transgenic Sweet Potato (*Ipomoea batatas*) Expressing Spinach Betaine aldehyde Dehydrogenases. *PLoS One* **2012**, *7*, e37344.

48. FAO. *How to Feed the World in 2050*. http://www.fao.org/fileadmin/templates/wsfs/docs/expert_paper/How_to_Feed_the_World_in_2050.pdf (accessed Oct 15, 2016).

49. Fatehi, F.; Hosseinzadeh, A.; Alizadeh, H.; Brimavandi, T.; Struik, P. C. The Proteome Response of Salt-resistant and Salt-sensitive Barley Genotypes to Long Term Salinity Stress. *Mol. Biol. Rep.* **2012**, *39*, 6387–6397.

50. Fernandez-Garcia N.; Hernandez M.; Casado-vela J.; Bru R.; Elortza F.; Hedden P. Changes to the Proteome and Targeted Metabolites of Xylem Sapin *Brassica oleracea* in Response to Salt Stress. *Plant Cell Environ.* **2011**, *34*, 821–836.

51. Finkelstein, R. R.; Gampala, S. S. L.; Rock, C. D. Abscisic Acid Signaling in Seeds and Seedlings. *Plant Cell* **2002**, *14*, S15–S45.

52. Flowers, T. J.; Colmer, T. D. Salinity Tolerance in Halophytes. *Tansley Rev.—New Phytol.* **2008**, *179*, 945–963.

53. Foyer, C. H.; Noctor, G. Ascorbate and Glutathione: The Heart of the Redoxhub. *Plant Physiol.* **2011**, *155*, 2–18.

54. Frary, A.; Gol, D.; Keles, D.; Okmen, B.; Pinar, H.; Sigva, H. O.; Yemenicioglu, A.; Doganlar, S. Salt Tolerance in *Solanum pennellii*: Antioxidant Response and Related Qtl. *BMC Plant Biol.* **2010**, *10*, 58–84.

55. Gapinska, M.; Sklodowska; Gabara, B. Effect of Short and Long Term Salinity on the Activities of Antioxidative Enzymes and Lipid Peroxidation in Tomato Roots. *Acta Physiol. Plant.* **2008**, *30*, 11–18.

56. Gemes, K.; Poor, P.; Horvath, E.; Kolbert, Z.; Szopko, D.; Szepesi, A.; Tari, I. Cross-talk between Salicylic Acid and NaCl-generated Reactive Oxygen Species and Nitric Oxide in Tomato during Acclimation to High Salinity. *Physiol. Plant.* **2011**, *142*, 179–192.

57. Ghanem, M. E.; Albacete, A.; Martínez-Andújar, C.; Acosta, M.; Romero-Aranda, R.; Dodd, I. C.; Lutts, S.; Pérez-Alfocea, F. Hormonal Changes during Salinity-induced leaf Senescence in Tomato (*Solanum lycopersicum* L.). *J. Exp. Bot.* **2008**, *59*, 3039–3050.

58. Gill, S. S.; Tuteja, N. Reactive Oxygen Species and Antioxidant Machinery in Abiotic Stress Tolerance in Crop Plants. *Plant Physiol. Biochem.* **2010,** *48*, 909–30.

59. Gill, T.; Sreenivasulu, Y.; Kumar, S.; Ahuja, P. S. Over-expression of Superoxide Dismutase Exhibits Lignifications of Vascular Structures in *Arabidopsis thaliana. J. Plant Physiol.* **2010,** *167*, 757–760.

60. Grassmann, J.; Hippeli, S.; Elstner, E. F. Plant's Defense and Its Benefits for Animals and Medicine: Role of Phenolics and Terpenoids in Avoiding Oxygen Stress. *Plant Physiol. Biochem.* **2002,** *40*, 471–478.

61. Gray, J.; Caparrós-Ruiz, D.; Grotewold, E. Grass Phenylpropanoids: Regulate before Using! *Plant Sci.* **2012,** *184*, 112–120.

62. Grigore, M. N.; Toma, C. Histo-anatomical Strategies of *Chenopodiaceae* Halophytes: Adaptive, Ecological and Evolutionary Implications. *WSEAS Trans. Biol. Biomed.* **2007,** *4*, 204–218.

63. Hasanuzzaman, M.; Alam, M. M.; Rahman, A.; Hasanuzzaman, M.; Nahar, K.; Fujita, M. Exogenous Proline and Glycine Betaine Mediated Upregulation of Antioxidant Defense and Glyoxalase Systems Provide Better Protection Against Salt-induced Oxidative Stress in Two Rice (*Oryza sativa* L.) Varieties. *Biomed. Res. Int.* **2014,** *2014,* Article ID757219, p 19.

64. Hasegawa, P. M.; Bressan, R. A.; Zhu, J. K.; Bohnert, H. J. Plant Cellular and Molecular Responses to High Salinity. *Annu. Rev. Plant Physiol. Plant Mol. Biol.* **2000,** *51*, 463–499.

65. He, T.; Cramer, G. R. Abscisic Acid Concentrations are Correlated with Leaf Area Reductions in Two Salt-stressed Rapid-cycling *Brassica* species. *Plant Soil* **1996,** *179*, 25–33.

66. Hernandez, J. A.; Ferrer, M. A.; Jimenez, A.; Barcelo, A. R.; Sevvilla, F. Antioxidant Systems and O_2/H_2O_2 Production in the Apoplast of Pea Leaves. Its Relation with Salt Induced Necrotic Lesions in Minor Veins. *Plant Physiol.* **2001,** *127*, 817–831.

67. Hoque, M. A.; Banu, M. N.; Nakamura, Y.; Shimoishi, Y.; Murata, Y. Proline and Glycinebataine Enhance Antioxidant Defense and Methylglyoxal Detoxification Systems and Reduce NaCl-induced Damage in Cultured Tobacco cells. *J. Plant Physiol.* **2008,** *165*, 813–824.

68. Hoque, M. A.; Banu, M. N. A.; Okuma, E.; Amako, K.; Nakamura, Y.; Shimoishi, Y.; Murata, Y. Exogenous Proline and Glycinebetaine Increase NaCl Induced Ascorbate Glutathione Cycle enzyme Activities, and Proline Improves Salt Tolerance More Than Glycinebetaine in Tobacco Bright Yellow-2 Suspension Cultured Cells. *J. Plant Physiol.* **2007,** *164*, 1457–1468.

69. Hossain, M. A.; Fujita M. Evidence for a Role of Exogenous Glycinebetaine and Proline in Antioxidant Defense and Methylglyoxal Detoxification Systems in Mung Bean Seedlings under Salt Stress. *Physiol. Mol. Biol. Plants* **2010,** *16*, 19–29.

70. Hu, L.; Li, H.; Chen, L.; Lou, Y.; Amombo, E.; Jinmin, F. RNA-seq for Gene Identification and Transcript Profiling in Relation to Root Growth of Bermuda Grass (*Cynodon dactylon*) under Salinity Stress. *BMC Genomics* **2015,** *16*, 575–587.

71. Hu, L.; Li, H.; Pang, H.; Fu, J. Responses of Antioxidant Gene, Protein and Enzymes to Salinity Stress in Two Genotypes of Perennial Ryegrass (*Lolium perenne*) Differing in Salt Tolerance. *J. Plant Physiol.* **2012,** *169*, 146–156.

72. Hu, L. X.; Huang, Z. H.; Liu, S. Q.; Fu, J. M. Growth Response and Gene Expression in Antioxidant Related Enzymes in Two Bermuda Grass Genotypes Differing in Salt Tolerance. *J. Am. Soc. Hortic. Sci.* **2012,** *137*, 134–143.

73. Hu, T.; Li, H. Y.; Zhang, X. Z.; Luo, H. J.; Fu, J. M. Toxic Effect of NaCl on Ion Metabolism, Antioxidative Enzymes and Gene Expression of Perennial Ryegrass. *Ecotoxicol. Environ. Saf.* **2011**, *74*, 2050–2056.

74. Hu, X.; Jiang, M.; Zhang, A.; Lu, J. Abscisic Acid-induced Apoplastic H_2O_2 Accumulation Up-regulates the Activities of Chloroplastic and Cytosolic Antioxidant Enzymes in Maize Leaves. *Planta* **2005**, *223*, 57–68.

75. Huang, C.; He, W.; Guo, J.; Chang, X.; Su, P.; Zhang, L. Increased Sensitivity to Salt Stress in an Ascorbate-deficient Arabidopsis Mutant. *J. Exp. Bot.* **2005**, *56*, 3041–3049.

76. Huang, G. T.; Ma, S. L.; Bai, L. P.; Zhang, L.; Ma, H.; Jia, P.; Liu, J.; Zhing, M.; Guo, Z. F. Signal Transduction during Cold, Salt, and Drought Stresses in Plants. *Mol. Biol. Rep.* **2012**, *39*, 969–987.

77. Iqbal, N.; Umar, S.; Khan, N. A. Nitrogen Availability Regulates Proline and Ethylene Production and Alleviates Salinity Stress in Mustard (*Brassica juncea*). *J. Plant Physiol.* **2015**, *178*, 84–91.

78. Iqbal, N.; Umar, S.; Khan, N. A.; Khan, M. I. R. A New Perspective of Phyto-hormones in Salinity Tolerance: Regulation of Proline Metabolism. *Environ. Exp. Bot.* **2014**, *100*, 34–42.

79. Ismail, A.: Takeda, S.; Nick, P. Life and Death under Salt Stress: Same Players, Different Timing? *J. Exp. Bot.* **2014**, *65*, 2963–2979.

80. Jaleel, C. A.; Manivannan, P.; Lakshmanan, G. M.; Sridharan, R.; Panneerselvam, R. NaCl As a Physiological Modulator of Proline Metabolism and Antioxidant Potential in *Phyllanthus amarus*. *Comptes Rendus Biol.* **2007**, *330*, 806–813.

81. Jebara, S.; Jebara, M.; Limam, F.; Aouani, M. E. Changes in Ascorbate Peroxidase, Catalase Guaiacol Peroxidase and Superoxide Dismutase Activities in Common Bean (*Phaseolus vulgaris*) Nodules under Salt Stress. *J. Plant Physiol.* **2005**, *162*, 929–936.

82. Jiang, M.; Zhang, J. Involvement of Plasma-membrane NADPH Oxidase in Abscisic Acid- and Water Stress-induced Antioxidant Defense in Leaves of Maize Seedlings. *Planta* **2002**, *215*, 1022–1030.

83. Jiang, M.; Zhang, J. Water Stress-induced Abscisic Acid Accumulation Triggers the Increased Generation of Reactive Oxygen Species and Up-regulates the Activities of Antioxidant Enzymes in Maize Leaves. *J. Exp. Bot.* **2002**, *53*, 2401–2410.

84. Jubany-Mari, T.; Munne-Bosch, S.; Lopez-Carbonell, M.; Alegre, L. Hydrogen Peroxide is Involved in the Acclimation of the Mediterranean Shrub, *Cistus albidus* L., to Summer Drought. *J. Exp. Bot.* **2009**, *60*, 107–120.

85. Kaur, L.; Zhawar, V. K. Phenolic parameters under Exogenous ABA, Water Stress, Salt Stress in Two Wheat Cultivars Varying in Drought Tolerance. *Indian J. Plant Physiol.* **2015**, *20*, 151–156.

86. Kaur, L.; Zhawar, V. K. Antioxidant Parameters under Salt Stress in Drought Tolerant and Susceptible Wheat Cultivars. *Indian J. Plant Physiol.* **2015**, *21*, 101–106.

87. Kaur, M.; Gupta, A. K.; Zhawar, V. K. Antioxidant Response and *Lea* Genes Expression under Exogenous ABA and Water Deficit Stress in Wheat Cultivars Contrasting in Drought Tolerance. *J. Plant Biochem. Biotechnol.* **2014**, *23*, 18–30.

88. Kaur, N.; Kumar, A.; Kaur, K.; Gupta, A. K.; Singh, I. DPPH Radical Scavenging Activity and Contents of H_2O_2, Malondialdehyde and Proline in Determining Salinity Tolerance in Chickpea Seedlings. *Indian J. Biochem. Biophys.* **2014**, *51*, 407–415.

89. Kavi-Kishore, P. B.; Sangam, S.; Amrutha, R. N.; Laxmi, P. S.; Naidu, K. R.; Rao, K. R. S. S.; Rao, S.; Reddy, K. J.; Theriappan, P.; Sreenivasulu, N. Regulation of Proline

Biosynthesis, Degradation, Uptake and Transport in Higher Plants: its Implications in Plant Growth and Abiotic Stress Tolerance. *Curr. Sci.* **2005,** *88,* 424–438.

90. Khan, M. I. R.; Asgher, M.; Khan, N. A. Alleviation of Salt-induced Photosynthesis and Growth Inhibition by Salicylic Acid Involves Glycinebetaine and Ethylene in Mung Bean (*Vigna radiata* L.). *Plant Physiol. Biochem.* **2014,** *80,* 67–74.

91. Khan, M. I. R.; Iqbal, N.; Masood, A.; Khan, N. A. Variation in Salt Tolerance of Wheat Cultivars: Role of Glycinebetaine and Ethylene. *Pedosphere* **2012,** *22,* 746–754.

92. Kim, S. H.; Ahn, Y. O.; Ahn, M. J.; Lee, H. S.; Kwak, S. S. Down-regulation of β-carotene Hydroxylase Increases β-carotene and Total Carotenoids Enhancing Salt Stress Tolerance in Transgenic Cultured Cells of Sweet Potato. *Phytochemistry* **2012,** *74,* 69–78.

93. Kobayashi, F.; Takumi, S.; Egawa, C.; Ishibashi, M.; Nakamura, C. Expression Patterns of the Low Temperature Responsive Genes in a Dominant ABA-less-sensitive Mutant of Common Wheat. *Physiol. Plant.* **2006,** *127,* 612–623.

94. Kobayashi, F.; Takumi, S.; Nakata, M.; Ohno, R.; Nakamura, T.; Nakamura, C. Comparative Study of the Expression Profiles of the Cor/Lea Gene Family in Two Wheat Cultivars with Contrasting Levels of Freezing. *Physiol. Plant.* **2004,** *120,* 585–594.

95. Kostopoulou, Z.; Therios, I.; Molassiotis, A. Resveratrol and its Combination with α-tocopherol Mediate Salt Adaptation in Citrus Seedlings. *Plant Physiol. Biochem.* **2014,** *78,* 1–9.

96. Kostopoulou, Z.; Therios, I.; Roumeliotis, E.; Kanellis, A. K.; Molassiotis, A. Melatonin Combined with Ascorbic Acid Provides Salt Adaptation in *Citrus aurantium* L. Seedlings. *Plant Physiol. Biochem.* **2015,** *86,* 155–165.

97. Kovacik, J.; Klejdus, B.; Hedbavny, J.; Backor, M. Salicylic Acid Alleviates NaCl-induced Changes in the Metabolism of *Matricaria chamomilla* plants. *Ecotoxicology* **2009,** *18,* 544–554.

98. Kukreja, S.; Nandwal, A. S.; Kumar, N.; Sharma, S. K.; Sharma, S. K.; Unvi, V.; Sharma, P. K. Plant Water status, H_2O_2 Scavenging Enzymes, Ethylene Evolution and Membrane Integrity of *Cicer arietinum* Roots as Affected by Salinity. *Biol. Plant.* **2005,** *49,* 305–308.

99. Kumari, A.; Das, P.; Parida, A. K.; Agarwal, P. K. Proteomics, Metabolomics and Ionomics Perspectives of Salinity Tolerance in Halophytes. *Front. Plant Sci.* **2015,** *6,* 537–542.

100. Lauchi, A.; Epstein, E. Mechanisms of Salt tolerance in Plants. *Calif. Agric.* **1984,** *38,* 18–20.

101. Li, M.; Guo, S.; Xu, Y.; Meng, Q.; Li, G.; Yang, X. Glycine Betaine-mediated Potentiation of HSP Gene Expression Involves Calcium Signaling Pathways in Tobacco Exposed to NaCl Stress. *Physiol. Plant.* **2014,** *150,* 63–75.

102. Lin, C. C.; Kao, C. H. Cell Wall Peroxidase Activity, Hydrogen Peroxide Level and NaCl-inhibited Root Growth of Rice Seedlings. *Plant Soil* **2001,** *230,* 135–143.

103. Lin, C. C.; Kao, C. H. Cell Wall Peroxidase Against Ferulic Acid, Lignin and NaCl-reduced Root Growth of Rice Seedlings. *J. Plant Physiol.* **2001,** *158,* 667–671.

104. Liu, Z.; Bao, H.; Cai, J.; Han, J.; Zhou, L. A Novel Thylakoid Ascorbate Peroxidase from *Jatropha curcas* Enhances Salt Tolerance in Transgenic Tobacco. *Int. J. Mol. Sci.* **2013,** *15,* 171–185.

105. Lokhande, V. H.; Suprasanna, P. Prospects of Halophytes in Understanding and Managing Abiotic Stress Tolerance. In *Environmental Adaptations and Stress Tolerance of Plants in the Era of Climate Change*; Ahmad, P., Prasad, M. N. V., Eds., Berlin: Springer, 2012; pp 26–59.

106. Lopez-Orenes, A.; Ros-Marin, A. F.; Ferrer, M. A.; Calderon AA. Antioxidant Capacity as a Marker for Assessing the in Vitro Performance of the Endangered *Cistus heterophyllus*. *Sci. World J.* **2013**, *2013*, Article ID 176295, p 10.

107. Lu, S.; Su, W.; Li, H.; Guo, Z. Abscisic Acid Improves Drought Tolerance of Triploid Bermuda Grass and Involves H_2O_2 and NO-induced Antioxidant Enzyme Activities. *Plant Physiol. Biochem.* **2009**, *47*, 132–138.

108. Luo, X.; Wu, J.; Li, Y.; Nan, Z.; Guo, X.; Wang, Y.; Zhang, A.; Wang, Z.; Xia, G.; Tian, Y. Synergistic Effects of GhSOD1 and GhCAT1 over Expression in Cotton Chloroplasts on Enhancing Tolerance to Methyl Viologen and Salt Stresses. *PLoS One* **2013**, *8*. https://doi.org/10.1371/journal.pone.0054002 (accessed May 18, 2017).

109. Manaa, A.; Ben Ahmed, H.; Valot, B.; Bouchet, J. P.; Aschi-Smiti, S.; Causse, M.; Faurobert, M. Salt and Genotype Impact on Plant Physiology and Root Proteome Variations in Tomato. *J. Exp. Bot.* **2011**, *62*, 2797–2813.

110. Mazid, M.; Khan, T. A.; Mohammad F. Role of Nitric Oxide in Regulation of H_2O_2 Mediating Tolerance of Plants to Abiotic Stress: a Synergistic Signaling Approach. *J. Stress Physiol. Biochem.* **2011**, *7*, 34–74.

111. Mhadhbi, H.; Fotopoulos, V.; Mylona, P. V.; Jebara, M.; Elarbi-Aouani, M.; Polidoros, A. N. Antioxidant Gene–Enzyme Responses in *Medicago truncatula* Genotypes with Different Degree of Sensitivity to Salinity. *Physiol. Plant.* **2011**, *141*, 201–214.

112. Miller, G.; Suzuki, N.; Ciftci-Yilmaz, S.; Mittler, R. Reactive Oxygen Species Homeostasis and Signaling During Drought and Salinity Stresses. *Plant Cell Environ.* **2010**, *33,* 453–467.

113. Mishra, P.; Bhoomika, K.; Dubey, R. S. Differential Responses of Antioxidative Defense System to Prolonged Salinity Stress in Salt Tolerant and Salt Sensitive Indica rice (*Oryza sativa* L.) Seedlings. *Protoplasma* **2013**, *250*, 3–19.

114. Missihoun, T. D.; Schmitz, J.; Klug, R.; Kirch, H. H.; Bartels, D. Betaine Aldehyde Dehydrogenase Genes from Arabidopsis with Different Sub Cellular Localization Affect Stress Responses. *Planta* **2011**, *233*, 369–382.

115. Molassiotis, A.; Tanou G.; Diamantidis G. NO Says More Than 'Yes' to Salt Tolerance: Salt Priming and Systemic Nitric Oxide Signaling in Plants. *Plant Signal. Behav.* **2010**, *5*, 209–212.

116. Moller, I. M.; Jensen, P. E.; Hansson, A. Oxidative Modifications to Cellular Components in Plants. *Annu. Rev. Plant Biol.* **2007**, *58*, 459–481.

117. Montero, E.; Cabot, C.; Poschenrieder, C. H.; Barcelo J. Relative Importance of Osmotic Stress and Ion-specific Effects on ABA-mediated Inhibition of Leaf Expansion Growth in *Phaseolus vulgaris*. *Plant Cell Environ.* **1998**, *21*, 54–62.

118. Moura, J. C.; Bonine, C. A.; De Oliveira Fernandes Viana, J.; Dornelas, M. C.; Mazzafera, P. Abiotic and Biotic Stresses and Changes in the Lignin Content and Composition in Plants. *J. Integr. Plant Biol.* **2010**, *52*, 360–376.

119. Munns, R.; Tester, M. Mechanisms of Salinity Tolerance. *Annu. Rev. Plant Biol.* **2008**, *59*, 651–681.

120. Naliwajski, M. R.; Sklodowska, M. Proline and its Metabolism Enzymes in Cucumber Cell Cultures During Acclimation to Salinity. *Protoplasma* **2015**, *251*, 201–209.

121. Nazar, R.; Umar, S.; Khan, N. A. Exogenous Salicylic Acid Improves Photosynthesis and Growth Through Increase in Ascorbate–Glutathione Metabolism and S Assimilation in Mustard under Salt Stress. *Plant Signal. Behav.* **2015**, *10*, 1–10.

122. Noreen, Z.; Ashraf, M. Assessment of Variation in Antioxidative Defense System in Salt Treated Pea (*Pisum sativum*) Cultivars and Its Putative Use as Salinity Tolerance Markers. *J. Plant Physiol.* **2009**, *166*, 1764–1774.

123. Nounjan, N.; Nghia, P. T.; Theerakulpisut, P. Exogenous Proline and Trehalose Promote Recovery of Rice Seedlings from Salt-stress and Differentially Modulate Antioxidant Enzymes and Expression of Related Genes. *J. Plant Physiol.* **2012**, *169*, 596–604.

124. Omoto, E.; Nagao, H.; Taniguchi, M.; Miyake, H. Localization of Reactive Oxygen Species and Change of Antioxidant Capacities in Mesophyll and Bundle Sheath Chloroplasts of Maize under Salinity. *Physiol. Plant.* **2013**, *149*, 1–12.

125. Ozturk, L.; Demir, Y. Effects of Putrescine and Ethephon on Some Oxidative Stress Enzyme Activities and Proline Content in Salt Stressed Spinach Leaves. *Plant Growth Regul.* **2003**, *40*, 89–95.

126. Parvin, S.; Lee, O. R.; Sathiyaraj, G.; Khorolragchaa, A.; Kim, Y. J.; Yang, D. C. Spermidine Alleviates the Growth of Saline-stressed Ginseng Seedlings through Antioxidative Defense System. *Gene* **2014**, *537*, 70–78.

127. Perez-Clemente, R. M.; Montoliu, A.; Zandalinas, S. I.; De Ollas, C.; Gomez-Cadenas, A. *Carrizo citrange* Plants Do Not Require the Presence of Roots to Modulate the Response to Osmotic Stress. *Sci. World J.* **2012**, *2012*, Article ID 795396, 1–13.

128. Perez-Lopez, U.; Robredo, A.; Lacuesta, M.; Sgherri, C.; Muñoz-Rueda, A.; Navari-Izzo, F.; Mena-Petite, A. The oxidative Stress Caused by Salinity in Two Barley Cultivars Is Mitigated by Elevated CO_2. *Physiol. Plant.* **2009**, *135*, 29–42.

129. Pessarakli, M.; Touchane H. Growth Responses of Bermudagrass and Seashore Paspalum Under Various Levels of Sodium Chloride Stress. *J. Food Agric. Environ.* **2006**, *4*, 240–243.

130. Qioa, W. H.; Xiao, S. H.; Yu, L.; Fan, L. M. Expression of a Rice Gene OsNOA1 Re-establishes Nitric Oxide Synthesis and Stress Related Gene Expression for Salt Tolerance in Arabidopsis Nitric Oxide-associated 1 Mutant Atnoa 1. *Environ. Exp. Bot.* **2009**, *65*, 90–98.

131. Qiu, Z. B.; Wang, Y. F.; Zhu, A. J.; Peng, F. L.; Wang, L. S. Exogenous Sucrose Can Enhance Tolerance of *Arabidopsis thaliana* Seedlings to Salt Stress. *Biol. Plant.* **2014**, *58*, 611–617.

132. Quesada, V.; Ponce, M. R.; Micol, J. L. Genetic Analysis of Salt-tolerant Mutants in *Arabidopsis thaliana*. *Genetics* **2000**, *154*, 421–436.

133. Ranjit, S. L.; Manish, P.; Penna, S. Early Osmotic, Antioxidant, Ionic and Redox Responses to Salinity in Leaves and Roots of Indian Mustard (*Brassica juncea* L.). *Protoplasma* **2016**, *253*, 792–797.

134. Rasool, S., Ahmad, A., Siddiqi, T. O. and Ahmad, P. Changes in Growth, Lipid Peroxidation and Some Key Antioxidant Enzymes in Chickpea Genotypes under Salt Stress. *Acta Physiol. Plant.* **2013**, *35*, 1039–1050.

135. Rasoulnia, A.; Bihamta, M. R.; Peyghambari, S. A.; Alizadeh, H.; Rahnama, A. Proteome Response of Barley Leaves to Salinity. *Mol. Biol. Rep.* **2011**, *38*, 5055–5063.

136. Ren, H.; Gao, Z.; Chen, L.; Wei, K.; Liu, J.; Fan, Y.; Davies, W. J.; Jia, W.; Zhang, J. Dynamic Analysis of ABA Accumulation in Relation to the Rate of ABA Catabolism in Maize Tissues under Water Deficit. *J. Exp. Bot.* **2007**, *58*, 211–219.

137. Ren, Z.; Zheng, Z.; Chinnusamy, V.; Zhu, J.; Cui, X.; Lida, K.; Zhu, J. K. RAS1, a Quantitative Trait Locus for Salt Tolerance and ABA Sensitivity in Arabidopsis. *Proc. Natl. Acad. Sci. U.S.A.* **2010**, *107*, 5669–5674.

138. Richter, J. A.; Erban, A.; Kopka, J.; Zorb, C. Metabolic Contribution to Salt Stress in Two Maize Hybrids with Contrasting Resistance. *Plant Sci.* **2015,** *233,* 107–115.

139. Robinson, S. P.; Jones, G. P. Accumulation of Glycinebetaine in Chloroplasts Provides Osmotic Adjustment during Salt Stress. *Aust. J. Plant Physiol.* **1986,** *13,* 659–668.

140. Roxas, V. P.; Lodhi, S. A.; Garrett, D. K.; Mahan, J. R.; Allen, R. D. Stress Tolerance in Transgenic Seedlings that Overexpress Glutathione-*S*-transferase/Glutathione Peroxidase. *Plant Cell Physiol.* **2000,** *41,* 1229–1234.

141. Roychoudhury, A.; Basu, S.; Sengupta, D. N. Amelioration of Salinity Stress by Exogenously Applied Spermidine or Spermine in Three Varieties of Indica Rice Differing in Their Level of Salt Tolerance. *J. Plant Physiol.* **2011,** *168,* 317–328.

142. Sairam, R. K.; Tyagi, A. Physiology and Molecular Biology of Salinity Stress Tolerance in Plants. *Curr. Sci.* **2004,** *86,* 407–421.

143. Sairam, R. K.; Rao, K. V.; Srivastava, G. C. Differential Response of Wheat Genotypes to Long Term Salinity Stress in Relation to Oxidative Stress, Antioxidant Activity and Osmolyte Concentration. *Plant Sci.* **2002,** *163,* 1037–1046.

144. Sakr, M. T.; El-Metwally, M. A. Alleviation of the Harmful Effects of Soil Salt Stress on Growth, Yield and Endogenous Antioxidant Content of Wheat Plant by Application of Antioxidants. *Pak. J. Biol. Sci.* **2009,** *12,* 624–630.

145. Seckin, B.; Sekmen, A. H.; Turkan, I. An Enhancing Effect of Exogenous Mannitol on the Antioxidant Enzyme Activities in Roots of Wheat under Salt Stress. *J. Plant Growth Regul.* **2009,** *28,* 12–20.

146. Sekman, A. H.; Turkan, I.; Takio, S. Differential Responses of Antioxidative Enzymes and Lipid Peroxidation to Salt Stress in Salt Tolerant *Plantago maritima* and Salt Sensitive *Plantago media. Physiol. Plant.* **2007,** *131,* 399–411.

147. Semida, W. M.; Taha, R. S.; Abdelhamid, M. T.; Rady, M. M. Foliar Applied α-tocopherol Enhances Salt Tolerance in *Vicia faba* L. Plants Grown under Saline Conditions. *S. Afr. J. Bot.* **2014,** *95,* 24–31.

148. Shah, S. H.; Tobita, S.; Shono, M. Supplemental Calcium Regulates Proline Accumulation in NaCl-stressed Suspension Culture of *Oryza sativa* L. at the Level of MRNA Translation. *Pak. J. Biol. Sci.* **2001,** *4,* 707–710.

149. Shankar, V.; Kumar, D.; Agarwal, V. Assessment of Antioxidant Enzyme Activity and Mineral Nutrients in Response to NaCl Stress and Its Amelioration through Glutathione in Chickpea. *Appl. Biochem. Biotechnol.* **2016,** *178,* 267–284.

150. Shao, H. B.; Chu, L. Y.; Lu, Z. H.; Kang, C. M. Primary Antioxidant Free Radical Scavenging and Redox Signaling Pathways in Higher Plant Cells. *Int. J. Biol. Sci.* **2008,** *4,* 8–14.

151. Shao, H. B.; Liang, Z. S.; Shao, M. A.; Sun, Q. Dynamic Changes of Anti-oxidative Enzymes of 10 Wheat Genotypes at Soil Water Deficits. *Colloids Surf. B: Biointerfaces* **2005,** *42,* 187–195.

152. Shao, H. B.; Liang, Z. S.; Shao, M. A.; Wang, B. C. Changes of Anti-oxidative Enzymes and Membrane Peroxidation for Soil Water Deficits Among 10 Wheat Genotypes at Seedling Stage. *Colloids Surf. B: Biointerfaces* **2005,** *42,* 107–113.

153. Shu, S.; Yuan, L. Y.; Guo, S. R.; Sun, J.; Yuan, Y. H. Effects of Exogenous Spermine on Chlorophyll Fluorescence, Antioxidant System and Ultrastructure of Chloroplasts in *Cucumis sativus* L. Under Salt Stress. *Plant Physiol. Biochem.* **2013,** *63,* 209–216.

154. Siddiqi, E. H.; Ashraf, M.; AL-Qurainy, F.; Akram, N. A. Salt Induced Modulation in Inorganic Nutrients, Antioxidant Enzymes, Proline Content and Seed Oil Composition in Safflower (*Carthamus tinctorius* L.). *J. Sci. Food Agric.* **2011,** *91,* 2785–2793.

155. Silva-Ortega, C. O.; Ochoa-Alfaro, A. E.; Reyes-Aguero, J. A.; Aguado-Santacruz, G. A.; Jimenez-Bremont, J. F. Salt Stress Increases the Expression of p5cs Gene and Induces Proline Accumulation in Cactus Pear. *Plant Physiol. Biochem.* **2008**, *46*, 82–92.

156. Sneha, S.; Rishi, A.; Dadhich, A.; Chandra, S. Effect of Salinity on Seed Germination, Accumulation of Proline and Free Amino Acids in *Pennisetum Glaucum* (L.) R. Br. *Pak. J. Biol. Sci.* **2013**, *16*, 877–881.

157. Sobahan, M. A.; Arias, C. R.; Okuma, E.; Shimoishi, Y.; Nakamura, Y.; Hirai, Y.; Mori, I. C.; Murata, Y. Exogenous Proline and Glycine Betaine Suppress Apoplastic Flow to Reduce Na$^+$ Uptake in Rice Seedlings. *Biosci. Biotechnol. Biochem.* **2009**, *73*, 2037–2042.

158. Sung, M. S.; Hsu, Y. T.; Hsu, Y. T.; Wu, T. M.; Lee, T. M. Hyper Salinity and Hydrogen Peroxide Upregulation of Gene Expression of Antioxidant Enzymes in *Ulva fasciata* against Oxidative Stress. *Mar. Biotechnol.* **2009**, *11*, 199–209.

159. Surender, Reddy P.; Jogeswar, G.; Rasineni, G. K.; Maheswari, M.; Reddy, A. R.; Varshney, R. K.; Kavi Kishor, P. B. Proline over Accumulation Alleviates Salt Stress and Protects Photosynthetic and Antioxidant Enzyme Activities in Transgenic Sorghum (*Sorghum bicolor* (L.) Moench). *Plant Physiol. Biochem.* **2015**, *94*, 104–113.

160. Szalai, G.; Kellos, T.; Galiba, G.; Kocsy, G. Glutathione as an Antioxidant and Regulatory Molecule in Plants under Abiotic Stress Conditions. *J. Plant Growth Regul.* **2009**, *28*, 66–80.

161. Tang, W.; Sun, J.; Liu, J.; Liu, F.; Yan, J.; Gou, X.; Lu, B. R.; Liu, Y. RNAi Directed down Regulation of Betaine Aldehyde Dehydrogenase 1 (OsBADH1) Results in Decreased Stress Tolerance and Increased Oxidative Markers without Affecting Glycine Betaine Biosynthesis in Rice (*Oryza sativa*). *Plant Mol. Biol.* **2014**, *86*, 443–454.

162. Tanou, G.; Molassiotis, A.; Diamantidis, G. Hydrogen Peroxide and Nitric Oxide Induced Systemic Antioxidant Prime Like Activity under NaCl Stress and Stress Free Conditions in Citrus Plants. *J. Plant Physiol.* **2009**, *166*, 1904–1913.

163. Thipyapong, P.; Stout, M. J.; Attajarusit, J. Functional Analysis of Polyphenol-oxidases by Antisense/Sense Technology. *Molecules* **2007**, *12*, 1569–1595.

164. Tsai, Y. C.; Hong, C. Y.; Liu, L. F.; Kao, C. H. Expression of Ascorbate Peroxidase and Glutathione Reductase in Roots of Rice Seedlings in Response to NaCl and H$_2$O$_2$. *J. Plant Physiol.* **2005**, *162*, 291–299.

165. Uchida, A.; Jagendorf, A. T.; Hibino, T.; Takabe, T.; Takabe, T. Effects of Hydrogen Peroxide and Nitric Oxide on Both Salt and Heat Stress Tolerance in Rice. *Plant Sci.* **2002**, *163*, 515–523.

166. Upadhyaya, C. P.; Venkatesh, J.; Gururani, M. A.; Asnin, L.; Sharma, K.; Ajappala, H.; Park, S. W. Transgenic Potato Overproducing L-ascorbic Acid Resisted an Increase in Methyl Glyoxal under Salinity Stress via Maintaining Higher Reduced Glutathione Level and Glyoxalase Enzyme Activity. *Biotechnol. Lett.* **2011**, *33*, 2297–2307.

167. Velikova, V.; Yordanov, I.; Edreva, A. Oxidative Stress and Some Antioxidant Systems in Acid Rain-treated Bean Plants. Protective Role of Exogenous Polyamines. *Plant Sci.* **2000**, *151*, 59–66.

168. Verslues, P. E.; Bray, E. A. Role of Abscisic Acid (ABA) and *Arabidopsis thaliana* ABA-insensitive Loci in Low Water Potential-induced ABA and Proline Accumulation. *J. Exp. Bot.* **2006**, *57*, 201–212.

169. Verslues, P. E.; Agarwal, M.; Katiyar-Agarwal, S.; Zhu, J.; Zhu, J. K. Methods and Concepts in Quantifying Resistance to Drought, Salt and Freezing, Abiotic Stresses that Affect Plant Water Status. *Plant J.* **2006**, *45*, 523–539.

170. Verslues, P. E.; Kim, Y. S.; Zhu, J. K. Altered ABA, Proline and Hydrogen Peroxide in an Arabidopsis Glutamate: Glyoxylate Aminotransferase Mutant. *Plant Mol. Biol.* **2007,** *64,* 205–217.

171. Wahid, A.; Ghazanfar, A. Possible Involvement of Some Secondary Metabolites in Salt Tolerance of Sugarcane. *J. Plant Physiol.* **2006,** *163,* 723–730.

172. Wahid, A.; Perveen, M.; Gelani, S.; Basra, S. M. A. Pretreatment of Seed with H_2O_2 Improves salt Tolerance of Wheat Seedlings by Alleviation of Oxidative Damage and Expression of Stress Proteins. *J. Plant Physiol.* **2007,** *164,* 283–294.

173. Wang, H.; Liang, X.; Wan, Q.; Wang, X.; Bi, Y. Ethylene and Nitric Oxide are Involved in Maintaining Ion Homeostasis in Arabidopsis Callus Under Salt Stress. *Planta* **2009,** *230,* 293–307.

174. Wang, L.; Liu, X.; Liang, M.; Tan, F.; Liang, W.; Chen, Y.; Lin, Y.; Huang, L.; Xing, J.; Chen, W. Proteome Analysis of Salt Responsive Proteins in the Leaves of Mangrove *Kandelia candel* during Short Term Stress. *PLoS One* **2014,** *e83141.* https://doi.org/10.1371/Journal.Pone.0083141 (accessed May 18, 2017).

175. Wang, W. B.; Kim, Y. H.; Lee, H. S.; Kim, K. Y.; Deng, X. P.; Kwak, S. S. Analysis of Antioxidant Enzyme Activity during Germination of Alfalfa under Salt and Drought Stresses. *Plant Physiol. Biochem.* **2009,** *47,* 570–577.

176. Wang, X.; Ma, Y.; Huang, C.; Wan, Q.; Li, N.; Bi, Y. Glucose-6-phosphate Dehydrogenase Plays a Central Role in Modulating Reduced Glutathione Levels in Reed Callus under Salt Stress. *Planta* **2008,** *227,* 611–623.

177. Widodo, I. H.; Patterson, E.; Newbigin, M.; Tester, M.; Bacic, A.; Roessner, U. Metabolic Responses to Salt Stress of Barley (*Hordeum vulgare* L.) Cultivars, Sahara and Clipper, which Differ in Salinity Tolerance. *J. Exp. Bot.* **2009,** *60,* 4089–4103.

178. Wu, G.; Wang, G.; Ji, J.; Gao, H.; Guan, W.; Wu, J.; Guan, C.; Wang, Y. Cloning of a Cytosolic Ascorbate Peroxidase Gene from *Lycium chinense* Mill. and Enhanced Salt Tolerance by Over Expressing in Tobacco. *Gene* **2014,** *543,* 85–92.

179. Xiong, L.; Schumaker, K. S.; Zhu, J. K. Cell Signaling during Cold, Drought, and Salt Stress. *Plant Cell* **2002,** *14*(Suppl.), 165–183.

180. Xu, C.; Sibicky, T.; Huang, B. Protein Profile Analysis of Salt Responsive Proteins in Leaves and Roots in Two Cultivars of Creeping Bent Grass Differing in Salinity Tolerance. *Plant Cell Rep.* **2010,** *29,* 595–615.

181. Xu, H.; Zhai, J.; Liu, Y.; Cheng, X.; Xia, Z.; Chen, F.; Cui, D.; Jiang, X. The Response of Mo-hydroxylases and Abscisic Acid to Salinity in Wheat Genotypes with Differing Salt Tolerances. *Acta Physiol. Plant.* **2012,** *34,* 1767–1778.

182. Xu, J.; Yin, H.; Yang, L.; Xie, Z.; Liu, X. Differential Salt Tolerance in Seedlings Derived from Dimorphic Seeds of *Atriplex centralasiatica*: from Physiology to Molecular Analysis. *Planta* **2011,** *233,* 859–871.

183. Yamaguchi, M.; Sharp, R. E. Complexity and Coordination of Root Growth at Low Water Potentials: Recent Advances from Transcriptomic and Proteomic Analyses. *Plant Cell Environ.* **2010,** *33,* 590–603.

184. Yang, X.; Liang, Z.; Wen, X.; Lu, C. Genetic Engineering of the Biosynthesis of Glycine Betaine Leads to Increased Tolerance of Photosynthesis to Salt Stress in Transgenic Tobacco Plants. *Plant Mol. Biol.* **2008,** *66,* 73–86.

185. Yildiztugay, E.; Sekmen, A. H.; Turkan, I.; Kucukoduk, M. Elucidation of Physiological and Biochemical Mechanisms of an Endemic Halophyte *Centaurea tuzgoluensis* under Salt Stress. *Plant Physiol. Biochem.* **2011,** *49,* 816–824.

186. Yoo, J. H.; Park, C. Y.; Kim, J. C.; Heo, W. D.; Cheong, M. S.; Park, H. C.; Kim, M. C.; Moon, B. C.; Choi, M. S.; Kang, Y. H.; Lee, J. H.; Kim, H. S.; Lee, S. M.; Yoon, H. W.; Lim, C. O.; Yun, D. J.; Lee, S. Y.; Chung, W. S.; Cho, M. J. Direct Interaction of a Divergent CaM Isoform and the Transcription Factor, MYB2, Enhances Salt Tolerance in Arabidpsis. *J. Biol. Chem.* **2005**, *280*, 3697–3706.

187. Yusuf, M.; Fariduddin, Q.; Varshney, P.; Ahmad, A. Salicylic Acid Minimizes Nickel and/ or Salinity-induced Toxicity in Indian Mustard (*Brassica juncea*) through an Improved Antioxidant System. *Environ. Sci. Pollut. Res. Int.* **2012**, *19*, 8–18.

188. Zhai, C. Z.; Zhao, L.; Yin, L. J.; Chen, M.; Wang, Q. Y.; Li, L. C.; Xu, Z. S.; Ma, Y. Z. Two Wheat Glutathione Peroxidase Genes Whose Products are Located in Chloroplasts Improve Salt and H_2O_2 Tolerances in Arabidopsis. *PLoS One* **2013**, *8*, e73989. https://doi.org/10.1371/Journal.Pone.0073989 (accessed May 18, 2017).

189. Zhang, A.; Jiang, M.; Zhang, J.; Ding, H.; Xu, S.; Hu, X.; Tan, M. Nitric Oxide Induced Hydrogen Peroxide Mediates Abscisic Acid-induced Activation of the Mitogen-activated Protein Kinase Cascade Involved in Antioxidant Defense in Maize Leaves. *New Phytol.* **2007**, *175*, 36–50.

190. Zhang, A.; Jiang, M.; Zhang, J.; Tan, M.; Hu, X. Mitogen-activated Protein Kinase is Involved in Abscisic Acid-induced Antioxidant Defense and Acts Downstream of Reactive Oxygen Species Production in Leaves of Maize Plants. *Plant Physiol.* **2006**, *141*, 475–487.

191. Zhang, H.; Dong, H.; Li, W.; Sun, Y.; Chen, S.; Kong, X. Increased Glycine Betaine Synthesis and Salinity Tolerance in *AhCMO* Transgenic Cotton Lines. *Mol. Breed.* **2009**, *23*, 289–298.

192. Zhang, J.; Jia, W.; Yang, J.; Ismail, A. M. Role of ABA in Integrating Plant Responses to Drought and Salt Stresses. *Field Crops Res.* **2006**, *97*, 111–119.

193. Zhang, J.; Tan, W.; Yang, X. H.; Zhang, H. X. Plastid Expressed Choline Mono-oxygenase Gene Improves Salt and Drought Tolerance Through Accumulation of Glycine Betaine in Tobacco. *Plant Cell Rep.* **2008**, *27*, 1113–1124.

194. Zhang, L.; Liu, N.; Ma, X.; Jiang L. The Transcriptional Control Machinery as well as the Cell Wall Integrity and Its Regulation are Involved in the Detoxification of the Organic Solvent Dimethyl Sulfoxide in *Saccharomyces cerevisiae*. *FEMS Yeast Res.* **2013**, *13*, 200–218.

195. Zhao, M.; Zhao, X.; Wu, Y.; Zhang, L. Enhanced Sensitivity to Oxidative Stress in Arabidopsis Nitric Oxide Synthase Mutant. *J. Plant Physiol.* **2007**, *164*, 737–745.

196. Zhao, M. G.; Tian, Q. Y.; Zhang, W. H. Nitric Oxide Synthase Dependent Nitric Oxide Production is Associated with Salt Tolerance in Arabidopsis. *Plant Physiol.* **2007**, *144*, 206–217.

197. Zhou, S.; Chen, X.; Zhang, X.; Li, Y. Improved Salt Tolerance in Tobacco Plants by Co-transformation of a Betaine Synthesis Gene BADH and a Vacuolar Na^+/H^+ Antiporter Gene SeNHX1. *Biotechnol. Lett.* **2008**, *30*, 369–376.

198. Ziaebrahimi, L.; Khavari-Nejad, R. A.; Fahimi, H.; Nejadsatari, T. Effects of Aqueous Eucalyptus Extracts on Seed Germination, Seedling Growth and Activities of Peroxidase and Polyphenoloxidase in Three Wheat Cultivar Seedlings (*Triticum aestivum* L.). *Pak. J. Biol. Sci.* **2007**, *10*, 3415–3419.

CHAPTER 5

BIOCHEMICAL, PHYSIOLOGICAL, AND MOLECULAR APPROACHES FOR IMPROVING SALT TOLERANCE IN CROP PLANTS: A REVIEW

ARCHANA SINGH, VEDA KRISHNAN, VINUTHA THIMMEGOWDA, and SURESH KUMAR

ABSTRACT

Salt tolerance is a multigenic complex trait; hence only limited success has been achieved so far in developing salt-tolerant crop varieties through conventional breeding methods. Ion homeostasis is considered to be crucial for normal growth and development of plants under salt stress. Osmoprotectants, enzymatic, and non-enzymatic antioxidants, and other antistress molecules constitute the defense system of plants. Identification of the genes associated with the stress tolerance has provided new insights into the biological mechanism to cope up with the stress. Genomic studies promise to complement the use of existing genetic diversity to accelerate plant improvement programs toward the development of salt tolerant, high yielding, multistress tolerant plants. Development of transgenic plant with ion transporters, compatible organic solutes, and enhanced antioxidant production appears to be a promising approach to engineer biosynthetic pathways. Combined together, these may integrate ion homeostasis, reactive oxygen species scavenging, reduced lipid peroxidation, and maintaining protein structure and functions to efficiently counteract the stress. Evidence also suggests that genomic DNA not only provides the genetic information for a trait but also serves as chromatin modulator to affect the expression of the associated genes. Epigenetic (DNA methylation, posttranslational histone modifications, and regulatory/noncoding RNAs biogenesis) changes are induced by environmental perturbations. Hence, the ultimate success of manipulating genes for stress tolerance in plant will only emerge if the transferred genes contribute to the stress tolerance not only at a particular stage of plant growth but also on the appearance of the stressful condition. Since the epigenetic state of chromatin is dynamic, transfer of a gene from one species to another would not only require the transfer of the gene(s) associated with a trait of interest but also the epigenetic state so that the genes can express optimally. In this chapter, we present newer insights for the successful development of salt-tolerant plants while utilizing the benefits of genetic engineering approach. The scope, challenges, and future prospects of genetic engineering, as a supplementary strategy to conventional breeding, have also been described.

5.1 INTRODUCTION

The global population is estimated to reach 9 billion by the year 2050. This would result in a predicted increase in food demands by 70% regardless of the further decrease in the quantity and quality of arable land, irrigation water, and increasingly variable weather conditions associated with the climate change.[111] Besides, urbanization and seawater intrusion are likely to increase over time; hence, any gain in agricultural production has to come from increased production and productivity of marginal lands especially the saline soils. Plants being sessile in nature need to cope up with various environmental stresses such as drought, heat, salinity, insect pests and diseases. Amongst these abiotic stresses are the major environmental constraints that affect growth, development, and productivity of crop plants, salinity being the most destructive stress. In fact, it has become a serious problem in several parts of the world, especially in the arid and semiarid regions.[139] It is assuming a serious dimension in these ecologies because inherently saline sodic irrigation waters are getting further degraded because of natural and anthropogenic reasons.[61] It has been estimated that 51–82% of the potential yield of annual crops is lost due to abiotic stresses.[30] It is assessed that irrigated lands world over contribute one-third of the world's food production but nearly 20% of these lands are affected by various degrees of soil salinity.[40] No exact assessment of salt affected soils is available for India, yet the most quoted figure states that about 6.73 million hectare (M ha) are salt affected and poses threat to the food security of India.[135,176]

5.1.1   PLANT RESPONSE TO SOIL SALINITY

Salinity induces osmotic and ionic imbalance in plant cells and exerts severe stress on the plants to affect plant growth and metabolism.[1] Higher concentration of salt imposes hyperosmotic shock by lowering the water potential causing turgor reduction that restricts cell expansion.[233] Higher concentration of salt in apoplast of cells generates primary and secondary effects that negatively affect growth, development, and survival of the plant. Primary effects include ionic toxicity, disequilibrium, and hyperosmolality.

Elevated cellular Na^+ and Cl^- concentrations in saline soils or saline water irrigations adversely affects crop yields by inhibiting cell expansion, photosynthetic activity and due to cytotoxic effects that accelerate leaf senescence, reduced carbon assimilation, and partitioning to reproductive structures.[139] These ions have inhibitory effects on cytosolic and organellar processes; however, Na^+ at > 0.1 M concentration becomes cytotoxic and directly affects certain biochemical and physiological processes. Secondary effects of salt stress also include disturbance in K^+ acquisition, membrane dysfunction, impairment of photosynthesis, generation of reactive oxygen species (ROS) and programed cell death.[70,82,167,206,234] Under salt stress, ROS is generated due to impaired electron transport pathway in chloroplast, mitochondrion and during photorespiration which may damage membranes, photosynthetic pigments, proteins, nucleic acids, and lipids.[41,51]

5.1.2   STRESS AVOIDANCE AND TOLERANCE

Plants show stress avoidance or tolerance through acclimatization and adaptation mechanisms that have evolved through natural selection process. Osmotic balance in the cytoplasm is maintained by the accumulation of organic solutes (compatible osmolytes) such as sugars (sucrose, fructose, trehalose), ions (K^+), charged metabolites (glycine betaine), and amino acids (proline) which do not inhibit metabolic processes. The function of compatible solutes is not only limited to osmotic balance, but being hydrophilic in nature, they replace water at the surface of proteins or membranes acting as low molecular weight chaperones.[82] In addition, they also function to protect cellular structures through scavenging ROS. Besides, plants produce low molecular weight antioxidants such as

ascorbic acid, reduced glutathione, tocopherol, carotenoids, and detoxifying enzymes to scavenge ROS[41,82]

Following stress recognition, plant's adaptation to salinity stress involves
complex physiological, biochemical, molecular networks, and metabolic pathways. Regulatory responses reestablish cellular and organellar
homeostasis and reduce episodic shock effects. So far, the use of natural
variations for improving abiotic stress tolerance in crop plants has generally been without the knowledge of specific causal genes and associated
biological mechanisms. The recent molecular characterization of the genes
associated with stress survival has enabled more expedient approaches for
crop improvement. In many instances, identification of the genes associated with the abiotic stress tolerance in one plant species led to the insights
into the specific molecular mechanism and translation of the knowledge to
other species. To exploit the existing genetic diversity for stress tolerance,
investigations into the molecular, biochemical and physiological processes
integral to the environmental adaptation and acclimatization are very much
desirable, which may also foster biotechnologically engineered solutions for
stress tolerance in crop plants.

A comprehensive understanding of how plants respond to salinity stress
at different levels and integrated approach of combining molecular tools with
biochemical and physiological techniques is essential for the development of
salt-tolerant crop varieties.[78] Although recent studies have identified various
adaptive responses to salinity stress at molecular, cellular, metabolic, and
physiological levels, yet mechanisms underlying salt tolerance in plant are
still far from being completely understood.

The focus of this chapter is to discuss the ways by which plants respond
to salt stress, utilize and integrate many common signals and subsequent
pathways to cope with the stress. The authors also discuss the aberrations in
metabolism and the strategies/approaches to increase salt tolerance through
traditional and biotechnological approaches to improve crop performance
in stressful environments. They also propose to discuss other mechanisms
involved in imparting salt stress tolerance such as the role of various regulatory genes, signal transduction involving hormones, genomic, proteomic,
epigenetic aspects, and strategies to improve salt stress tolerance. Therefore, this chapter provides a comprehensive review of the major research
advances in biochemical, physiological, and molecular mechanisms regulating plant's adaptation to salt stress.

5.2   BIOCHEMICAL AND PHYSIOLOGICAL ASPECTS OF SALT TOLERANCE

Plants have developed several physiological and biochemical mechanisms to cope with salinity stress. Some of the mechanisms include, but are not limited to

a)   Ion uptake, transport and compartmentalization,
b)   Biosynthesis of osmoprotectants and compatible solutes,
c)   Antioxidant enzyme activity and synthesis of antioxidant compounds,
d)   Synthesis of polyamines (PAs),
e)   Generation of nitric oxide, and
f)   Hormone-mediated regulation of salt stress tolerance.

Certain biochemical and physiological parameters have been identified as indicators of the ability of a plant to tolerate salt stress. ROS produced during stressed environment cause chlorophyll degradation and membrane-lipid peroxidation, the percent loss in chlorophyll content being widely accepted as an important attribute to assess salt tolerance ability of a genotype.[189] Salt tolerance level in plants can also be estimated by determining malondialdehyde (MDA) level[86] because MDA is accumulated due to membrane-lipid peroxidation in plants under stress. Total phenolic content in the plants also increases significantly as reports indicate its accumulation is triggered by salt stress in salvia plant grown under stress.[205] A better understanding of these and such other plant defense mechanisms under abiotic stress can help in engineering more tolerant crop plants to enhance yield and productivity, which is crucial in the present times.

5.2.1   ION HOMEOSTASIS AND SALT TOLERANCE

Salt stress inhibits plant growth through two major effects: (1) osmotic effect, where salt in soil water reduces plant's ability to take up nutrients through water leading to reduced growth and (2) ion-toxicity effect due to accumulation of high level of toxic ions. Since sodium chloride (NaCl) is the major salt present in the soils, accumulation of Na^+ and Cl^- in the cytoplasm with reduced uptake of beneficial ions (e.g., K^+ and Ca^{2+}) leads to oxidative stress. Negative effects caused by Na^+ and Cl^- ions accumulation include membrane damage, nutrient imbalance, altered levels of growth regulators, enzymatic inhibition, and improper metabolic functions, which ultimately lead to plant death. Therefore, the main focus of research has been to study the transport mechanisms and compartmentalization of Na$^+$ ions vis-à-vis the other ions. Since neither halophytes nor glycophytes can tolerate high salt concentration in their cytoplasm; hence, excessive salts are either compartmentalized into vacuoles or sequestered in older tissues which are sacrificed, thereby protecting the plant from salt stress.[234] Ion uptake, compartmentalization, and homeostasis are thus crucial for the normal growth and development of plants under salt stress (Fig. 5.1). Depending upon the plant's ability to do this, crop plants differ significantly in their ability to tolerate salts.

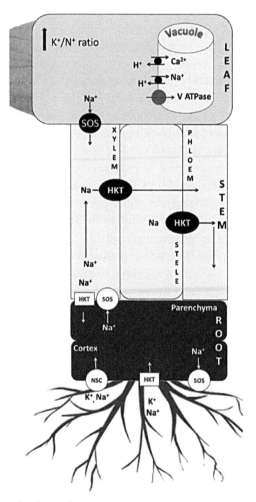

FIGURE 5.1 **(See color insert.)** Ionic influx and efflux during salt stress.

Adaptation of plants to salt stress requires cellular ion homeostasis involving net intracellular Na^+ and Cl^- uptake and subsequent vacuolar compartmentalization without toxic ion accumulation in the cytosol. Sodium is compartmentalized into vacuoles in preference to K^+ resulting in maintenance of high K^+/Na^+ ratio in the cytosol of shoot cells. Transport systems necessary for ion homeostasis include plasma membrane and tonoplast H^+-pumps, Na^+/H^+ antiporters, and Na^+ influx systems.[27,222] Na^+ that enters the cytoplasm is transported to the vacuole via Na^+/H^+ antiporter. Vacuolar type H^+-ATPase (V-ATPase) is the most prevailing H^+-pump in the plant cell. V-ATPase maintains solute homeostasis during unstressed condition; however, under stressed condition, survival of the plant depends upon the activity of V-ATPase.[52] In halophytes, vacuolar pyrophosphatase (V-PPase) H^+ pump has been reported to play important role.[209]

In many species, a large component of tolerance to high Na^+ concentration can be attributed to the ability of plants to exclude Na^+ from shoot. Most of the plants show a high degree of K^+/Na^+ discrimination for uptake, and plant genome contains a number of genes encoding potassium transporter and channels. Sodium exclusion is achieved through low net uptake by cells in the root cortex and tight control of net unloading of the xylem by parenchyma cells in the stele.[141]

The K^+ is an essential element for plants and Na^+ competes with K^+ for uptake into cells, particularly when the external concentration is substantially greater.[166] In contrast to marine organisms (which live in a medium with an inexhaustible supply of K^+), terrestrial life evolved in oligotrophic environments where low supply of K^+ limited the growth of colonizing plants. Under these limiting conditions, Na^+ could substitute for K^+ in some cellular functions but in others it is toxic. In vacuole, accumulated Na^+ is not toxic and can undertake osmotic functions, reducing the total K^+ requirements and improving plant growth when K^+ is a limiting factor. Because of these physiological requirements, the terrestrial plants depend on high-affinity K^+ uptake and high-affinity Na^+ uptake systems.

The phylogenetic analysis of available HKT sequences revealed two major subfamilies depending on whether the specific transporter has a highly conserved serine (subfamily 1) or glycine (subfamily 2) residue in the first pore-loop of the protein, and on the extracellular Na^+/K^+ ratio.[90] The HKTs have been shown to function as Na^+/K^+ symporters and as Na^+ selective uniporters.[90] It is suggested that HKT1 transporters are permeable to Na^+ only, whereas HKT2 refers to the transporters that are permeable to both Na^+ and K^+. Generally, HKT1 has relatively higher Na^+ to K^+ selectivity than HKT2 transporters. High-affinity potassium transporters (HKTs) have two important functions: (1) to take up Na^+ from the soil to reduce K^+ requirements when K^+ is a limiting factor and (2) to reduce Na^+ accumulation in leaves by both removing Na^+ from the xylem sap and loading Na^+ into the phloem sap (Fig. 5.1). Differential expression of HKTs in contrasting genotypes supports the idea that Na^+ membrane transporters may contribute to the observed differential stress tolerance at both biochemical and molecular levels.[46,186]

Soybean plant showed enhanced salt stress tolerance through maintenance of ion homeostasis, osmotic adjustment, restoration of oxidative balance, and other metabolic and structural adaptations.[152] Besides, Ca^{2+} has been long implicated as a decoder in stress signaling cascades.[178,234] Ca^{2+} alleviates Na^+ toxicity through different mechanisms, including the control of K^+/Na^+ selective accumulation and others that are being elucidated.[174]

Studies on the effects of salinity on growth and photosynthesis in *Brassica* sp., variability in salt tolerance during germination in *Medicago*, accumulation and ion distribution in shoot and root of maize plants,[202] biochemical and mineral element analysis in maize cultivars,[37] and salt stress induced changes in germination, sugars, starch, and enzyme of carbohydrate metabolism[53]—indicated the complexity of salt stress tolerance at metabolic levels. In addition, the presence of putrescine (Put) differentially influences the effect of salt stress through PA metabolism and ethylene synthesis in rice cultivars[161] and thus supports regulation at metabolic level through ion homeostasis.

5.2.1.1   SALINITY CAUSED IONIC AND OSMOTIC STRESS

Na^+ sequestration into vacuole depends on expression and activity of Na^+/H^+ antiporter driven by electrochemical gradient of protons generated by vacuolar H^+-ATPase and H^+-pyrophosphatase. Sodium extrusion at the root–soil interface is presumed to be crucial for salt tolerance. A rapid efflux of Na^+ from roots must occur to control its rate of influx. Salt overly sensitive 1 (*SOS1*), a Na^+/H^+ antiporter, localized in the plasma membrane is the only Na^+ efflux protein characterized so far from plants.[31] Ion accumulation in vacuole facilitates osmotic adjustment with minimal deleterious impact on cytosolic and organellar machinery. Osmotic homeostasis is accomplished by accumulation of compatible osmolytes in the cytosol, a process that is less deciphered.

Establishment of ionic and osmotic homeostasis under salt stress is essential not only to prevent cellular death but also for the physiological and biochemical steady states necessary for growth and completion of the life cycle. Elevated concentrations of salts are built in the apoplast, and eventually inside the cell, as water evaporates. The accumulation of ions in plant tissues results in progressive damage and ion specific stress supplements the effects caused by hyperosmolarity. Among the various ions, Na^+ reaches to a toxic concentration before the other ions do; therefore most of the studies have concentrated on control of Na^+ homeostasis. Maintenance of appropriate intracellular K^+/Na^+ balance is critical for metabolic activities as Na^+ cytotoxicity is largely due to its competition with K^+ for binding sites on the enzymes essential for cellular functions.[182] Na^+ has a strong inhibitory effect on K^+ uptake by cells, presumably by interfering with transporters such as HAK/KUP/HKT that mediate high-affinity and low-affinity K^+ transport at the plasmalemma and tonoplast.[160,167]

The fine-tuned control of net ion accumulation in shoot involves precise *in planta* coordination between the mechanisms that are intrinsically cellular with those that are operational at the intercellular, tissue, or organ level.[61] Several processes are involved including regulation of Na^+ transport into shoot, preferential Na^+ accumulation in shoot cells that are metabolically not very active, and reduction of Na^+ content in shoot by recirculation through the phloem back to the root.[24,138]

Ions loaded into the root xylem are transported to the shoot largely by mass-flow driven by the size of the transpirational sink.[140] A control response is required to lower transpiration by reduction in stomata aperture; however, this is only effective as a short-term response because plants need to maintain water status, carbon fixation, and solute transport.[82] Controlling ion load into the root xylem restricts accumulation in the shoot to a level where cells can act as effective ion repositories by vacuolar compartmentalization.[82,140] Since Na^+ content is greater in older than in younger leaves, it is presumed that the former are ion sinks that restrict cytotoxic ion accumulation in cells that are crucial for development.[138] Ionic and osmotic stresses may be recognized by distinct sensors followed by the activation of signaling cascades that exert transcriptional and/or posttranscriptional control over determinants that mediate adaptation.[234] However, substantial complexity in the functioning is envisaged because of overlapping stress effects, output diversity, and pathway networking required to coordinate the processes.

5.2.2   OSMOPROTECTANTS FOR SALT STRESS TOLERANCE

Osmolytes are a group of chemically diverse organic compounds that are uncharged, polar, water-soluble in nature and do not interfere with the cellular metabolism even at high concentration. They mainly include proline, glycine betaine, sugars, and polyols. Organic osmolytes are synthesized and accumulated in varying amounts in different plants. For example, quaternary ammonium compound betaine (β-alanine) accumulation has been reported among a few members of *Plumbaginaceae*,[80] whereas accumulation of amino acid proline has been reported in diverse sets of plants.[180] The concentration of compatible solutes within the cell is maintained either by irreversible synthesis of the compounds or by a combination of synthesis and degradation. As their accumulation is proportional to the external osmolarity, the major functions of these osmolytes are to protect the structure and to maintain osmotic balance within the cell via continuous water influx.[82]

Glycine betaine is a nontoxic cellular osmolyte ubiquitously found in microorganisms, higher plants, and animals and is electrically neutral over a wide range of pH. It is highly soluble in water but also contains nonpolar moiety. Because of its unique structural features, it interacts both with hydrophobic and hydrophilic domains of macromolecules such as enzymes and protein complexes. Glycine betaine raises osmolarity of cell during stress period and plays an important function in stress mitigation. It is synthesized within the cell from either choline or glycine. Its accumulation has been reported in a wide variety of plants belonging to different taxonomical background.

Rahman et al.[162] reported positive effects of glycine betaine on rice seedlings exposed to salt stress. Amino acids such as cysteine, arginine, and methionine, which constitute about 55% of the total free amino acids, decrease when exposed to salinity stress, whereas proline concentration rises in response to salinity stress.[55] Intracellular proline accumulation not only provides tolerance to salt stress but also serves as an organic nitrogen reserve during stress recovery. It functions as an O_2 quencher, thereby revealing its antioxidant capability. It has been reported that proline improves salt tolerance in *Nicotiana tabacum* by increasing the activity of enzymes involved in antioxidant defense system.[89]

Deivanai et al.[49] demonstrated that rice seedlings from the seeds pretreated with proline exhibited improved salt stress tolerance. The proteins accumulated under salt stress conditions provide a storage form of nitrogen that is reutilized in post stress recovery and also play a role in osmotic adjustments.[116] Fine-tuning of protein levels and activity by post translational modifications has recently emerged as essential regulatory mechanisms, thereby constituting interesting intervention targets for a tighter control of plant defense responses.[127]

Sugars, a class of polyols, function as compatible solutes, as low molecular weight chaperones and as ROS scavenging compounds. Mannitol, acyclic sugar synthesis is induced in plant during stress via the action of NADPH dependent mannose-6-phosphate reductase. These compatible solutes function as a protector/stabilizer of enzymes/membrane structures that are sensitive to dehydration/ionically induced damage. Pinitol, a cyclic sugar, is accumulated in plant cell when the plant is subjected to salinity stress. Inositol methyl transferase enzyme plays major role in the synthesis of pinitol that also plays a significant role in stress alleviation. Accumulation of polyols, either straight-chain metabolites such as mannitol and sorbitol or cyclic polyols such as myo-inositol and its methylated derivatives, has been correlated with tolerance to drought and/or salinity, based on polyol distribution in many species, including microbes, plants, and animals.[28]

Accumulations of glucose, fructose, fructans, trehalose, and starch occur under salt stress.[149] The major role played by these carbohydrates in stress mitigation involves osmoprotection, carbon storage, and scavenging of ROS. Besides being a carbohydrate reserve, trehalose accumulation protects organisms against several physical and chemical stresses including salinity stress.

5.2.3   ANTIOXIDANT ENZYMATIC ACTIVITY AND SALT STRESS

Singlet oxygen ($^1O_2$), hydroxyl radical ($^{\bullet}OH$), superoxide radical ($O_2^{\bullet-}$), and hydrogen peroxide (H_2O_2) are strong oxidizing compounds and are potentially harmful for cell integrity.[75] Antioxidant metabolism, including antioxidant enzymes and non-enzymatic compounds, plays critical role in detoxifying ROS produced during the stresses. Salinity tolerance has been positively correlated with the activity of antioxidant enzymes such as superoxide dismutase (SOD), catalase (CAT), peroxidase (POX), ascorbate POX (APX), glutathione reductase (GR), and accumulation of non-enzymatic antioxidant compounds.[11,79]

Singh et al.[186] observed increase in specific activity of several antioxidant enzymes in the contrasting wheat genotypes with increasing concentration (50–200 mM) of NaCl. Maximum activity of the antioxidant enzymes was observed 6 days after salt treatment, particularly in the salt-tolerant genotype (Kharchia-65). SOD activity was observed to be higher in untreated salt-sensitive genotype than that in salt tolerant genotype. Salt tolerant genotype showed high GR activity, and a distinct isoform of GR was observed in salt tolerant genotype. Related studies suggest that tolerance to salt stress can be correlated with a more efficient antioxidant system, with high enzymatic activities and better biomass production in salt tolerant genotype.

Ascorbate is another antioxidant present within the cell. Exogenous application of ascorbate mitigated adverse effects of salinity stress in different plants and promoted plant recovery from the stress.[4,137] Another antioxidant in stress mitigation is glutathione, which reacts with $O_2^{\cdot-}$, $^\cdot OH$, and H_2O_2, and function as free radical scavenger. It also participates in regeneration of ascorbate via ascorbate–glutathione cycle.[64] On exogenous application, glutathione helped maintaining plasma membrane permeability and cell viability during salinity stress.[8]

5.2.4   POLYAMINES AND SALT STRESS TOLERANCE

PAs are ubiquitous, low molecular weight, polycationic aliphatic, molecules. Put, spermidine (Spd), and spermine (Spm) are three commonly found PAs in plants. These are well known for antisenescence and antistress effects due to their acid neutralizing as well as antioxidant properties. PAs are considered as biochemical markers for abiotic stress tolerance, particularly for salinity. Though physiological relationship between salinity tolerance and PAs has been reported for more than four decades, ever since a debate also exists whether increased PAs levels protect plant (due to their ability to deal with free radicals) or cause damage due to the H_2O_2 produced on their catabolism.[5,118,184]

PA affects a large spectrum of cellular activities through physiological, biochemical, and molecular mechanisms. Increase in endogenous PA level has been reported when the plant is exposed to salinity stress and a wide range of physiological alterations have been reported. Intracellular PA level is well regulated by PA catabolism by amine oxidases which include copper binding diamine oxidases and FAD binding PA oxidases. To understand whether PAs actually protect cells from stress-induced damages, exogenous application of PAs was investigated before or during salinity stress. PA priming was found to increase the level of endogenous PAs during the stress.[171,173]

PA accumulation under salt stress has been reported to make the tonoplast more cation (K^+) selective, leading to higher vacuolar Na^+ sequestration and improved cytosolic K^+/Na^+ homeostasis.[204] Absence of Spm in an *Arabidopsis* mutant caused imbalance in Ca^{2+} homeostasis and showed hypersensitivity to salinity, suggesting its involvement in modulating the activity of certain Ca^{2+}-permeable channels and changes in Ca^{2+} allocation compared to unstressed state, which may prevent Na^+ and K^+ entry into the cytosol, enhance Na^+ and K^+ influx into the vacuole, or suppress Na^+ and K^+ release from the vacuole. However, further research would be required to understand these intricate interactions between PAs and vacuolar transport systems.[155]

5.2.5 NITRIC OXIDE AND SALT STRESS TOLERANCE

Nitric oxide (NO) is a volatile gaseous molecule involved in regulation of various plant growth and developmental processes and stress signaling.[119,120,230] NO reacts with lipid radicals, thus preventing lipid oxidation, exerting a protective effect by scavenging superoxide radical and formation of peroxynitrite that can be neutralized by other cellular processes. It also helps in activation of antioxidant enzymes.[20,50] Exogenous NO application has been reported to play roles in stress mitigation,[91] but the effects depend on NO concentration. Positive effects of NO on salinity tolerance have been attributed to antioxidant activities and modulation of ROS detoxification system.[134]

Improved plant growth under salinity stress by exogenous application of NO was associated with increases in antioxidant enzymes such as SOD, CAT, GPX, APX, and GR[229] and suppression of MDA production or lipid peroxidation.[143] Effects of NO on salinity tolerance are also related to its regulation of plasma membrane H^+-ATPase and Na^+/K^+ ratio.[45] NO stimulates H^+-ATPase (H^+-PPase), thereby producing a H^+ gradient and offering the force for Na^+/H^+ exchange. Although NO acts as a signal molecule under salt stress and induces salt tolerance by increasing PM H^+-ATPase activity, yet research on *Populus euphratica* indicated that NO cannot activate purified PM H^+-ATPase activity, at least in vitro.[223] Their results suggested that H_2O_2 induced increased PM H^+-ATPase activity which resulted in an increased K^+/Na^+ ratio leading to salt stress adaptation.

5.2.6 HORMONE-MEDIATED REGULATION OF SALT STRESS TOLERANCE

The responses of plants to abiotic stresses depend on the multiple factors, among them plant hormones (phytohormones) are considered as one of the active players of signal cascade. Phytohormones are key regulators of plant physiology, development and growth, as well as mediators of the response to environmental stress by integrating environmental stimuli with regulatory networks. Phytohormone includes not only the so-called classical hormones (viz., auxins, cytokinins, gibberellic acid (GA), abscisic acid (ABA), and ethylene) but also comprises of several other plant growth regulators, namely, jasmonic acid (JA), salicylic acid (SA), brassinosteroids (BRs). The phyto-hormones are structurally diverse compounds which play important roles as signaling compounds to stimulate physiological responses, modulate plant growth regulations, and induce stress responses to various environmental factors at nanomole concentration.[151]

Hormone signaling pathways are linked to various defense signaling path-ways and play important role in metabolism and developmental processes of plants. Decline in endogenous level of phytohormones has been related to the repressive effect of salt stress.[47] Exogenous applications of cytokinin, auxin, GA, ethylene, JA and SA, etc. have been reported to alleviate the negative effect of salt stress.[99,144] Seed priming with BR, GA, and SA have been reported to reduce inhibitory effects of salt stress.[15,74,123,150]

Salinity stress causes osmotic and water deficit stresses, increasing the production of ABA in shoot and root tissues.[89] ABA is an important phyto-hormone which ameliorates the effect of abiotic stresses by mitigating the inhibitory effect of salinity on photosynthesis, growth, and translocation of assimilates.[49,89] It is a vital cellular signal that modulates expression of a number of salt and water deficit-responsive genes. Positive relationship between ABA accumulation and salinity tolerance has been attributed to the accumulation of K^+, Ca^{2+}, proline, and sugars in vacuoles of root, which counteract with the uptake of Na^+ and Cl^-.[67,130] Fukuda and Tanaka[66] demon-strated the effects of ABA on expression of genes for vacuolar H^+-inorganic pyrophosphatase and H^+-ATPase in *Hordeum vulgare* under salinity stress. ABA treatment induced expression of the stress associated genes in wheat under salinity stress.[104]

GA not only plays crucial role in regulating various plant growth and development processes but also mitigates deleterious effects of salinity stress. Exogenous application of GA_3 was reported to minimize the adverse effects of NaCl stress through increased germination percentage and seed-ling growth of barley and lettuce seeds.[38] Ethylene signaling modulates salt response at different levels, including membrane receptors, components in cytoplasm, and nuclear transcription factors in the pathway.[3,36] Down-regulated expression of ETR1 gene encoding ethylene receptor was reported under salt and osmotic stress in *Arabidopsis* and loss-of-function mutant (*etr1-7*) showed increased tolerance to salt stress.[231]

Aminocyclopropane-carboxylic acid (an ethylene precursor) suppressed salt sensitivity in NTHK1 transgenic *Arabidopsis* plants, suggesting that ethylene is required for counteraction of receptor function to improve salt tolerance.[35] Ethylene stimulates modulation of ion homeostasis, upregulates activity of PM H[+]-ATPase which in turn improves salt tolerance in *Arabidopsis*.[210] A novel ortholog of ethylene response-factor (MsERF11) was isolated from *Medicago sativa* and its overexpression was reported to enhance salt stress tolerance in transgenic *Arabidopsis* plants.[39]

Auxin and cytokinin affect various aspects of plant growth and development as well as regulation of gene expression under abiotic stress.[58,100] Seed priming of wheat with auxins (IAA, IBA) diminished the negative implications of salt stress on endogenous ABA level.[94] Fang and Yang[59] showed up-regulated expression of NIT1 (Nitrilase1) and NIT2 genes involved IAA-biosynthesizes and down-regulation of ABA responsive genes under salinity stress suggesting that under salt stress the increased level of IAA regulates expression of salt responsive genes. Nevertheless, little information is available on the relationship between IAA levels in plants under salinity and the role of IAA in mitigating salt stress.[99] It has been reported that under salt stress the level of cytokinin decreases.[71] Overexpression of cytokinin biosynthetic gene might serve as a strategy to mitigate the effects of salt stress. Increased tolerance to salinity stress was observed in wheat plants raised from cytokinin primed seeds.[95] Exogenous application of cytokinin has shown improved salinity tolerance because of the increased proline contents in eggplant.[216]

SA has received significant attention because of its ability to respond to biotic and abiotic stresses through extensive signaling crosstalk with other growth substances.[12] It plays important role in defense responses against different abiotic stresses including osmotic and salinity stress.[83] Foliar application of SA significantly reduced toxic effects of NaCl by decreasing Na[+] and increasing K[+] and Mg[2+] in root and shoot of tomato plants.[85]

Pretreatment of tomato with SA in hydroponic culture was able to trigger the accumulation of ABA, leading to improved acclimation to salt stress.[191] Application of SA improved plant growth by promoting protective reactions and maintaining membrane integrity in barley.[56] SA increased antioxidant activity in maize plants under salinity stress through decreased lipid peroxidation and membrane permeability.[77] Recently, JA has been the focus of much attention because of its ability to provide protection against salinity stress.[201]

Exogenous application of JA improved recovery of salt-stressed rice seedlings. Barley leaf segments exposed to osmotic stress exhibited a sharp increase in endogenous JA contents.[109] Methyl jasmonate (MeJA) level in rice root was significantly enhanced in response to 200 mM NaCl.[136] Similarly, BRs, a group of naturally occurring novel steroidal phytohormones, regulate plant growth and development by producing an array of physiological changes.[102,106] Rice seedlings exposed to salinity stress and treated with BRs showed a significant increase in the activities of CAT, SOD, GR, and a slight increase in the APX activity.[147] Application of BRs increased accumulation of proline and enhanced activities of antioxidant enzymes in salt stressed *Cicer arietinum* and *Vigna radiata*.[6,84]

Thus, manipulation of hormone metabolism and signaling processes represents a promising alternative toward obtaining enhanced stress tolerance in crop plants. However, a tight spatio-temporal regulation would be necessary in order to minimize possible negative effects of hormonal imbalance on plant growth and development.[34]

5.2.7 MULTILEVEL SIGNAL TRANSDUCTION

Plants deploy multilevel signal transduction to induce stress responses. Interactions between PAs, ROS, hormones, antioxidants along with other signaling molecules like Ca^{2+}, cyclic nucleotides, and reactive nitrogen species such as NO form a complex signaling network, and induce diverse, apparently contradictory physiological effects.[72,155,206] Increased level of PAs during salt stress has been shown to have dual effects (Fig. 5.2). While exogenous application of PA was correlated with improved tolerance to abiotic stress (due to increased ability to inactivate oxidative radicals), it was reported to decrease plant's capacity to withstand stress, possibly due to the increased levels of H_2O_2 resulted from Pas' catabolism.[133]

Both anabolism and catabolism of PAs were reported to increase during abiotic stress with the net effect of increased cellular levels of ROS and antioxidants.[133,156] Interestingly, ABA was involved in regulation of both anabolic and catabolic pathways of PAs in *Arabidopsis*.[92] It has been reported that PAs can induce production of NO, thus serving as signal for salt tolerance by increasing K^+ to Na^+ ratio through stimulating expression of plasma membrane H^+-ATPase and Na^+/H^+ antiporters in tonoplast.[226,228,231] However, further research is needed to determine the exact nature of the intricate signaling network in context of salt stress tolerance.

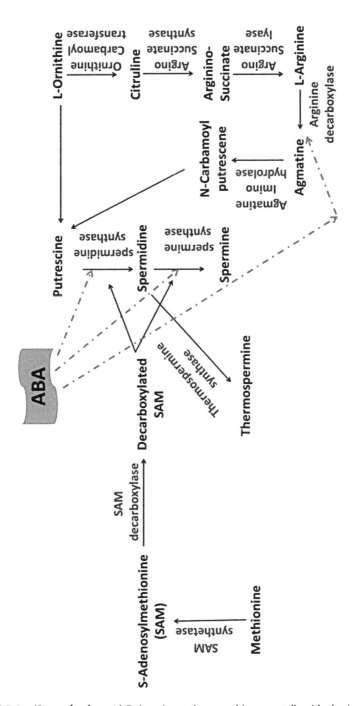

FIGURE 5.2 **(See color insert.)** Polyamine pathway and its cross talk with abscisic acid.

5.2.8   GENETIC AND GENOMIC APPROACHES FOR SALT STRESS TOLERANCE

Some plants are salt avoiders due to the adaptive morphological structures such as glands bladders or leaves as salt sinks.[198] However, the plants that possess such avoidance capabilities utilize conserved physiological processes to escape the unfavorable environment. There has been a limited success in developing salt-tolerant cultivars of different potential crops using conventional breeding approaches. A major problem conventional plant breeders have been facing is the low magnitude of genetic variability in the gene pools of crop species. Hence, salt-tolerant wild relatives of crop plants can be utilized as a source of genes for crop improvement. But, transferring salt-tolerant genes from wild relatives to the crop plant is not so easy because of reproductive barriers.[14]

Chinnusamy et al.[41] have described some of the reasons for limited success in improvement of crops for salt tolerance through conventional breeding trait. Advances in identification of genes associated with the stress (Fig. 5.3) have led to new insights into the biological mechanism and translation of the knowledge to other species. Genetic and genomic studies promise to complement the use of existing genetic diversity to accelerate plant improvement programs toward development of climate smart crops.[38]

Proteome analysis of salt stressed tomato plant suggested that improvement in salt tolerance might be achieved through the exogenous application of compatible soluble solutes like glycinebetain.[39] An adaptation to high salt may involve hydrolysis of some proteins,[149] modifications in gene expression, accumulation and depletion of certain metabolites carrying forward an imbalance of proteins at cellular level.[107] The total soluble protein contents of tomato cultivars were significantly decreased by salt stress depending upon time intervals.

Research with tobacco and sorghum plants established that glycophytes could adapt to high levels of salinity, provided that stress imposition occurs in moderate increments.[9] Genetic determinants for salt stress tolerance can be identified through quantitative trait locus (QTL) mapping, association mapping and screening by recurrent selection.[193] Detailed molecular characterization of QTLs and the associated genes that facilitate stress survival has enabled more expedient transfer of genes in breeding programs by using molecular markers associated with the molecular marker-based breeding enabled identification of loci responsible for salt tolerance and facilitated separation of physically linked loci that negatively influence the yield.[62,63]

Monogenic introgression of salt tolerance determinant directly into high-yielding crop varieties would simplify the breeding efforts to improve crop yield stability and salt concentrations.[54]

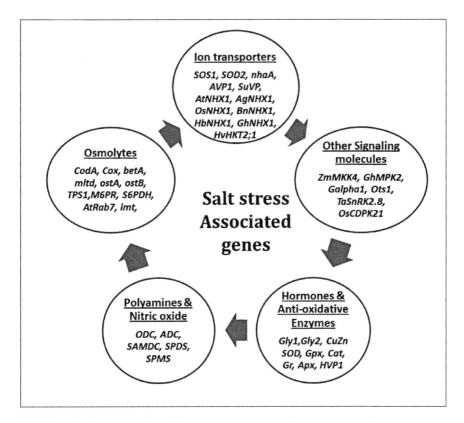

FIGURE 5.3   **(See color insert.)** Interaction networks of stress-associated genes in plant.

Kawasaki et al.[103] proposed that genes implicated in the recovery process are those that continue to have elevated expression after seven days of salt stress because plants at this stage have resumed faster growth and can be considered to be adapted. Many genes that are induced in *Arabidopsis* by salt stress were constitutively expressed including SOS1 and AtP5CS. Constitutive expression of AtP5CS (a key enzyme in proline biosynthesis) resulted in higher levels of this amino acid under salt cress, even in the absence of salt.[192] Use of molecular markers and bioinformatics tools have led to the identification of genes with major contributions to Na^+ and K^+ homeostasis.[203]

5.2.8.1   GENETIC MODELS AND SALT TOLERANCE DETERMINANTS

One of the goals of many researchers has been to identify the genetic determinants that can enhance yield stability of crops under saline conditions.[10,28,57] Salt tolerance determinants can be categorized into two functional groups: effecter and regulatory genes.[82]

The effectors include proteins that function by protecting cells from high cytoplasmic Na[+], such as transporters, enzymes required for biosynthesis of various compatible osmolytes, LEA proteins, chaperones, and detoxification enzymes.

Regulatory molecules include transcription factors and signaling molecules like protein kinases and enzymes involved in phosphoinositide metabolism.[233] Subtractive libraries from soybean and pigeon pea under multiple abiotic stresses led to identification of differentially expressed genes and were characterized for their functions.[110,159] Isolation and characterization of pigeon pea *CcHyPRP*, *CcCYP*, and *CcCDR* genes, and their overexpression in *Arabidopsis* had been reported to impart multiple abiotic stress tolerance in the plants.[159,181,194] *CYP* genes from maize, bean, *Solanum commersonii*, and *Solanum tuberosum* were also found to confer multiple stress tolerance during drought, salinity, and extreme temperatures.[73]

Genetic and physiological evidences support the likelihood that *HKT*1 mediates Na[+] influx across the plasma membrane.[117,175] Involvement of *AtHKT*1 in Na[+] homeostasis was established when dysfunctional alleles were determined to suppress Na[+]-hypersensitivity of SOS3 plants, implicating *AtHKT*1 as a Na[+] influx system in root.[173] Antisense wheat lines with reduced *TaHKT*1 activity exhibited improved salt tolerance and reduced Na[+] uptake into root.[117,121] The *HKT*1 mutations resulted in higher Na[+] content in shoot and lower Na[+] content in root.[131] SOS1 functions in the loading of Na[+] into the xylem for transport to the shoot where the ion is loaded into leaf cells.[185] However, under severe salt stress (100 mM NaCl), it is proposed that SOS1 functions to restrict Na[+] accumulation in shoot by limiting net Na[+] uptake at the root tip and loading into the xylem sap.[185] A model for *AtHKT*1 function indicates its involvement in the redistribution of Na[+] from the shoot to root, apparently by facilitating Na[+] loading into the phloem in the shoot and unloading in the root.[24]

5.2.8.2 ALLELIC VARIATION FOR SODIUM TRANSPORTERS

Studies on model plants identified *HKT1*, *SOS1*, and *NHX* (*Na⁺/H⁺ exchanger*) as crucial determinants of cellular Na^+ homeostasis, with HKT1 and SOS1 controlling net flux across the plasma membrane and NHX controlling movement across the tonoplast membrane into the vacuole.[48,172] HKT1;5 is a plasma membrane-localized Na^+ transporter that partitions Na^+ from root xylem vessels to adjacent parenchyma cells and perhaps into the cortex.[81,153] Commercial durum varieties (*Triticum turgidum* ssp. *durum*) lack *HKT1;5* and are salt sensitive because of high concentrations of leaf Na^+ and disturbed leaf K^+/Na^+ homeostasis.[121,142] Introgression of the *Nax2* locus (containing *TmHKT1;5-A*) from the wild wheat relative *Triticum monococcum* into durum wheat substantially reduced leaf Na^+ concentration owing to effective root xylem vessel exclusion, leading to enhanced grain yield by 25% on saline soils.[97] *Nax2* was used by ~30 wheat breeding programs world over for introgression into local high-yielding lines. Salt tolerance of modern hexaploid bread wheat (AABB genomes from *T. turgidum* and DD genome from *Aegilops tauschii*) is linked to the *Kna1* locus78 (harboring *TaHKT1;5D*79) contributed by *A. tauschii*. *TaHKT1;5D* has neofunctionalized from the *A. taushcii* gene *AtHKT1;5D* during hybridization to have NaCl-induced expression. The *TaHKT1;5D* allele benefit is associated with reduced leaf Na^+ concentration and higher fitness of hexaploid *Triticum aestivum* relative to *T. turgidum* under salt stress.[33]

Distinctions in salt tolerance in rice seedlings are connected with allelic variation at *OsHKT1;5*. Some *OsHKT1;5* alleles effectively maintain shoot K^+/Na^+ homeostasis owing to specific amino acid variations that enhance K^+ over Na^+ transport into root xylem sap, that alter protein transmembrane stability or that alter phosphorylation that affects function.[163] Systematic evaluation of *OsHKT1;5* allelic variation and tissue Na^+ and K^+ concentrations indicates that *OsHKT1;5* and additional as yet uncharacterized genetic determinants function together to alleviate damage from Na^+ in *O. sativa* and *Oryza glaberrima* seedlings.[17] K^+/Na^+ homeostasis and salt tolerance have also been linked to *HKT1* paralogues in tomato, further highlighting evolutionary variation in this transporter.

Although considerable progress was made during the 20th century to improve crop yield and quality through conventional breeding, yet there is not enough work on improving abiotic stress tolerance, especially salinity tolerance in crop plants. Conventional breeding approaches are time-consuming and labor-intensive; undesirable genes are often transferred in combination with the desirable ones; and reproductive barriers limit transfer of favorable alleles from interspecific and intergeneric sources. Due to these limitations, genetic engineering is being employed, as a supplementary strategy to conventional breeding, and for selective gene transfer across the reproductive barriers.

5.2.9 EPIGENETICS OF SALT STRESS TOLERANCE IN PLANT

Most crop plants are salt-sensitive and their sensitivity varies from one genotype to another genotype.[96] Some genotypes possess unique ability to adapt to a toxic level of salt stress, whereas others are highly sensitive, indicating their uniqueness in genetic makeup and regulatory architecture. Unfortunately, epigenetic response of contrasting genotypes under salinity stress has yet been underexplored.

Epigenetic processes involve chromatin remodeling by DNA methylation and histone modifications and play important roles in modulating gene expression in response to environmental stimuli. Epigenetic regulation of plant growth and development, particularly under environmental stresses, has created a newer basis of genetic variation for improving crop productivity and adaptation to environmental stresses. Plant epigenetics refers to heritable variation in gene expression resulting from covalent modifications of DNA and its associated chromatin proteins without changing the underlying nucleotide sequences.[59] Such epigenetic modification, which often reverts to its earlier state, may show transgenerational inheritance and can alter the phenotype.[23]

Methylated cytosine (5-methylcytosine, 5-mC), also known as the fifth base of DNA, was identified long before the DNA was recognized as the genetic material of living organism. Genetic information in DNA is concealed in nucleotide (A, T, C, G) sequence, but epigenetic mechanisms including noncoding RNA activities may alter the gene activity without making any alteration in the DNA sequence. One of the well-studied epigenetic mechanisms is DNA methylation, which refers to the addition of a methyl group to C_5 of cytosine residue. Cytosine methylation is a post replicative event by the action of DNA methyltransferase enzyme in plant. Deamination of C residue converts it to U, which either gets repaired by the action of Uracil DNA glycosylase enzyme in due course of DNA repair process or gets replaced by T during the DNA replication process.[113]

Cytosine methylation is a conserved epigenetic mechanism involved in many important biological processes, including movement of transposable elements, genome imprinting, and regulation of gene expression.[212] DNA methylation is usually associated with inactivation of genes, while demethylation results in gene activation.[221] Genome-wide high-resolution mapping and functional analysis of DNA methylation in rice revealed that 8% of the active genes were methylated within their promoter, while methylation of coding regions was observed in 31% of the expressed genes.[220]

Analysis of stress related genes and their regulation in response to environmental stresses have been commonly employed for enhancing plants productivity under the stress. However, growing evidences implicate epigenetic mechanisms in regulating gene expression under environmental stresses.[29] Demethylation of functionally inactive genes has also been reported due to abiotic stresses.[42] Abiotic stresses may cause heritable changes in cytosine methylation, forming novel epialleles.[108]

Mangrove plants—growing in contrasting natural habitats (riverside and salt marsh)—showed considerable variation with respect to cytosine methylation despite having little genetic variation.[124] In spite of some exceptions, it is widely accepted that DNA methylation of promoter sequence is conversely related to the gene expression level.[235] Genome-wide high-resolution analysis of DNA methylation in *Arabidopsis thaliana* revealed that methylation of coding region of gene may occasionally have a positive effect on gene expression.[225]

Effect of salinity stress on variation in cytosine methylation has been studied in crop plants.[114,211] Such epigenetic variations need to be explored so that useful variability can be identified and exploited in crop breeding programs to enhance crop adaptability to unfavorable environmental conditions.[115] Reduced level of DNA methylation often has a negative effect on the plant's ability to tolerate environmental stresses, reduced DNA methylation having decreased the ability of *Arabidopsis* plant to cope with salt stress.[19] In addition, methylation levels may also be used to distinguish salt-tolerant or salt sensitive varieties. A salt tolerant wheat variety showed higher level of methylation after 10 days of salt stress imposition compared to a salt-sensitive variety.[232]

Karan et al.[101] studied the extent and pattern of cytosine methylation in four contrasting rice genotypes under salinity stress using MSAP technique. They reported methylation and demethylation patterns to be genotype and tissue specific. Differential expression of stress-related genes influenced by methylation status under salinity stress revealed the possible role of epigenetic mechanisms in stress adaptation. Gene body methylation may have an important role in regulating gene expression in tissue and genotype specific manner. Association between salt tolerance and observed methylation changes in some cases suggest that many of the methylation changes are not directed. The natural genetic variation for salt tolerance observed in crop plants may be independent of the DNA methylation induced by abiotic stress followed by accumulation through the natural selection process.[101]

The studies by authors on changes in cytosine methylation in contrasting wheat genotypes on salt stress imposition revealed that basic methylation level in shoot of both the genotypes was higher (5.4–6.6%) compared to that in the root (3.4–4.0%), and salt stress caused further increase (by 10%) in methylation in the tolerant genotype. Quantitative analyses of 5-mC revealed that *HKT2;3* gene in tolerant genotype was hypermethylated (88%) in both root and shoot compared to its methylation status (74%) in the sensitive genotype. Salt stress caused increase in cytosine methylation even in susceptible genotypes (from 74% to 79%) and the increase was more prominent in root than that in the shoot. Most of the cytosine in CG context was methylated in both the genotypes, but a significant variation in methylation level was observed in the CHH context. Expression analysis of *HKT2;3* gene indicated that hypermethylation of the gene in shoot of salt-tolerant genotype down-regulated its expression.[114] Thus, understanding epigenetic regulation of gene expression in different tissues under varying environmental conditions may be helpful in modulating salt stress tolerance in crop plants.

Histone modification plays equally important role in epigenetic regulation of gene expression. Histone proteins are associated with nuclear DNA and their N-terminal residues undergo posttranslational modifications, specifically acetylation, methylation, phosphorylation, ubiquitination, glycosylation, carbonylation, ADP-ribosylation, sumoylation, and biotination. Such histone modifications cause alternations in the nucleosomal structure and affect gene expression activity. Modifications within the histones alter the ability of plants to tolerate adverse environmental conditions such as high salt, drought, and bacterial infections. Recently, DNA methylation analysis of plants grown under the stress showed that DNA methylation is associated with H3K9me2 enrichment and H3K9ac depletion in the histones of salt-stressed progenies.[26] The study also showed that DNA methylation and histone modification is associated with global gene repression and salt tolerance.

Small noncoding RNAs, such as microRNAs (miRNAs) and endogenous small interfering RNAs (siRNAs), are 20–28 nucleotides single-stranded RNAs that regulate expression of targeted (complementary) genes by affecting mRNA build-up, chromatin remodeling, DNA methylation, histone modification or by mediating translational repression[105] and epigenetic transgenerational memory of the progenies.[29] In addition, miRNA is able to direct

DNA methylation to a specific locus using the RNA-directed DNA methylation (RdDM) mechanism[132] by binding miRNA to the target genes. In rice, global gene expression analysis showed the role of miRNAs in controlling gene expression when plants were exposed to environmental stresses such as cold, drought, and high-salt.[183] MiRNAs isolated from *Arabidopsis* showed their critical role in salt tolerance in plants. For example, miR159 regulated expression of MYB101 and MYB33 transcription factors by controlling their cleavage,[164] and miR160 was found to be a potential ABA regulator.[124]

Genetic variability and phenotypic diversity are basic materials for the selection and breeding programs. Alterations in DNA methylation produce epialleles which may also result into phenotypic variations. Epiallele can be defined as any two or more genetically identical genes that are epigenetically distinct due to epigenetic modification(s). Epialleles may be utilized in plant breeding to improve plant's tolerance to stresses. However, DNA methylation within the genome can be stable, unstable, or stochastic.

Stable DNA methylation takes place when a consistent and heritable variation in DNA methylation occurs at specific loci due to an environmental factor for several generations and persists even in the absence of the triggering environmental factor. Unstable DNA methylation occurs when the variation disappears once the plants return to normal environmental conditions or when it is not transmittable to the next generation. Therefore, it is difficult to use unstable DNA methylation in crop improvement programs. Traits associated with methylated or unmethylated loci can be identified based on epigenetic mapping of the corresponding QTLs. Despite the importance of epigenetic maps in plant breeding, the use of such maps for improving plant traits is still uncommon. Moreover, molecular control of DNA methylation and the inheritance of epialleles are not currently manageable using the available knowledge and technologies.

Recent studies have identified several stress-regulated miRNAs, depicting their role as master regulators of gene-regulatory networks.[125,190] Transgenic approach has been recently applied to overexpress or knock-down specific miRNAs or their targets.[33] For instance, overexpression of miR319 yielded rice plants with enhanced cold, salt, and drought tolerance.[233] However, a more thorough understanding of the networks of miRNA gene regulation is essential to avoid off-targets and minimize pleiotropic effects. Several siRNAs accumulating under stress conditions have been recently identified in plants, and their role in stress signaling networks are starting to be unveiled.[190] Future research should focus on better understanding the effect of histone modifiers and miRNAs on DNA methylation and their inheritance.

5.3   TRANSGENIC APPROACHES FOR SALT STRESS TOLERANCE

Although considerable progress has been made during the 20th century to improve crop yield and quality through conventional breeding, yet limited studies have been conducted on improving abiotic stress tolerance, especially salinity tolerance in crop plants. Conventional breeding approaches are time-consuming and labor-intensive; undesirable genes are often transferred in combination with the desirable ones; and reproductive barriers limit transfer of favorable alleles from interspecific and intergeneric sources. Due to these issues, genetic engineering is being employed emphatically as a supplementary strategy to conventional breeding for selective gene transfer across the reproductive barriers. Genetic engineering and transformation technologies have enabled scientists to tailor make genes and have them expressed precisely in a predictable manner. Hence, genetic engineering can be a promising approach to manipulate osmoprotectant biosynthetic pathways for accumulating such molecules that act by scavenging ROS, reducing lipid peroxidation, maintaining protein structure, and functions.[2,13] For improving salt tolerance in crop plants through genetic engineering, biotechnologists have focused much on the gene that encodes ion transporter, compatible organic solute, antioxidant, heat-shock protein, late embryogenesis abundant protein, and transcription factor (Table 5.1).

5.3.1   TRANSGENIC PLANTS FOR ION TRANSPORTERS

Evidences suggest that Na$^+$ moves passively through a general cation channel from saline growth medium into the cytoplasm of plant cells,[16] but there is also evidence that Na$^+$ can move actively through Na$^+$/H$^+$ antiporters.[146,185] The regulation of Na$^+$ uptake and accumulation into plant cells relies on the regulation of proton pumps and antiporters operating at both plasma membrane and tonoplast.[13]

Engineering plants for overexpression of genes for different types of antiporters have recently gained interest for controlling the uptake of toxic ions and hence improving salt tolerance. Verma et al.[207] reported 81% improvement in shoot and root lengths in transgenic rice line expressing *PgNHX1* from *Pennisetum glaucum*, under salt stress. Tobacco transgenic lines for *GhNHX1*[215] and *AlNHX*[224] have shown 100 and 150% higher plant dry-weight compared to nontransgenic plant when tested at 300 and 400 mM NaCl, respectively. A remarkable improvement (215%) in shoot dry weight of transgenic alfalfa plant overexpressing vacuolar H$^+$-pyrophosphatase gene was observed when

TABLE 5.1 Some of the Successful Examples of Genetic Engineering in Plants for Improving Salt Stress Tolerance.

Transgene	Transgenic plant	Source of gene	Improvement	References
		Transgenic plant for osmoprotectants		
ADC	Solanum melongena	Avena sativa	Significant (3–5-fold) increase in Put, Spd, Spm, and salt tolerance	[157]
Choline dehydrogenase (betA)	Cabbage (Brassica oleracea L.)	Escherichia coli	Improvement osmotic potential of transgenic plants at 150 and 300 mM NaCl	[25]
L-Myo-inositol 1-phosphate synthase from P. coarctata (Roxb.) designated as PcINO1	Tobacco (N. tabacum L.)	Porteresia coarctata (Roxb.)	2–7-fold increase in the inositol contents and about 2-fold higher photosynthetic competence at 300 mM NaCl	[129]
Mt1D and GutD	Loblolly pine (Pinus taeda L.)	Agrobacterium tumefaciens	Synthesis and accumulation of mannitol and glucitol	[197]
NADP-dependent S6PDH	Diospyros kaki Thunb.	Malus domestica Borkh.	Higher sorbitol contents (14.5–61.5 μmol/g fresh wt.) at 50 mM NaCl in transgenic plants	[68]
ODC	Nicotiana tabacum	Mouse	2–3-fold increase in Put & Spd, and increased salt tolerance	[117]
SAMDC	Oryza sativa	Tritordeum	Significant (2–3-fold) increase in Put, Spd, Spm, and enhanced salt tolerance	[170]
SPMS	Lycopersicon esculentum	Malus sylvestris	Significant (1–5-fold) increase in Spd, Spm and improved salt tolerance	[145]
TPS1	Tomato (Lycopersicon esculentum L.)	Saccharomyces cerevisiae	Higher trehalose, chlorophyll and starch contents in transgenic plants	[44]
p5cs	Rice (Oryza sativa L.)	Vigna aconitifolia	120–315% more proline content in transgenic plants	[188]
P5CS	Potato (Solanum tuberosum L.)	A. thaliana L.	3–10 times increase in proline	[87]
P5CS	Wheat (T. aestivum L.)	V. aconitifolia	Higher level of P5CS protein and accumulated 2.5-fold higher proline	[179]

TABLE 5.1 *(Continued)*

Transgene	Transgenic plant	Source of gene	Improvement	References
TaP5CR	Arabidopsis thaliana L.	Triticum aestivum L.	2–4-fold higher proline at 150 mM NaCl stress	[128]
Transgenic plant for ion transporters				
GST and GPX	Tobacco (Nicotiana tabacum L.)	Nicotiana tabacum	Higher levels of glutathione and ascorbate	[169]
Gly I and II	Tobacco (Nicotiana tabacum L.)	Nicotiana tabacum	Reduced accumulation of methylglyoxal (MG)	[219]
Mn-SOD	Rice (Oryza sativa L.)	S. pombe	1.7-fold increase in SOD and 1.5-fold in APX	[195]
OPR1 cDNA designated as ZmOPR1	A. thaliana L.	Zea mays L.	Altered transcription of RD22, RD29B and ABF2	[76]
Plasma membrane Na^+/H^+ antiporter sodium2 (SOD2)	Rice (O. sativa L.)	S. pombe	Accumulated higher K^+ (29.4%), Ca^{2+} (22.2%), Mg^{2+} (53.8%) and lower Na^+ (26.6%) contents and decreased MDA level (57.10%)	[227]
Transgenic plant for enhanced production of antioxidants				
Vacuolar H^+-pyrophosphatase AVP1	Alfalfa (M. sativa L.)	A. thaliana L.	High Na^+, K^+, and Ca^{2+} in leaves and roots	[21]
Vacuolar Na^+/H^+ antiporter MsNHX1	Arabidopsis thaliana L.	Medicago sativa L.	Increased osmotic adjustment and MDA content	[22]
Vacuolar Na^+/H^+ antiporter AtNHX1	Tall fescue (Festuca arundinacea Schreb.)	A. thaliana L.	Higher root sodium contents and activity of vacuolar Na^+/H^+ antiporter	[229]
Vacuolar Na^+/H^+ antiporter AgNHX1	Rice (Oryza sativa L.)	Atriplex gmelini	Eight-fold higher activity of the vacuolar-type Na^+/H^+ antiporter	[148]
Vacuolar Na^+/H^+ antiporter PgNHX1	Rice (O. sativa L.)	Pennisetum glaucum (L.) R. Br.	Extensive root system	[207]

TABLE 5.1 *(Continued)*

Transgene	Transgenic plant	Source of gene	Improvement	References
Vacuolar Na⁺/H⁺ antiporter PgNHX1	Wheat (*Triticum aestivum* L.)	*A. thaliana* L.	Improved biomass production, grain yield, and leaf K⁺ accumulation, and reduced leaf Na⁺	[217]
Vacuolar Na⁺/H⁺ antiporter gene AtNHX1	Tall fescue (*F. arundinacea* Schreb.)	*A. thaliana* L.	Better growth at 200 mM NaCl	[199]
Vacuolar Na⁺/H⁺ antiporter AtNHX1	Maize (*Z. mays* L.)	*A. thaliana* L.	80 % germination of transgenics at 0.5% NaCl, compared to < 57% of nontransgenics	[217]
Vacuolar Na⁺/H⁺ antiporter GhNHX1	Tobacco (*N. tabacum* L.)	*Gossypium hirsutum* L.	Na⁺ vacuolar sequestration	[215]
Vacuolar Na⁺/H⁺ antiporter AlNHXI	Tobacco (*N. tabacum* L.)	*Aeluropus littoralis*	Compartmentalized more Na⁺ in the roots and kept a relative high K⁺/Na⁺ ratio in the leaves	[224]
Vacuolar Na⁺/H⁺ antiporter cDNA gene, HbNHX1	Tobacco (*N. tabacum* L.)	*Hordeum brevisubulatum*	Improved growth, leaf Na⁺ and K⁺ contents, Na⁺/K⁺ ratio and proline contents	[126]
Transgenic plant for transcription factor				
AtSKIP	*Arabidopsis thaliana*	*Arabidopsis thaliana*	Conferred salt and osmotic tolerance	[122]
BjDREB1 (dehydration responsive element binding protein) gene	Tobacco (*Nicotiana tabacum* L.)	*Brassica juncea* L.	2–3-fold higher proline at 150 mM NaCl stress	[43]
JcDREB	*Arabidopsis thaliana*	*Jatropha curcas*	Enhanced salt and freezing stress tolerance	[196]

Mt1D, Mannitol-1-Phosphate Dehydrogenase; GutD, Glucitol-6-Phosphate Dehydrogenase; S6PDH, Sorbitol-6-Phosphate Dehydrogenase; TPS1, Trehalose-6-Phosphate Synthase; p5cs, Δ1-Pyrroline-5-Carboxylate Synthetase; P5CS, Δ1-Pyrroline-5-Carboxylate Synthetase; P5CS, Δ1-Pyrroline-5-Carboxylate Synthetase; TaP5CR, Δ1-Pyrroline-5-Carboxylate Reductase; GST, Glutathione S-Transferase; GPX, Glutathione Peroxidase; Gly, Glyoxalase; SOD, Superoxide Dismutase; APX, Ascorbate POX.

tested under saline conditions.[21] Xue et al.[218] reported enhanced salt tolerance of transgenic wheat overexpressing vacuolar Na^+/H^+ antiporter (AtNHX1).

5.3.2   TRANSGENIC PLANTS FOR COMPATIBLE ORGANIC SOLUTES

The organic solutes, which play active role in salt tolerance in plant, include proline, trehalose, sucrose, polyols, and quaternary ammonium compounds such as glycine betaine, proline betaine, alanine betaine, hydroxyproline betaine, pipecolate betaine, and choline O-sulfate.[165] Among the various compatible organic solutes, proline, glycine betaine, and trehalose have been the most favored for transgenic experiments. Overproduction of Δ1-pyrroline-5-carboxylate synthetase (P5CS) gene in rice,[188] potato,[87] wheat,[179] and tobacco[88] resulted in accumulation of proline many fold higher than that in wild type plants and showed a marked increase in growth under saline conditions. Genes encoding key enzymes of glycine betaine biosynthesis (such as choline dehydrogenase, choline oxidase, and choline mono-oxygenase) have been engineered. Cabbage,[25] rice,[177] and *Brassica juncea*[158] plants transformed with glycine betaine encoding genes showed enhanced salt tolerance. Transgenic rice line expressing trehalose gene exhibited enhanced tolerance not only to salinity stress, but also to drought and cold stress.[69,98]

Different plants have also been transformed for overproduction of mannitol, myo-inositol, sorbitol, etc. Most of transgenic plants have exhibited improved salt tolerance. Transgenic manipulation of PA biosynthesis pathway genes (ADC and ODC) resulted in a significant increase (3–10-fold) in Put content with relatively smaller (2-fold) changes in Spd and Spm. Transgenic manipulation of SPDS, SPMS, and SAMDC genes caused 2–3-fold increase in Spd and Spm contents.[117,157,171] Unfortunately, most of these studies were carried out under controlled conditions and none of the lines showing extraordinary performance in terms of overproduction of a specific osmoprotectant have been tested under field conditions so far.

5.3.3   TRANSGENIC PLANTS FOR ENHANCED ANTIOXIDANT PRODUCTION

Although ROS are also generated in plants under optimal growth conditions yet their concentration is very low.[119] Most environmental adversaries trigger enhanced production of ROS. Plants have their innate systems for scavenging/detoxifying ROS, which are commonly referred to as antioxidants. Studies have revealed that improvement in salt tolerance of plant is possible through engineering the genes for enzymes which can effectively scavenge ROS.[65,154]. Ushimaru et al.[204] demonstrated that expression of rice DHAR in transgenic *Arabidopsis* enhanced the degree of salt tolerance.

By engineering with genes for different types of SOD in rice,[195] tobacco,[18] and Chinese cabbage,[200] a marked improvement in salt tolerance was achieved. Similarly, transgenic plants overexpressing glyoxalase pathway enzymes that do not allow accumulation of methylglyoxal have been developed in different plants such as tobacco,[187,219] Japonica and indica rice.[208] Genes for glutathione *S*-transferase[169] and CAT[7] were also engineered in tobacco. It is suggested that not only one antioxidant enzyme activity needs to be enhanced to counteract oxidative stress, but also more than one related enzymes may need to be enhanced for better performance of transgenic plant under salt stress.

Application of genetic engineering principles has emerged as synthetic biology approach for developing stress tolerant plants toward: rational designing and development of new biological modules from the naturally existing components, to facilitate de novo genetic switches and circuits, and manipulation of signaling processes.[213,214] This emerging field has already shown its potential applications in bacterial, yeast, and mammalian cell systems for the production of chemicals, metabolic engineering, and biomedical applications.[214]. Thus, synthetic biology approach for engineering plant for overexpression of genes encoding antiporters, antioxidant enzyme(s), and osmoprotectant(s) may provide effective method for generating salt-tolerant plant (Fig. 5.4). Gene expression studies using constitutive promoters provide limited benefits compared to the use of inducible promoters or tissue-specific promoters. The choice of promoters can significantly affect performance of the transgenic plants under salt stress. Thus, genetic manipulation approaches that must supplement engineering for salt-tolerant crops plant include[32]:

a) Applying plant synthetic biology approaches toward rational designing of genetic switches, circuits and signaling processes;

b) Fine-tuning of the stress response by modulating novel regulatory targets;

c) Maintaining hormone homeostasis to avoid pleiotropic effects under the stress;

d) Modulating the stress-associated miRNA expression; and

e) Post translational modifications that regulate plant's performance under the stress.

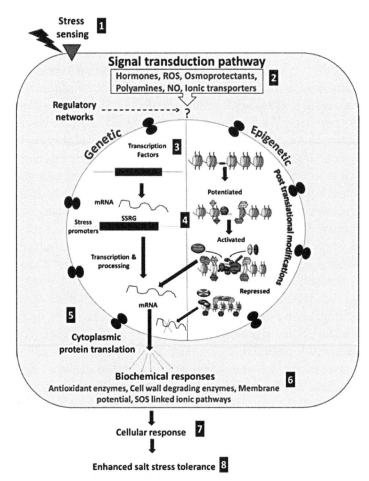

FIGURE 5.4 (See color insert.) Model depicting various possible biochemical and molecular responses in salt stress: (1) Stress sensing; (2) Signal transduction through various inducers; (3) Transcription factors; (4) Transcriptional activation of SSRG (salt stress related genes) through genetic and epigenetic interactions; (5) Synthesis and accumulation of stress proteins, resulting finally in (6) biochemical responses, (7) cellular responses, and (8) physiological response, that is, enhanced salt stress tolerance.

5.4 CHALLENGES AND FUTURE PROSPECTS

Reduced crop yield as a consequence of increasing soil salinization is one of the major threats to global food security. Salinity affected area is gradually spreading world over because of the anthropogenic interventions and increasing use of poor quality water for irrigation. Prolonged exposure to abiotic stresses especially salinity, results in altered metabolism and damage to the cellular biomolecules.

In general, salt stress causes a series of morphological, physiological, biochemical, and molecular changes that adversely affect plant's growth, development, and productivity. Substantial efforts have been made during the last century to develop salt tolerant lines and varieties of various crops using conventional plant breeding methods. Although conventional breeding has been effectively used in the fight against abiotic stresses especially salinity stress, it is time-consuming and labor-intensive. Moreover, undesirable genes are often transferred in combination with the desirable ones and reproductive barriers limit transfer of favorable alleles from interspecific and intergeneric sources.

Due to these reasons, genetic engineering, as a supplementary strategy to conventional breeding, is being employed emphatically for selective gene transfer across the reproductive barriers. Several siRNAs accumulating under stress conditions have been recently identified in plants and their role in stress signaling networks are starting to be unveiled. Epigenetic processes involve chromatin remodeling by DNA methylation and histone modifications and play important roles in modulating gene expression in response to environmental stimuli. However, a more thorough understanding of the networks of miRNA gene regulation is essential to avoid off-targets and minimize pleiotropic effects. Future research should focus on better understanding the effect of histone modifiers and miRNAs on DNA methylation and their inheritance.

Modern tools and techniques of genetic engineering are being increasingly employed currently yet these have met with limited success in achieving the desired goals. The main reasons for the limited success in producing salt tolerant genotypes through genetic engineering is that salt tolerance is a multigenic trait and several physiological, biochemical, and molecular processes are involved in the mechanism of salt tolerance.[233]

Many of transformation experiments so far have been carried out with single gene in model plants such as *Arabidopsis* and tobacco. No doubt, there have been amazing successes in terms of enhanced salt tolerance, yet to take full advantage of the advances made, we need to introduce multiple salt-tolerance genes into a crop plant of our interest. Since several physiological and biochemical processes interact with one another to resist salt stress at molecular, cellular or whole plant levels, genetic engineering must exploit multifarious metabolic pathways involving multiple genes. For example, in case of salt stress, if genes for ion homeostasis, osmotic adjustment, and antioxidant production are all incorporated into a single plant, they are likely to work in coordination to efficiently counteract the stress. Thus, genetic manipulation for many candidate genes for salt stress tolerance would be crucial to address complex traits like salt stress tolerance.

Plants growing on salt affected soils particularly in arid and semiarid regions often simultaneously encounter multiple abiotic stresses such as high temperature, high light intensity, water deficit stress, etc. Thus, it is imperative to elucidate how defense mechanisms are regulated during a combination of abiotic stresses. Epigenomic studies have been successful in identifying epigenetic variations under environmental stresses, but evidence also points toward the role of transposable elements in epigenetic regulation of gene expression.[60] Thus, an inclusive knowledge of all metabolic phenomena related directly or indirectly to the mechanisms of salt tolerance is vital for generating salt-tolerant plants through molecular breeding. However, further advances in genomics, proteomics and epigenomics research would make this multigenic trait engineering more effective and precise in future.

Though genetic engineering has resulted into relatively little success particularly in terms of enhanced stress tolerance in crop plants in spite of the use of huge resources in generating transgenic lines, yet it has generated lots of basic knowledge required for effective genetic manipulation of crop plants. Also, the performance of transgenic plants has so far only been assessed under controlled conditions (laboratory or glasshouse) that too using only single species of salts, mostly sodium chloride. A number of earlier studies have shown that the response of a plant to chloride or sulfate salinity varies to a great extent.[93,168]

None of the experimental material has been released yet as a field-tested salt-tolerant cultivar except the one or two.[217,218] This needs to be extended to natural stress environments, because the interplay of a number of edaphic and climatic factors and the interaction of multispecies of salts that may influence overall plant growth and development.[16] It is expected that future advances in transformation technology; RNAi mediated gene silencing, epigenomics, and comprehensive knowledge of signaling pathways would enable us to generate climate smart stress tolerant crop plants in future.[112,114] Engineering plants for overexpression of genes for different types of antiporters has recently gained interest for controlling the uptake of toxic ions and hence improving salt tolerance.

Genetic engineering and transformation technologies have enabled scientists now to tailor make genes and have them expressed precisely in a predictable manner. Hence, genetic engineering would be a promising approach to manipulate osmoprotectant biosynthetic pathways for accumulating such molecules that act by scavenging ROS, reducing lipid peroxidation, maintaining protein structure and functions. For improving salt tolerance in crop plants through genetic engineering, biotechnologists have focused much on the gene that encodes ion transporter, compatible organic solute, antioxidant, heat-shock protein, late embryogenesis abundant protein, and transcription factor.

Ultimate aim of manipulating genes for salt tolerance in important crops should be that the genes incorporated should contribute to tolerance not only at a particular stage of plant growth but throughout the plant growth and development. Another key challenging issue is that many of the stress-related metabolic phenomena are not well understood yet. Work to unfold these mechanisms should be initiated. The increased understanding of the complex defense mechanisms of salt stress can facilitate engineering of more stress tolerant plants for better yield and productivity, and to develop climate smart crop varieties.

5.5 CONCLUSION

Although the genes/loci that safeguard growth, development, and productivity of plant under environmental stresses do exist in the nature, yet they need to be identified and tagged for utilization in the breeding programs toward development of the stress tolerant crop varieties. Therefore, the subject of salt tolerance in plant's metabolism, productivity, and sustainability is gaining considerable significance in the present scenario.

Plants deploy multilevel signal transduction to induce stress responses. Interactions between PAs, ROS, hormones, antioxidants along with other signaling molecules like Ca^{2+}, cyclic nucleotides, and reactive nitrogen species such as NO form a complex signaling network, and induce diverse, apparently contradictory physiological effects. Engineering plants for over-expression of genes for different types of antiporters has recently gained interest for controlling the uptake of toxic ions and hence improving salt tolerance. Although considerable progress was made during the 20th century to improve crop yield and quality through conventional breeding, we find relatively little work on improving abiotic stress tolerance, especially salinity tolerance in crop plants.

Conventional breeding approaches are time-consuming and labor-intensive, undesirable genes are often transferred in combination with the desirable ones, and reproductive barriers limit transfer of favorable alleles from interspecific and intergeneric sources. Due to these reasons genetic engineering, as a supplementary strategy to conventional breeding, is being employed emphatically for selective gene transfer across the reproductive barriers. Several siRNAs accumulating under stress conditions have been recently identified in plants, and their role in stress signaling networks is starting to be unveiled. Role of epigenetics in regulation of stress-associated genes have begun to be understood.

5.6   SUMMARY

The chapter describes the role of a major abiotic stress, soil salinity, in adversely impacting crop productivity. It dwells upon the plant response to salinity stress highlighting that salt stress tolerance is a complex trait. Following a comprehensive detailed analysis of how plants respond to salinity stress at different levels, a case for an integrated approach of combining molecular tools with biochemical and physiological techniques for the development of salt tolerant crop varieties is build-up. It is argued that many of the stress-related metabolic phenomena are not well understood yet. Therefore, conventional breeding has met with limited success in the development of improved crop varieties for salt tolerance.

The authors argue that advances in identification of genes associated with the stress have led to new insights into the biological mechanism to cope up with the salt stress. Genetic and genomic studies, therefore, promise to complement the use of existing genetic diversity to accelerate plant improvement programs toward developing salt tolerant high yielding multistress tolerant plants. While dwelling upon transgenic approaches for salt stress tolerance, issues related to transgenic plants for ion transporters, transgenic plants for compatible organic solutes, and transgenic plants for enhanced antioxidant production are described. Put together, genetic engineering appears to be a promising approach to manipulate osmoprotectant biosynthetic pathways for accumulating molecules that act by scavenging ROS, reducing lipid peroxidation, and maintaining protein structure and functions.

Ultimate success of manipulating genes for salt tolerance in important crops will only emerge, if the genes incorporated contribute to tolerance not only at a particular stage of plant growth but throughout the plant growth and development. Toward the end of the chapter, while advocating genetic engineering for developing salt tolerance in plants, scope, challenges, and future prospects of genetic engineering are included. It is argued that genetic engineering must exploit multifarious metabolic pathways involving multiple genes. For example, if genes for ion homeostasis, osmotic adjustment, and antioxidant production are all incorporated into a single plant, they are likely to work in coordination to efficiently counteract the stress.

KEYWORDS

- **crop improvement**
- **DNA methylation**
- **epigenetics**
- **gene regulation**
- **high-affinity potassium transporter**
- **ion homeostasis**
- **reactive oxygen species**
- **salt tolerance**

REFERENCES

1. Abdel-Hamid, A. M. E. Physiological and Molecular Markers for Salt Tolerance in Four Barley Cultivars. *Eur. Sci. J.* **2014,** *10*, 252–272.
2. Abogadallah, G. M. Antioxidative Defense under Salt Stress. *Plant Signal Behav.* **2010,** *5*, 369–374.
3. Achard, P.; Cheng, H.; De Grauwe, L. Integration of Plant Responses to Environmentally Activated Phytohormonal Signals. *Science* **2006,** *311*, 91–94.
4. Agarwal, S.; Shaheen, R. Stimulation of Antioxidant System and Lipid Peroxidation by Abiotic Stresses in Leaves of *Momordica charantia. Braz. J. Plant Physiol.* **2007,** *19*, 149–161.
5. Alet, A. I.; Sanchez, D. H.; Cuevas, J. C.; Del, V. S.; Altabella, T.; Tiburcio, A. F. Putrescine Accumulation in *Arabidopsis thaliana* Transgenic Lines Enhances Tolerance to Dehydration and Freezing Stress. *Plant Signal Behav.* **2011,** *6*, 278–286.
6. Ali, B.; Hayat, S.; Ahmad, A. 28-homobrassinolide Ameliorates the Saline Stress in Chickpea (*Cicer arietinum* L). *Environ. Exp. Bot.* **2007,** *59*, 217–223.
7. Al-Taweel, K.; Iwaki, T.; Yabuta, Y.; Shigeoka, S.; Murata, N.; Wadano, A. A Bacterial Transgene for Catalase Protects Translation of D1 Protein During Exposure of Salt-stressed Tobacco Leaves to Strong Light. *Plant Physiol.* **2007,** *145*, 258–265.
8. Aly-Salama, K. H.; Al-Mutawa, M. M. Glutathione-triggered Mitigation in Salt-induced Alterations in Plasma Lemma of Onion Epidermal Cells. *Int. J. Agri. Biol.* **2009,** *5*, 639–642.
9. Amzallag, G. N.; Lemer, H. R.; Poljakoff-Mayber, A. Induction of Increased Salt Tolerance in Antiporter Gene Improves Salt Tolerance in *Arabidopsis thaliana. Nat. Biotechnol.* **1990,** *21*, 81–85.

10. Apse, M. P.; Blumwald, E. Engineering Salt Tolerance In Plants. *Curr. Opin. Biotechnol.* **2002,** *13,* 146–150.

11. Asada, K. The Water–Water Cycle in Chloroplasts: Scavenging of Active Oxygens and Dissipation of Excess Photons. *Ann. Rev. Plant Biol.* **1999,** *50,* 601–639.

12. Asensi-Fabado, M.; Munné-Bosch, S. The aba3-1 Mutant of *Arabidopsis thaliana* Withstands Moderate Doses of Salt Stress by Modulating Leaf Growth and Salicylic Acid Levels. *J. Plant Growth Regul.* **2011,** *30,* 456–466.

13. Ashraf, M. Biotechnological Approach of Improving Plant Salt Tolerance Using Antioxidants as Markers. *Biotechnol. Adv.* **2009,** *27,* 84–93.

14. Ashraf, M.; Akram, N. A. Improving Salinity Tolerance of Plants through Conventional Breeding and Genetic Engineering: An Analytical Comparison. *Biotechnol. Adv.* **2009,** *27,* 744–752.

15. Ashraf, M.; Akram, N.; Arteca, R. The Physiological, Biochemical and Molecular Roles of Brassino-steroids and Salicylic Acid in Plant Processes and Salt Tolerance. *Crit. Rev. Plant Sci.* **2010,** *3,* 162–90.

16. Ashraf, M.; Athar, H. R.; Harris, P. J. C.; Kwon, T. R. Some Prospective Strategies for Improving Crop Salt Tolerance. *Adv. Agron.* **2008,** *97,* 45–110.

17. Asins, M. J.; Villalta, I.; Aly, M. M.; De Morales, P. A.; Huertas, R.; Li, J.; Jaime-Pérez, N.; Haro, R.; Raga, V.; Carbonell, E. A.; Belver, A. Two Closely Linked Tomato HKT Coding Genes are Positional Candidates for the Major Tomato QTL Involved in Na^+/K^+ Homeostasis. *Plant Cell Environ.* **2013,** *36,* 1171–1191.

18. Badawi, G. H.; Yamauchi, Y.; Shimada, E.; Sasaki, R.; Kubo, A.; Tanaka, K. Enhanced Tolerance to Salt Stress and Water Deficit by Overexpressing Superoxide Dismutase in Tobacco (*Nicotiana tabacum*) Chloroplasts. *Plant Sci.* **2004,** *166,* 919–928.

19. Baek, D.; Jiang, J.; Chung, J. S.; Wang, B.; Chen, J.; Xin, Z.; Shi, H. Regulated *AtHKT1* Gene Expression by a Distal Enhancer Element and DNA Methylation in the Promoter Plays an Important Role in Salt Tolerance. *Plant Cell Physiol.* **2011,** *52,* 149–161.

20. Bajguz, A. Nitric oxide: role in plants under abiotic stress. In *Physiological Mechanisms and Adaptation Strategies in Plants under Changing Environment*; Ahmad, P., Wani, M. R., Eds.; Springer Science + Business Media: New York, 2014; pp 137–159.

21. Bao, A. K.; Wang, S. M.; Wu, G. O.; Xi, J. J.; Zhang, J. L.; Wang, C. M. Overexpression of the Arabidopsis H^+-PPase Enhanced Resistance to Salt and Drought Stress in Transgenic Alfalfa (*Medicago sativa* L.). *Plant Sci.* **2009,** *176,* 232–240.

22. Bao-Yan, A. N.; Yan, L.; Jia-Rui, L.; Wei-Hua, Q.; Xian-Sheng, Z.; Xin-Qi, Z. Expression of a Vacuolar Na^+/H^+ Antiporter Gene of Alfalfa Enhances Salinity Tolerance in Transgenic Arabidopsis. *Acta Agron. Sin.* **2008,** *34,* 557–564.

23. Bender, J. DNA Methylation and Epigenetics. *Annu. Rev. Plant Biol.* **2004,** *55,* 41–68.

24. Berthomieu, P.; Conejero, G.; Nublat, A. et al. Functional Analysis of AtHKT1 in *Arabidopsis* Shows that Na^+ Recirculation by the Phloem is Crucial for Salt Tolerance. *EMBO J.* **2003,** *22,* 2004–2014.

25. Bhattacharya, R. C.; Maheswari, M.; Dineshkumar, V.; Kirti, P. B.; Bhat, S. R.; Chopra, V. L. Transformation of *Brassica oleracea* var. Capitata with Bacterial *betA* Gene Enhances Tolerance to Salt Stress. *Sci. Hortic.* **2004,** *100,* 215–227.

26. Bilichak, A.; Ilnystkyy, Y.; Hollunder, J.; Kovalchuk, I. The Progeny of *Arabidopsis thaliana* Plants Exposed to Salt Exhibit Changes in DNA Methylation, Histone Modifications and Gene Expression. *PLoS One* [Online] **2012,** *7.* https://doi.org/10.1371/journal.pone.0030515 (accessed on May 18, 2017).

27. Blumwald, E. Engineering Salt Tolerance in Plants. *Biotechnol. Genet. Eng. Rev.* **2003,** *20*, 261–275.

28. Bohnert, H. J.; Nelson, D. E.; Jensen, R. G. Adaptations to Environmental Stresses. *Plant Cell* **1995,** *7*, 1099–1111.

29. Boyko, A.; Kovalchuk, I. Genome Instability and Epigenetic Modification-heritable Responses to Environmental Stress? *Curr. Opin. Plant Biol.* **2011,** *14,* 260–266.

30. Bray, E. A.; Bailey-Serres, J.; Weretilnyk, E. Responses to Abiotic Stresses. In *Biochemistry and Molecular Biology of Plants*; Buchanan, B., Gruissem, W., Jones, R., Eds.; American Society of Plant Physiologists: USA; 2000; pp 1158–1249.

31. Brini, F.; Masmoudi, K. Ion Transporters and Abiotic Stress Tolerance in Plants. *International Scholarly Research Network.* (*ISRN*) *Mol. Biol.* **2012,** p 13, ID 927436; www.hindawi.com/journals/isrn/2012/927436/.

32. Byrt, C. S.; Xu, B.; Krishnan, M.; Lightfoot, D. J.; Athman, A.; Jacobs, A. K.; Watson-Haigh, N. S.; Munns, R.; Tester, M.; Gilliham, M. The Na^+ Transporter, TaHKT1;5-D, Limits Shoot Na^+ Accumulation in Bread Wheat. *Plant J.* **2014,** *80,* 516–526.

33. Cabello, J. V.; Lodeyro, A. F.; Zurbriggen, M. D. Novel Perspectives for the Engineering of Abiotic Stress Tolerance in Plants. *Curr. Opin. Biotechnol.* **2014,** *26*, 62–70.

34. Cao, W. H.; Liu, J.; Zhou, Q. Y.; Cao, Y. R.; Zheng, S. F.; Du, B. X.; Zhang, J. S.; Chen, S. Y. Expression of Tobacco Ethylene Receptor NTHK1 Alters Plant Responses to Salt Stress. *Plant Cell Environ.* **2006,** *29*, 1210–1219.

35. Cao, W. H. Liu, J.; He, X. J.; Mu, R. L.; Zhou, H. L.; Chen, S. Y.; Zhang, J. S. Modulation of Ethylene Responses Affects Plant Salt-stress Responses. *Plant Physiol.* **2007,** *143*, 707–719.

36. Carpici, E. B.; Celika, N.; Bayram, .; Asik, B. B. The Effect of Salt Stress on the Growth, Biochemical Parameter and Mineral Element Content of Some Maize (*Zea mays* L.) Cultivars. *Afr. J. Biotechnol.* **2010,** *9*, 6937–6942.

37. Chakrabarti, N.; Mukherji, S. Alleviation of NaCl Stress by Pretreatment with Phytohormones in *Vigna radiata*. *Biol. Plant.* **2003,** *46*, 589–594.

38. Chandra, A.; Tiwari, K. K.; Nagaich, D.; Dubey, N.; Kumar, S.; Bhatt, R. K.; Roy, A. K. Development and Characterization of Microsatellite Markers from Tropical Forage *Stylosanthes* Species, Analysis of Genetic Variability and Cross-species Transferability. *Genome* **2011,** *54*, 1016–1028.

39. Chen, L.; Ren, F.; Zhou, L.; Wang, Q. Q.; Zhong, H.; Li, B. X. The *Brassica napus* Calcineurin B-like 1/CBL-interacting Protein Kinase 6 (CBL1/CIPK6) Components is Involved in the Plant Response to Abiotic Stress and ABA Signaling. *J. Exp. Bot.* **2012,** *63*, 6211–6222.

40. Chinnusamy, V.; Jagendorf, A.; Zhu, J.-K. Understanding and Improving Salt Tolerance in Plants. *Crop Sci.* **2005,** *45*, 437–448.

41. Chinnusamy, V.; Schumaker, K.; Zhu, J.-K. Molecular Genetic Perspectives on Cross Talk and Specificity in Abiotic Stress Signaling in Plants. *J. Exp. Bot.* **2004,** *55*, 225–236.

42. Choi, C. S.; Sano, H. Abiotic-stress Induces Demethylation and Transcriptional Activation of a Gene Encoding a Glycerophosphodiesterase-like Protein in Tobacco Plants. *Mol. Genet. Genomics* **2007,** *277*, 589–600.

43. Cong, L.; Chai, T. Y.; Zhang, Y. X. Characterization of the Novel Gene BjDREB1B Encoding a DRE-binding Transcription Factor from *Brassica juncea* L. *Biochem. Biophys. Res. Commun.* **2008,** *371*, 702–706.

44. Cortina, C.; Culianez-Macia, F. A. Tomato Abiotic Stress Enhanced Tolerance by Trehalose Biosynthesis. *Plant Sci.* **2005,** *169*, 75–82.

45. Crawford, N. M. Mechanisms for Nitric Oxide Synthesis in Plants. *J. Exp. Bot.* **2006,** *57*, 471–478.

46. Cuin, T. A.; Shabala, S. Potassium Transport and Plant Salt Tolerance. *Physiol. Plant.* **2008,** *133*, 651–669.

47. Debez, A.; Chaibi, W.; Bouzid, S. Effect of NaCl and Growth Regulators on the Germination of *Atriplex halimus* L. (*Efecto du NaCl et de regulatoeurs de croissance sur la germination d'Atriplex halimus* L.) *Can. Agric.* **2001,** *10*, 135–138.

48. Deinlein, U.; Stephan, A. B.; Horie, T.; Luo, W.; Xu, G.; Schroeder, J. I. Plant Salt Tolerance Mechanisms. *Trends Plant Sci.* **2014,** *19*, 371–379.

49. Deivanai, S.; Xavier, R.; Vinod, V.; Timalata, K.; Lim, O. F. Role of Exogenous Proline in Ameliorating Salt Stress at Early Stage in Two Rice Cultivars. *J. Stress Physiol. Biochem.* **2011,** *7*, 157–174.

50. Delledonne, M.; Xia, Y.; Dixon, R. A.; Lamb, C. Nitric Oxide Functions as a Signal in Plant Disease Resistance. *Nature* **1998,** *394*, 585–588.

51. Demiral, T.; Turkan, I. Comparative Lipid Peroxidation, Antioxidant Defence Systems and Proline Content in Roots of Two Rice Cultivars in Salt Tolerance. *Environ. Exp. Bot.* **2005,** *53*, 247–257.

52. Dietz, K. J.; Tavakoli, N.; Kluge, C. Significance of the Vtype ATPase for the Adaptation to Stressful Growth Conditions and Its Regulation on the Molecular and Biochemical Level. *J. Exp. Bot.* **2001,** *52*, 1969–1980.

53. Dkhil, B. B.; Denden, M. Salt Stress Induced Changes in Germination, Sugars, Starch and Enzyme of Carbohydrate Metabolism in *Abelmoschus esculentus* (L.) Moench Seeds. *Afr. J. Agric. Res.* **2010,** *5*, 408–415.

54. Doganlar, Z. B.; Demir, K.; Basak, H.; Gul, I. Effect of Salt Stress on Pigment and Total Soluble Protein Contents of Three Different Tomato Cultivars. *Afr. J. Agric. Res.* **2010,** *5*, 2056–2065.

55. El-Shintinawy, F.; El-Shourbagy, M. N. Alleviation of Changes in Protein Metabolism in NaCl-stressed Wheat Seedlings by Thiamine. *Biol. Plant.* **2001,** *44*, 541–545.

56. El-Tayeb, M. A. Response of Barley Grains to the Interactive Effect of Salinity and Salicylic Acid. *Plant Growth Regul.* **2005,** *45*, 215–224.

57. Epstein, E.; Norlyn, J. D.; Rush, D. W.; Kingsbury, R.; Kelly, D. B.; Wrana, A. F. Saline Culture of Crops: A Genetic Approach. *Science* **1980,** *210*, 399–404.

58. Fahad, S.; Hussain, S.; Matloob, A.; Khan, F. A.; Khaliq, A.; Saud, S. Phytohormones and Plant Responses to Salinity Stress: A Review. *Plant Growth Regul.* **2015,** *75*, 391–404.

59. Fang, B.; Yang, L. J. Evidence that the Auxin Signaling Pathway Interacts with Plant Stress Response. *Acta Bot. Sin.* **2002,** *44*, 532–536.

60. Fazzari, M. J.; Greally, J. M. Epigenomics: Beyond CpG Islands. *Nat. Rev. Genet.* **2004,** *5*, 446–455.

61. Flowers, T. J. Improving Crop Salt Tolerance. *J. Exp. Bot.* **2004,** *55*, 307–319.

62. Flowers, T. J.; Koyama, M. L., Flowers, S. A.; Sudhakar, C.; Singh, K. P.; Yeo, A. R. QTL: Their Place in Engineering Tolerance of Rice to Salinity. *J. Exp. Bot.* **2000,** *51*, 99–106.

63. Foolad, M. R.; Zhang, L. P.; Lin, G. Y. Identification and Validation of QTLs for Salt Tolerance During Vegetative Growth in Tomato by Selective Genotyping. *Genome* **2001,** *44*, 444–454.

64. Foyer, C. H.; Lopez-Delgado, H.; Dat, J. F.; Scott, I. M. Hydrogen Peroxide- and Glutathione-associated Mechanisms of Acclimatory Stress Tolerance and Signaling. *Physiol. Plant.* **1997,** *100*, 241–254.

65. Foyer, C. H.; Souriau, N.; Perret, S.; et al. Overexpression of Glutathione Reductase but not Glutathione Synthetase Leads to Increases in Antioxidant Capacity and Resistance in Photoinhibition in Poplar Trees. *Plant Physiol.* **1995,** *109,* 1047–57.

66. Fukuda, A.; Tanaka, Y. Effects of ABA, Auxin, and Gibberellin on the Expression of Genes for Vacuolar H^+-inorganic Pyrophosphatase, H^+-ATPase Subunit A and Na^+/H^+ Antiporter in Barley. *Plant Physiol. Biochem.* **2006,** *44,* 351–358.

67. Gadallah, M. A. A. Effects of Proline and Glycinebetaine on *Vicia faba* Responses to Salt Stress. *Biol. Plant.* **1999,** *42,* 249–257.

68. Gao, M.; Tao, R.; Miura, K. A.; Dandekar, M.; Sugiura, A. Transformation of Japanese Persimmon (*Diospyros kaki* Thunb.) with Apple cDNA Encoding NADP-dependent Sorbitol-6-phosphate Dehydrogenase. *Plant Sci.* **2001,** *160,* 837–845.

69. Garg, A. K.; Kim, J. K.; Owens, T. G.; et al. Trehalose Accumulation in Rice Plants Confers High Tolerance Levels to Different Abiotic Stresses. *Proc. Natl. Acad. Sci. U.S.A.* **2002,** *9,* 15898–15903.

70. Gechev, T. S.; Breusegem, F. V.; Stone, J. M.; Denev, I.; Laloi, C. Reactive Oxygen Species as Signals that Modulate Plant Stress Responses and Programmed Cell Death. *BioEssays* **2006,** *28,* 1091–1101.

71. Ghanem, M. E.; Albacete, A.; Martínez-Andújar, C.; Acosta, M.; Romero-Aranda, R.; Dodd, I. C.; Lutts, S.; Pérez-Alfocea, F. Hormonal Changes During Salinity-induced Leaf Senescence in Tomato (*Solanum lycopersicum* L.). *J. Exp. Bot.* **2008,** *59,* 3039–3050.

72. Gill, S. S.; Tuteja, N. Reactive Oxygen Species and Antioxidant Machinery in Abiotic Stress Tolerance in Crop Plants. *Plant Physiol. Biochem.* **2010,** *48,* 909–930.

73. Godoy, A. V.; Lazzaro, A. S.; Casalongue, C. A.; San, S. B. Expression of a *Solanum tuberosum* Cyclophilin Gene is Regulated by Fungal Infection and Abiotic Stress Conditions. *Plant Sci.* **2000,** *152,* 123–134.

74. Gomez, C. A.; Arbona, V.; Jacas, J.; PrimoMillo, E.; Talon, M. Abscisic Acid Reduces Leaf Abscission and Increases Salt Tolerance in Citrus Plants. *J Plant Growth Regul.* **2002,** *21,* 234–240.

75. Groß, F.; Durner, J.; Gaupels, F. Nitric Oxide, Antioxidants and Prooxidants in Plant Defence Responses. *Front. Plant Sci.* **2013,** *4,* 419. http://journal.frontiersin.org/article/10.3389/fpls.2013.00419/full (accessed on May 18, 2017).

76. Gu, D.; Liu, X.; Wang, M.; et al. Overexpression of ZmOPR1 in Arabidopsis Enhanced the Tolerance to Osmotic and Salt Stress During Seed Germination. *Plant Sci.* **2008,** *174,* 124–30.

77. Gunes, A.; Inal, A.; Alpaslan, M.; Eraslan, F.; Bagci, E. G.; Cicek, N. Salicylic Acid Induced Changes on Some Physiological Parameters Symptomatic for Oxidative Stress and Mineral Nutrition in Maize (*Zea mays* L.) Grown under Salinity. *J. Plant Physiol.* **2007,** *164,* 728–736.

78. Gupta, B.; Huang, B. Mechanism of Salinity Tolerance in Plants: Physiological, Biochemical, and Molecular Characterization. *Int. J. Genomics* **2014,** *2014,* 1–18.

79. Gupta, K. J.; Stoimenova, M.; Kaiser, W. M. In Higher Plants, Only Root Mitochondria, but not Leaf Mitochondria Reduce Nitrite to NO, In Vitro and In Situ. *J. Exp. Bot.* **2005,** *56,* 2601–2609.

80. Hanson, A. D.; Rathinasabapathi, B.; Rivoal, J.; Burnet, M.; Dillon, M. O.; Gage, D. A. Osmoprotective compounds in the Plumbaginaceae: A Natural Experiment in Metabolic Engineering of Stress Tolerance. *Proc. Natl. Acad. Sci. U.S.A.* **1994,** *91,* 306–310.

81. Hasegawa, P. M. Sodium (Na$^+$) Homeostasis and Salt Tolerance of Plants. *Environ. Exp. Bot.* **2013,** *92,* 19–31.

82. Hasegawa, P. M.; Bressan, R. A.; Zhu, J.-K.; Bohnert, H. J. Plant Cellular and Molecular Responses to High Salinity. *Ann. Rev. Plant Physiol. Plant Mol. Biol.* **2000,** *51,* 463–499.

83. Hayat, Q.; Hayat, S.; Irfan, M.; Ahmad, A. Effect of Exogenous Salicylic Acid under Changing Environment: A Review. *Environ. Exp. Bot.* **2010,** *68,* 14–25.

84. Hayat, S.; Hasan, S. A.; Yusuf, M.; Hayat, Q.; Ahmad, A. Effect of 28-homobrassinolide on Photosynthesis, Fluorescence and Antioxidant System in the Presence or Absence of Salinity and Temperature in *Vigna radiata. Environ. Exp. Bot.* **2010, 69,** 105–112.

85. He, Y.; Zhu, Z. Exogenous Salicylic Acid Alleviates NaCl Toxicity and Increases Antioxidative Enzyme Activity in *Lycopersicon esculentum. Biol. Plant.* **2008,** *52,* 792–795.

86. Hernandez, J. A.; Almansa, M. S. Short Term Effects of Salt Stress on Antioxidant Systems and Leaf Water Relations of Pea Leaves. *Physiol. Plant.* **2002,** *115,* 251–257.

87. Hmida-Sayari, A.; Gargouri-Bouzid, R.; Bidani, A.; Jaoua, L.; Savoure, A.; Jaoua, S. Overexpression of Δ1-pyrroline-5-carboxylate Synthetase Increases Proline Production and Confers Salt Tolerance in Transgenic Potato Plants. *Plant Sci.* **2005,** *169,* 746–752.

88. Hong, Z.; Lakkineni, K.; Zhang, Z.; Pal, D.; Verma, S. Removal of Feedback Inhibition of Δ1-pyrroline-5-Carboxylate Synthetase Results in Increased Proline Accumulation and Protection of Plants from Osmotic Stress. *Plant Physiol.* **2000,** *122,* 1129–1136.

89. Hoque, M. A.; Banu, M. N. A.; Nakamura, Y.; Shimoishi, Y.; Murata, Y. Proline and glycinebetaine Enhance Antioxidant Defense and methylglyoxal detoxification Systems and Reduce NaCl-induced Damage in Cultured Tobacco Cells. *J. Plant Physiol.* **2008,** *165,* 813–824.

90. Horie, T.; Hauser, F.; Schroede, J. I. HKT transporter-mediated Salinity Resistance Mechanisms in *Arabidopsis* and Monocot Crop Plants. *Trends Plant Sci.* **2009,** *14,* 660–668.

91. Hossain, K. K.; Itoh, R. D.; Yoshimura, G. Effects of Nitric Oxide Scavengers on Thermoinhibition of Seed Germination in *Arabidopsis thaliana. Russian J. Plant Physiol.* **2010,** *57,* 222–232.

92. Hussain, S. S.; Ali, M.; Ahmad, M.; Siddique, K. H. M. Polyamines: Natural and Engineered Abiotic and Biotic Stress Tolerance in Plants. *Biotechnol. Adv.* **2011,** *29,* 300–311.

93. Inal, A. Growth, Proline Accumulation and Ionic Relations of Tomato (*Lycopersicon esculentum* L.) as Influenced by NaCl and Na$_2$SO$_4$ Salinity. *Turk. J. Bot.* **2002,** *26,* 285–290.

94. Iqbal, M.; Ashraf, M. Salt Tolerance and Regulation of Gas Exchange and Hormonal Homeostasis by Auxin-priming in Wheat. *Pesq. Agropec. Bras. Brasília* **2013,** *48,* 1210–1219.

95. Iqbal, M.; Ashraf, M.; Jamil, A. Seed Enhancement with Cytokinins: Changes in Growth and Grain Yield in Salt Stressed Wheat Plants. *Plant Growth Regul.* **2006,** *50,* 29–39.

96. Ismail, A. M.; Heuer, S.; Thomson, M. J.; Wissuwa, M. Genetic and Genomic Approaches to Develop Rice Germplasm for Problem Soils. *Plant Mol. Biol.* **2007, 65,** 547–570.

97. James, R. A.; Zwart, A. B.; Hare, R. A.; Rathjen, A. J.; Munns, R. Impact of Ancestral Wheat Sodium Exclusion Genes *Nax1 and Nax2* on Grain Yield of Durum Wheat on Saline Soils. *Funct. Plant Biol.* **2012,** *39,* 609–618.

98. Jang, I. C.; Oh, S. J.; Seo, J. S.; et al. Expression of a Bifunctional Fusion of the *Escherichia coli* Genes for Trehalose-6-phosphate Synthase and Trehalose-6-phosphate Phosphatase in Transgenic Rice Plants Increases Trehalose Accumulation and Abiotic Stress Tolerance Without Stunting Growth. *Plant Physiol.* **2003**, *131*, 516–524.

99. Javid, M. G.; Sorooshzadeh, A.; Moradi, F.; Sanavy, S. A. M. M.; Allahdadi, I. The Role of Phytohormones in Alleviating Salt Stress in Crop Plants. *Aust. J. Crop Sci.* **2011**, *5*, 726–734.

100. Jung, J. H.; Park, C. M. Auxin Modulation of Salt Stress Signaling in Arabidopsis Seed Germination. *Plant Signal Behav.* **2011**, *6*, 1198–1200.

101. Karan, R.; DeLeon, T.; Biradar, H.; Subudhi, P. K. Salt Stress Induced Variation in DNA Methylation Pattern and Its Influence on Gene Expression in Contrasting Rice Genotypes. *PLoS One* **2012**, *7*. http://journals.plos.org/plosone/article?id=10.1371/journal.pone.0040203 (accessed on May 16, 2017).

102. Kartal, G.; Temel, A.; Arican, E.; Gozukirmizi, N.; Effects of Brassinosteroids on Barley Root Growth, Antioxidant System and Cell Division. *Plant Growth Regul.* **2009**, *58*, 261–267.

103. Kawasaki, S.; Borchert, C.; Deyholos, M.; Wang, H.; Brazille, S.; Kawai, K.; Galbraith, D.; Bohnert, H. J. Gene Expression Profiles During the Initial Phase of Salt Stress in Rice. *Plant Cell* **2001**, *13*, 889–905.

104. Keskin, B. C.; Sarikaya, A. T.; Yuksel, B.; Memon. A. R. Abscisic Acid Regulated Gene Expression in Bread Wheat (*Triticum aestivum* L.). *Aust. J. Crop Sci.* **2010**, *4*, 617–625.

105. Khraiwesh, B.; Zhu, J.; Zhu, J. Role of miRNAs and siRNAs in Biotic and Abiotic Stress Responses of Plants. *Biochim. Biophys. Acta* **2012**, *1819*, 137–148.

106. Khripach, V.; Zhabinskii, V. N.; deGroot, A. E. Twenty years of Brassino-steroids: Steroidal Plant Hormones Warrant Better Crops for the XXI Century. *Ann. Bot.* **2000**, *86*, 441–447.

107. Kong-Ngern, K.; Daduang, S.; Womggkham, C.; Bunnag, S.; Kosittrakun, M.; Theerakulpisu, P. Protein Profiles in Response to Salt Stress in Leaf Sheaths of Rice Seedlings. *Sci. Asia* **2005**, *31*, 403–408.

108. Kou, H. P.; Li, Y.; Song, X. X.; et al. Heritable Alteration in DNA Methylation Induced by Nitrogen-deficiency Stress Accompanies Enhanced Tolerance by Progenies to the Stress in Rice (*Oryza sativa* L.). *J. Plant Physiol.* **2011**, *168*, 1685–1693.

109. Kramell, R.; Miersch, O.; Atzorn, R.; Parthier, B.; Wasternack, C. Octadecanoid-derived Alteration of Gene Expression and the Oxylipin Signature in Stressed Barley Leaves. Implications for Different Signaling Pathways. *Plant Physiol.* **2000**, *123*, 177–188.

110. Kudapa, H.; Bharti, A. K.; Cannon, S. B.; et al. A Comprehensive Transcriptome Assembly of Pigeonpea (*Cajanus cajan* L.) Using Sanger and Second-generation Sequencing Platforms. *Mol. Plant* **2012**, *5*, 1020–1028.

111. Kumar, S. The Role of Biopesticides in Sustainably Feeding the Nine Billion Global Populations. *J. Biofertilizer Biopesticides* **2013**, *4*, 1–3.

112. Kumar, S. RNAi (RNA interference) Vectors for Functional Genomics Study in Plants. *Natl. Acad. Sci. Lett.* **2014**, *37*, 289–294.

113. Kumar, S.; Bhat, V. Application of Omics Technologies in Forage Crop Improvement. In *Omics Application in Crop Science*; Debmalya, B., Ed.; CRC Press, Taylor & Francis Group: USA, 2014; pp 523–547.

114. Kumar, S.; Beena, A. S.; Awana, M.; Singh, A. Salt-induced Tissue-specific Cytosine Methylation Downregulates Expression of *HKT* Genes in Contrasting Wheat (*Triticum aestivum* L.) Genotypes. *DNA Cell Biol.* **2017**, *36*, 283–294.

115. Kumar, S.; Singh, A. Epigenetic Regulation of Abiotic Stress Tolerance in Plants. *Adv. Plants Agric. Res.* **2016**, *5*, 1–6.

116. Kumar, M.; Hasan, M.; Arora, A.; Gaikwad, K.; Kumar, S.; Rai, R. D.; Singh, A. Sodium Chloride-induced Spatial and Temporal Manifestation in Membrane Stability Index and Protein Profiles of Contrasting Wheat (*Triticum aestivum* L.) Genotypes under Salt Stress. *Indian J. Plant Physiol.* **2015**, *20*, 271–275.

117. Kumria, R.; Rajam, M. V. Ornithine Decarboxylase Transgene in tobacco Affects Polyamines, *In Vitro*-morphogenesis and Response to Salt Stress. *J. Plant Physiol.* **2002**, *159*, 983–990.

118. Kusano, T.; Yamaguchi, K.; Berberich, T.; Takahashi, Y. Advances in Polyamine Research. *J. Plant Research* **2007**, *120*, 345–350.

119. Laloi, C.; Apel, K.; Danon, A. Reactive Oxygen Signaling: The Latest News. *Curr. Opin. Plant Biol.* **2004**, *7*, 323–328.

120. Lamattina, L.; García-Mata, C.; Graziano, M.; Pagnussat, G. Nitric Oxide: The Versatility of an Extensive Signal Molecule. *Annu. Rev. Plant Biol.* **2003**, *54*, 109–136.

121. Laurie, S.; Feeney, K. A.; Maathuis, F. J.; Heard, P. J.; Brown, S. J.; Leigh, R. A. A Role for HKT1 in Sodium Uptake by Wheat Roots. *Plant J.* **2002**, *32*, 139–149.

122. Lim, G.-H.; Zhang, X.; Chung, M.-S.; et al. A Putative Novel Transcription Factor, AtSKIP, Is Involved in Abscisic Acid Signaling and Confers Salt and osmotic Tolerance in *Arabidopsis*. *New Phytol.* **2010**, *185*, 103–113.

123. Liopa-Tsakalidi, A.; Kaspiris, G.; Salahas, G.; Barouchas, P. Effect of Salicylic Acid (SA) and Gibberellic Acid (GA3) Pre-soaking on Seed Germination of Stevia (*Stevia rebaudiana*) under Salt Stress. *J. Med. Plants Res.* **2012**, *6*, 416–423.

124. Lira-Medeiros, C. F.; Parisod, C.; Fernandes, R. A.; Mata, C. S.; Cardoso, M. A. Epigenetic Variation in Mangrove Plants Occurring in Contrasting Natural Environment. *PLoS One* **2010**, *5*. http://journals.plos.org/plosone/article?id=10.1371/journal.pone.0010326 (accessed on May 16, 2017).

125. Liu, P. P.; Montgomery, T. A.; Fahlgren, N.; Kasschau, K. D.; Nonogaki, H.; Carrington, J. C. Repression of Auxin Response Factor10 by MicroRNA160 is Critical for Seed Germination and Post-germination Stages. *Plant J.* **2007**, *52*, 133–146.

126. Lu, S. Y.; Jing, Y. X.; Shen, S. H.; et al. Antiporter Gene from *Hordum brevisubulatum* (Trin.) Link and Its Overexpression in Transgenic Tobaccos. *J. Integr. Plant Biol.* **2005**, *47*, 343–349.

127. Lyzenga, W. J.; Stone, S. L. Abiotic Stress Tolerance Mediated by Protein Ubiquitination. *J. Exp. Bot.* **2012**, *63*, 599–616.

128. Ma, L.; Zhou, E.; Gao, L.; Mao, X.; Zhou, R.; Jia, J. Isolation, Expression Analysis and Chromosomal Location of P5CR Gene in Common Wheat (*Triticum aestivum* L.). *S. Afr. J. Bot.* **2008**, *74*, 705–712.

129. Majee, M.; Maitra, S.; Dastidar, G. K.; et al. A Novel Salt Tolerant L-myo-inositol-1-phosphate Synthase from *Porteresia coarctata* (Roxb.) Tateoka, A Halophytic Wild Rice. *J. Biol. Chem.* **2004**, *279*, 28539–28552.

130. Makela, P.; Karkkainen, J.; Somersalo, S. Effect of Glycinebetaine on Chloroplast Ultrastructure, Chlorophyll and Protein Content, and RuBPCO Activities in Tomato Grown under Drought or Salinity. *Biol. Plant.* **2000**, *43*, 471–475.

131. Maser, P.; Eckelman, B.; Vaidyanathan, R.; et al. Altered Shoot/Root Na^+ Distribution and Bifurcating Salt Sensitivity in Arabidopsis by Genetic Disruption of the Na^+ Transporter AtHKT1. *FEBS Lett.* **2002**, *53*, 157–161.

132. Matzke, M.; Kanno, T.; Huettel, B.; Daxinger, L.; Matzke, A. J. Targets of RNA-directed DNA Methylation. *Curr. Opin. Plant Biol.* **2007,** *10,* 512–519.

133. Minocha, R.; Majumdar, R.; Minocha, S. C. Polyamines and Abiotic Stress in Plants: A Complex Relationship. *Front. Plant Sci.* **2014,** *5.* https://doi.org/10.3389/fpls.2014.00175 (accessed on May 18, 2017).

134. Mishra, S.; Jha, A. B.; Dubey, R. S. Arsenite Treatment Induces Oxidative Stress, Upregulates Antioxidant System, and Causes Phytochelatin Synthesis in Rice Seedlings. *Protoplasma* **2011,** *248,* 565–577.

135. Mondal, A. K.; ObiReddy, G. P; Ravisankar, T. Digital Database of Salt Affected Soils in India Using Geographic Information System. *J. Soil Salinity Water Qual.* **2011,** *3,* 16–29.

136. Moons, A.; Prinsen, E.; Bauw, G.; Montagu, M. V. Antagonistic Effects of Abscisic Acid and Jasmonates on Salt Stress-inducible Transcripts in Rice Roots. *Plant Cell* **1997,** *9,* 2243–2259.

137. Munir, N.; Aftab, F. Enhancement of Salt Tolerance in Sugarcane by Ascorbic Acid Pretreatment. *Afr. J. Biotechnol.* **2011,** *10,* 18362–18370.

138. Munns, R. Comparative Physiology of Salt and Water Stress. *Plant Cell Environ.* **2002,** *25,* 239–250.

139. Munns, R.; Tester. M. Mechanisms of Salinity Tolerance. *Ann. Rev. Plant Biol.* **2008,** *59,* 651–681.

140. Munns, R.; Husain, S.; Rivelli, A. R; et al. Avenues for Increasing Salt Tolerance of Crops, and the Role of Physiologically-based Selection Traits. *Plant Soil* **2002,** *247,* 93–105.

141. Munns, R.; James, R. A.; Lauchli, A. Approaches to Increasing the Salt Tolerance of Wheat and Other Cereals. *J. Exp. Bot.* **2006,** *57,* 1025–1043.

142. Munns, R.; James, R. A.; Xu, B.; et al. Wheat Grain Yield on Saline Soils is Improved by an Ancestral Na^+ Transporter Gene. *Nat. Biotechnol.* **2012,** *30,* 360–364.

143. Nalousi, A. M.; Ahmadiyan, S.; Hatamzadeh, A.; Ghasemnezhad, M. Protective Role of Exogenous Nitric Oxide Against Oxidative Stress Induced by Salt Stress in Bell-pepper (*Capsicum annum* L.). *American-Eurasian J. Agric. Environ. Sci.* **2012,** *12,* 1085–1090.

144. Nazar, R.; Iqbal, N.; Syeed, S.; Khan, N. A. Salicylic Acid Alleviates Decreases in Photosynthesis under Salt Stress by Enhancing Nitrogen and Sulfur Assimilation and Antioxidant Metabolism Differentially in Two Mungbean Cultivars. *J. Plant Physiol.* **2011,** *168,* 807–815.

145. Neily, M. H.; Baldet, P.; Arfaoui, I.; Saito, T.; Li, Q. L.; Asamizu, E. Overexpression of Apple Spermidine Synthase 1 (*MdSPDS1*) Leads to Significant Salt Tolerance in Tomato Plants. *Plant Biotechnol. J.* **2011,** *28,* 33–42.

146. Niu, X.; Bressan, R. A.; Hasegawa, P. M. Halophytes Up-regulate Plasma Membrane Hþ-ATPase Gene More Rapidly than Glycophytes in Response to Salt Stress. *Plant Physiol.* **1993,** *102,* 130–133.

147. Nunez, M.; Mazzafera, P.; Mazorra, L. M.; Siqueira, W. J.; Zullo, M. A. T. Influence of a Brassinosteroid Analog on Antioxidant Enzymes in Rice Grown in Culture Medium with NaCl. *Biol. Plant.* **2003,** *47,* 67–70.

148. Ohta, M.; Hayashi, Y.; Nakashima, A.; et al. Introduction of a Na^+/H^+ Antiporter Gene from the Halophyte *Atriplex gmelini* Confers Salt Tolerance to Rice. *FEBS Lett.* **2002,** *532,* 279–282.

149. Parida, A. K. Das, A. B.; Mohanty, P. Investigations on the Antioxidative Defence Responses to NaCl Stress in a Mangrove, *Bruguiera parviflora*: Differential Regulations of Isoforms of Some Antioxidative Enzymes. *Plant Growth Regul.* **2004,** *42,* 213–226.

150. Parida, A. K.; Das, A. B. Salt Tolerance and Salinity Effects on Plants: A Review. *Ecotoxicol. Environ. Saf.* **2005**, *60*, 324–349.

151. Pedranzani, H.; Racagni, G.; Alemano, S.; et al. Salt Tolerant Tomato Plants Show Increased Levels of Jasmonic Acid. *Plant Growth Regul.* **2003**, *41*, 149–158.

152. Phang, T.; Shao, G.; Lam, H. Salt Tolerance in Soybean. *J. Integr. Plant Biol.* **2008**, *50*, 1196–1212.

153. Platts, B. E.; Grismer, M. E. Chloride Levels Increase after 13 Years of Recycled Water Use in the Salinas Valley. *Calif. Agric.* **2014**, *68*, 68–74.

154. Polidoros, A. N.; Scandalios, J. G. Role of Hydrogen Peroxide and Different Classes of Antioxidants in the Regulation of Catalase and Glutathione *S*-transferase Gene Expression in Maize (*Zea mays* L.). *Physiol. Plant* **1999**, *106*, 112–120.

155. Pottosin, I.; Shabala, S. Polyamines Control of Cation Transport Across Plant Membranes: Implications for Ion Homeostasis and Abiotic Stress Signaling. *Front. Plant Sci.* **2014**, *5*, 154–158.

156. Pottosin, I.; Velarde-Buendía, A. M.; Bose, J.; Zepeda-Jazo, I.; Shabala, S.; Dobrovinskaya, O. Spermidine Treatment to Rice Seedlings Recovers Salinity Stress-induced Damage of Plasma Membrane and PM-bound H^+-ATPase in Salt-tolerant and Salt-sensitive Rice Cultivars. *Plant Sci.* **2014**, *168*, 583–591.

157. Prabhavathi, V. R.; Rajam, M. V. Polyamine Accumulation in Transgenic Eggplant Enhances Tolerance to Multiple Abiotic Stresses and Fungal Resistance. *Plant Biotechnol.* **2007**, *24*, 273–282.

158. Prasad, K. V. S. K.; Sharmila, P.; Kumar, P. A.; Saradhi, P. P. Transformation of *Brassica juncea* (L.) Czern with Bacterial codA Gene Enhances its Tolerance to Salt Stress. *Mol. Breed.* **2000**, *6*, 489–499.

159. Priyanka, B.; Sekhar, K.; Sunita, T.; Reddy, V. D.; Rao, K. V. Characterization of Expressed Sequence Tags (ESTs) of Pigeonpea (*Cajanus cajan* L.) and Functional Validation of Selected Genes for Abiotic Stress Tolerance in *Arabidopsis thaliana*. *Mol. Genet. Genomics* **2010**, *283*, 273–287.

160. Qi, Z.; Spalding, E. P. Protection of Plasma Membrane K^+ Transport by the Salt Overly Sensitive Na^+/H^+ Antiporter During Salinity Stress. *Plant Physiol.* **2004**, *136*, 2548–2555.

161. Quinet, M.; Ndayiragije, A.; Lefevre, I.; Lambillotte, B.; Gillain, C. C. D.; Lutts, S. Putrescine Differently Influences the Effect of Salt Stress on Polyamine Metabolism and Ethylene Synthesis in Rice Cultivars Differing in Salt Resistance. *J. Exp. Bot.* **2010**, *61*, 2719–2733.

162. Rahman, S.; Miyake, H.; Takeoka, Y. Effects of Exogenous Glycinebetaine on Growth and Ultrastructure of Salt-stressed Rice Seedlings (*Oryza sativa* L.) *Plant Prod. Sci.* **2002**, *5*, 33–44.

163. Ren, Z. H.; Gao, J. P.; Li, L. G.; et al. A Rice Quantitative Trait Locus for Salt Tolerance Encodes a Sodium Transporter. *Nat. Genet.* **2005**, *37*, 1141–1146.

164. Reyes, J. L.; Chua, N. H. ABA Induction of miR159 Controls Transcript Levels of Two MYB Factors During Arabidopsis Seed Germination. *Plant J.* **2007**, *49*, 592–606.

165. Rhodes, D.; Hanson, A. D. Quaternary ammonium and Tertiary Sulfonium Compounds in Higher-plants. *Annu. Rev. Plant Physiol. Plant Mol. Biol.* **1993**, *44*, 357–84.

166. Rodriguez-Navarro, A. Potassium Transport in Fungi and Plants. *Biochim. Biophys. Acta* **2000**, *146*, 1–30.

167. Rodriguez-Navarro, A.; Rubio, F. High-affinity Potassium and Sodium Transport Systems in Plants. *J. Exp. Bot.* **2006**, *57*, 1149–1160.

168. Rogers, M. E.; Grieve, C. M.; Shannon, M. C. The Response of Lucerne (*Medicago sativa* L.) to Sodium Sulphate and Chloride Salinity. *Plant Soil* **1998**, *202*, 271–280.

169. Roxas, V. P.; Lodhi, S. A.; Garrett, D. K.; Mahan, J. R.; Allen, R. D. Stress Tolerance in Transgenic Tobacco Seedlings that Overexpress Glutathione *S*-transferase/Glutathione Peroxidase. *Plant Cell Physiol.* **2000**, *41*, 1229–1234.

170. Roy, M.; Wu, R. Overexpression of *S*-adenosyl Methionine Decarboxylase Gene in Rice Increases Polyamine Level and Enhances Sodium Chloride-stress Tolerance. *Plant Sci.* **2002**, *163*, 987–992.

171. Roy, P.; Niyogi, K.; Sengupta, D. N.; Ghosh, B. Spermidine Treatment To Rice Seedlings Recovers Salinity Stress-induced Damage of Plasma Membrane and PM-bound H^+-ATPase in Salt-tolerant and Salt-sensitive Rice Cultivars. *Plant Sci.* **2005**, *168*, 583–591.

172. Roy, S. J.; Negra, S.; Tester, M. Salt Resistant Crop Plants. *Curr. Opinion Plant Biology* **2014**, *26*, 115–124.

173. Roychoudhury, A.; Basu, S.; Sengupta, D. N. Amelioration of Salinity Stress by Exogenously Applied Spermidine or Spermine in Three Varieties of Indica Rice Differing in Their Level of Salt Tolerance. *J. Plant Physiol.* **2011**, *168*, 317–328.

174. Rubio, L.; Rosado, A.; Linares-Rueda, A.; et al. Regulation of K^+ Transport in Tomato Roots by the TSS1 Locus. Implications in Salt Tolerance. *Plant Physiol.* **2004**, *134*, 452–459.

175. Rus, A.; Yokoi, S.; Sharkhuu, A.; et al. AtHKT1 is a salt Tolerance Determinant that Controls Na^+ Entry into Plant Roots. *Proc. Natl. Acad. Sci. U.S.A.* **2001**, *98*, 14150–14155.

176. Sahi, C.; Singh, A.; Kumar, K.; Blumwald, E.; Grover, A. Salt Stress Response in Rice: Genetics, Molecular Biology, and Comparative Genomics. *Funct Integ Genomics* **2005**, *6*, 263–284.

177. Sakamoto, A.; Murata, A.; Murata, N. Metabolic Engineering of Rice Leading to Biosynthesis of Glycinebetaine and Tolerance to Salt and Cold. *Plant Mol. Biol.* **1998**, *38*, 1011–1019.

178. Sanders, D.; Pelloux, J.; Brownlee, C.; Harper, J. F. Calcium at the Crossroads of Signaling. *Plant Cell* **2002**, *14*, 401–417.

179. Sawahel, W. A.; Hassan, A. H. Generation of transgenic Wheat Plants Producing High Levels of the Osmoprotectant Proline. *Biotechnol. Lett.* **2002**, *24*, 721–725.

180. Saxena, S. C.; Kaur, H.; Verma, P.; et al. Osmo-protectants: Potential for Crop Improvement under Adverse Conditions. In *Plant Acclimation to Environmental Stress*; Springer: New York, NY, USA, 2013; pp 197–232.

181. Sekhar, K.; Priyanka, B.; Reddy, V. D.; Rao, K. V. Isolation and Characterization of a Pigeon Pea Cyclophilin (*CcCYP*) Gene, and Its Over-expression in *Arabidopsis* Confers Multiple Abiotic Stress Tolerance. *Plant Cell Environ.* **2010**, *33*, 655–669.

182. Serrano, R. Salt Tolerance in Plants and Microorganisms: Toxicity Targets and Defense Responses. *Int. Rev. Cytology* **1996**, *165*, 1–52.

183. Shen, J.; Xie, K.; Xiong, L.; Global Expression Profiling of Rice MicroRNAs by One-tube Stem-loop Reverse Transcription Quantitative PCR Revealed Important Roles of MicroRNAs in Abiotic Stress Responses. *Mol. Genet. Genomics* **2010**, *284*, 477–488.

184. Shi, H.; Chan, Z. Improvement of Plant Abiotic Stress Tolerance through Modulation of the Polyamine Pathway. *J. Integr. Plant Biol.* **2014**, *56*, 114–121.

185. Shi, H.; Wu, S. J.; Zhu, J.-K. Overexpression of a Plasma Membrane Na^+/H^+ Antiporter Improves Salt Tolerance in Arabidopsis. *Nat. Biotechnol.* **2003**, *21*, 81–85.

186. Singh, A.; Bhushan, B.; Gaikwad, K.; Yadav, O. P.; Kumar, S.; Rai, R. D.; Induced Defence Response of Contrasting Bread Wheat Genotype under Differential Salt Stress Imposition. *Ind. J. Biochem. Biophysics* **2015**, *52*, 78–85.

187. Singla-Pareek, S. L.; Reddy, M. K.; Sopory, S. K. Genetic Engineering of the Glyoxalase Pathway in Tobacco Leads to Enhanced Salinity Tolerance. *Proc. Natl. Acad. Sci. U.S.A.* **2003**, *100*, 14672–14677.

188. Su, J.; Wu, R. Stress-inducible Synthesis of Proline in transgenic Rice Confers Faster Growth under Stress Conditions than that with Constitutive Synthesis. *Plant Sci.* **2004**, *166*, 941–948.

189. Sultana, N.; Ikeda, T.; Itoh, R. Effect of NaCl Salinity on Photosynthesis and dry Matter Accumulation in Developing Rice Grains. *Environ. Exp. Bot.* **2000**, *42*, 211–220.

190. Sunkar, R.; Li, Y.; Jagadeeswaran, G. Functions of MicroRNAs in Plant Stress Responses. *Trends Plant Sci.* **2012**, *17*, 196–203.

191. Szepesi, Á.; Csiszár, J.; Gémes, K.; Horváth, E.; Horváth, F.; Simon, M. L. Salicylic Acid Improves Acclimation to Salt Stress by Stimulating Abscisic Aldehyde Oxidase Activity and Abscisic Acid Accumulation, and Increases Na^+ Content in leaves without Toxicity Symptoms in *Solanum lycopersicum* L. *J. Plant Physiol.* **2009**, *166*, 914–925.

192. Taji, T.; Seki, M.; Satou, M.; et al. Comparative Genomics in Salt Tolerance between *Arabidopsis* and *Arabidopsis*—Relative Halophyte Salt Cress Using *Arabidopsis* microarray. *Plant Physiol.* **2004**, *135*, 1697–1709.

193. Takeda, S.; Matsuoka, M. Genetic Approaches to crop Improvement: Responding to Environmental and Population Changes. *Nat. Rev. Genet.* **2008**, *9*, 444–457.

194. Tamirisa, S.; Vudem, D. R.; Khareedu, V. R. Overexpression of Pigeon Pea Stress-induced Cold and Drought Regulatory Gene (*CcCDR*) Confers Drought, Salt, and Cold Tolerance in *Arabidopsis*. *J. Exp. Biol.* **2014**, *65*, 4769–4781.

195. Tanaka, Y.; Hibino, T.; Hayashi, Y.; et al. Salt Tolerance of Transgenic Rice Overexpressing Yeast Mitochondrial Mn–SOD in Chloroplasts. *Plant Sci.* **1999**, *148*, 131–138.

196. Tang, M.; Liu, X.; Deng, H.; Shen, S. Over-expression of JcDREB, A Putative AP2/EREBP Domain-containing Transcription Factor Gene in Woody Biodiesel Plant *Jatropha curcas*, Enhances Salt and Freezing Tolerance in Transgenic *Arabidopsis thaliana*. *Plant Sci.* **2011**, *181*, 623–631.

197. Tang, W.; Peng, X.; Newton, R. J. Enhanced Tolerance to salt Stress in Transgenic Loblolly Pine Simultaneously Expressing Two Genes Encoding Mannitol-1-phosphate Dehydrogenase and Glucitol-6-phosphate Dehydrogenase. *Plant Physiol. Biochem.* **2005**, *43*, 139–146.

198. Tester, M.; Davenport, R. Na^+ Tolerance and Na^+ Transport in Higher Plants. *Ann. Bot., London* **2003**, *91*, 503–527.

199. Tian, L.; Huang, C.; Yu, R.; et al. Overexpression AtNHX1 Confers Salt Tolerance of Transgenic Tall Fescue. *Afr. J. Biotechnol.* **2006**, *5*, 1041–1044.

200. Tseng, M. J.; Liu, C. W.; Yiu, J. C. Enhanced Tolerance to Sulfur Dioxide and Salt Stress of Transgenic Chinese Cabbage Plants Expressing Both Superoxide Dismutase and Catalase in Chloroplasts. *Plant Physiol. Biochem.* **2007**, *745*, 822–833.

201. Tsonev, T. D.; Lazova, G. N.; Stoinova, Z. G.; Popova, L. P. A Possible Role for Jasmonic Acid in Adaptation of Barley Seedlings to Salinity Stress. *J. Plant Growth Regul.* **1998**, *17*, 153–159.

202. Turan, M. A.; Elkarim, A. H. A.; Taban, N.; Taban, S. Effect of Salt Stress on Growth and Ion Distribution and Accumulation in Shoot and Root of Maize Plant. *Afr. J. Agric. Res.* **2010,** *5,* 584–588.

203. Tyagi, A.; Singh, A.; Rai, R. D. Abiotic stress responses in crop plants. In *Crop Biology and Agriculture in Harsh Environments*; LAP Publishing: Germany, 2013; pp 81–130.

204. Ushimaru, T.; Nakagawaa, T.; Fujiokaa, Y.; et al. Transgenic Arabidopsis Plants Expressing the rice Dehydroascorbate Reductase Gene are Resistant to Salt Stress. *J. Plant Physiol.* **2006,** *163,* 1179–1184.

205. Valifard, M.; Mohsenzadeh, S.; Kholdebarin, B.; Rowshan, V. Effects of Salt Stress on Volatile Compounds, Total Phenolic Content and Antioxidant Activities of *Salvia mirzayanii*. *S. Afr. J. Bot.* **2014,** *93,* 92–97.

206. Velarde-Buendía, A. M.; Shabala, S.; Cvikrova, M.; Dobrovinskaya, O.; Pottosin, I. Salt-sensitive and Salt-tolerant Barley Varieties Differ in the Extent of Potentiation of the ROS-induced K^+ Efflux by Polyamines. *Plant Physiol. Biochem.* **2012,** *61,* 18–23.

207. Verma, D.; Singla-Pareek, S. L.; Rajagopal, D.; Reddy, M. K.; Sopory, S. K. Functional Validation of a Novel Isoform of Na^+/H^+ Antiporter from *Pennisetum glaucum* for Enhancing Salinity Tolerance in Rice. *J. Biosci.* **2007,** *32,* 621–628.

208. Verma, M.; Verma, D.; Jain, R. K.; Sopory, S. K.; Wu, R. Overexpression of Glyoxalase I Gene Confers Salinity Tolerance in Transgenic Japonica and Indica Rice Plants. *Rice Genet. Newslett.* **2005,** *22,* 58–62.

209. Wang, B.; Luttge, U. L.; Ratajczak, R. Effects of Salt Treatment and Osmotic Stress on V-ATPase and V-PPase in Leaves of the Halophyte Suaeda Salsa. *J. Exp. Bot.* **2001,** *52,* 2355–2365.

210. Wang, H.; Liang, X.; Huang, J.; Lu, H.; Liu, Z.; Bi, Y. Involvement of Ethylene and Hydrogen Peroxide in Induction of Alternative Respiratory Pathway in Salt-treated *Arabidopsis* Calluses. *Plant Cell Physiol.* **2010,** *51,* 1754–1765.

211. Wang, W. S.; Zhao, X.; Pan, Y.; et al. DNA Methylation Changes Detected by Methylation-sensitive Amplified Polymorphism in Two Contrasting Rice Genotypes under Salt Stress. *J. Genet. Genomics* **2011,** *38,* 419–424.

212. Wang, X.; Li, Q.; Yuan, W.; Kumar, S.; Li, Y.; Qian, W. The Cytosolic Fe–S Cluster Assembly Component MET18 is Required for the Full Enzymatic Activity of ROS1 in Active DNA Demethylation. *Sci. Rep.* **2016,** *6,* 1–15.

213. Weber, W.; Fussenegger, M. Molecular Diversity—The Toolbox for Synthetic Gene Switches and Networks. *Curr. Opin. Chem. Biol.* **2011,** *15,* 414–420.

214. Weber, W.; Fussenegger, M. Emerging Biomedical Applications of Synthetic Biology. *Nat. Rev. Genet.* **2012,** *13,* 21–35.

215. Wu, C. A.; Yang, G. D.; Meng, Q. W.; Zheng, C. C. The Cotton GhNHX1 Gene Encoding a Novel Putative Tonoplast Na^+/H^+ Antiporter Plays an Important Role in Salt Stress. *Plant Cell Physiol.* **2004,** *45,* 600–607.

216. Wu, X.; He, J.; Chen, J.; Yang, S.; Zha, D. Alleviation of Exogenous 6-benzyladenine on Two Genotypes of Egg-plant (*Solanum melongena* Mill.) Growth under Salt Stress. *Protoplasma* **2013,** *251,* 169–176.

217. Xiao-Yan, Y.; Ai-Fang, Y.; Ke-Wei, Z.; Ju-Ren, Z. Production and Analysis of Transgenic Maize with Improved Salt Tolerance by the Introduction of AtNHX1 Gene. *Acta Bot. Sin.* **2004,** *46,* 854–861.

218. Xue, Z. Y.; Zhi, D. Y.; Xue, G. P.; Zhao, Y. X. Xia, G. M. Enhanced Salt Tolerance of Transgenic Wheat (*Triticum aestivum* L.) Expressing a Vacuolar Na^+/H^+ Antiporter

Gene with Improved Grain Yield in Saline Soils in the Field and a Reduced Level of Leaf Na$^+$. *Plant Sci.* **2004**, *167*, 849–59.

219. Yadav, S. K.; Singla-Pareek, S. L.; Reddy, M. K.; Sopory, S. K. Transgenic Tobacco Plants Overexpressing Glyoxalase Enzymes Resist an Increase in Methylglyoxal and Maintain Higher Reduced Glutathione Levels under Salinity Stress. *FEBS Lett.* **2005**, *579*, 6265–71.

220. Yan, H. H.; Kikuchi, S.; Neumann, P. Genome Wide Mapping of Cytosine Methylation Revealed Dynamic DNA Methylation Patterns Associated with Genes and Centromeres in Rice. *Plant J.* **2010**, *63*, 353–365.

221. Zemach, A.; McDaniel, I. E.; Silva, P.; Zilberman, D. Genome-wide Evolutionary Analysis of Eukaryotic DNA Methylation. *Science* **2010**, *328*, 916–919.

222. Zepeda-Jazo, I.; Shabala, S.; Chen, Z.; Pottosin, I. I. Na-K Transport in Roots under Salt Stress. *Plant Signal Behav.* **2008**, *3*, 401–403.

223. Zhang, F.; Wang, Y.; Yang, Y.; Wu, H.; Wang, D.; Liu, J. Involvement of Hydrogen Peroxide and Nitric Oxide in Salt Resistance in the Calluses from *Populus euphratica*. *Plant Cell Environ.* **2007**, *30*, 775–785.

224. Zhang, G. H.; Su, Q.; An, L. J.; Wu, S. Characterization and Expression of a Vacuolar Na$^+$/H$^+$ Antiporter Gene from the Monocot Halophyte *Aeluropus littoralis*. *Plant Physiol. Biochem.* **2008**, *46*, 117–126.

225. Zhang, X.; Yazaki, J.; Sundaresan, A.; et al. Genome-wide High-resolution Mapping and Functional Analysis of DNA Methylation in Arabidopsis. *Cell* **2006**, *126*, 1189–1201.

226. Zhang, Y.; Wang, L.; Liu, Y.; Zhang, Q.; Wei, Q.; Zhang, W. Nitric Oxide Enhances Salt Tolerance in Maize Seedlings through Increasing Activities of Proton-pump and Na$^+$/H$^+$ Antiport in the Tonoplast. *Planta* **2006**, *224*, 545–555.

227. Zhao, F.; Guo, S.; Zhang, H.; Zhao, Y. Expression of Yeast SOD2 in Transgenic Rice Results in Increased Salt Tolerance. *Plant Sci.* **2006**, *170*, 216–224.

228. Zhao, J.; Zhi, D.; Xue, Z.; Liu, Z.; Xia, G. Enhanced Salt Tolerance of Transgenic Progeny of Tall Fescue (*Festuca arundinacea*) Expressing a Vacuolar Na$^+$/H$^+$ Antiporter Gene from Arabidopsis. *J. Plant Physiol.* **2007**, *164*, 1377–1383.

229. Zhao, L.; Zhang, F.; Guo, J.; Yang, Y.; Li, B.; Zhang, L. Nitric Oxide Functions as a Signal in Salt Resistance in the Calluses from Two Ecotypes of Reed. *Plant Physiol.* **2004**, *134*, 849–857.

230. Zhao, M. G.; Chen, L.; Zhang, L. L.; Zhang, W. H. Nitric Reductase-dependent Nitric Oxide Production is Involved in Cold Acclimation and Freezing Tolerance in *Arabidopsis*. *Plant Physiol.* **2009**, *151*, 755–767.

231. Zhao, X. C.; Schaller, G. E. Effect of Salt and Osmotic Stress upon Expression of the Ethylene Receptor ETR1 in *Arabidopsis thaliana*. *FEBS Lett.* **2004**, *562*, 189–192.

232. Zhong, L.; Wang, J. B. The Role of DNA Hypermethylation in Salt Resistance of *Triticum aestivum* L. *Wuhan Zhiwuxue Yanjiu.* **2007**, *25*, 102–104.

233. Zhu, J.-K. Salt and Drought Stress Signal Transduction in Plants. *Annu. Rev. Plant Physiol. Plant Mol. Biol.* **2002**, *53*, 247–273.

234. Zhu, J.-K. Regulation of Ion Homeostasis under Salt Stress. *Curr. Opinion Plant Biology* **2003**, *6*, 441–445.

235. Zilberman, D.; Gehring, M.; Tran, R. K.; Ballinger, T.; Henikoff, S. Genome-wide Analysis of Arabidopsis Thaliana DNA Methylation Uncovers an Interdependence between Methylation and Transcription. *Nat. Genet.* **2007**, *39*, 61–69.

CHAPTER 6

GENOMICS TECHNOLOGIES FOR IMPROVING SALT TOLERANCE IN WHEAT

AMIT KUMAR SINGH, RAKESH SINGH, RAJESH KUMAR, SHIKSHA CHAURASIA, SHEEL YADAV, SUNDEEP KUMAR, and DHARMMAPRAKASH P. WANKHEDE

ABSTRACT

Salinity tolerance in plants is a complex trait. Besides being controlled by many genes, it is also influenced by environmental conditions. Large and complex genome in wheat also poses many difficulties in gene discovery/ genetic mapping of salinity tolerance associated traits. On the other hand, recent advances in genomics and availability of high-throughput genome-wide analysis tools can greatly assist in deciphering molecular basis of salt tolerance in wheat. Nowadays, "omics" approaches such as transcriptomics, proteomics, metabolomics ionomics, and phenomics are increasingly being applied to get a finer level understanding of mechanisms underlying salt tolerance in the salt-tolerant accessions of different crops including wheat. Association mapping, which exploits diverse set of germplasm lines have allowed fine high-resolution mapping to identify loci controlling abiotic stress tolerance associated traits in cereals. Further, advanced breeding approaches such as marker-assisted selection and transgenic are being exploited for fast tracking the development of crop varieties. The, genomic selection is another promising approach for breeding salt-tolerant varieties. In this chapter, we have discussed and reviewed these approaches in grater details, with the focus on their potential in either underpinning molecular basis of salt tolerance or developing salt-tolerant wheat varieties.

6.1   INTRODUCTION

Wheat is a staple food crop supplying about 20% of the total calorie intake for people worldwide. Ensuring food security to the ever-growing population is a major challenge for the agriculture researchers and policy makers around the world. In order to meet the food demand in 2050, the global food grain production must increase by 70% over the 2006 level.[43] A substantial part of this increase would have to come from wheat as it occupies the largest area under cultivated crops. However, in the past few years, there has not been a considerable increase in the global wheat production primarily due to losses because of increasing incidences of biotic and abiotic stresses. Salinity stress is one of the major factors limiting wheat productivity in major wheat producing regions of the world; especially so in the irrigated arid and semiarid regions.[117]

High levels of salt in root zone adversely affect plant growth and development in wheat causing severe yield reductions. Wheat productivity in the salt-affected soils (SAS) can be enhanced by two ways:

- Soil reclamation using amendments and
- Technology intervention in the form of salt-tolerant high yielding varieties.

Since soil reclamation using chemical amendments is practically difficult for the resource poor farming communities in the developing countries, cultivation of salt-tolerant genotypes giving high yields under salt stress conditions is increasingly being considered an effective approach to augment the productivity of SAS. The approach is particularly relevant to the needs of resource poor farmers with dismal adaptive capacity which restricts the choice of conventionally used reclamation practices often involving higher initial as well as recurring costs ascribed to, for example, repeated use of amendments such as gypsum.[143] Consequently, wheat breeders face a tough task in developing salt tolerant and high yielding wheat varieties capable of producing acceptable yields under saline and alkali conditions with limited or no use of amendments.

Salt stress response in wheat is a highly complex trait determined by various factors such as genotype, level of the stress (i.e., concentration of soluble salts and duration of exposure) and the developmental stage of plant. For example, wheat plants adopt different physiological and biochemical mechanisms at seedling and maturity stages to safeguard plant metabolisms and growth from the anticipated salt injury. Molecular studies in various crops have shown that salinity tolerance is a highly complex trait governed by multiple genes,[27,63] the genes associated with salinity tolerance being transcription factors, microRNAs, Na^+ transporters and those involved in osmolyte biosynthesis.[33] Some of these genes have also been shown to be associated with salinity tolerance in wheat. However, a comprehensive knowledge of molecular processes associated with salt tolerance in various wheat genotypes is still lacking. Therefore, it is imperative to have a detailed understanding of molecular mechanisms/genes responsible for salt tolerance in available wheat germplasm for developing wheat varieties with durable salinity tolerance. With the technological advancements in the field of omics, it is now possible to obtain comprehensive snapshot of the transcript, protein and metabolite level changes in plants associated with any events, intrinsic or extrinsic.[95] Concerted application of these tools is likely to accelerate the research aimed at deciphering the complex regulatory networks determining salinity responses in wheat.

This chapter discusses the scope and potential of genomics technologies in

a) Deciphering the molecular basis of salt tolerance and
b) Development of salt-tolerant wheat varieties.

6.2   PHYSIOLOGICAL RESPONSES OF WHEAT TO SALINITY STRESS

Salinity stress severely affects seed germination and growth in wheat, and higher salt levels may completely inhibit the seed germination, seedling emergence and survival.[137] Although higher levels of salt virtually affect all the plant growth stages, reproductive stage is particularly extremely sensitive to salt injury. The harmful effects of salinity stress are mostly attributed to[7,12]

• Inability to absorb water due to low soil osmotic potential,
• Toxic effect of various ions, mainly Na^+ and Cl^-,
• Nutritional stress due to reduced uptake and transport of essential mineral ions to the aerial tissues, and
• Oxidative stress caused by the accumulation of reactive oxygen species (ROS).[7,121]

The deleterious effects of salinity stress are like that of water deficit. Under salinity stress, plant cannot absorb water efficiently which causes reduction in turgor potential of leaf cells affecting vital physiological processes including photosynthesis, stomatal conductance and the rate of transpiration.[142] The excess accumulation of Na^+ in leaves perturbs the photosynthetic assimilation that in turn affects energy production for plant growth. Na^+ causes damage to the components of photosystem-II (PS-II) and reduces the activity of enzymes involved in CO_2 fixation.[8] It has been observed that photosynthetic capacity of salinized wheat genotypes can be determined by cellular and subcellular partitioning of Na^+, K^+, and Cl^- ions with salt-tolerant genotypes showing significantly higher photosynthetic rate ascribed to high K^+/Na^+ ratio in leaves.[80] Salinity stress also affects leaf membrane stability, lipid peroxidation, chlorophyll and carotenoid contents.[137,171]

6.3   GENOMICS APPROACHES TO UNRAVEL MECHANISMS OF SALT TOLERANCE

6.3.1   GENOME SEQUENCE AND MOLECULAR MARKER RESOURCES

The genome sequence of a crop is the ultimate genomic resource for performing any kind of genome-wide analysis. Next-generation sequencing (NGS) technologies have drastically reduced the time and cost involved in the whole genome sequencing. Without NGS, it would not have been possible to sequence a large and complex genome like wheat in a short period.[77] Besides bread wheat, the draft genomes of *Triticum uratu* and *Aegilops tauschii* considered to be A- and D-genome progenitors of bread wheat, respectively, are now available in the public domain.[83,99] These genomes can be scanned to identify abiotic stress related orthologues or syntenic genes, their copies and chromosomal locations. Annotation of *Aegilops tauschii* genomes revealed that it has an invaluable repertoire of adaptation related genes.[83] The predicted abiotic stress related genes can be analyzed using functional genomics tools to identify their potential role in salt tolerance. Moreover, analysis of whole genome sequence data of wheat and their relatives can also shed light on evolutionary path of various salt tolerance mechanisms. Such studies are extremely important as the hexaploid (AABBDD) and diploid wheat species (AA and DD) are relatively more tolerant to salinity than tetraploid wheat (AABB) barring few accessions of wild emmer wheat.[37,163] Molecular markers are other major valuable genomic resource for performing genomic-wide analysis in crops.

For high-resolution mapping of the salt tolerance traits, it is important to generate genotyping data for mapping populations/association panel with large number of genome wide markers. In the last couple of years, many high-throughput genotyping tools have been developed in wheat.[26,97] A 90K iSelect SNP array has been developed and used for characterizing genetic diversity of durum and bread wheat populations.[159] Besides, some studies have also applied DArT (diversity array technology) and GBS (genotyping by sequencing) for genetic mapping.[97,166] The availability of wheat genome sequence is expected to accelerate development of more number of SNP (single nucleotide polymorphism) assays and enhanced application of GBS in gene mapping studies. Application of advanced genotyping tools is likely to fast-track fine mapping and cloning of salt tolerance associated loci/genes in wheat as well as in their progenitors.

6.3.2   DISCOVERING QTLS ASSOCIATED WITH SALINITY TOLERANCE

6.3.2.1   QTL MAPPING

Salinity tolerance in wheat is a complex trait involving osmotic adjustment, partitioning of the ions and changes at the morphological level.[117] These salt tolerance traits are quantitatively inherited, controlled by many genes designated as a QTL (quantitative trait locus). QTL mapping approach has been widely applied to understand the genetic and molecular basis of salt tolerance in several crop plants. In wheat, a good number of QTL mapping studies have been performed to identify loci/genomic regions associated with salinity tolerance. Initially, salt tolerance in wheat genotypes was solely attributed to their ability to exclude Na^+ from leaves.[39,57,140] Consequently, genetic studies in 1990s mainly focused to identify loci/chromosomal regions responsible of high K^+/Na^+ ratio in leaves under saline conditions.

Gorham et al.[57] were first to report that Na^+ exclusion trait on chromosome 4D, using a set of disomic durum wheat substitution lines in which each of the A- or B-chromosome was replaced by the corresponding D-genome homologue (Chinese Spring). Dubcovsky et al.[38] identified a major salt tolerance associated locus, *Kna1* on the long arm of the chromosome 4D. The *Kna1* locus controls ability of wheat genotypes to discriminate K^+ from Na^+ in roots which enables preferential transport of K^+ from roots to leaves. Subsequently, another major salt tolerance locus *Nax 1 (Sodium Exclusion locus 1)* was identified on chromosome 2AL using a F_2 mapping population derived from cross between Line 169 (salt-tolerant durum line) × Tamro (a salt sensitive Australian durum wheat cultivar).[98] Durum line 169 was a selection from cross between durum wheat cv Marrocos and a salt-tolerant accession of *Triticum monococcum* (C68-10).[117] The *Nax1* locus explained only 38% variation in leaf Na^+ concentration and thus it was suggested that more than one locus may be responsible for sodium exclusion.[98] This observation was in line with an earlier study which predicted at least two major loci for Na^+ exclusion in Line 169 based on genetic analysis.[117] The second Na^+ exclusion locus from line 169 was found to be present on chromosome 5AL and designated as *Nax2*.[23] Interestingly, the *Nax2*-containing region on 5AL and the *Kna1 locus containing region on* 4DL were *ancestrally related to the 4AL* suggesting that *Nax2* of durum wheat essentially corresponds to *Kna 1* of bread wheat.[24] In the subsequent studies, *Nax1* and *Nax2* were cloned and their role in salinity tolerance was clearly established at the molecular and physiological levels.[5,75,23,81] Both the *Nax* genes encode for high affinity K transporter (HKT) proteins; HKT1; 4 (*Nax1*) and HKT 1;5 (*Nax2*) which are involved in regulating transport of Na^+ from root to shoot tissues by restricting its loading in root xylem vessels.[23,75]

Additionally, the HKT1; 4 is capable of withdrawing Na^+ ions from shoot to leaf sheath tissues, preventing its accumulation in leaf lamina. Recently, it was reported that the bread wheat *Kna* locus also encodes for a HKT family transporter, HKT1;5D.[24] The role of HKT family transporters in salt tolerance is well established in other species as well such as *Arabidopsis*,[109] barley,[113] and rice.[130]

Although salt tolerance in wheat genotypes is mostly linked with their ability to maintain low Na^+ concentration in the leaves, recent studies show that it is not always the case.[51,133] There are few wheat genotypes like Mahuti, Berkut, GBA Ruby, etc., which accumulate significantly higher amount of Na^+ in leaves yet they can tolerate high levels of salinity.[9,51] These wheat genotypes may have other salinity tolerance mechanisms.[51,133] Therefore, it is equally imperative to focus on other morpho-physiological traits as well, because few of them may be either directly or indirectly associated with salinity tolerance in wheat genotypes. Consequently, in the past few years, QTL mapping studies in wheat for salinity tolerance have included other traits as well such as chlorophyll content, seedling biomass, shoot and root dry weights, yield, spikelet number, seed germination, tiller number and leaf injury (death), etc.[34,52,53,105,123] Moreover, QTL mapping studies have been performed for both seedling and adult plant stage traits because plants could have different sets of salt tolerance genes/mechanisms at these two stages. Important QTL mapping studies in wheat for salinity tolerance associated traits are listed in Table 6.1.

TABLE 6.1 List of Studies Performed to Identify QTLs Associated with Salinity Stress.

Trait	Species	Method of salinity treatment	Chromosome/linkage groups	Mapping population/association panel	References
Shoot Na+ concentration	Bread wheat	Hydroponics	2A, 2B, 2B, 6A, and 7A	DH population (Berkut X Krichauff)	[52]
		Gravel culture with provision of automatic sub irrigation	2	F2 population (Line149XTamaroi)	[98]
	Bread wheat	Hydroponics and field	7	Two DH populations (Cranbrook/Halberd) and (Excalibur/Kukri)	[40]
		Hydroponics	2, 3, 4, 6, and 7	Synthetic backcross lines derived from Aus29639 X Yitpi	[125]
		Field (six field trials)	2A, 5A, 5D, 6A, and 7D (detected in more than two field trials)	DH population (Berkut X Krichauff)	[53]
Na+/K+ discrimination	Bread wheat	Hydroponics	4D	RIL population for chromosome 4B/4D	[38]
Shoot K concentration	Bread wheat	Hydroponics	1D, 3B, 3D, 4A, 4D, and 5A	DH population (Berkut X Krichauff)	[52]
	Bread wheat	Field (six field trials)	5D, 5A, 1A, 2A (detected in more than 2 field trials)	DH population (Berkut X Krichauff)	[53]
Seedling biomass	Bread wheat	Hydroponics	2A, 4B, 5A, 5B, 6A, 6D, and 7A	DH population (Berkut X Krichauff)	[52]
Shoot dry matter	Bread wheat		1 and 3	RIL population (Opata85XW7984)	[105]
% Death leaves	Durum wheat		4B	119 durum varieties	[152]
Chlorophyll content	Bread wheat		2D, 5A, 5B, and 5D	DH population Berkut X Krichauff	[52]
			3 and 7	RIL (Opata85XW7984)	[105]
	Durum wheat	Hydroponics (150 mM NaCl)	3A	119 durum varieties	[152]
Tiller number	Bread wheat	Hydroponics	1A, 4B, 5A, 5B, and 5D	DH population (Berkut X Krichauff)	[52]

TABLE 6.1 (Continued)

Trait	Species	Method of salinity treatment	Chromosome/linkage groups	Mapping population/association panel	References
Grain yield	Bread wheat	Field irrigated with saline water	1, 2, 3, 4, 5, 6, and 7	DH population {Chinese Spring X SQ1 (a high abscisic acid expressing line)}	[131]
Germination	Bread wheat	Filter paper	3, 4, 5, and 7	RIL population (Opata85XW7984)	[105]
Grain yield and agronomical traits	Bread wheat	Field condition	29 stable QTLs on nine different chromosomes 1A, 1B, 2A, 2B, 3B, 5B, 5D, 6B, and 7A	RIL population (Attila/Kauz X Kharchia)	[123]
Agronomical traits	Bread wheat	Field condition	2D, 4A, 2A, 5A, 6A, 7A, 1B, 4B, 3B, 6B, 7B, and 6D	RIL (Opata85XW7984)	[34]

RIL, recombinant inbred line.

Ma et al.[105] evaluated 114 recombinant inbred lines (RILs) of ITMI (International triticeae mapping initiative) population (Opata 85 × W7984) at germination and seedling stages under salinity stress and detected a total of 47 QTLs associated with different agronomical traits. Another independent study with the same ITMI population, detected as many as 36 salinity associated QTLs for various agronomical traits under saline conditions. Of these, 15 QTLs were present on single chromosome 2D.[34] Genc et al.[52] evaluated a set of 152 double haploid (DH) lines (Brekut X Krichauff) under hydroponics saline condition and identified 40 QTLs for seven different physiological traits including leaf symptoms, seedling biomass, tiller number, chlorophyll content, and shoot Na^+ and K^+ concentration. A total of five such QTLs were linked with Na^+ concentration and two of the colocalized with seedling biomass (2A and 6A). They were first to report an association between a Na^+ exclusion QTL interval (wPt-3114-wmc170) and increased seedling biomass.

6.3.2.2   GENOME WIDE ASSOCIATION MAPPING

In recent years, there is a tremendous interest in using genome wide association (GWA)/linkage disequilibrium based mapping for discovering QTLs/ genomic regions for economically important traits in the food crops.[68,101,132,148] In contrast to the classical QTL mapping, GWA mapping relies on the historical recombination events among the diverse germplasm lines to identify associations between marker and the targeted trait. Currently, association mapping (AM) is preferred over QTL mapping due to following advantages:

- High-resolution genetic mapping enhances the chances of identifying very close association between marker and traits because large parts of the genome of the selected individuals may have undergone multiple round of recombination events over historical time,
- Highly cost effective as the genotyping information of an association panel can be used for genetic mapping of different traits, and
- No need to generate mapping population, thus saving time and labor.

AM in crops has become common with the availability of high throughput, reliable and cost effective genotyping tools such as DArT, SNP assays and GBS.[65,97] It has been increasingly used to identify QTLs for biotic and abiotic stress tolerance in wheat.[4,10] However, so far there is just one report of AM in wheat for salinity tolerance associated QTLs.

Turki et al.[152] evaluated 119 durum wheat varieties at seedling and adult plant stages under salinity stress (150 mM NaCl) and mapped 12 QTLs for eight different traits distributed to chromosomes 4B, 3A, 5A, 5B, 6A, and 7A.[152] They identified dead leaves (%), as a highly reliable trait for measuring salinity tolerance in durum wheat and it was found to be controlled by a major QTL on long arm of chromosome 4B.

6.3.2.3   ROLE OF MICRO-RNAs IN SALT TOLERANCE

MicroRNA (miRNA), a class of small endogenous RNAs (18–24 nucleotides long), are considered key regulators of plant growth and development.[13] They bind to complementary region in the target transcripts directing their degradation or translational inhibition in a sequence dependent manner.[86] MicroRNAs are considered master regulators of gene expression in eukaryotes because each miRNA can regulate expression of many genes, and together they constitute a miRNA-gene(s) network. Plants have several such miRNA-gene regulatory networks associated with one or more physiological processes. In plants, miRNA essentially represents an endogenous gene silencing mechanism evolved to fine tune the developmental processes and responses to biotic and abiotic stresses.

Environmental stresses such as salinity and drought can significantly alter the miRNA profiles of plants in a tissue and genotype dependent manner.[169] The two wheat genotypes possessing different levels of salt tolerance may also vary with respect to their miRNA profiles. It is generally observed that some miRNA genes are significantly up-regulated in salt susceptible genotypes. These miRNAs negatively regulate transcription factors involved in various signaling pathways (ABA, ethylene, and auxin) and also stress responsive proteins (Fig. 6.1). Microarray and small RNA-sequencing are the two most widely followed approaches for miRNA identification and have facilitated discovery of salt responsive miRNA in many plant species including wheat, *Arabidopsis*, rice, maize, barley, etc.[11,35,36,41,82,108,127]

Salt responsive miRNAs were identified from the roots of two wheat genotypes contrasting for salt tolerance; Seri-82 (salt tolerant) and Bezostaza (susceptible), using microarray containing all known plant miRNAs as probes (11862).[41] A total of 44 miRNAs were salt responsive, of which three (hvu-miR5049a, ppt-miR1074, and osa-miR444b.2) had significantly higher expression in Bezostaza as compared to Seri-82.

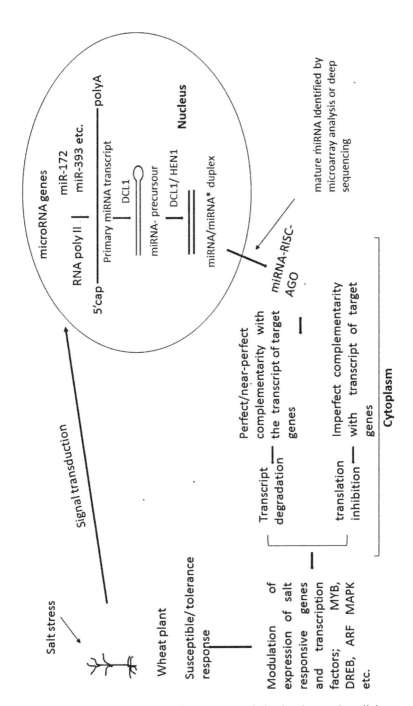

FIGURE 6.1 Overview of miRNA-mediated gene regulation in wheat under salinity stress.

The availability of NGS technologies has greatly accelerated discovery of abiotic stress responsive miRNAs in crop plants.[12] Salt responsive miRNAs can be easily identified by sequencing small RNA pool in different tissues of salt stressed and control plants. Novel and conserved salt responsive miRNA were identified in wheat *cv* "PBW343" using deep sequencing.[127] Several salt responsive miRNAs were predicted to target genes encoding for transcription factors (NAC, ARF, Class III HD-ZIP protein), proteins involved in hormone signaling (MAPK) and stress related proteins (Cu-Zn SOD) (Table 6.2). Recently, Agharbaoui et al.[2] performed a comprehensive miRNAome analysis in wheat to identify abiotic stress responsive (salt, cold, and aluminum toxicity) miRNAs using deep transcriptome sequencing approach. Their analysis identified a total of 198 miRNAs in root and vegetative tissues, out of which 55 were differentially expressed under salt stress.

Salt responsive microRNAs can also be identified by applying bioinformatics tools. Since majority of miRNA families are conserved across plant lineages, their members can be used as query to search the recently sequenced wheat genome as well as expressed sequenced tags and genome survey sequence databases for identifying corresponding orthologues. The role of these identified miRNA orthologues in salinity stress can be validated by expression analysis. Gupta et al.[64] analyzed expression pattern of some conserved miRNAs in wheat under salt, drought and cold stresses.[64] Two miRNAs (miR168 and miR397) were upregulated in all the three analyzed stresses, suggesting their involvement in the responses, which may be commonly associated with abiotic stresses.

Generally, the miRNAs which negatively regulate expression of genes and transcription factors associated with salt tolerance are up-regulated in salt sensitive varieties. Silencing of these miRNAs by RNAi or genome editing technologies could significantly improve salt tolerance of wheat genotypes.

Salinity stress is perceived at the level of plasma membrane and subsequently transduced to nucleus causing up or down regulation of stress responsive miRNA genes. Salt responsive miRNA genes are transcribed by RNA Polymerase II into long primary transcripts which are then processed in two different steps into the corresponding miRNA/miRNA* duplex by the action of various proteins including Dicer Like 1 (DCL1), HYL1, and HEN1. The miRNA/miRNA* duplex is translocated to cytosol where it gets separated and miRNA is incorporated into RISC (RNA induced silencing complex) assembly, a complex of proteins including AGO-1 and directs transcript cleavage or translational inhibition (Fig. 6.1).

TABLE 6.2 Conserved miRNAs Differentially Expressed Under Salinity Stress in Wheat.

miRNA	Expression under salinity stress	Predicted targets	References
miR156	Down-regulated	Squamosa promoter-binding-like (SPL) protein	[127]
miR164	Down-regulated	NAC transcription factors, mitogen-activated protein kinase (MAPK), harpin-induced protein 1 containing protein	[64,127]
miR166a	Down-regulated	Class III HD-ZIP protein	[127]
miR167a	Down-regulated	ARF transcription factor	[127]
miR168	Down-regulated	Argonaut	[64]
miR171a	Down-regulated	Superoxide dismutase [Cu–Zn], pentatricopeptide repeat-containing protein-like protein, cysteine proteinase-like protein	[127]
miR172	Down-regulated	AP2-domain transcription factor, ARF, floral homeotic protein, ethylene-responsive TF RAP2-7-like	[41,64]
miR397	Down-regulated	Laccase, Ice1 (inducer of CBF expression 1) transcription factor	[64]
miR393	Up-regulated	bHLH TF, transport inhibitor and response 1/auxin F-box	[64,103]
miR529	Down-regulated	AP2-like transcription factor, squamosa-promoter binding protein-like	[64]
miR1029	Down-regulated	AP2-like transcription factor, DREB transcription factor	[64]
miR396d	Down-regulated	Growth regulating factor 3 (GRF-3) transcription factor	[127]
miR444b.2	Up-regulated	MADS-box transcription factor-27-like ABC transporter C family member 10	[41]
miR1074	Up-regulated	Malate dehydrogenase, receptor protein kinase TMK1	[41]
miR5049a	Up-regulated	60S ribosomal protein 136	[41]

DREB, dehydration responsive element binding. (Modified from Ref. 93.)

6.4 OTHER OMICS-BASED APPROACHES

Plant salt tolerance is a genetically complex quantitative trait governed by different soil, environmental and management factors. Conventionally, plant response to salt stress is evaluated by different morphological and physiological indices. However, the use of such indices often fails to provide the detailed understanding of biochemical and molecular basis of salt stress regulation at the cellular level. This has necessitated the integrated use of different approaches to analyze the expression and role of genes and gene products in plants exposed to saline conditions. Usefulness of such approaches, collectively referred to as "omics" (including transciptomics, proteomics, metabolomics, ionomics, and phenomics) in the enhanced understanding of molecular basis of salt tolerance in wheat is discussed in this section.

6.4.1   TRANSCRIPTOME PROFILING

Transcriptome profiling is routinely applied to study global changes in the gene expression to identify key genes and metabolic processes involved in plant adaptation to environmental stresses.[59] In the initial years, gene expression studies in wheat were mainly performed using methods like suppression-subtractive-cDNA library, differential display reverse transcription polymerase chain reaction, serial amplification of gene amplification, etc. These methods failed to give in-depth understanding of molecular processes involved in abiotic stresses because they could measure only a small fraction of differentially expressed genes under abiotic stresses. It was only after the availability of wheat microarrays with large number of c-DNA probes that made possible the detailed study of the abiotic stresses on global gene expression.[31,88,115]

Kawaura et al.[88] analyzed root and shoot transcriptomes of 2-week-old seedlings of wheat which were subjected to salt stress (150 mM of NaCl for 1, 6, and 24 h) using a 32K-oligo DNA microarray. They observed a clear shift in global gene expression pattern during the course of salinity treatment. The early salt stress responsive genes mainly included transcription factors, transcription-regulator and those with DNA-binding functions. By contrast, the late stress responsive genes predominately belonged to transferase and transporter categories. Additionally, it was shown that multiple signaling pathways were activated under salinity stress. In another study, root and leaf transcriptomes of five different wheat genotypes; Chinese Spring (salt sensitive) two salt-tolerant genotypes (W4909 and W4910), and their parental lines AJDAj5 and ph1 were analyzed using Affymetrix wheat GeneChip1 array. Salt-tolerant genotypes showed significantly higher expression of tonoplast aquaporin (from ph1line) and a K channel protein (from AJDAj5). It was suggested that tonoplast aquaporin and K channel protein might have improved salt tolerance of wheat genotypes W4909 and W4910 by maintaining ion homeostasis and osmotic adjustment.[115] Other studies have also showed involvement of aquaporin and K transport proteins in salt tolerance.[73,74,172]

Lately, the NGS technologies have revolutionized the field of transcriptomics. RNA sequencing (RNA-seq) also known as transcriptome sequencing is considered more accurate and reliable than microarray as the relative abundance of transcript in different tissue samples is measured by sequencing the whole transcripts. It is highly suitable for complex genome like wheat which has three different but very similar genomes, because the sequenced transcripts can be easily assigned to the respective homologous chromosomes. Ma et al.[106] analyzed transcriptome of two wheat mutants—RH8706-49 (salt tolerant) and RH8706-34 (salt susceptible)—in response to salt stress using RNA seq. The mutants mainly differed in the expression levels of genes associated with oxidation phosphorylation and ribosome pathway. Significantly higher expression of the oxidative phosphorylation genes may have enabled salt-tolerant mutant, RH8706-49 to meet energy required for normal metabolic activities during salinity stress.

Zhang et al.[168] compared root and leaf transcriptomes of salt tolerant (QM) and susceptible wheat genotype (Chinese Spring) subjected to salinity stress using RNA-Seq. Their study revealed reprograming as well as expression partitioning of salt responsive genes to homologous chromosomes (paralogues located on A, B, and D genomes) during the course of salt treatment. Furthermore, salt tolerance genes were found to be tandem arrayed on various chromosomes suggesting a role for tandem duplication in their evolution. It was suggested that expression of the salt-tolerant gene located on the three subgenomes might have diverged during polyploidization event.[168] More recently, RNA-seq was applied to generate root transcriptome landscape of the famous Indian salt tolerant land race "Kharchia local" in response to salinity stress.[59] The "Kharchia local" transcripts were mainly distributed to five metabolic pathways; ribosome, oxidative phosphorylation, spliceosome, carbon metabolism, and RNA transport. Expectedly, they noted higher expression of V-ATPase (V-type proton-transporting ATPase) in the roots of salt stressed seedlings indicating its involvement in conferring salinity tolerance. The V-ATPase are known to generate strong H^+ electrochemical gradient across the tonoplast, thereby facilitating sequestration of excess of cytosolic Na^+ into vacuole by the Na^+/H^+ antiporter.[114,162] Additionally, transcriptomics can be also combined with QTL approach to identify the candidate salt tolerance genes.[126]

Besides generating a wealth of knowledge on metabolic and regulatory pathway operating in various tissues under salt stress, transcriptomic analysis can also provide a list of putative candidate genes for salt tolerance. The detailed characterization of these candidates using functional genomics approaches like RNAi (RNA interference) technology, genome editing and transgenic would make their possible role in salt tolerance. A putative salt-tolerant gene *TaSR* (*Triticum aestivum* salt responsive) was identified and characterized from the microarray data of a salt-tolerant wheat mutant using transgenic (rice and *Arabidopsis*) as well as RNAi approach.[107]

Transgenic rice and *Arabidopsis* plants expressing *TaSR* were salt tolerant and demonstrated lower concentration of Na^+ and higher concentrations of K^+ and Ca^{++} in leaves and other aerial tissues in salt stress as compared to unstressed plants as well as those in which this gene was silenced using RNAi. In another study, a putative salt-tolerant candidate gene (*TaZNF*) encoding for a C2-H2 type Dof zinc finger domain protein was characterized in *Arabidopsis*. *Arabidopsis* transgenic plants expressing *TaZNF* demonstrated higher Na^+ exclusion activity in roots and were more tolerant to salt stress compared to wild types. The TaZNF protein was localized into the nucleus which indicated that it may be regulating the expression of other salinity stress responsive genes at the transcriptional level.[106]

6.4.2 PROTEOMICS

Proteomics deals with the systematic study of all proteins expressed in a cell, tissue, or organ of an organism. It has been pointed out in various studies that transcript level changes do not always results into corresponding changes at the level of protein.[93,145] Therefore, it is not always possible to make a valid conclusion about the involvement of a gene(s) and or biochemical pathway(s) in abiotic stress tolerance only based on expression studies. Proteins are integral component of abiotic stress signaling and are also involved in various physiological and metabolic processes associated with plant acclimation to abiotic stresses. Moreover, some proteins can regulate stress responses by altering the transcription and translation of various stress related genes. Therefore, proteome studies are extremely useful in reconstructing the metabolic networks (pathways) operative in crop plants under salinity stress. Already, some research groups have used proteomics approach to decipher the physiological and biochemical basis of salt tolerance in wheat (Table 6.3).[25,60,65,76,87,92,94,95,129]

TABLE 6.3 Proteomics Studies in Wheat for Salt Tolerance.

Genotypes	Tissue	Salinity treatment method and duration	Proteomics tool	Major metabolic pathways affected under salt stress	Important differentially expressed proteins	References
Bread wheat						
Shanrong-3 (salt tolerant) and Jinan 77 (salt susceptible)	Leaf	Half Hoagland solution, 200 mM NaCl (24 h)	2DE, MALDI-TOF and MALDI-TOF/TOF	Signaling proteins, Energy generation (glycolysis), osmoregulation	Up-regulated in Shanrong-3: BTF3 (NAC transcription factor), DEAD-box RNA helicase involved in modulation of abiotic stress inducible transcription factor DREB, V-ATPase subunit E, translation initiation factor eIF5A2, Up-regulated in Jinan 77: G-protein β subunit, ethylene receptor ETR1	[158]
Wyalkatchem (salt tolerant), Clinger and Janz (salt susceptible)	Shoot (mitochondrial fraction)	200 mM NaCl concentration was raised gradually (50 mM NaCl / day)	2D-DIGE, LC–MS/MS	ROS detoxification	Up-regulated: Mn-SOD, voltage-dependent anion channel (VDAC), alternate oxidase (AOX) Down-regulated: citrate synthase (CS), nucleotide diphosphate kinase (NDPK), outer mitochondrial membrane porin Genotype differences: Mn-SOD and AOX had higher expression in Wyalkatchem as compared to Calingiri, Janz (SS)	[79]

TABLE 6.3 *(Continued)*

Genotypes	Tissue	Salinity treatment method and duration	Proteomics tool	Major metabolic pathways affected under salt stress	Important differentially expressed proteins	References
Zhengmai 9023	Leaf	1, 1.5, 2, 2.5% NaCl in Hoagland solution, 2 days	2D-DIGE, Q-TOF-MS	Carbohydrate metabolism, protein folding, ROS detoxification	Up-regulated: ATPases, TPI, ferritin and glutathione-S-transferase Down-regulated: Rubisco LSU, RubisCO activase, 23 kDa oxygen evolving protein	[61]
Keumkang	Leaf (chloroplast fraction)	150 mM NaCl (1, 2, and 3 days)	2DE, LTQ-FTICR-MS	Amino acid and nitrogen metabolism, energy metabolism (photosynthesis)	Up-regulated: fructose bisphosphate aldolase, glycerol-3-phospahte dehydrogenase eIF5 A-1/2, eIF5 A-3, glutamine dehydrogenase (GDH), GS, pyridoxal phosphate proteins (PDX1.2 and PDX1.3), CA, SAMS, cytb-6, germin like protein; Down-regulated: RubisCO, ATP synthase subunits (α, β, γ) and V-type proton ATPase, oxygen evolving protein OEE1 and OEE2	[87]
Chinese Spring (salt susceptible) and *T. aestivum* X *Lophopyrum elongatum* amphiploid (salt tolerant)	Shoot and root (mitochondrial fraction)	200 mM NaCl	2D-DIGE, MALDI-TOF/TOF and HPLC-Q-TOF MS/MS	ROS detoxification	Up-regulated in *T. aestivum* × *Lophopyrum elongatum* amphiploid:: Mn-SOD, malate dehydrogenase (MDH), aconitase, serine hydroxyl methyl transferase (SHMT), beta cynoalanine synthase (β-CAS)	[79]

TABLE 6.3 *(Continued)*

Genotypes	Tissue	Salinity treatment method and duration	Proteomics tool	Major metabolic pathways affected under salt stress	Important differentially expressed proteins	References
				Durum wheat		
Ofanto	Leaf	100 mM NaCl, 2 days	2DE, MALDI-TOF	Carbohydrate metabolism, amino acid and nitrogen metabolism, energy metabolism (glycolysis), ROS detoxification	Up-regulated: triose phosphate isomerase (TPI), glucose-6-phosphate dehydrogenase, Rubisco activase, glutamine synthetase (GS), carbonic anhydrase (CA); glycine dehydrogenase and SAMS (glycine betaine biosynthesis) APX, Cu/Zn-SOD, thiol-specific antioxidant protein, RAB-related COR protein Down-regulated: RubisCO SSU, phosphoribulokinase, ATP synthase CF1 α subunit, OEEE-1 precursor, β-glucosidase	[25]

DREB, dehydration responsive element binding; ROS, reactive oxygen species.

Under salinity stress, wheat plants accumulate significantly higher concentration of key enzymes of glycolysis and carbohydrate metabolism including triose phosphate isomerase, fructose bisphosphate aldolase, and enolase indicating metabolic reprograming to meet the energy required for metabolic activities.[25,87] The other major categories of proteins consistently up-regulated in salt-tolerant genotypes are those involved in ROS detoxification. Some of the important ROS scavenging enzymes commonly observed in salinity stressed wheat plants include Cu/Zn-SOD, Mn-SOD, glutathione-S-transferase, ascorbate peroxidase and peroxiredoxin.[49,78] Proteins involved in osmolyte biosynthesis, mainly those of glycine betaine (glycine dehydrogenase; S-adenosylmethionine synthetase, SAMS) and proline synthesis (glutamine synthetase) have been shown to be up-regulated by salt stress.[3,22] Moreover, the proteins involved in nucleotide metabolism (nucleoside diphosphate kinase), signal transduction (G-proteins, MAP kinases, ethylene signaling), and ion homeostasis (vacuolar ATPase subunits, ABC transporters) are often accumulated in enhanced quantity in the salt-tolerant wheat genotypes.

Expression and functionality of chloroplast proteins are also affected in wheat plants under salinity stress.[76,87] Huo et al.[76] analyzed salinity induced changes in leaf proteome of two wheat mutants having contrasting salt tolerance responses. The mutants demonstrated qualitative and quantitative differences for key chloroplast proteins such as glutamine synthetase 2, putative 33 kDa oxygen evolving protein of PS-II, ribulose-1,5-bisphosphate carboxylase/oxygenase small subunit and H$^+$-ATPase. Since these proteins are essential for the proper functioning of chloroplast associated processes, they are mostly functionally active in salt-tolerant wheat genotypes under salinity stress.[76]

Proteome profiling of a wheat cv. "Zhengmai 9023" identified a large number of salinity induced proteins. These proteins were grouped into following major categories; membrane transport associated proteins, carbon metabolism, ATP synthesis, ROS detoxification enzymes and protein folding. Some key proteins which were up-regulated under salt stress include H$^+$-ATPases, ferritin, triose phosphate isomerase and glutathione S-transferase.[61] In addition, the general observation has been that salinity reduces expression of key proteins of CO_2 fixation pathway; RuBisco small sub units (SSU), RuBisco large subunit (LSU) and phosphoglyceratekinase. It is a significant metabolic shift considering the fact that CO_2 fixation is an energy intensive process and by doing so plants might save energy (ATP and NADPH2) and channelize it for normal metabolic activities.[25,87,93]

In recent times, emphasis is shifting on analyzing chloroplast and mito-chondria proteomes as majority of salinity stress responses are associated to these organelles.[79] Kamal et al.[87] reported variation in the chloroplast proteome during the course of salt treatment. Several proteins which were down-regulated 1 day after salt treatment got up-regulated on second or third day. This appears to be a strategy by plants to counterbalance the negative effects associated with salt stress. On the other hand, some proteins were steadily up-regulated during entire period of salinity stress. These included Cytb6-f, GLP, glutamine synthetase, fructose-bisphosphate aldolase, ATP synthase subunit ω, SAMS and carbonic anhydrase. Studies are also focusing on posttranslation modifications and protein–protein interactions as they play crucial role in framing the plant responses to abiotic stresses.[161,167]

Phosphorylation state of proteins is commonly affected during abiotic stresses. Indeed, some proteins become functionally active only after they are phosphorylated. Recently, a proteomic and phoshoproteomic study in *Tritium monococcum* identified a set of biomarkers associated with salt tolerance. These included betaine-aldehyde dehydrogenase, Cu/Zn super-oxide dismutase, 2-Cys peroxiredoxin BAS1, leucine aminopeptidase 2, and cp31BHv.[104] Several such biomarkers may be identified and targeted for improvement of salt tolerance in wheat. Moreover, proteome analysis can be combined with other genomics tools to unearth finer details of molecular processes associated with salinity tolerance of wheat genotypes.

6.4.3 METABOLOMICS

Plant responses to abiotic stresses are essentially the manifestation of the changes at the level of metabolites. Salt stress can affect several metabolic pathways including photosynthesis, osmotic solute and hormone biosynthesis pathways and there by altering metabolite composition in different tissues.[122] Therefore, a comparative analysis of total metabolite pool of salt stresses and control wheat plants is expected to provide better understanding of the mechanisms underlying salt tolerance. Metabolite profiling can also shed light on salt tolerance mechanisms (ion homeostasis, osmotic stress adjustment, detoxification of reactive oxygen species) present in different wheat genotypes. Several studies have combined metabolomics and other genomics approaches to decipher the mechanisms responsible for abiotic stress tolerance in plants.[6,17,96] So far metabolomics studies in wheat have mainly focused on analyzing grain quality related traits.[18,170] Nevertheless, lately some studies have used metabolic profiling to understand abiotic stress responses in wheat.[20,62]

Comparative metabolome profiling of alkali and salt stressed wheat seedlings revealed that the alkali stress has more intense effect on root and leaves than salt stress. Salt stressed seedlings showed a metabolic shift toward gluconeogenesis (increased synthesis of sugars from fatty acids) which helped them maintain osmotic balance and energy required to support normal metabolic activities. By contrast, alkali stressed wheat seedlings accumulated low amount of sugars and amino acids and were also poor in nitrogen metabolism.[62] This study established that the salt and alkali stresses have entirely different impact on the plant metabolism. Therefore, to develop alkali stress tolerant wheat varieties a different strategy should be adopted which can focus on enhancing sugar and amino acid biosynthesis.

6.4.4   IONOMICS

The term "ionome" refers to complete mineral nutrient pool of a tissue or whole organism. Ionomics is the systematic quantitative analysis of elemental composition of living systems and changes therein with response to environmental aberrations, developmental states, and genetic modifications.[14,138] In the past 10 years, significant advances have taken place in the instrumentation for ionome profiling in plants; it is possible to simultaneously measure ionome profiles of a large number of genotypes in short time and at reasonable cost. The most commonly used ionomics technologies include, inductively coupled plasma–mass spectrometry, inductively coupled plasma–optical emission spectrometry, and neutron activation analysis. Ionomics in combination with other genomic analysis tool can help to identify genes responsible for salinity tolerance in plants.

Several studies have reported ionome profiling to decipher the physiological and molecular basis of salinity stress responses in food crops including wheat, rice, *Arabidopsis*, and barley.[72,160,163] Ionome profiling of two barley genotypes contrasting in salt tolerance which were grown in NaCl (150 mM) showed that salt-tolerant genotype had lower amount of Zn and Cu in roots and Ca, Mg, and S in shoots as compared to salt susceptible genotype. Moreover, the Na^+ accumulation in shoot was negatively correlated with metabolites involved in energy metabolism pathways, glycolysis and Kreb's cycle. Therefore, it was clearly evident that the rearrangement at the level of metabolome and ionome play role in determining physiological responses to salinity stress.[161]

Various physiological parameters including mineral profiles of different wheat species diploid (*Ae. tauschi*), tetraploid (*T. turgidum*), natural hexaploid, and synthetic hexaploid (synthetic) were analyzed under salinity stress, and it was concluded that higher salt tolerance in hexaploid was attributable to inheritance of favorable traits from tetraploid (higher seed germination rate) and diploid (Na^+ exclusion) parents.[163] Guo et al.[62] analyzed metabolome and ionome responses of the wheat seedlings to both salinity and alkali stress. They observed a sharp rise in root Ca content immediately after imposing the alkali stresses. Calcium accumulation activated salt overly sensitive pathway (SOS) mediated Na^+ exclusion mechanism and improved alkali tolerance capacity of wheat seedlings. The above described studies clearly show potential of ionomics in underpinning salinity responses in wheat. The ionome profiling of large number of wheat genotypes may help in identifying novel genes/mechanisms associated with salt tolerance.

6.4.5   PHENOMICS

Phenomics is the discipline of biology concerned with systematic study of phenome on genome-wide scale.[16,48] Collecting accurate and precise phenotyping data continues to be the major limitation in performing genome wide studies for quantitative traits. In traditional phenotyping approach, plants must be harvested for measuring shoot biomass so it is practically impossible to measure all the salinity induced morpho-physiological responses from the same plant. Although with the availability of high-throughput, NGS-based genotyping technologies generation of genotyping data has become fast and cost effective, yet the high-throughput precision phenotyping is still lagging behind. Consequently, currently the greater emphasis is placed on developing sophisticated phenotyping facilities equipped with high-end imaging and robotics based tools for large-scale precision phenotyping.[42,44,47] Imaging-based phenotyping tools can provide real time measurement of changes in shoot biomass in response to salinity.[29,54] Even the osmotic and ionic changes caused by salt stress can be measured from the live plant.[118,133,146]

Thermal imaging tools are quite often used for measuring leaf surface temperature to make an inference about the usually stomatal conductance which is generally affected by salt stress. The salt-tolerant varieties are known to have lower reduction in stomatal conductance than susceptible ones under salinity stress. Sirault et al.[146] developed a highly sensitive infrared based imaging based system to quantify osmotic stress responses of barley and wheat genotypes to salt stress.[146] It may detect small differences (as low as 1°C) between the leaf surface temperature of salt treated (150 mM NaCl) and control plant.

Rajendran et al.[133] developed imaging-based assays for high-throughput measurement of Na^+ exclusion, tissue tolerance and osmotic tolerance in a set of 12 genotypes of *T. monococcum*. Using these assays, they concluded that the salinity tolerance in *T. monococcum* genotypes was an outcome of different combinations of all the three salt-tolerant mechanisms; Na^+ exclusion, osmotic adjustment and tissue tolerance. Large-scale phenomics facilities can accommodate hundreds of individuals, and thus, it can be used to evaluate entire wheat core-collection, mapping population, or wheat association panel for salinity associated traits. Moreover, real-time phenomics data may also give detailed insights into the salt tolerance mechanisms present in various wheat genotypes. Phenomics facilities have been set up by various institutions across the world. One of the most advanced state of the art phenotyping facility is the Australian Plant Phenomics Facility, located at the University of Adelaide. It is extensively used by researchers for genome wide analysis of stress tolerance in crops including wheat.[150]

In India, phenomics facilities have been established at ICAR-Indian Agricultural Research Institute, New Delhi and ICAR-National Institute of Abiotic Stress Management, Maharashtra to cater to the need of Indian researchers. These facilities would be extremely useful in accelerating genomics research for abiotic stress tolerances in important agricultural crops including wheat and rice.

6.4.6   SYSTEM BIOLOGY

The whole idea of system biology is based on the very premise that a biological system is more than the sum of its parts. System biology approach involves an integrative analysis of the large biological data set generated using genomics tools for gaining holistic understanding of biological process.[46,103] The crux of system biology is to use different levels of biological information and their interactions to construct metabolic networks for better understanding of gene(s) to phenotype relationships and to answer biological questions such as molecular basis of tolerance to biotic and abiotic stresses, grain yield and grain quality, etc.

The integration of "omics" data has allowed reconstruction of drought tolerance regulatory network in model plants (*Arabidopsis thaliana*, *Medicago sativa*) and some food crops such as rice, soybean, and maize.[154,85,164] An integrated transcriptomic and metabolic study in grapevine showed differences in metabolic responses to salt and drought stress. Drought stressed caused up-regulation of processes needed for osmoregulation and protection against oxidative stresses and photoinhibition. Salinity stressed induced mainly the processes involved in ion transport, energy metabolism and protein synthesis.[32] It clearly shows the need for adopting system biology approach for dissecting the complex mechanisms of salt tolerance in wheat genotypes.

6.5 GENOMICS-BASED APPLICATIONS TOWARD DEVELOPMENT OF SALT-TOLERANT WHEAT VARIETIES

6.5.1 TRANSGENIC APPROACH

Recombinant DNA technology provides an alternative strategy for genetic improvement of crops. It enables researchers to exploit genes from a totally unrelated species such as bacteria, fungi, animals, etc. The importance of this approach has begun to be realized with release of abiotic stress transgenics in major crops such as maize, soybean, and sugarcane (http://www.isaaa.org/). In wheat, abiotic stress tolerant transgenics are at different stages of development and yet to be released for commercial cultivation. Salt tolerance of wheat could be improved by introduction of potential salt tolerance genes. The genes so far deployed for improving salt tolerance in wheat perform either of the functions, such as (1) salt stress signaling, (2) ion homeostasis, and (3) osmolyte biosynthesis.[117] Some of these are briefly discussed in this section.

6.5.1.1 TRANSCRIPTION FACTORS

Genes encoding for transcription factors are considered good candidates for developing stress tolerant transgenic plants as introduction of just one key transcription factor gene may lead to altered expression of several downstream genes involved in stress tolerance.[155,156] Transcription factors bind to certain conserve motif in promoters of their targeted genes to regulate their expression at the transcriptional level. For example, genes regulated by dehydration element binding (DREB) transcription factors contain a DRE (dehydration responsive elements)/CRT (C-repeat) motif in their promoters. Salt-tolerant wheat has been developed with transcription factor genes belonging to NAC and AP2-ERBP family.[50,84,135,136,144]

Shiqing et al.[144] developed two sets of transgenic wheat plants with soybean *DREB*gene *(GmDREB1)* driven under rd-29 and ubiquitin promoters, respectively. Both the sets of *GmDREB1 wheat transgenic plants were* tolerant to salt as well as drought. The salt and drought tolerance of transgenic *GmDREB1* plants was attributed to enhanced accumulation of oxidative and osmotic stress related proteins and osmolytes (mainly proline and glycine betaine). In another independent study, introduction of cotton *DREB* gene *(GDREB)* in wheat improved tolerance to various abiotic stresses (salt, drought as well as frost) which was attributed to enhanced accumulation of chlorophyll and proline in leaves.[50] Transformation of wheat with *SNAC1* gene from rice improved both salt and drought tolerance. Transgenic wheat lines expressing *SNAC,* exhibited higher RWC, biomass and chlorophyll content. Moreover, these lines showed upregulation of genes involved in ABA signaling under salinity stress.[136] Wheat plants *(cv. Yangmai 12)* overexpressing *TaERF3* gene showed enhanced tolerance to salt stress which was ascribed to higher leave chlorophyll and protein content.[135] Besides, several other transcription factors for abiotic stress tolerance have been reported in rice, *Arabidopsis* and various other species can be also exploited improving salt tolerance in wheat.[155,156]

6.5.1.2   SODIUM TRANSPORTERS

Plants have different type of Na^+ transporters responsible for uptake and transport Na^+ from soil to various tissues. They have been shown to play important role in maintaining ion homeostasis during salinity stress and exploited for developing salt-tolerant transgenic plants. The Na^+/K^+ vacuolar class transporters which facilitate sequestration of excess Na^+ into vacuoles have been widely used to engineer salinity tolerance in crops. Transgenic wheat plants containing *Arabidopsis AtNHX* gene demonstrated increased biomass production and grain yield under saline conditions.[114,162] In addition, the vacuolar Na^+/H^+ antiporter genes from salt-tolerant wheat genotypes can be also used to improve salt tolerance capacity of salt sensitive wheat varieties. *Arabidopsis* plants transformed with wheat *TNHX1* and H^+-PPase TVP1 genes, exhibited enhanced salt tolerance and could grow in 200 mM $NaCl$.[21]

In another study, Wang et al.[157] developed transgenic *Arabidopsis* plants with V-ATPase subunit B (vacuolar H^+-ATPase) isolated from a salt-tolerant wheat mutant line, RH8706-49. Transgenic *Arabidopsis* plants exhibited better germination rate and seedling growth under salt stress than untransformed *Arabidopsis* plants. Additionally, the HKT family genes which encode for Na^+/K^+ transporters could be also exploited for developing salt-tolerant transgenic wheat. HKT family members with potential role in salt tolerance have been identified in many plant species including rice and *Arabidopsis*.[109,130] Interestingly, the *salol* locus which is well known to confer higher levels of salt tolerance in some rice land races also encodes for a HKT transporter (*Os HKT1;5*).[130]

6.5.1.3 OSMOLYTE BIOSYNTHESIS

Osmolytes/compatible solutes are the diverse groups of organic compounds which provide protection to plants against abiotic stresses via osmotic adjustment and also in some cases by stabilizing membrane proteins and/or scavenging reactive oxygen species. Some common osmolytes are sugars (sorbitol, mannitol, trehalose), glycine betaine, polyols, polyamines, and amino acid (proline). As the excess accumulation of osmolytes does not cause any harm to plant cellular machinery, gene encoding for their biosynthesis have been targeted for engineering abiotic stress tolerant crops.[89,147,155] Wheat plants transformed with *"P5CS,"* a gene encoding for Δ^1-pyrroline-5-carboxylate synthetase from *Vigna aconitifolia* accumulated large amount of proline and showed higher tolerance than control plants.[139]

Transgenic wheat plants expressing *Escherichia coli mtlD* (mannitol dehydrogenase), a key gene of mannitol biosynthesis pathway, exhibited relatively lower reduction in shoot fresh weight, dry weight, plant height, and flag leaf length than the untransformed plants.[1] In another independent study, transgenic plants expressing *mtlD* gene had a modified sugar profile; higher content of various soluble sugars such as fructose glucose and galacturonic acid and lower contents of insoluble sugars. Moreover, the transgenic wheat plants expressing *mtlD* gene gave higher yield per plant as compared to wild type wheat plants.[134]

Glycine betaine is another most widely targeted osmolyte for improvement of salt tolerance in crops. In general, cereals including wheat do not accumulate significantly higher amount of glycine betaine which is attributed to truncated transcript of GB synthesis enzyme betaine aldehyde dehydrogenase. Transgenic wheat plant expressing BetA gene encoding for choline dehydrogenase enzyme from *E. coli* could accumulate significantly higher glycine betaine and demonstrated higher chlorophyll content and lower electrolyte leakage as compared to wild type plants. It was suggested that glycine betaine could have improved salt tolerance capacity of transgenic wheat lines via maintaining osmotic balance and protecting the integrity of cell membranes.[69]

6.5.1.4 OTHER SALT TOLERANCE GENES

In addition to the above described genes, those encoding for salt stress tolerance associated proteins such as dehydrins, protease inhibitors can be also exploited for developing salt-tolerant transgenic wheat. Dehydrins or class II LEA proteins protect plants against abiotic stresses primarily via preventing aggregation of proteins and membrane associated complexes.[66] In many crops, transgenics have been developed with dehydrin genes for improving tolerance to abiotic stresses.[95,128] However, till date, there is not a single report of transgenic wheat with dehydrin gene. Protease inhibitors take part in various plant physiological processes by regulating the activity of proteases. Transformation of *Arabidopsis* plants with *WRSI5*, a protease inhibitor gene from the salt-tolerant wheat genotype Shanrong No. 3 (SR3), significantly enhanced their capacity to tolerate salt stress.[141]

6.5.2 MARKER-ASSISTED SELECTION (MAS) FOR BREEDING SALT-TOLERANT WHEAT

Nowadays, it is increasingly common to use molecular markers to track introgression of desired traits in crop breeding programs. Marker assisted backcrossing approach has been successfully applied to introgress potential biotic and abiotic stress tolerant gene/alleles from ancestors or land races into the high yielding varieties in various crops including rice and wheat.[33,112] In rice, a large number of abiotic stress tolerance varieties have been developed using MAS.[100] One of the successful stories of application of MAS in rice is the development of salt-tolerant rice varieties with favorable alleles of salinity locus *saltol*, originally discovered from a salt tolerant rice land race, Pokkali.[151] By contrast, there are not many reports of MAS for breeding abiotic stress tolerant wheat varieties. It is ascribed to non-availability of abiotic stress tolerance associated major validated QTLs.

Recently, Munns et al.[119] achieved a major breakthrough in breeding salt-tolerant wheat variety using MAS. They developed a salt-tolerant variety of durum wheat by introgressing an ancestral salt tolerance gene *Nax 2* (*TmHKT1;5-A*). When grown in SAS, this variety produced 25% higher grain yield than near isogeneic line lacking *Nax2* gene. Besides, MAS can also be used to pyramid salt tolerance genes from diverse sources. In the foregoing sections, we have discussed that genomic regions/genes for salt tolerance can be present on more than one chromosomes both in wheat as well as in its progenitors. Moreover, some of the salt tolerance component phenotype (such as Na^+ exclusion, Na^+/K^+ ratio, biomass, etc.) is encoded by more than one gene. Therefore, stacking of multiple salt tolerance genes from diverse sources may prove very useful in improving salt tolerance of wheat varieties. Wheat varieties with multiple salt tolerance genes are expected to perform better than those with single salt-tolerant genes under diverse saline field conditions.

6.5.3 GENOMIC SELECTION

Conventional MAS approach uses one or few markers (linked with major genes) for selecting improved genotypes for the targeted traits. This approach has met limited success with complex traits which are controlled by large number of genes each with small effects. In contrast, genomic selection (GS) or genome-wide selection is an advanced marker-based selection strategy in which markers distributed across the genomes are used to estimate the breeding value of genotypes.[71] GS involves two steps. In the first step, marker effects are estimated based on the genotyping and phenotypic data of a set of individuals, designated as training or reference population. Subsequently, a prediction model derived from the marker effects estimated in the training population is used to predict genomic breeding values of individuals in another population solely based on their genotyping data. GS selection is considered very promising for improvement of complex quantitative traits such as drought, yield, plant height, etc.[102] Already, researchers have used GS for improvement of quantitative traits in major crops including wheat,[70] sugarcane,[58] maize,[15,110] etc.

It is generally considered that salt tolerance is a less complex trait, governed by few major genes and therefore, it is possible to improve salt tolerance by introgressing one or two major tolerance QTLs. Even salt tolerance rice varieties were developed by introgressing a major QTL, *saltol*. However, recent studies have showed that salt tolerance in rice is controlled by large number of loci distributed across the genome. In wheat also, many salinity tolerance QTLs have been identified which are located on different chromosomes. In such a scenario, GS can be attempted for selecting salt-tolerant wheat varieties. Furthermore, application of GS would also reduce time involved in development of salt-tolerant wheat varieties because it does not require phenotyping data to select candidate genotypes possessing desired salt tolerance associated traits.

6.6   MINING WHEAT GERMPLASM COLLECTIONS FOR NOVEL SALT-TOLERANT GENES/ALLELES

Over the past decades, a large number of explorations were made to collect wheat germplasm including their wild relatives and land races from diverse habitat and their seeds have been conserved in various national and international gene banks. Some gene banks have several thousands of wheat accessions (http://www.fao.org/docrep/013/i1500e/i1500e12.pdf): CIMMYT (110281), NSGC, USA (57348 accessions), ICGAR-CAAS, China (43039 accession), and NBPGR, India (35889). Taken together, these wheat collections represent substantial part of global wheat genetic diversity and are thus repertoire of genes and alleles for economically important traits including abiotic stresses. However, so far the real worth of these collections could not be realized largely due to unavailability of their evaluation data. It is nearly impossible for breeders to evaluate several thousands of wheat accessions for every trait, particularly for abiotic stresses. Therefore, greater emphasis is placed on identification of smaller set of wheat genotypes designated as core (approximately 10% accessions of the original collection) or mini-core collections (approximately 1% accessions of the original collection) which are representative of total diversity present in the original wheat collection.[45,91,153]

Many studies have identified wheat core and mini-core sets and large number of them are country specific.[19,67,68] Wheat core collections once identified are valuable genetic resource for genome wide and candidate based association studies for discovering novel genes and alleles for abiotic stress tolerance. With the availability of high-throughput maker systems such as DART GBS and SNP assays, core collections can be genotyped with greater marker density for high-resolution LD mapping of abiotic stress tolerance traits. Additionally, the wheat core collections can be mined for novel allelic variants of the abiotic stress tolerance genes (*DREB, MYB, HKT,* etc.) for their association with salinity tolerance. However, to obtain reliable marker trait association, it is important that core collections are precisely phenotyped under controlled as well multilocation saline-field conditions.

6.7   GENOMIC-BASED EXPLORATION OF SALT TOLERANCE GENES FROM WILD AND RELATED WHEAT SPECIES

Wheat wild relatives and progenitors grow in diverse habitats and thus possess considerable variability for abiotic stress tolerance traits. They represent valuable genetic resource for improving abiotic stress tolerance capacity of cultivated wheats. The tribe *Triticeae* possesses many salt-tolerant species including some halophytes such as different species of wheat grasses (E and J genome species); *Thinopyrum* spp., *Lophopyrum elongatum*, etc.[30,56,111] Some of the accession of *Thinopyrum ponticum* can *survive in 750* mM NaCl.[111] In addition, various diploid A-genome wheat species such as *Tritium uratu*, *T. monococcum*, and *T. boeticum* are considered highly efficient in Na⁺ exclusion as compared to the tetraploid durum wheat.[117,120] Even some of the accessions *Aegilops tauschi*, the DD genome progenitor of bread wheat[55,144] and wild emmer wheat (*Triticum dicoccoides*) have been also shown to possess relatively higher Na⁺ exclusion capacity than cultivated wheat.[28,124]

Physiological studies have shown that Na⁺ exclusion trait to hexaploid wheat was mainly contributed by the D-genome progenitor.[55] Consequently, breeders have always looked toward them for broadening range of variations for salt tolerance traits in cultivated wheats. Several attempts have been made to transfer salt tolerance genes from wild relatives to cultivated wheat.[39,56,90]

The crosses between cultivated wheat and salt-tolerant wild relatives have produced amphidiploids, recombinant lines, and disomic addition and substitution lines having varying level of salt tolerance.[30] However, the value of wheat wild relative and progenitors as donor of salt tolerance genes has not yet been fully realized due to undesirable affects (linkage drag) associated with the transfer of salt-tolerant locus/genomic region.[116] For example, King et al. developed a novel salt-tolerant cereal *Tritipyrum* by crossing *Thinopyrum* with *Triticum* spp.[90] But, *it could* not match to bread wheat with respect to grain yield and considered suitable for forage purpose. Therefore, greater emphasis is placed on minimizing linkage drag while transferring salt tolerance associated traits from wheat wild relatives and progenitors.[149,165] Toward this end, molecular markers may prove very useful.

Using high-throughput marker techniques, salt tolerance traits in wild relatives and progenitors can be fine mapped that will facilitate identification of closely linked flanking markers to the genes/genomic regions controlling these traits. These flanking markers may be used to track the size of the genomic fragments introgressed from the wild relatives and progenitors in each backcross generation. Besides, genomic-based approach can also be applied to identify key genes and molecular mechanisms responsible for high level of salinity tolerance in wild relatives and progenitors. The molecular understanding of salt tolerance in wheat wild relatives and progenitors would accelerate their utilization in wheat improvement program for development of salt-tolerant varieties. The key salt genes identified by genomics approaches can be used to develop salt-tolerant transgenic wheat. The one added advantage of the transgenic approach is that, it helps in overcoming the problem of linkage drag, a major limitation in exploiting wild relatives for breeding salt-tolerant wheat varieties. An integrated genomics approach for development of salt-tolerant wheat varieties is summarized in Figure 6.2.

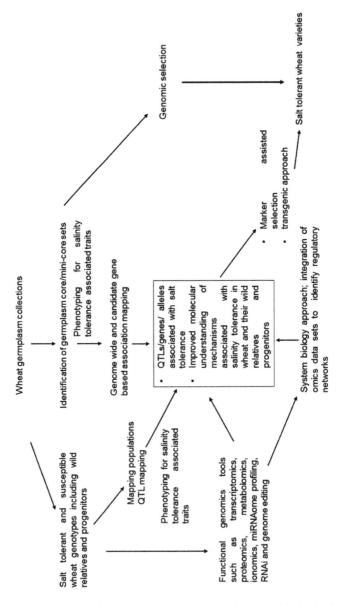

FIGURE 6.2   Flow diagram depicting genomics and related approaches for the development of salt-tolerant wheat varieties.

6.8 CONCLUSION

Genomics-based approaches offer immense opportunity to wheat researchers to perform targeted analysis and molecular dissection of salt tolerance traits and their utilization in designing wheat varieties which can withstand higher level of salinity stress. The need of the hour is to apply these approaches in systematic and integrative fashion to identify salt tolerance associated genetic elements. The key genes/alleles contributing to significantly higher level of salt tolerance can be transferred to elite varieties through markers assisted different breeding approaches. Alternatively, the validated salt tolerance genes/markers can also be successfully introgressed through transgenic approach in the popular wheat cultivars to accelerate wheat breeding program.

6.9 SUMMARY

In the beginning of the chapter, role of wheat crop in global food security is described. Considering the ever growing food demands, it is suggested to expand wheat cultivation to marginally productive saline areas. A summary of the physiological responses of wheat to excess salts is included. It is followed by a review of genomics and related approaches in broadening the current knowledge on physiological and genetic basis of salt tolerance in wheat. This discussion is proceeded to describe the precise genetic manipulation tools such as genetic transformation, marker-assisted selection, and GS for developing salt-tolerant wheat cultivars. An integrated genomics approach for accelerating the breeding for salt tolerance in wheat is described. While discussing the means and ways for enhancing salt tolerance in wheat, it is emphasized that development of salt-tolerant cultivars is an economically viable approach to harness the productivity of SAS.

ACKNOWLEDGMENT

The authors acknowledge financial support by SERB, DST (Research grant No.SB/YS/LS-31/2014).

KEYWORDS

- association mapping
- breeding salt-tolerant wheat varieties
- genome-wide analysis
- HKT genes
- sodium transporters
- marker-assisted selection
- omics approaches
- salt tolerance

REFERENCES

1. Abebe, T.; Guenzi, A. C.; Martin, B.; Cushman, J. C. Tolerance of Mannitol-accumulating Transgenic Wheat to Water Stress and Salinity. *Plant Physiol.* **2003,** *31*, 1748–1755.
2. Agharbaoui, Z.; Leclercq, M.; Remita, M. A.; Badawi, M. A.; Lord, E.; Houde, M.; Danyluk, J.; Diallo, A. B.; Sarhan, F. An Integrative Approach to Identify Hexaploid Wheat miRNAome Associated with Development and Tolerance to Abiotic Stress. *BMC Genom.* **2015,** *16*, 1–17.
3. Ahmad, R.; Lim, C. J.; Kwon, S. Y. Glycine Betaine: A Versatile Compound with Great Potential for Gene Pyramiding to Improve Crop Plant Performance Against Environmental Stresses. *Plant Biotechnol. Rep.* **2013,** *7*, 49–57.
4. Ain, Q.; Rasheed, A.; Anwar, A.; Mahmood, T.; Imtiaz, M.; Mahmood, T.; Xia, X.; He, Z.; Quraishi, U. M. Genome-wide Association for Grain Yield Under Rainfed Conditions in Historical Wheat Cultivars from Pakistan. *Front. Plant Sci.* **2015,** *6*, 1–15.
5. Amar, S. B.; Brini, F.; Sentenac, H.; Masmoudi, K.; Véry, A. A. Functional Characterization in *Xenopus* Oocytes of Na+ Transport Systems from Durum Wheat Reveals Diversity among Two HKT1; 4 Transporters. *J. Exp. Bot.* **2014,** *65*, 213–222.
6. Arbona, V.; Manzi, M.; Ollas, C. D.; Gómez-Cadenas. A. Metabolomics as a Tool to Investigate Abiotic Stress Tolerance in Plants. *Int. J. Mol. Sci.* **2013,** *14*, 4885–4911.
7. Arzani, A. Improving Salinity Tolerance in Crop Plants: A Biotechnological View. *In Vitro Cellul. Dev. Biol. Plants* **2008,** *44*, 373–383.
8. Ashraf, M.; Harris, P. J. Photosynthesis Under Stressful Environments: An Overview. *Photosynthetica* **2013,** *51*, 163–190.
9. Babgohari, M. Z.; Niazi, A.; Moghadam, A. A.; Deihimi, T.; Ebrahimie, E. Genome-wide Analysis of Key Salinity-tolerance Transporter (HKT1; 5) in Wheat and Wild Wheat Relatives (A & D Genomes). *Vitro Cellul. Dev. Biol. Plants* **2013,** *49*, 97–106.
10. Bajgain, P.; Rouse, M. N.; Bulli, P.; Bhavani, S.; Gordon, T.; Wanyera, R.; Njau, P. N.; Legesse, W.; Anderson, J. A.; Pumphrey, M. O. Association mapping of North American Spring Wheat Breeding Germplasm Reveals Loci Conferring Resistance to Ug99 and Other African Stem Rust Races. *BMC Plant Biol.* **2015,** *15*, 1–19.

11. Barciszewska-Pacak, M.; Milanowska, K.; Knop, K.; Bielewicz, D.; Nuc, P.; Plewka, P.; Pacak, A. M.; Vazquez, F.; Karlowski, W.; Jarmolowski, A.; Szweykowska-Kulinska, Z. *Arabidopsis* microRNA Expression Regulation in a Wide Range of Abiotic Stress Responses. *Front. Plant Sci.* **2015**, *6*, 1–14.

12. Barrera-Figueroa, B. E.; Gao, L.; Wu, Z.; Zhou, X.; Zhu, J.; Jin, H.; Liu, R.; Zhu, J. K. High Throughput Sequencing Reveals Novel and Abiotic Stress-regulated microRNAs in the Inflorescences of Rice. *BMC Plant Biol.* **2012**, *12*, 1–11.

13. Bartel, D. P. MicroRNAs: Genomics, Biogenesis, Mechanism, and Function. *Cell* **2004**, *16*, 281–297.

14. Baxter, I. Ionomics: The Functional Genomics of Elements. *Briefing Funct. Genom.* **2010**, *9*, 149–156.

15. Beyene, Y.; Semagn, K.; Mugo, S.; Tarekegne, A.; Babu, R.; Meisel, B.; Sehabiague, P.; Makumbi, D.; Magorokosho, C.; Oikeh, S.; Gakunga, J. Genetic Gains in Grain Yield Through Genomic Selection in Eight Bi-parental Maize Populations Under Drought Stress. *Crop Sci.* **2015**, *55*, 154–156.

16. Bilder, R. M.; Sabb, F. W.; Cannon, T. D.; London, E. D.; Jentsch, J. D.; Parker, D. S.; Poldrack, R. A.; Evans, C.; Freimer, N. B. Phenomics: The Systematic Study of Phenotypes on a Genome-wide Scale. *Neuroscience* **2009**, *164*, 30–42.

17. Bino, R. J.; Hall, R. D.; Fiehn, O.; Kopka, J.; Saito, K.; Draper, J.; Nikolau, B. J.; Mendes, P.; Roessner-Tunali, U.; Beale, M. H.; Trethewey, R. N. Potential of Metabolomics as a Functional Genomics Tool. *Trends Plant Sci.* **2004**, *9*, 418–425.

18. Bonte, A.; Neuweger, H.; Goesmann, A.; Thonar, C.; Mäder, P.; Langenkämper, G.; Niehaus, K. Metabolite Profiling on Wheat Grain to Enable a Distinction of Samples From Organic and Conventional Farming Systems. *J. Sci. Food Agric.* **2014**, *94*, 2605–2612.

19. Bordes, J.; Branlard, G.; Oury, F. X.; Charmet, G.; Balfourier, F. Agronomic Characteristics, Grain Quality and Flour Rheology of 372 Bread Wheats in a Worldwide Core Collection. *J. Cereal Sci.* **2008**, *48*, 569–579.

20. Bowne, J. B.; Erwin, T. A.; Juttner, J.; Schnurbusch, T.; Langridge, P.; Bacic, A.; Roessner, U. Drought Responses of Leaf Tissues from Wheat Cultivars of Differing Drought Tolerance at the Metabolite Level. *Mol. Plant* **2012**, *5*, 418–429.

21. Brini, F.; Hanin, M.; Mezghani, I.; Berkowitz, G. A.; Masmoudi, K. Overexpression of Wheat Na+/H+ Antiporter TNHX1 and H+-pyrophosphatase TVP1 Improve Salt- and Drought-stress Tolerance in *Arabidopsis thaliana* Plants. *J. Exp. Bot.* **2007**, *58*, 301–308.

22. Bruździak, P.; Panuszko, A.; Stangret, J. Influence of Osmolytes on Protein and Water Structure: A Step to Understanding the Mechanism of Protein Stabilization. *J. Phys. Chem. B* **2013**, *117*, 11502–11508.

23. Byrt, C. S.; Platten, J. D.; Spielmeyer, W.; James, R. A.; Lagudah, E. S.; Dennis, E. S.; Tester, M.; Munns, R. HKT1; 5-like Cation Transporters Linked to Na+ Exclusion Loci in Wheat, Nax2 and Kna1. *Plant Physiol.* **2007**, *143*, 1918–1928.

24. Byrt, C. S.; Xu, B.; Krishnan, M.; Lightfoot, D. J.; Athman, A.; Jacobs, A. K.; Watson Haigh, N. S.; Plett, D.; Munns, R.; Tester, M.; Gilliham, M. The Na+ Transporter, *TaHKT1; 5D*, Limits Shoot Na$^+$ Accumulation in Bread Wheat. *Plant J.* 2014, *80*, 516–526.

25. Caruso, G.; Cavaliere, C.; Guarino, C.; Gubbiotti, R.; Foglia, P.; Laganà, A. Identification of Changes in *Triticum durum* L. Leaf Proteome in Response to Salt Stress by Two-dimensional Electrophoresis and MALDI-TOF Mass Spectrometry. *Anal. Bioanal. Chem.* **2008**, *391*, 381–390.

26. Cavanagh, C. R.; Chao, S.; Wang, H. B. E.; Stephen. S.; Kiani, S.; Forrest, K.; Saintenac, C.; Brown-Guedira, G. L.; Akhunova, A.; See, D.; Bai, G.; Pumphrey, M.; Tomar, L.; Wong, D.; Kong, S.; Matthew Reynolds, M.; Lopez, M. S.; Bockelman, H.; Talbert, L.; Anderson, J. A.; Dreisigacker, S.; Baenziger, S.; Carter, A.; Korzun, V.; Morrell, P. L.; Dubcovsky, J.; Morell, M. K.; Sorrells, M. E.; Hayden, M. J.; Akhunov, E. Genome-wide Comparative Diversity Uncovers Multiple Targets of Selection for Improvement in Hexaploid Wheat Landraces and Cultivars. *Proc. Natl. Acad. Sci.* **2013**, *110*, 8062–8067.

27. Chinnusamy, V.; Jagendorf, A.; Zhu, J. K. Understanding and Improving Salt Tolerance in Plants. *Crop Sci.* **2005**, *45*, 437–448.

28. Chen, L.; Ren, J.; Shi, H.; Chen, X.; Zhang, M.; Pan, Y.; Fan, J.; Nevo, E.; Sun, D.; Fu, J.; Peng, J. Physiological and Molecular Responses to Salt Stress in Wild Emmer and Cultivated Wheat. *Plant Mol. Biol. Rep.* **2013**, *31*, 1212–1219.

29. Cobb, J. N.; DeClerck, G.; Greenberg, A.; Clark, R; McCouch, S. Next-generation Phenotyping: Requirements and Strategies for Enhancing our Understanding of Genotype – Phenotype Relationships and its Relevance to Crop Improvement. *Theoret. Appl. Genet.* **2013**, *126*, 867–887.

30. Colmer, T. D.; Flowers, T. J.; Munns, R. Use of Wild Relatives to Improve Salt Tolerance in Wheat. *J. Exp. Bot.* **2016**, *57*, 1059–1078.

31. Coram, T. E.; Brown-Guedira, G.; Chen, X. M. Using Transcriptomics to Understand the Wheat Genome. *CAB Rev.: Perspect. Agric., Vet. Sci. Nutr. Natl. Resour.* **2008**, *83*, 1–9.

32. Cramer, G. R.; Ergül, A.; Grimplet, J.; Tillett, R. L.; Tattersall, E. A.; Bohlman, M. C.; Vincent, D.; Sonderegger, J.; Evans, J.; Osborne, C.; Quilici, D. Water and Salinity Stress in Grapevines: Early and Late Changes in Transcript and Metabolite Profiles. *Funct. Integr. Genom.* **2007**, *7*, 111–134.

33. Das, G.; Rao, G. J. Molecular Marker Assisted Gene Stacking for Biotic and Abiotic Stress Resistance Genes in an Elite Rice Cultivar. *Front. Plant Sci.* **2015**, *6*, 1–18.

34. De León, J. L.; Escoppinichi, R.; Geraldo, N.; Castellanos, T.; Mujeeb-Kazi, A.; Röder, M. S. Quantitative Trait Loci Associated with Salinity Tolerance in Field Grown Bread Wheat. *Euphytica* **2011**, *181*, 371–383.

35. Deng, P.; Wang, L.; Cui, L.; Feng, K.; Liu, F.; Du, X.; Tong, W.; Nie, X.; Ji, W.; Weining, S. Global Identification of microRNAs and Their Targets in Barley Under Salinity Stress. *PLoS One*, **2015**, *10*, http://dx.doi.org/10.1371/journal.pone.0137990 (accessed on May 18, 2017).

36. Ding, D.; Zhang, L.; Wang, H.; Liu, Z.; Zhang, Z.; Zheng, Y. Differential Expression of miRNAs in Response to Salt Stress in Maize Roots. *Annal. Bot.* **2009**, *103*, 29–38.

37. Dubcovsky, J.; Dvorak, J. Genome Plasticity a Key Factor in the Success of Polyploid Wheat under Domestication. *Science* **2007**, *316*, 1862–1866.

38. Dubcovsky, J.; Santa Maria, G.; Epstein, E.; Luo, M. C.; Dvořák, J. Mapping of the K^+/Na^+ Discrimination Locus Kna1 in Wheat. *Theoret. Appl. Genet.* **1996**, *92*, 448–454.

39. Dvořák, J.; Noamann, M. M.; Goyal, S.; Gorham J. Enhancement of the Salt Tolerance of *Triticum turgidum* L. by the *Kna1* Locus Transferred from the *Triticum aestivum* L. Chromosome 4D by Homologous Recombination. *Theoret. Appl. Genet.* **1994**, *87*, 872–877.

40. Edwards, J.; Shavrukov, Y.; Ramsey, C.; Tester, M.; Langridge, P.; Schnurbusch, T. In *Identification of a QTL on Chromosome 7AS for Sodium Exclusion*, Proceedings of 11th International Wheat Genetics Symposium; Appels, R.; Eastwood, R.; Lagudah,

E.; Langridge, P.; Mackay, M.; McIntyre, L.; Sharp, P. Eds; Sydney University Press: Australia, 2008; Vol. III, pp 891–893.

41. Eren, H.; Pekmezci, M. Y.; Okay, S.; Turktas, M.; Inal, B.; Ilhan, E.; Atak, M.; Erayman, M.; Unver, T. Hexaploid wheat (*Triticum aestivum*) root miRNome Analysis in Response to Salt Stress. *Ann. Appl. Biol.* **2015**, *167*, 208–216.

42. Fahlgren, N.; Gehan, M. A. Baxter, I. Lights, Camera, Action: High-throughput Plant Phenotyping is Ready for a Close-up. *Curr. Opin. Plant Biol.* **2015**, *24*, 93–99.

43. FAO. *How to Feed the World in 2050.* Food and Agriculture Organization, Rome, Italy, 2009.

44. Fiorani, F.; Schurr, U. Future Scenarios for Plant Phenotyping. *Annu. Rev. Plant Biol.* **2013**, *64*, 267–291.

45. Frankel, O. H.; Brown, A. H. D. Plant Genetic Resources Today: A Critical Appraisal. In *Crop Genetic Resources: Conservation & Evaluation*; Holden, J. H. W. and Williams, J. T. Eds.; George Alien & Unwin Ltd.: London, 1984; pp 249–257.

46. Fukushima, A.; Kusano, M.; Redestig, H.; Arita, M.; Saito, K. Integrated Omics Approaches in Plant Systems Biology. *Curr. Opin. Chem. Biol.* **2009**, *13*, 532–538.

47. Furbank, R. T. Plant Phenomics: From Gene to Form and Function. *Funct. Plant Biol.* **2009**, *36*, 5–6.

48. Furbank R. T.; Tester M. Phenomics–Technologies to Relieve the Phenotyping Bottleneck. *Trends Plant Sci.* **2011**, *16*, 635–644.

49. Gao, L.; Yan, X.; Li, X.; Guo, G.; Hu, Y.; Ma, W.; Yan, Y. Proteome Analysis of Wheat Leaf Under Salt Stress by Two-dimensional Difference Gel Electrophoresis (2D-DIGE). *Phytochemistry* **2011**, *72*, 1180–1191.

50. Gao, S. Q.; Chen, M.; Xia, L. Q.; Xiu, H. J.; Xu, Z. S.; Li, L. C.; Zhao, C. P.; Cheng, X. G.; Ma, Y. Z. A Cotton (*Gossypium hirsutum*) DRE-binding Transcription Factor Gene, GhDREB, Confers Enhanced Tolerance to Drought, High Salt, and Freezing Stresses in Transgenic Wheat. *Plant Cell Rep.* **2009**, *28*, 301–311.

51. Genc, Y.; Mcdonald, G. K.; Tester, M. Reassessment of Tissue Na+ Concentration as a Criterion for Salinity Tolerance in Bread Wheat. *Plant Cell Environ.* **2007**, *30*, 1486–1498.

52. Genc, Y.; Oldach, K.; Verbyla, A. P.; Lott, G.; Hassan, M.; Tester, M.; Wallwork, H.; McDonald, G. K. Sodium Exclusion QTL Associated with Improved Seedling Growth in Bread Wheat Under Salinity Stress. *Theoret. Appl. Genet.* **2010**, *121*, 877–894.

53. Genc, Y.; Oldach, K.; Gogel, B.; Wallwork, H.; McDonald, G. K.; Smith, A. B. Quantitative Trait Loci for Agronomic and Physiological Traits for a Bread Wheat Population Grown in Environments with a Range of Salinity Levels. *Mol. Breed.* **2013**, *32*, 39–59.

54. Golzarian, M. R.; Frick, R. A.; Rajendran, K.; Berger, B.; Roy, S.; Tester, M.; Lun, D. S. Accurate Inference of Shoot Biomass from High-throughput Images of Cereal Plants. *Plant Methods* **2011**, *7*, e-article; DOI: 10.1186/1746-4811-7-2.

55. Gorham, J. Salt Tolerance in the Triticeae: K/Na Discrimination in *Aegilops* Species. *J. Exp. Bot.* **1990**, *41*, 615–621.

56. Gorham, J.; Jones, R. G. Utilization of *Triticeae* for Improving Salt Tolerance in Wheat. In *Towards the Rational Use of High Salinity Tolerant Plants*; Lieth, H., Al-Massom, A. A. Eds.; Springer: Netherlands, 1993; pp 27–33.

57. Gorham, J.; Hardy, C.; Jones, R. W.; Joppa, L. R.; Law C. N. Chromosomal Location of a K/Na Discrimination Character in the D Genome of Wheat. *Theoret. Appl. Genet.* **1987**, *74*, 584–588.

58. Gouy, M.; Rousselle, Y.; Bastianelli, D.; Lecomte, P.; Bonnal, L.; Roques, D.; Efile, J. C.; Rocher, S.; Daugrois, J.; Toubi, L.; Nabeneza, S. Experimental Assessment of the Accuracy of Genomic Selection in Sugarcane. *Theoret. Appl. Genet.* **2013**, *126*, 2575–2586.

59. Goyal, E.; Amit, S. K.; Singh, R. S.; Mahato, A. K.; Chand, S.; Kanika, K. Transcriptome Profiling of the Salt-stress Response in *Triticum aestivum* cv. *Kharchia Local. Sci. Rep.*, **2016**, *6*, 1–14.

60. Guo, G.; Ge, P.; Ma, C.; Li, X.; Lv, D.; Wang, S.; Ma, W.; Yan, Y. Comparative Proteomic Analysis of Salt Response Proteins in Seedling Roots of Two Wheat Varieties. *J. Prot.* **2012**, *75*, 1867–1885.

61. Gao, L.; Yan, X.; Li, X.; Guo, G.; Hu, Y.; Ma, W.; Yan, Y. Proteome Analysis of Wheat Leaf under Salt Stress by Two-dimensional Difference Gel Electrophoresis (2D-DIGE). *Phytochemistry* **2011**, *72*, 1180–1191.

62. Guo, R.; Yang, Z.; Li, F.; Yan, C.; Zhong, X.; Liu, Q.; Xia, X.; Li, H.; Zhao, L. Comparative Metabolic Responses and Adaptive Strategies of Wheat (*Triticum aestivum*) to Salt and Alkali Stress. *BMC Plant Biol.* **2015**, *15*, 1–13.

63. Gupta, B.; Huang, B. Mechanism of Salinity Tolerance in Plants: Physiological, Biochemical, and Molecular Characterization. *Int. J. Genom.* **2014**, *2014*, 1–18.

64. Gupta, O. P.; Meena, N. L.; Sharma, I.; Sharma, P. Differential Regulation of microRNAs in Response to Osmotic, Salt and Cold Stresses in Wheat. *Mol. Biol. Rep.* **2014**, *41*, 4623–4629.

65. Gupta, P. K.; Kulwal, P. L.; Jaiswal, V. Association Mapping in Crop Plants: Opportunities and Challenges. *Adv. Genet.* **2013**, *85*, 109–147.

66. Hanin, M.; Brini, F.; Ebel, C.; Toda, Y.; Takeda, S.; Masmoudi, K. Plant Dehydrins and Stress Tolerance: Versatile Proteins for Complex Mechanisms. *Plant Signal. Behav.* **2011**, *6*, 1503–1509.

67. Hao, C.; Dong, Y.; Wang, L.; You, G.; Zhang, H.; Ge, H.; Jia, J.; Zhang, X. Genetic Diversity and Construction of Core Collection in Chinese Wheat Genetic Resources. *Chinese Sci. Bull.* **2008**, *53*, 1518–1526.

68. Hao, C.; Wang, L.; Ge, H.; Dong, Y.; Zhang, X. Genetic Diversity and Linkage Disequilibrium in Chinese Bread Wheat (*Triticum aestivum* L.) Revealed by SSR Markers. *PLoS One*, **2011**, *6*, e17279.

69. He, C.; Yang, A.; Zhang, W.; Gao, Q.; Zhang, J. Improved Salt Tolerance of Transgenic Wheat by Introducing *bet-A* Gene for Glycine Betaine Synthesis. *Plant Cell Tissue Organ Cult.* **2010**, *101*, 65–78.

70. He, S.; Schulthess, A. W.; Mirdita, V.; Zhao, Y.; Korzun, V.; Bothe, R.; Ebmeyer, E.; Reif, J. C.; Jiang, Y. Genomic Selection in a Commercial Winter Wheat Population. *Theoret. Appl. Genetics* **2016**, *129*, 641–651.

71. Heffner, E. L.; Sorrells, M. E.; Jannink, J. L. Genomic Selection for Crop Improvement. *Crop Sci.* **2009**, *49*, 1–2.

72. Hill, C. B.; Jha, D.; Bacic, A.; Tester, M.; Roessner, U. Characterization of Ion Contents and Metabolic Responses to Salt Stress of Different *Arabidopsis* AtHKT1; 1 Genotypes and Their Parental Strains. *Mol. Plant* **2013**, *6*, 350–368.

73. Horie, T.; Karahara, I.; Katsuhara, M. Salinity Tolerance Mechanisms in Glycophytes: An Overview with the Central Focus on Rice Plants. *Rice*, **2012**, *5*, 1–18.

74. Hu, W.; Yuan, Q.; Wang, Y.; Cai, R.; Deng, X.; Wang, J.; Zhou, S.; Chen, M.; Chen, L, Huang, C. Ma, Z. Overexpression of a Wheat Aquaporin Gene, TaAQP8, Enhances Salt Stress Tolerance in Transgenic Tobacco. *Plant Cell Physiol.* **2012**, *53*, 2127–2141.

75. Huang, S.; Spielmeyer, W.; Lagudah, E. S.; James, R. A.; Platten, J. D.; Dennis, E. S. Munns, R. A Sodium Transporter (HKT7) is a Candidate for Nax1, a Gene for Salt Tolerance in Durum Wheat. *Plant Physiol.* **2006,** *142,* 1718–1727.

76. Huo, C. M.; Zhao, B. C.; Ge, R. C.; Shen, Y. Z.; Huang, Z. J. Proteomic Analysis of The Salt Tolerance Mutant of Wheat Under Salt Stress. *Yi Chuan Xue Bao* **2004,** *12,* 1408–1414.

77. International Wheat Genome Sequencing Consortium. A Chromosome-based Draft Sequence of the Hexaploid Bread Wheat (*Triticum aestivum*) Genome. *Science* **2014,** *345.* https://www.ncbi.nlm.nih.gov/pubmed/25035500 (accessed on May 18, 2017).

78. Jacoby, R. P.; Millar, A. H.; Taylor, N. L. Wheat Mitochondrial Proteomes Provide New Links Between Antioxidant Defense and Plant Salinity Tolerance. *J. Prot. Res.,* **2010,** *12,* 6595–6604.

79. Jacoby, R. P.; Millar, A. H.; Taylor, N. L. Investigating the Role of Respiration in Plant Salinity Tolerance by Analyzing Mitochondrial Proteomes from Wheat and a Salinity-tolerant Amphiploid (Wheat × *Lophopyrum elongatum*). *J. Prot. Res.* **2013,** *12,* 4807–4829.

80. James, R. A.; Davenport, R. J.; Munns, R. Physiological Characterization of Two Genes for Na$^+$ Exclusion in Durum Wheat, Nax1 and Nax2. *Plant Physiol.* **2006,** *142,* 1537–1547.

81. James, R. A.; Munns, R.; Von Caemmerer, S.; Trejo, C.; Miller, C.; Condon, T. A. Photosynthetic Capacity is Related to the Cellular and Subcellular Partitioning of Na$^+$, K$^+$ and Cl$^-$ in Salt Affected Barley and Durum Wheat. *Plant Cell Environ.* **2006,** *29,* 2185–2197.

82. Jeong, D. H.; Green, P. J. The Role of Rice microRNAs in Abiotic Stress Responses. *J. Plant Biol.* **2013,** *56,* 187–197.

83. Jia, J.; Zhao, S.; Kong, X.; Li, Y.; Zhao, G.; He, W.; Appels, R.; Pfeifer, M.; Tao, Y.; Zhang, X.; Jing, R. *Aegilops tauschii* Draft Genome Sequence Reveals a Gene Repertoire for Wheat Adaptation. *Nature* **2013,** *496,* 91–95.

84. Jiang, Q.; Hu, Z.; Zhang, H.; Ma, Y. Overexpression of GmDREB1 Improves Salt Tolerance in Transgenic Wheat and Leaf Protein Response to High Salinity. *Crop J.* **2014,** *2,* 120–131.

85. Jogaiah, S.; Govind, S. R.; Tran, L. S. Systems Biology-based Approaches Toward Understanding Drought Tolerance in Food Crops. *Crit. Rev. Biotechnol.* **2013,** *33,* 23–39.

86. Jones-Rhoades, M. W.; Bartel, D. P.; Bartel, B. MicroRNAs and Their Regulatory Roles in Plants. *Annu. Rev. Plant Biol.* **2006,** *57,* 19–53.

87. Kamal, A. H.; Cho, K.; Kim, D. E.; Uozumi, N.; Chung, K. Y.; Lee, S. Y.; Choi, J. S.; Cho, S. W.; Shin, C. S.; Woo S. H. Changes in Physiology and Protein Abundance in Salt-stressed Wheat Chloroplasts. *Mol. Biol. Rep.* **2012,** *39,* 9059–9074.

88. Kawaura, K.; Mochida, K.; Ogihara, Y. Genome-wide Analysis for Identification of Salt-responsive Genes in Common Wheat. *Funct. Integr. Genom.* **2008,** *8,* 277–286.

89. Khan, M. S.; Ahmad, D.; Khan, M. A. Utilization of Genes Encoding Osmoprotectants in Transgenic Plants for Enhanced Abiotic Stress Tolerance. *Electr. J. Biotechnol.* **2015,** *18,* 257–266.

90. King, I. P.; Forster, B. P.; Law, C. C.; Cant, K. A.; Orford, S. E.; Gorham, J.; Reader, S.; Miller, T. E. Introgression of Salt Tolerance Genes from *Thinopyrum bessarabicum* into Wheat. *New Phytol.* **1997,** *137,* 75–81.

91. Kobayashi, F.; Tanaka, T.; Kanamori, H.; Wu, J.; Katayose, Y.; Handa, H. Characterization of a Mini Core Collection of Japanese Wheat Varieties Using Single-nucleotide Polymorphisms Generated by Genotyping-by-sequencing. *Breed. Sci.* **2016,** *66,* 213–225.

92. Komatsu, S.; Kamal, A. H.; Hossain, Z. Wheat Proteomics: Proteome Modulation and Abiotic Stress Acclimation. *Front. Plant Sci.* **2014,** *5,* 1–19.

93. Kosová, K.; Vítámvás, P.; Prášil, I. T.; Renaut, J. Plant Proteome Changes Under Abiotic Stress-contribution of Proteomics Studies to Understanding Plant Stress Response. *J. Proteom.* **2011,** *74,* 1301–1322.

94. Kosová, K.; Vítámvás, P.; Prášil, I. T. Proteomics of Stress Responses in Wheat and Barley—Search for Potential Protein Markers of Stress Tolerance. *Front. Plant Sci.* **2014,** *5,* e-article. DOI: 10.3389/fpls.2014.00711.

95. Kumar, S.; Kumari, P.; Kumar, U.; Grover, M.; Singh, A. K.; Singh, R.; Sengar, R. S. Molecular Approaches for Designing Heat Tolerant Wheat. *J. Plant Biochem. Biotechnol.* **2013,** *22,* 359–371.

96. Kumari, A.; Das, P.; Parida, A. K.; Agarwal, P. K. Proteomics, Metabolomics, and Ionomics Perspectives of Salinity Tolerance in Halophytes. *Front. Plant Sci.* **2015,** *6,* e-article. DOI: ORG/10.3389/FPLS.2015.00537.

97. Li, H.; Vikram, P.; Singh, R. P.; Kilian, A.; Carling, J.; Song, J.; Burgueno-Ferreira, J. A.; Bhavani, S.; Huerta-Espino, J.; Payne, T.; Sehgal, D. A High Density GBS Map of Bread Wheat and Its Application for Dissecting Complex Disease Resistance Traits. *BMC Genom.* **2015,** *16,* e-article. DOI: 10.1186/s12864-015-1424-5.

98. Lindsay, M. P.; Lagudah, E. S.; Hare, R. A.; Munns, R. A Locus for Sodium Exclusion (Nax1), A Trait for Salt Tolerance, Mapped in Durum Wheat. *Funct. Plant Biol.* **2004,** *31,* 1105–1114.

99. Ling, H. Q.; Zhao, S.; Liu, D.; Wang, J.; Sun, H.; Zhang, C.; Fan, H.; Li, D.; Dong, L.; Tao, Y.; Gao, C. Draft Genome of The Wheat A-genome Progenitor *Triticum urartu. Nature* **2013,** *496,* 87–90.

100. Linh, L. H.; Linh, T. H.; Xuan, T. D.; Ham, L. H.; Ismail, A. M.; Khanh, T. D. Molecular Breeding to Improve Salt Tolerance of Rice (*Oryza sativa* L.) in the Red River Delta of Vietnam. *Int. J. Plant Genom.* **2012,** *13,* 1–9.

101. Liu, Y.; Wang, L.; Mao, S.; Liu, K.; Lu, Y.; Wang, J.; Wei, Y.; Zheng, Y. Genome-wide Association Study of 29 Morphological Traits in *Aegilops tauschii. Sci. Rep.* **2015,** *5,* 1–12.

102. Lorenzana, R. E.; Bernardo, R. Accuracy of Genotypic Value Predictions for Marker-based Selection in Biparental Plant Populations. *Theoret. Appl. Genet.* **2009,** *120,* 151–161.

103. Lucas, M.; Laplaze, L.; Bennett, M. J. Plant Systems Biology: Network Matters. *Plant Cell Environ.* **2011,** *34,* 535–553.

104. Lv, D. W.; Zhu, G. R.; Zhu, D.; Bian, Y. W.; Liang, X. N.; Cheng, Z. W.; Deng, X.; Yan, Y. M. Proteomic and Phosphoproteomic Analysis Reveals the Response and Defense Mechanism in Leaves of Diploid Wheat *T. monococcum* under Salt Stress and Recovery. *J. Proteom.* **2016,** *143,* 93–105.

105. Ma, L.; Zhou, E.; Huo, N.; Zhou, R.; Wang, G.; Jia. J. Genetic Analysis of Salt Tolerance in a Recombinant Inbred Population of Wheat (*Triticum aestivum* L.). *Euphytica* **2007,** *153,* 109–117.

106. Ma, X.; Liang, W.; Gu, P.; Huang, Z. Salt Tolerance Function of The Novel C2H2-type Zinc Finger Protein TaZNF in Wheat. *Plant Physiol. Biochem.* **2016,** *106,* 129–140.

107. Ma, X. L.; Cui, W. N.; Zhao, Q.; Zhao, J.; Hou, X. N.; Li, D. Y.; Chen, Z. L.; Shen, Y. Z.; Huang, Z. J. Functional Study of a Salt Inducible TaSR Gene in *Triticum aestivum*. *Physiol. Plant.* **2016**, *156*, 40–53.

108. Mangrauthia, S. K.; Agarwal, S.; Sailaja, B.; Madhav, M. S.; Voleti, S. R. MicroRNAs and Their Role in Salt Stress Response in Plants. In *Salt Stress in Plants*; Ahmad, P.; Azooz, M. M., Prasad, M. N. V., Eds.; Springer: New York, 2013; pp 15–46.

109. Mäser, P.; Eckelman, B.; Vaidyanathan, R.; Horie, T.; Fairbairn, D. J.; Kubo, M.; Yamagami, M.; Yamaguchi, K.; Nishimura, M.; Uozumi, N.; Robertson, W. Altered Shoot/Root Na$^+$ Distribution and Bifurcating Salt Sensitivity in *Arabidopsis* by Genetic Disruption of the Na$^+$ Transporter AtHKT1. *FEBS Lett.* **2002**, *531*, 157–161.

110. Massman, J. M.; Jung, H. J. G.; Bernardo, R. Genome-wide Selection Versus Marker-assisted Recurrent Selection to Improve Grain Yield and Stover-quality Traits for Cellulosic Ethanol in Maize. *Crop Sci.* **2013**, *53*, 58–66.

111. McGuire, G. E.; Dvořák, J. High Salt Tolerance Potential in Wheatgrasses. *Crop Sci.* **1981**, *21*, 702–705.

112. Merchuk-Ovnat, L.; Barak, V.; Fahima, T.; Ordon, F.; Lidzbarsky, G. A.; Krugman, T.; Saranga, Y. Ancestral QTL Alleles from Wild Emmer Wheat Improve Drought Resistance and Productivity in Modern Wheat Cultivars. *Front. Plant Sci.* **2016**, *7*, 1–14.

113. Mian, A.; Oomen, R. J.; Isayenkov, S.; Sentenac, H.; Maathuis, F. J.; Véry, A. A. Overexpression of an Na$^+$ and K$^+$ Permeable HKT Transporter in Barley Improves Salt Tolerance. *Plant J.* **2011**, *68*, 468–479.

114. Moghaieb, R. E.; Sharaf, A. N.; Soliman, M. H.; El-Arabi, N. I.; Momtaz, O. A. An efficient and reproducible protocol for the production of salt tolerant transgenic wheat plants expressing the *Arabidopsis* AtNHX1 gene. *GM Crops Food*, **2014**, *5*, 132–138.

115. Mott, I. W.; Wang, R. R. Comparative Transcriptome Analysis of Salt-tolerant Wheat Germplasm Lines using Wheat Genome Arrays. *Plant Sci.* **2007**, *173*, 327–339.

116. Mullan, D. J.; Mirzaghaderi, G.; Walker, E.; Colmer, T. D.; Francki, M. G. Development of Wheat–*Lophopyrum elongatum* Recombinant Lines for Enhanced Sodium 'exclusion' During Salinity Stress. *Theoret. Appl. Genet.* **2009**, *119*, 1313–1323.

117. Munns, R.; James, R. A.; Läuchli, A. Approaches to Increasing the Salt Tolerance of Wheat and Other Cereals. *J. Exp. Bot.* **2006**, *57*, 1025–1043.

118. Munns, R.; James, R. A.; Sirault, X. R.; Furbank, R. T.; Jones, H. G. New Phenotyping Methods for Screening Wheat and Barley for Beneficial Responses to Water Deficit. *J. Exp. Bot.* **2010**, *61*, 3499–3507.

119. Munns, R.; James, R. A.; Xu, B.; Athman, A.; Conn, S. J.; Jordans, C.; Byrt, C. S.; Hare, R. A.; Tyerman, S. D.; Tester, M.; Plett, D. Wheat Grain Yield on Saline Soils is Improved by an Ancestral Na$^+$ Transporter Gene. *Nat. Biotechnol.* **2012**, *30*, 360–364.

120. Munns, R.; Rebetzke, G. J.; Husain, S.; James, R. A.; Hare, R. A. Genetic Control of Sodium Exclusion in Durum Wheat. *Crop Pasture Sci.* **2003**, *54*, 627–635.

121. Munns, R.; Termaat, A. Whole-plant Responses to Salinity. *Funct. Plant Biol.* **1986**, *13*, 143–160.

122. Munns, R.; Tester, M. Mechanisms of Salinity Tolerance. *Annu. Rev. Plant Biol.* **2008**, *59*, 651–681.

123. Narjesi, V.; Mardi, M.; Hervan, E. M.; Azadi, A.; Naghavi, M. R.; Ebrahimi, M.; Zali, A. A. Analysis of Quantitative Trait Loci (QTL) for Grain Yield and Agronomic Traits in Wheat (*Triticum aestivum* L.) under Normal and Salt-stress Conditions. *Plant Mol. Biol. Rep.* **2015**, *33*, 2030–2040.

124. Nevo, E. Evolution of Wild Emmer Wheat and Crop Improvement. *J. Syst. Evol.* **2014,** *52*, 673–696.

125. Ogbonnaya, F. C.; Huang, S.; Steadman, E.; Emebiri, L. C.; Dreccer, M. F.; Lagudah, E.; Munns, R. In *Mapping Quantitative Trait Loci Associated with Salinity Tolerance in Synthetic Derived Backcrossed Bread Lines.* Proceedings of 11th International Wheat Genetics Symposium; Appels, R.; Eastwood, R.; Lagudah, E., Langridge, P., Mackay, M., McIntyre, L., Sharp, P., Eds.; Sydney University Press: Australia, 2008; pp 943–945.

126. Pandit, A.; Rai, V.; Bal, S.; Sinha, S.; Kumar, V.; Chauhan, M.; Gautam, R. K.; Singh, R.; Sharma, P. C.; Singh, A. K.; Gaikwad, K.; Sharma, T. R.; Mohapatra, T.; Singh, N. K. Combining QTL Mapping and Transcriptome Profiling of Bulked RILs for Identification of Functional Polymorphism for Salt Tolerance Genes in Rice (*Oryza sativa* L.). *Mol. Genet. Genomics* **2010,** *284*, 121–136.

127. Pandey, R.; Joshi, G.; Bhardwaj, A. R.; Agarwal, M.; Katiyar-Agarwal, S. A Comprehensive Genome-wide Study on Tissue-specific and Abiotic Stress-specific miRNAs in *Triticum aestivum*. *PLoS One*, **2014,** *9*, 1–15.

128. Peng, Y.; Reyes, J. L.; Wei, H.; Yang, Y.; Karlson, D.; Covarrubias, A. A.; Krebs, S. L.; Fessehaie, A.; Arora, R. RcDhn5, A Cold Acclimation Responsive Dehydrin from *Rhododendron catawbiense* Rescues Enzyme Activity from Dehydration Effects In Vitro and Enhances Freezing Tolerance in RcDhn5 Overexpressing *Arabidopsis* Plants. *Physiol. Plant.* **2008,** *134*, 583–597.

129. Peng, Z.; Wang, M.; Li, F.; Lv, H.; Li, C.; Xia, G. A Proteomic Study of the Response to Salinity and Drought Stress in an Introgression Strain of Bread Wheat. *Mol. Cell. Proteomics* **2009,** *12*, 2676–2686.

130. Platten, J. D.; Egdane, J. A.; Ismail, A. M. Salinity Tolerance, Na^+ Exclusion and Allele Mining of HKT1; 5 in *Oryza sativa* and *O. glaberrima*: Many Sources, Many Genes, One Mechanism? *BMC Plant Biol.* **2013,** *13*. DOI: 10.1186/1471-2229-13-32.

131. Quarrie, S. A.; Steed, A.; Calestani, C.; Semikhodskii, A.; Lebreton, C.; Chinoy, C.; Steele, N.; Pljevljakusić, D.; Waterman, E.; Weyen, J.; Schondelmaier, J. A High-density Genetic Map of Hexaploid Wheat (*Triticum aestivum* L.) from the Cross Chinese Spring × SQ1 and Its Use to Compare QTLs for Grain Yield Across a Range of Environments. *Theoret. Appl. Genet.* **2005,** *110*, 865–880.

132. Rafalski, J. A. Association Genetics in Crop Improvement. *Curr. Opin. Plant Biol.* **2010,** *13*, 174–180.

133. Rajendran, K.; Tester, M.; Roy, S. J. Quantifying the Three Main Components of Salinity Tolerance in Cereals. *Plant Cell Environ.* **2009,** *32*, 237–249.

134. Ramadan, A. M.; Eissa, H. F.; Hassanein, S. E.; Azeiz, A. A.; Saleh, O. M.; Mahfouz, H. T.; El-Domyati, F. M. Increased Salt Stress Tolerance and Modified Sugar Content of Bread Wheat Stably Expressing the mtlD Gene. *Life Sci. J.* **2013,** *10*, 2348–2362.

135. Rong, W.; Qi, L.; Wang, A.; Ye, X.; Du, L.; Liang, H.; Xin, Z.; Zhang, Z. The ERF Transcription Factor TaERF3 Promotes Tolerance to Salt and Drought Stresses in Wheat. *Plant Biotechnol. J.* **2014,** *12*, 468–479.

136. Saad, A. S.; Li, X.; Li, H. P.; Huang, T.; Gao, C. S.; Guo, M. W.; Cheng, W.; Zhao, G. Y.; Liao, Y. C. A Rice Stress-responsive NAC Gene Enhances Tolerance of Transgenic Wheat to Drought and Salt Stresses. *Plant Sci.* **2013,** *203*, 33–40.

137. Sairam, R. K.; Rao, K. V.; Srivastava, G. C. Differential Response of Wheat Genotypes to Long Term Salinity Stress in Relation to Oxidative Stress, Antioxidant Activity and Osmolyte Concentration. *Plant Sci.* **2002,** *163*, 1037–1046.

138. Salt, D. E.; Baxter, I.; Lahner, B. Bionomics and The Study of The Plant Ionome. *Annu. Rev. Plant Biol.* **2008,** *59*, 709–733.

139. Sawahel, W. A.; Hassan, A. H. Generation of Transgenic Wheat Plants Producing High Levels of The Osmoprotectant Proline. *Biotechnol. Lett.* **2002,** *4*, 721–725.

140. Shah, S. H.; Gorham, J.; Forster, B. P.; Jones, R. W. Salt Tolerance in the Triticeae: The Contribution of the D Genome to Cation Selectivity in Hexaploid Wheat. *J. Exp. Bot.* **1987,** *38*, 254–269.

141. Shan, L.; Li, C.; Chen, F.; Zhao, S.; Xia, G. A Bowman Birk Type Protease Inhibitor is Involved in The Tolerance to Salt Stress in Wheat. *Plant Cell Environ.* **2008,** *31*, 1128–1137.

142. Sharma, N.; Gupta, N. K.; Gupta, S.; Hasegawa, H. Effect of NaCl Salinity on Photosynthetic Rate, Transpiration Rate, and Oxidative Stress Tolerance in Contrasting Wheat Genotypes. *Photosynthetica* **2005,** *43*, 609–613.

143. Sharma, D. K.; Singh, A. Salinity Research in India: Achievements, Challenges and Future Prospects. *Water Energy Int.* **2005,** *58*, 35–45.

144. Shiqing, G. A.; Huijun, X. U.; Xianguo, C.; Ming, C.; Zhaoshi, X. U.; Liancheng, L.; Xingguo, Y. E.; Lipu, D. U.; Xiaoyan, H. A.; Youzhi, M. A. Improvement of Wheat Drought and Salt Tolerance by Expression of a Stress-inducible Transcription Factor GmDREB of Soybean (*Glycine max*). *Chinese Sci. Bull.* **2005,** *50*, 2714–2723.

145. Singh, A. K.; Rajkumar, S.; Kumar, R.; Singh, R. Molecular Approaches to Understand Nutritional Potential of Coarse Cereals. *Curr. Genomics* **2016,** *17*, 177–192.

146. Sirault, X. R.; James, R. A.; Furbank, R. T. A New Screening Method for Osmotic Component of Salinity Tolerance in Cereals Using Infrared Thermography. *Funct. Plant Biol.* **2009,** *36*, 970–977.

147. Slama, I.; Abdelly, C.; Bouchereau, A.; Flowers, T.; Savouré, A. Diversity, Distribution and Roles of Osmoprotective Compounds Accumulated in Halophytes Under Abiotic Stress. *Ann. Bot.* **2015,** *115*, 433–447.

148. Sorkheh, K.; Malysheva-Otto, L. V.; Wirthensohn, M. G.; Tarkesh-Esfahani, S.; Martínez-Gómez, P. Linkage Disequilibrium, Genetic Association Mapping and Gene Localization in Crop Plants. *Genet. Mol. Biol.* **2008,** *31*, 805–814.

149. Suiyun, C.; Guangmin, X.; Taiyong, Q.; Fengnin, X.; Yan, J.; Huimin, C. Introgression of Salt-tolerance from Somatic Hybrids Between Common Wheat and *Thinopyrum ponticum*. *Plant Sci.* **2004,** *167*, 773–779.

150. Takahashi, F.; Tilbrook, J.; Trittermann, C.; Berger, B.; Roy, S. J.; Seki, M.; Shinozaki, K.; Tester, M. Comparison of Leaf Sheath Transcriptome Profiles with Physiological Traits of Bread Wheat Cultivars Under Salinity Stress. *PLoS One*, **2015,** *10*, 1–23.

151. Thomson, M. J.; de Ocampo, M.; Egdane, J.; Rahman, M. A.; Sajise, A. G.; Adorada, D. L.; Tumimbang-Raiz, E.; Blumwald, E.; Seraj, Z. I.; Singh, R. K.; Gregorio, G. B. Characterizing the Saltol Quantitative Trait Locus for Salinity Tolerance in Rice. *Rice* **2010,** *3*, 148–160.

152. Turki, N.; Shehzad, T.; Harrabi, M.; Okuno, K. Detection of QTLs Associated with Salinity Tolerance in Durum Wheat Based on Association Analysis. *Euphytica* **2015,** *201*, 29–41.

153. Van Hintum, T. J. L.; Brown, A. H. D.; Spillane, C Hodgkin, T. *Core Collection of Plant Genetic Resources*. Technical Bulletin. No. 3; International Plant Genetic Resources Institute (IPGRI), Rome, 2000; p 51.

154. Van Norman, J. M.; Benfey, P. N. *Arabidopsis thaliana* as a Model Organism in Systems Biology. *Wiley Interdiscip. Rev. Syst. Biol. Med.* **2009,** *1*, 372–379.

155. Vinocur, B.; Altman, A. Recent Advances in Engineering Plant Tolerance to Abiotic Stress: Achievements and Limitations. *Curr. Opin. Biotechnol.* **2005,** *116,* 123–132.

156. Wang, H.; Wang, H.; Shao, H.; Tang, X. Recent Advances in Utilizing Transcription Factors to Improve Plant Abiotic Stress Tolerance by Transgenic Technology. *Front. Plant Sci.* **2016,** *7,* 1–13.

157. Wang, L.; He, X.; Zhao, Y.; Shen, Y.; Huang, Z. Wheat Vacuolar H⁺-ATPase Subunit B Cloning and its Involvement in Salt Tolerance. *Planta* **2011,** *234,* 1–7.

158. Wang, M. C.; Peng, Z. Y.; Li, C. L.; Li, F.; Liu, C.; Xia, G. M. Proteomic Analysis on a High Salt Tolerance Introgression Strain of *Triticum aestivum/Thinopyrum ponticum.* *Proteomics* **2008,** *8,* 1470–1489.

159. Wang, S.; Wong, D.; Forrest, K.; Allen, A.; Chao, S.; Huang, B. E.; Maccaferri, M.; Salvi, S.; Milner, S. G.; Cattivelli, L.; Mastrangelo, A. M. Characterization of Polyploid Wheat Genomic Diversity using a High Density 90,000 Single Nucleotide Polymorphism Array. *Plant Biotechnol. J.* **2014,** *12,* 787–796.

160. Wu, D.; Shen, Q.; Cai, S.; Chen, Z. H.; Dai, F.; Zhang, G. Ionomic Responses and Correlations Between Elements and Metabolites Under Salt Stress in Wild and Cultivated Barley. *Plant Cell Physiol.* **2013,** *54,* 1976–1988.

161. Wu, X.; Gong, F.; Cao, D.; Hu, X.; Wang, W. Advances in Crop Proteomics: PTMs of Proteins Under Abiotic Stress. *Proteomics* **2016,** *16,* 847–865.

162. Xue, Z. Y.; Zhi, D. Y.; Xue, G. P.; Zhang, H.; Zhao, Y. X.; Xia, G. M. Enhanced Salt Tolerance of Transgenic Wheat (*Tritivum aestivum* L.) Expressing a Vacuolar Na⁺/H⁺ Antiporter Gene with Improved Grain Yields in Saline Soils in the Field and a Reduced Level of Leaf Na⁺. *Plant Sci.* **2004,** *167,* 849–859.

163. Yang, C.; Zhao, L.; Zhang, H.; Yang, Z.; Wang, H.; Wen, S.; Zhang, C.; Rustgi, S.; von Wettstein, D.; Liu, B. Evolution of Physiological Responses to Salt Stress in Hexaploid Wheat. *Proc. Natl Acad. Sci.* **2014,** *111,* 11882–11887.

164. Yoshida, T.; Mogami, J.; Yamaguchi-Shinozaki, K. Omics Approaches Toward Defining the Comprehensive Abscisic Acid Signaling Network in Plants. *Plant Cell Physiol.* **2015,** *56,* 1043–1052.

165. Yuan, W. Y.; Tomita, M. *Thinopyrum ponticum* Chromatin-integrated Wheat Genome Shows Salt-tolerance at Germination Stage. *Int. J. Mol. Sci.* **2015,** *16,* 4512–4517.

166. Zhang, L.; Liu, D.; Guo, X.; Yang, W.; Sun, J.; Wang, D.; Sourdille, P.; Zhang, A. Investigation of Genetic Diversity and Population Structure of Common Wheat Cultivars in Northern China using DArT Markers. *BMC Genetics* **2011,** *12,* 1–11.

167. Zhang, M.; Lv, D.; Ge, P.; Bian, Y.; Chen, G.; Zhu, G.; Li, X.; Yan, Y. Phosphoproteome Analysis Reveals New Drought Response and Defense Mechanisms of Seedling Leaves in Bread Wheat (*Triticum aestivum* L.). *J. Proteom.* **2014,** *109,* 290–308.

168. Zhang, Y.; Liu, Z.; Khan, A. W.; Lin, Q.; Han, Y.; Mu, P.; Liu, Y.; Zhang, H.; Li, L.; Meng, X.; Ni, Z.; Zin, M. Expression Partitioning of Homeologs and Tandem Duplications Contribute to Salt Tolerance in Wheat (*Triticum aestivum* L.). *Sci. Rep.* **2016,** *6,* 1–10.

169. Zhang, B. MicroRNA: A New Target for Improving Plant Tolerance to Abiotic Stress. *J. Exp. Bot.* **2015,** *66,* 1749–1761.

170. Zhen, S.; Dong, K.; Deng, X.; Zhou, J.; Xu, X.; Han, C.; Zhang, W.; Xu, Y.; Wang, Z.; Yan, Y. Dynamic Metabolome Profiling Reveals Significant Metabolic Changes During Grain Development of Bread Wheat (*Triticum aestivum* L.). *J. Sci. Food Agri.* **2014,** *96,* 3731–3740.

171. Zheng, C.; Jiang, D.; Liu, F.; Dai, T.; Jing, Q.; Cao, W. Effects of Salt and Waterlogging Stresses and Their Combination on Leaf Photosynthesis, Chloroplast ATP Synthesis, and Antioxidant Capacity in Wheat. *Plant Sci.* **2009,** *176*, 575–582.
172. Zhou, S.; Hu, W.; Deng, X.; Ma, Z.; Chen, L.; Huang, C.; Wang, C.; Wang, J.; He, Y.; Yang, G.; He, G. Overexpression of the Wheat Aquaporin Gene, TaAQP7, Enhances Drought Tolerance in Transgenic Tobacco. *PLoS One,* **2012,** *7*, 1–14.

(A)   (B)

FIGURE 1.1   A saline (A) and a sodic (B) soil in Haryana (India).

FIGURE 1.3   A view of the rice tolerant variety CSR 30 (A) and wheat variety KRL 210 (B).

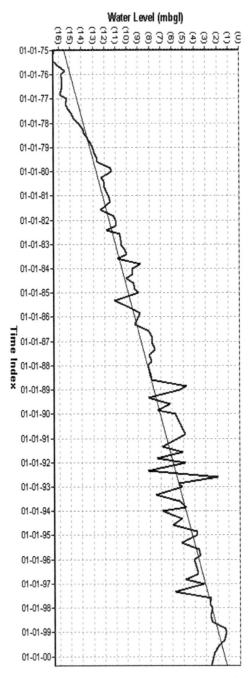

FIGURE 2.1   Rise in water table (0.6 m per annum) upon introduction of irrigation in Bhakra Canal Command in Haryana, India. *Note*: mbgl, meter below ground level.

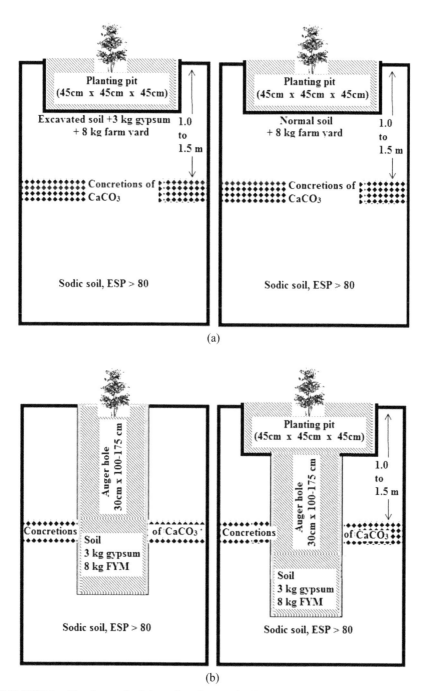

FIGURE 2.2 Planting methods in sodic soils: (a) planting methods in sodic soils: planting pit technique and (b) planting methods in sodic soils: auger hole technique (Courtesy: Authors[70]).

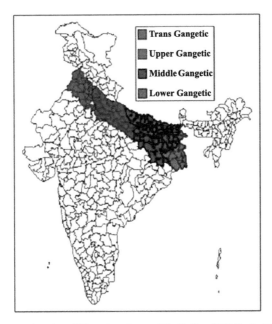

FIGURE 3.1   A map showing different regions of the Indian IGP (Indo-Gangetic plains).

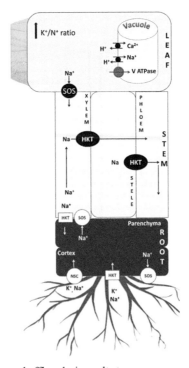

FIGURE 5.1   Ionic influx and efflux during salt stress.

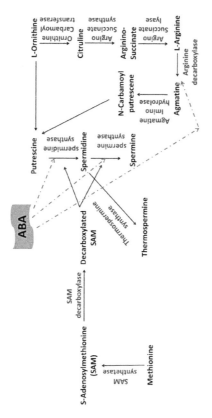

FIGURE 5.2   Polyamine pathway and its cross talk with abscisic acid.

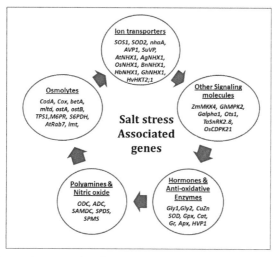

FIGURE 5.3   Interaction networks of stress-associated genes in plant.

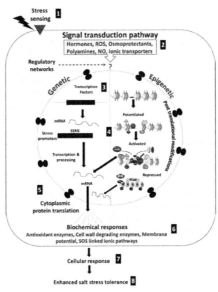

FIGURE 5.4   Model depicting various possible biochemical and molecular responses in salt stress: (1) Stress sensing; (2) Signal transduction through various inducers; (3) Transcription factors; (4) Transcriptional activation of SSRG (salt stress related genes) through genetic and epigenetic interactions; and (5) Synthesis and accumulation of stress proteins, resulting finally in (6) biochemical responses, (7) cellular responses, and (8) physiological response, that is, enhanced salt stress tolerance.

FIGURE 7.1   A view of the screening facilities and screening trials at CSSRI.

FIGURE 8.1 Water stagnation in mustard fields due to heavy rainfall in February (left) and nonuniform growth and yellowing of wheat crop in improperly leveled lands (right).

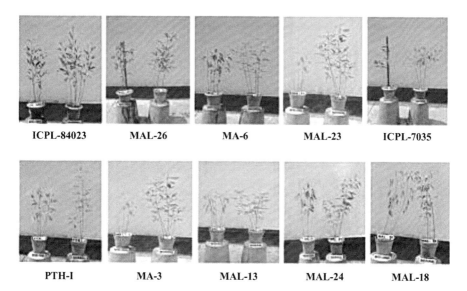

FIGURE 8.3 Effects of waterlogging stress on pigeonpea genotypes (waterlogged plants are to the left and nonwaterlogged plants to the right for each genotype). Stress was imposed after 40 days of sowing and photographs were taken after 20 days of imposing waterlogging stress during 2007–2008.

Source: Adapted from Ref. [22].

FIGURE 10.1 Sparse and stunted plant growth in fenugreek in a saline water irrigated field at Hisar, Haryana, India (Courtesy: Rameshwar Lal Meena).

FIGURE 10.2 Adverse effects of salinity on vegetative growth (left) and seed formation (right) in cumin crop (Jaisalmer, Rajasthan, India).

PART III

Adaptations and Screening of Plants Under Water Logging and Salinity Stresses

CHAPTER 7

MORPHO-BIOCHEMICAL AND MOLECULAR MARKERS FOR SCREENING AND ASSESSING PLANT RESPONSE TO SALINITY

MD. NASIM ALI, LUCINA YEASMIN, VIBHA SINGH, and BHASWATI GHOSH

ABSTRACT

Salinity, the second major factor that limits crop yields after drought, is now a global threat posing several challenges to the crop breeders. Breeding for introgression of increased salt tolerance into existing crop varieties is more energy efficient and cost effective than other available temporary curative and managerial hydrochemical measures commonly adopted to reduce soil salinity. Fundamental understanding of crop response to salinity is required for breeding crops against abiotic stresses. Germplasm collection from widely varying agro-conditions including problem environments and their quick and accurate screening for salt tolerance is considered as an essential prerequisite of any breeding strategy. Comprehensive screening of plants for diverse morphological, biochemical, physiological, and molecular traits will enhance the understanding of relative contribution of different attributes underlying salt tolerance. This chapter emphasizes the importance of appropriate techniques as well as descriptors for screening the crop germplasm under saline conditions to identify elite lines to be utilized for future breeding for salinity tolerance. The pros and cons of each category of descriptors are outlined in the present chapter. Marker linked with quantitative trait locis for salt tolerance and their possible utilization in marker-assisted breeding along with robust phenotyping of germplasm will help to develop salt tolerant varieties of different crops leading to a food secure globe.

7.1   INTRODUCTION

To meet the ever-growing food requirements of a burgeoning global population, food production must increase by 70% by 2050 compared to the current level.[32] It appears to be a herculean task as diverse biotic and abiotic stresses continue to suppress the global crop yields. Abiotic stress in the form of soil salinity, a threat to agriculture known for over 3000 years, has increased manifold both in the extent and severity in recent times.[33,89] The United Nations Environmental Program has estimated that about 20% of the global agricultural land and ~50% of the crop land are salt stressed[35] with at least 20% of all irrigated land being salt affected.[112] Some reports have assessed the extent of damage to varying degrees to exceed 1100 million ha.[146] Reports have also emerged that salinity not only diminishes the potential yields but also reduces the caloric and nutritional value of agricultural products.[152]

Cultivation of salt tolerant crops is getting increasing attention being an effective approach to enhance the productivity of saline soils and water. Salt tolerant varieties are capable of giving consistent yields even with little or no use of chemical amendments. Breeding for introgression of increased salt tolerance into existing crop varieties is more energy efficient and cost effective than other available hydro-chemical approaches. This has significant favorable implications especially for the resource poor farmers in marginal environments often having dismal adaptive capacity and poor policy support. Nonetheless, development of promising salt tolerant cultivars is not an easy task *per se*. It calls for germplasm collection from widely varying agro-conditions including problem environments and their quick and accurate screening for salt tolerance.

Using relevant examples, this chapter underscores the importance of appropriate techniques for screening the crop germplasm to identify the novel lines that can be used as parents in the future breeding programs under salinity conditions.

7.2   SALINITY AND PLANT RESPONSE

7.2.1   DEFINITION OF SALINITY

Salinity as a collective term refers to the problems arising due to presence of excess salts in the crop root zone. Based on the nature and properties, salt-affected soils (SAS) are categorized as

- *Saline soils* have high electrical conductivity of the soil saturation extract (EC_e >4 dS m⁻¹) with Cl⁻ and SO_4^- as the dominant ions. The exchangeable sodium percentage (ESP) in these soils is below 15 and soil pH remains below 8.2.[2] Although saline soils predominantly occur in arid regions and coastal tracts, yet they have been reported from virtually every climatic region of the world.[119]
- *Sodic soils*, widely distributed in semiarid and sub-humid regions, have high concentrations of free carbonates and bicarbonates and excess Na⁺ on the soil exchange complex. They are deficient in nutrients such as nitrogen and zinc. They have high pH_s (>8.2; sometimes up to 10.7) and high ESP (>15). Excess Na⁺ on exchange sites disperses the clay particles resulting in poor aggregate stability. Clay dispersion coupled with low organic matter make these soils sticky when wet and hard when dry. Sodic soils often exhibit quite poor hydraulic conductivity that causes water accumulation on surface after rains or irrigation.
- *Saline-sodic soils* have both high EC_e (>4 dS m⁻¹) and high ESP (>15).[2,99]

7.2.2 RESPONSE OF PLANTS TO SOIL SALINITY

Under saline conditions, plant growth is affected in a number of ways such as those caused by the osmotic stress and specific ion toxicities.[73,152] In most cases, initial injury is attributed to the alterations in plant water relations with subsequent damage being caused by the excess Na⁺ and Cl⁻ ions present in the soil solution. According to the biphasic model of salinity induced growth reduction,

- The initial rapid decrease in plant growth is attributed to *osmotic effect*, that is, changes outside the plant which reduce water uptake by the roots impairing the growth.[97,98,99] Such a condition, often described as physiological drought, is ascribed to the lowering of soil water potential. Interestingly, virtually no genotypic differences are observed under osmotic stress.[109] After several minutes from the initial response, the growth gradually recovers to reach a steady state depending on the outside salt concentration.[98]
- In the next phase, a much slower effect occurs over a period of days or weeks or even months due to accumulation of salt in the leaf (primarily in the older leaves) leading to *salt toxicity*. Heavy leaf injury, and in extreme cases plant mortality, may occur when salt concentration exceeds the capacity of salt compartmentalization in cytoplasm.[101,102] Eventually, membrane disorganization, accumulation

of reactive oxygen species, photosynthetic inhibition, poor nutrient acquisition, and other metabolic impairments prove detrimental to the plant growth.[48]

7.2.3 MANAGEMENT OPTIONS TO MITIGATE SALINITY STRESS

Augmenting the productivity of the existing arable lands has necessitated appropriate agronomic interventions to raise the productivity of cultivated lands as well as to bring marginal and degraded lands (e.g., salt affected soils: SAS) under crop production.[74,131] As far as SAS are concerned, two approaches are suggested to increase their productivity:

- The first approach essentially revolves around the use of chemicals and amendments to alleviate the salt stress such that a soil environment conducive to normal crop growth is created.[127]
- The second approach, in contrast, is based on the development of appropriate crops/cultivars capable of giving higher yields in SAS with no or little use of amendments virtually reducing the costs involved in chemical use.[72]

It is due to these reasons that breeding for salinity tolerance has gradually emerged as the focal point of salinity research.[128]

7.3   SCREENING OF GERMPLASM FOR SALT TOLERANCE

Crop landraces are one of the best sources of novel genes,[44,53,79] since these have evolved over a period of thousands of years due to dynamic interactions between the farmers' and natural selection pressures.[17] As a result, traditional varieties are often better adapted to adverse growing conditions making it necessary to screen them to identify potentially novel alleles for broadening the genetic base so that selection and hybridization can give better results.[154] Since plant species and crop genotypes show differential response under saline conditions, the screening of existing germplasm is an essential prerequisite.

It helps to identify the potential donors for future crop improvement programs within the rich germplasm diversity that exists amongst the existing crops and their varieties. Alternatively, promising high yielding lines identified during screening may directly be released for commercial cultivation.

Salinity is a multigenic trait and even subtle environmental variations could significantly influence plant response to salt stress. For example, changes in atmospheric temperature and humidity may affect evapotranspiration that may in turn affect salt uptake and transport in plants.[129] Moreover, the key physiological processes, controlling the salt tolerance problems, are being modulated by soil heterogeneity and other environmental variables. Such observations reflect the shortcomings of traditional phenotypic evaluation under field conditions. As such, screening under laboratory or controlled condition has many advantages over field screening, although it cannot entirely replace the necessity of field screening.

Some of the screening methods commonly employed are germination trays, pot culture studies, microplots (Fig. 7.1), and long plots having gradient of salinity stress along the length. The screening in hydroponic system can take care of soil heterogeneity. In nutshell, reliable and rapid screening techniques will ensure timely characterization of available germplasm by greatly minimizing the adverse effects of assorted soil and climatic factors.[43]

Highly alkli plot
Note the muddy water,
a chracteristic of
alkali soils

FIGURE 7.1 **(See color insert.)** A view of the screening facilities and screening trials at CSSRI.

7.3.1   *DESCRIPTORS FOR SALINITY SCREENING*

The adverse effects of salinity on plant growth are often determined by timing of exposure, as the growth stage dependent differences in salinity tolerance significantly affect phenotypic expression that varies with the genotypes; indicates to the need to identify growth stage- and genotype-specific markers involved in salt tolerance.[20,31] It is now well established that salinity adversely affects the biometric, morpho-physiological, and biochemical processes in plants necessitating comprehensive screening of plants for diverse morphological, biochemical, physiological, and molecular traits to enhance the understanding of relative contribution of different attributes underlying salt tolerance.[134] The generation of this kind of information can help to identify promising genotypes for the potential use in breeding programs.

7.4   MORPHOLOGICAL DESCRIPTORS FOR SCREENING

Since the degree of plant injury is dependent on the growth stage at which a plant is exposed to salinity stress, screening studies must be performed at various growth stages. Seed germination and seedling establishment stages are usually the most critical stages, resulting in considerable delay and reduction in seed germination and seedling growth. For example, sugar beet is quite sensitive to salts at the germination stage,[90] whereas wheat pearl millet and *Sesbania* are affected more at early seedling growth than germination.[38] Prevention or delay in seed germination is attributed to osmotic stress that reduces water availability[25,145] or ion toxicity that often renders the seeds unviable.[54] Toxic level of Na^+ and Cl^- ions generate an outside osmotic potential that avoids water uptake or results in prolonged dormancy of seeds.[100] The overall response at this stage determines the final crop yield.[6,14] Consistent with the fact that salts also tend to accumulate in the upper soil layers where seeds are planted, studying the effects of salinity on seed germination assumes importance.[113] For some crops (such as rice), phenotypic screening is recommended at different growth stages, namely, germination, seedling, vegetative growth stage, and reproductive stage. It is because rice is reported to be tolerant at germination, becomes very sensitive during early seedling stage (2–3 leaf stage), gains tolerance during vegetative growth stage, becomes sensitive during pollination and fertilization, and then becomes increasingly more tolerant at maturity.[57,111]

7.4.1 SCREENING BASED ON GERMINATION TRAITS

For screening crops and varieties at germination and seedling stage, shallow depth wooden germination trays are commonly used. Seed germination indices such as germination percentage, germination index, germination rate coefficients, hypocotyl length, radical length, seedling fresh weight, seedling dry weight, radical dry weight, hypocotyl dry weight, protein concentration, and seedling vigor index are used to screen for salt tolerance at this stage.[24] As water and nutrient uptake from saline substrates is dependent on root traits, characteristics such as radical length also provide important clues to plant response to salinity stress.[26,95,96]

Relative salt harm rate (RSHR) has been used to grade salt tolerance at this stage, which can be determined using an approach proposed by Bagci et al.[13] The following formula is used to determine RSHR:

$$\text{RSHR}\,(\%) = \left[1 - \frac{(T1 + T2 + \cdots + Tn)/n}{(CK1 + CK2 + \cdots + CKn)/n}\right] \times 100 \qquad (7.1)$$

where CK is the germination rate under control condition (nonsaline) and T is the germination rate under salinity stress. The numbers 1, 2, and n depict the replications used in the trial.

The grading of various varieties/germplasm is made by using the guidelines wherein a grade of highly tolerant is given if the RSHR is equal to or less than 20, salt tolerant if RSHR is more than 20 but less than or equal to 40.[13] The other grades are moderately tolerant (40.1–60), salt sensitive (60.1–80), and highly salt sensitive, if the RSHR is more than 80 or less than or equal to 100.

Seedling stage in many crops is also reported to be a salt sensitive stage. At seedling stage, morphological screening is done using the indices like root length, shoot length, plant biomass, shoot Na^+/K^+ ratio, etc.[3,35,42,61,150] To screen rice germplasm at seedling stage, standard evaluation score (SES) based on visible salt injury has been given by Gregorio.[41] According to their report, visible salt injury to the green leaf area is a convenient and reliable index to assess the salt tolerance at vegetative stage of rice (Table 7.1).

For seedling tolerance of jute, an index of salt harm (ISH) has been proposed.[80] The seedlings in such tests are cultivated under salt stress and nonsaline environment. The seedlings are compared visually and categorized in five groups, namely,

- 0: Seedling growing normally without any injury.
- 1: The edge of one or two leaves of seedling turns yellow and presents some black spots or withers.
- 2: One whole leaf of seedling turning yellow withers, bestrewing with black spots, or falls off.
- 3: Seedling growth is restrained with 2 or 3 leaves severely withering, turning yellow or falling off.
- 4: Seedling growth is severely restrained with many leaves severely withering and falling off or the whole seedling on the verge of death.

TABLE 7.1 SES for Rice Screening at Seedling Stage.

Score	Observation	Tolerance
1	Normal growth, no leaf symptoms	Highly tolerant
3	Nearly normal growth, but leaf tips or few leaves whitish and rolled	Tolerant
5	Growth severely retarded; most leaves rolled and only a few elongating	Moderately tolerant
7	Complete cessation of growth; most leaves dry; some plants dying	Susceptible
9	Almost all plants dead or dying	Highly susceptible

The average ISH of each genotype is calculated by the following formula:

$$AISH(\%)=100 \times \left[\sum (\text{number of '0'} \times 0)+(\text{number of '1'} \times 1)+ \right.$$
$$\left(\text{number of '2'} \times 2\right)+\left(\text{number of '3'} \times 3\right)+\left(\text{number of '4'} \times 4\right)\big] /$$
$$\left[4 \times \text{number of test seedlings}\right] \qquad (7.2)$$

where, number of '0' represents the seedling number of Grade 0, and so forth. The grade guidelines for AISH are the same as discussed for RSHR.

7.4.2 SCREENING BASED ON WHOLE PLANT AND CROP YIELD

Large number of investigations on differential response of crops or varieties have been reported at germination stage and attempts made to extrapolate the tolerance limit for final performance. More than often, such attempts proved frustrating because tolerance characteristics differ from one stage of growth to another. Plant survival has been used as a criterion by ecologists, but without yield, plant survival alone has little value to the farmer. On the other hand, it can prove to be a useful criterion for plant breeders. Absolute plant growth or yield is of greatest interest to the farmers. Grain weight or number of grains per year, 1000-grain weight, yield per plant, yield per line or plot, and percent yield reduction are some of the parameters used in assessing the salt tolerance.

Gregorio et al.[42] have also mentioned that grain yield per plant or plot is the best score for tolerance. In legume crops, a total of 10 morphological characters—plant height, number of branches/plant, number of pods/plant, pod length (cm), number of seeds/plant, 100 seed weight (g), seed density, biomass (g), harvest index, and yield/plant—have been found to estimate the reproductive stage salinity tolerance.[37,138] However, these criteria do not permit comparison between crops because yields of different crops are not expressed in comparable terms. The relative growth or yield, defined as the yield on a saline soil as a fraction of the yield on a nonsaline soil under similar environment and nutritional conditions, seems to be the most desirable and popular criteria used in crop tolerance studies. This criterion is reliable provided the level of other essential factors such as nutritional factor or water does not affect yield reductions.

7.4.2.1   SALINITY AND RELATIVE YIELD

The adverse effects of salinity vary with the crop, variety and plant growth stage at which stress is applied. The relative yield of crops declines with increasing salinity as compared to nonsaline environment. Some crops and varieties are more tolerant to salinity compared to others. However, the decline in yield in most crops can be expressed in the form of a piece-wise linear model.[83] In this model, crop yield remains unaltered until a threshold (EC_t), which represents the salinity up to which there is no appreciable yield reduction. The yield thereafter decline linearly with increasing salinity. The slope (S) of this portion of the curve represents the decrease in yield per unit increase in salinity beyond the threshold (Fig. 7.2). The yield at some salinity level thereafter becomes zero. The salinity is often expressed as EC_0, the salinity of the soil saturation extract at which yield is zero. Mathematically, the model in terms of relative yield (Y) is described as follows[83]:

$$Y = 100, \quad 0 \leq EC \leq EC_t$$

$$Y = 100 - S\,(EC - EC_t), \quad EC_t \leq EC \leq EC_0$$

$$Y = 0 \qquad EC \geq EC_0 \tag{7.3}$$

Although the model is known to have some limitations in describing salinity response functions for some crops, nevertheless it has been widely tested and used. This model has also been validated to describe crop response to specific ion toxicity due to boron, ESP[46] and for salinity of the irrigation water.[91]

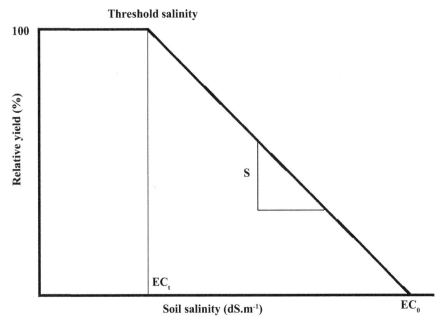

FIGURE 7.2   Crop response function based on piecewise linear model.

On this basis, crop response functions for various crops under different agro-climatic conditions have been derived and relevant parameters presented in Table 7.2.[45,46,91] More comprehensive data have been published by Maas and Hoffman[83] and Mass.[82] According to Mass, crops have been categorized into four groups:

- Sensitive ($EC_t < 1.2$ and $EC_0 < 8$);
- Moderately sensitive (EC_t 1.2–3 and EC_0 8–16);
- Moderately tolerant (EC_t 3–6 and EC_0 16–24); and
- Tolerant (EC_t 6–10 and EC_0 24–32).

The crops in first group are chickpea, beans, sesame, onion, carrot, green gram while in the second group are broad bean, maize, groundnut, sugarcane, alfalfa, *Berseem*, radish, paddy, cauliflower, chilly. Crops in the moderately tolerant and tolerant groups are: wheat, mustard, cowpea, oats, sorghum, soybean, spinach, pearl millet, muskmelon and barley, cotton, sugar beet, safflower, respectively.[82] Later studies in the tolerant crop barley have shown that within this crop, varietal differences exist to make it an ideal crop to grow under saline environment.[45,46,91] From the crop response functions for few grain crops as reported by Shannon,[129] it is revealed that threshold EC_t varied from 1.7 for corn to 8.0 for barley, while slope varied from 5.0 for barley to 16.0 for sorghum (Table 7.3). Any rating on this basis, however, may not provide an accurate absolute estimate of the expected yield as the yield depends not only upon level of salinity but also upon many other cultural and environmental factors. The interaction between salinity and various soil, water and climatic factors modifies the ability of the plant to tolerate salinity.

TABLE 7.2 Crop Tolerance to Soil Salinity.

Sites	Soil type	Crop	Threshold EC (dS m⁻¹)	Slope (%)
Sampla	Sandy loam	Wheat	4.0	29.0
		Mustard	6.0	75.0
		Barley	7.0	19.0
Karnal	Sandy loam	Green gram	1.8	20.7
		Mustard	3.8	6.9
		Sorghum	2.2	10.6
Agra	Sandy loam	Wheat	8.2	19.8
		Mustard	6.1	20.7
		Berseem	3.5	12.5
		Potato	4.4	16.1
		Tomato	1.3	6.5
Dharwad	Black clay	Wheat	2.3	20.5
		Safflower	2.8	20.7
		Sorghum (W)	2.1	3.9
		Italian millet	6.1	50.0
		Sataria	1.1	2.19
Indore	Black clay	*Berseem*	2.0	11.2
		Safflower	2.8	5.0
		Maize	0.5	7.9

TABLE 7.3   Threshold EC and Slope of the Piecewise Linear Model for Few Grain Crops.

Crops	Threshold EC (dS m^{-1})	Slope (%)
Barley	8.0	5.0
Corn	1.7	12.0
Rice	3.0	12.0
Sorghum	6.8	16
Wheat	6.0	7.1

7.5   PHYSIOLOGICAL AND BIOCHEMICAL PARAMETERS FOR SCREENING

A combined pleotropic effect of the low osmotic potential of soil solution, specific ion effects (salt stress), nutritional imbalances and/or a combination of these factors alters plant growth and development at physiological, biochemical and molecular level.[8,67,76] Accordingly, screening needs to be based on distinctive indicators at the whole plant, tissue or cellular levels.[9,21,40,60,98,130] Several physiological responses including alterations in ionic balance, water status, mineral nutrition, stomatal behavior, photosynthetic efficiency and allocation and/or utilization of carbon combine to adversely impact the plant growth due to salt stress.[34,40,102] The visual comparison of growth, survival and biomass production of locally adapted genotypes along with analysis of ionic concentration (Na^+ and Cl^- ions) in different plant organs are the good indicators and can be utilized as effective physiological markers.[36]

Noble and Shannon[108] reported that accumulation of specific ions in shoots or leaves or production of a specific metabolite are also effective indicators of salt tolerance. A study by Cicek and Cakirlar[23] showed that while total fresh and dry weights, shoot length and leaf area decreased, leaf proline, Na^+ and the Na^+/K^+ ratio increased and K^+ levels remained unaffected with increasing salinity leading to a concurrent decrease in leaf relative water content in two maize cultivars.

Singh et al.[133] observed that leaf area index plays an important role in contributing to salt tolerance. Twenty-one rice genotypes were screened for salinity tolerance based on leaf K^+/Na^+ ratio by Asch et al.,[7] higher the ratio, more the expected tolerance to salt stress. Natarajan et al.[105] have grouped rice genotypes for salinity tolerance based on grain yield and Na^+/K^+ ratio under coastal environment.

Osmotic adjustment by plants is one of the best protection mechanisms against salinity stress[134]. It is achieved either by the accumulation of high concentrations of inorganic ions or low molecular weight organic solutes.[124,143,144] Although both the groups of compounds play crucial roles as osmolytes, yet wide differences are noticed not only among the species and cultivars but also among different plant parts of even a single cultivar.[8,40,153] Chlorophyll concentration in stressed plants gives a reasonable estimate of tissue tolerance to NaCl. On the other hand, the mechanisms of chlorophyll degradation in salinized plants still remain obscure.

Yeo et al.[151] observed that the internal leaf Na^+ concentration required to decrease the chlorophyll content by 50% was higher in susceptible rice variety IR4630 than tolerant variety Nona Bokra. Chlorophyll concentration, however, is not necessarily an exclusive measure of tissue tolerance as Yeo et al.[149] demonstrated that in some rice genotypes, photosynthesis reduced by half at such a leaf Na^+ concentration that did not suppress the chlorophyll levels. The stomatal conductance, transpiration rate and photosynthesis were severely affected in common bean cultivars *Bassbeer*, *Beladi*, *Giza* 3, HRS 516, and RO21 exposed to 50 and 100 mM NaCl stress. Leaf turgor and leaf osmotic potential also significantly differed with the cultivars and salt levels.[36]

One of the major biochemical changes occurring in salt stressed plants is the excessive production of activated oxygen species (AOS) that can damage membrane lipids, proteins, and nucleic acids.[39, 93] Plants possess a number of antioxidant enzymes that protect them against the damaging effects of AOS. Superoxide dismutase (SOD) is a major scavenger of O_2, and its enzymatic action results in the formation of H_2O_2 and O_2.[28,51] The H_2O_2 produced is then scavenged by catalase (CAT) and a variety of peroxidases. Free radicals induce the lipid peroxidation that in turn causes the cell membrane instability and disorganization.[27,49,68,86] A significant decrease of SOD activity in the salt sensitive varieties of rice has been reported, while the tolerant variety showed slightly higher SOD activity in response to salt treatment. Singha and Choudhuri[136] observed that NaCl decreased SOD activity in rice seedlings. Salinity also decreased SOD activity in the leaves, chloroplasts and mitochondria of pea plants.[49,52] Hernandez et al.[50] reported that decrease in the catalytic activity of SOD isozymes in cowpea plants was a function of in vitro salt concentration.

Researchers dealing with rice[92] and other plants[65,81,132] have shown increased peroxidase activity in salt sensitive cultivars under salt stress. It is not clear whether the observed increase in salt induced peroxidase activity was due to increased activity of peroxidase encoding genes or increased activation of already existing enzymes. Mittal and Dubey[92] suggested that salinity mainly affects the *de novo* synthesis of enzyme(s) since inhibition under in vitro conditions and activation under in vivo conditions were observed in the salt sensitive cultivars. Besides their role in the metabolism of active oxygen, peroxidases in plants are also involved in the biosynthesis of cell wall constituents[106] and play important roles in cell wall lignification and suberization.[106,114] Available evidences show that high peroxidase activity is correlated with the reduction in plant growth.[75,84,155]

The accumulation of soluble carbohydrates has been widely reported as a response to salinity or drought stresses despite the significant decrease in net CO_2 assimilation rate.[103,115] Several studies have tried to establish correlations between plant responses to salinity with changes in carbohydrate concentrations. Ashraf and Tufail[12] reported significant increase in total sugar content in salt tolerant lines as compared to salt sensitive lines of sunflower. In another study in sunflower, differential patterns of soluble sugar accumulation were noted with one of the salt tolerant lines showing solute concentration equal to that observed in salt sensitive accessions.[10] These observations suggest considerable variation with regard to the accumulation of soluble sugars in response to salt stress both at interspecific and intraspecific levels. Despite many such experiments, precise information regarding the role(s) of soluble sugars in adaptation of plants to salinity is still insufficient to draw valid conclusions to designate it as a distinct indicator for salinity tolerance.

Proline, which widely occurs in higher plants, accumulates in larger amounts than other amino acids in salt stressed plants.[4,8,63,153] Its accumulation has been reported in response to both water deficit as well as salinity.[8] Proline regulates the accumulation of useable N, is osmotically very active,[8,63] contributes to membrane stability,[47,78,121] and mitigates the effect of NaCl on cell membrane disruption. Even at supra-optimal levels, proline does not suppress enzyme activity.[29,64] Maggio et al.[85] suggested that proline may act as a signaling/regulatory molecule able to activate multiple responses that are components of the adaptation process. In addition to acting as an osmoprotectant,[22] proline also serves as a sink for energy to regulate the redox potential,[18,125] as a hydroxy radical scavenger[137] and as a solute that protects macromolecules against denaturation.[126] Although proline accumulation is one of the common characteristics in many monocotyledonous plants under saline conditions,[62,139] yet barley seedlings did not show any appreciable changes in proline accumulation under NaCl stress.[148]

Proteins may be synthesized *de novo* in response to salt stress or may be present constitutively in low concentrations and increase when the plants are exposed to salt stress.[110] Proteins that accumulate in plants grown under saline conditions may provide a storage form of nitrogen that is reutilized when stress is relieved[135] and may play a role in osmotic adjustment. Several salt-induced proteins identified in plants have been classified into two distinct groups.[4,55,87,110] The first group comprises proteins that accumulate only under salinity. The second group consists of stress associated proteins which also accumulate in response to heat, cold, drought, waterlogging, and nutrient deficiency. A higher content of soluble proteins has been observed in salt tolerant than in salt sensitive cultivars of barley,[55] sunflower,[12] finger millet,[142] and rice.[80,110,117] During characterization of salt induced proteins in tobacco, a 26-kDa protein named as osmotin was detected.[135] In salt stressed *Mesembryanthemum crystallinum* osmotin-like protein increased relative to unstressed plants.[140] In barley, two 26-kDa polypeptides, immunologically unrelated to osmotin and identified as germin, increased in response to salt stress.[56]

Lopez et al.[79] found a 22-kDa protein in response to salt stress in radish. In wheat, Ashraf and O'Leary[11] reported that total soluble proteins increased due to salt stress in all the tested cultivars, the increase being marked in the salt sensitive cultivar Potohar but less pronounced in the salt tolerant line S24 compared with other lines. Ashraf and Fatima[10] found that salt tolerant and salt sensitive accessions of safflower did not differ significantly in soluble proteins present in the leaf.

7.6   MOLECULAR MARKERS FOR SCREENING

The phenotypic expression and the genetic diversity analysis traditionally based on differences in morphological characteristics are significantly influenced by the environment.[104] In comparison to screening under controlled conditions that greatly minimizes the environmental effects, the conventional methods of salinity screening often provide biased results due to environmental influences and narrow sense heritability of salt tolerance.[41] If environmental variation is large in comparison to genetic variation, diversity estimates based on morphological data may poorly reflect the actual level of genetic diversity among the accessions. These shortcomings substantially hinder the development of an accurate, rapid, and reliable screening technique. Use of molecular marker for screening the genotypes is a far more reliable technique than the use of previously described morphological and physiological descriptors.

Studies based on seed proteins[120] and isozymes[59] are also not much efficient owing to the lack of polymorphism and varied (environmental, source tissue, and developmental stage) effects on trait expression. Various molecular marker systems such as restriction fragment length polymorphism (RFLP), random amplification of polymorphic DNA (RAPD), inter simple sequence repeat (ISSR), simple sequence repeat (SSR), and amplified fragment length polymorphism (AFLP)[66] having advantages like growth stage neutrality, freedom from environmental effects, absence of epistasis, and abundant polymorphism can be adopted to overcome most of these shortcomings related to morphological and isozyme markers. DNA markers seem to be the best candidates for the efficient evaluation and selection of germplasm. Recent progress and technical advances in DNA marker technology permit substantial reduction in time while enhancing the precision of results.[70]

Zhang et al.[156] worked with 130 RFLP probes distributed throughout the rice genome using an F_2 population derived from a salt tolerant japonica rice mutant M-20 and the sensitive parent 77-170A. They mapped one important gene involved in salt tolerance, which can be utilized for evaluating the salt tolerance in rice. El-Kadi et al.[30] worked on Egyptian cotton (*Gossypium barbadense* L.) to develop molecular markers linked to salt tolerance. They utilized RAPD markers and identified three potential positive RAPD bands of 317 bp, 1507 bp, and 2137 bp linked to salt tolerance response using *OP-Z07* and *OP-Z15* primers which could be utilized further for evaluation of potential resistant genotypes. RAPD technique has been employed in rice for the evaluation of salt tolerant germplasm and polymorphic differential banding pattern was observed among tolerant and susceptible lines which further confirmed the phenotypic evaluation.[88] SSR markers are also an efficient tool for genetic mapping,[58,107] marker-assisted selection,[16] and marking the genetic diversity in crop plants. SSR markers can play an important role in gene identification for expediting the crop improvement programs for salt tolerance.

Lang et al.[71] detected a significant association of salt tolerance at SSR locus RM223 on chromosome 8 and validated it using 93 improved varieties of rice. Sabouri and Biabani[122] mapped quantitative trait loci (QTL) for salt tolerance in rice using 74 SSR markers on rice chromosomes 1, 2, 3, and 6. Mohammadi-Nejad et al.[94] screened 36 rice genotypes using 33 SSR markers linked to *Saltol* QTL. Yadav et al.[147] employed 28 ISSR primers for genetic diversity assessment of 9 rice cultivars grown under saline conditions. Their results showed high degree of polymorphism among the salt sensitive and salt tolerant cultivars.

Bhowmik et al.[15] screened rice genotypes using RM7075, RM336, and RM253 SSR markers for seedling stage salinity tolerance. The genotypes having similar banding pattern with Pokkali were considered as salt tolerant. Ali et al.[5] reported 4 landraces of rice as salt tolerant varieties using 11 SSR markers linked to salt tolerant QTL. Abbasi et al.[1] studied the association between SSR markers and morpho-physiological traits linked to salinity in *Beta vulgaris* and identified two potential SSR markers for screening. The markers were validated in 168 genotypes.

Several QTLs linked to salt tolerance have been mapped on different chromosomes of rice using RIL population.[123] Prasad et al.[116] identified seven QTLs linked to seedling stage salinity tolerance that were dispersed on 5 different chromosomes. Koyama et al.[69] reported the chromosomal location of ion transport and selectivity traits that are compatible with agronomic needs. They showed that QTL for Na^+ transport are likely to act through the control of root development and K^+ uptake are likely to act through structure and the regulation of membrane localized transport compartmentalization. A major QTL on rice chromosome 1 has been mapped which is flanked by SSR marker RM23 and RM140 and named as *Saltol*.[43] The QTL was mapped using a population generated from a cross between salt-sensitive IR29 and salt-tolerant Pokkali. This QTL accounted for more than 70% of the variation in salt uptake and is now being mapped within 1-cM distance using a large set of near isogenic lines.[19]

Lin et al.[77] identified two major QTLs linked to salt stress in rice with very large effect. The QTLs qSNC-7 and qSKC-1 for shoot Na^+ and shoot K^+ concentration, respectively, explained 48.5% and 40.1% of the total phenotypic variance. Ren et al.[118] suggested that *SKC1* is involved in regulating K^+/Na^+ homeostasis under salt stress, providing a potential tool for improving salt tolerance in crops. Thomson et al.[141] identified QTLs for seedling height and chlorophyll content on chromosome 4 and for root NA^+/K^+ ratios on chromosome 9 while they were investigating the *Saltol* QTL. Their experiment also confirmed the location of *Saltol* QTL on chromosome 1 and revealed that *Saltol* mainly acted to control shoot NA^+/K^+ homeostasis.

7.7 CONCLUSION

Cultivation of salt tolerant varieties has emerged as one of the major low cost intervention to manage and bio-reclaim salt affected lands. Contrarily, development of promising salt tolerant cultivars to suit various kinds and degrees of salt stressors is not an easy task. It is concluded that amongst the various morphological traits relative yield seems to be quite appropriate to screen strains against salt tolerance. Although few physiological and biochemical parameters are relevant yet more work is required to finalize the appropriate recommendations to ensure consistent results. To expedite the salt tolerant varieties development programs, use of molecular marker systems such as RFLP, RAPD, ISSR, SSR, and AFLP hold promise to screen germplasm being efficient and cost effective.

7.8   SUMMARY

Cultivation of salt tolerant crops is getting increasing attention being an effective approach to enhance the productivity of saline soils and water. Salt tolerant varieties are capable of giving consistent yields even with little or no use of chemical amendments virtually eliminating the need to procure and apply costly amendments that are also becoming scarce with time. Nonetheless, development of promising salt tolerant cultivars is not an easy task *per se*. Using relevant examples, this chapter tries to underscore the importance of appropriate techniques for screening the crop germplasm using morphological, physiological and biochemical parameters and molecular markers for screening of plant germplasm to identify the novel lines that can be used as parents in the future breeding programs.

Amongst the morphological traits, tests at germination, seedling stage, and crop yield are reviewed to conclude that relative yield seems to be quite appropriate trait for screening of crop cultivars. Similarly, physiological and biochemical parameters are reviewed to conclude that in few cases consistent results are not obtained. Still more work is required to finalize the appropriate recommendations. Consistent with the fact that conventionally used morphological and isozyme descriptors have limited use in germplasm screening, use of molecular markersystems such as RFLP, RAPD, ISSR, SSR, and AFLP are suggested to expedite the screening in a more efficient and cost-effective manner.

ACKNOWLEDGMENTS

The suggestions and modifications done by Dr. S K Gupta, EX-PC, and INAE Distinguished Professor, to finalize this chapter are duly acknowledged.

KEYWORDS

- biochemical parameters
- DNA markers
- morphological traits
- physiological descriptors
- screening-tolerant crops (salt)

REFERENCES

1. Abbasi, Z.; Majidi, M. M.; Arzani, A.; Rajabi, A.; Mashayekhi, P.; Bocianowski, J. Association of SSR Markers and Morpho-physiological Traits Associated With Salinity Tolerance in Sugar Beet (*Beta vulgaris* L.). *Euphytica* **2015,** *205,* 785–797.

2. Abrol, I. P.; Dargan, K. S.; Bhumbla, D. R. *Reclaiming Alkali Soils.* Central Soil Salinity Research Institute, Karnal, 1973; Bulletin 2; p 58.

3. Akbar, M.; Khush, G. S.; Hille Ris L. D. Genetics of salt tolerance in rice. *Proceedings of the Rice Genetics.* International Rice Research Institute, Manila, Philippines, 1986; pp 399–409.

4. Ali, G.; Srivastava, P. S.; Iqbal, M. Proline Accumulation, Protein Pattern and Photosynthesis in Regenerants Grown Under NaCl Stress. *Biol. Plant.* **1999,** *42,* 89–95.

5. Ali, M. N.; Yeasmin, L.; Gantait, S.; Goswami, R.; Chakraborty, S. Screening of Rice Landraces for Salinity Tolerance at Seedling Stage Through Morphological and Molecular Markers. *Physiol. Mol. Biol. Plants* **2014,** *20,* 411–423

6. Almansouri, M.; Kinet, J. M.; Lutts, S. Effect of Salt and Osmotic Stresses on Germination in Durum Wheat (*Triticum durum Desf.*). *Plant Soil* **2001,** *231,* 243–254.

7. Asch, F.; Dingkuhn, M.; Dörffling, K.; Miezan, K. Leaf K/Na Ratio Predicts Salinity Induced Yield Loss in Irrigated Rice. *Euphytica* **2000,** *113,* 109–118.

8. Ashraf, M. Breeding for Salinity Tolerance in Plants. *Crit. Rev. Plant Sci.* **1994,** *13,* 17–42.

9. Ashraf, M. Salt Tolerance of Cotton: Some New Advances. *Crit. Rev. Plant Sci.* **2002,** *21,* 1–30.

10. Ashraf, M.; Fatima, H. Responses of Some Salt Tolerant and Salt Sensitive Lines of Safflower (*Carthamus tinctorius* L.), *Acta Physiol. Plant.* **1995,** *17,* 61–71.

11. Ashraf, M.; O'Leary, J. W. Changes in Soluble Proteins in Spring Wheat Stressed with Sodium Chloride. *Biol. Plant.* **1999,** *42,* 113–117.

12. Ashraf, M.; Tufail, M. Variation in Salinity Tolerance in Sunflower (*Helianthus annuus* L.). *J. Agron. Soil Sci.* **1995,** *174,* 351–362.

13. Bagci, S. A.; Ekiz, H.; Yilmaz, A. Determination of the Salt Tolerance of Some Barley Genotypes and the Characteristics Affecting Tolerance. *Turk. J. Agric. For.* **2003,** *27,* 253–260.

14. Bhattacharjee, S. Triadimefon Pretreatment Protects Newly Assembled Membrane System and Causes Up-regulation of Stress Proteins in Salinity Stressed *Amaranthus lividus* L. During Early Germination. *J. Environ. Biol.* **2008,** *29*, 805–810.

15. Bhowmik, S. K.; Titov, S.; Islam, M. M.; Siddika, A.; Sultana, S.; Hoque, M. D. S. Phenotypic and Genotypic Screening of Rice Genotypes at Seedling Stage for Salt Tolerance. *Afr. J. Biotechnol.* **2009,** *8*, 6490–6494.

16. Bhuiyan M. A. R. *Efficiency in Evaluating Salt Tolerance in Rice using Phenotypic and Marker Assisted Selection.* M.Sc. dissertation; Department of Genetics and Plant Breeding, Bangladesh Agricultural University, Mymensingh, Bangladesh, 2005; p 96.

17. Bisht, V.; Singh, B.; Rao, K. S.; Maikhuri, R. K.; Nautiyal, A. R. Genetic Divergence of Paddy Landraces in Nanakosi Micro-watershed of Uttarakhand Himalaya. *J. Trop. Agric.* **2007,** *45*, 48–50.

18. Blum, A.; Ebercon, A. Genotypic Responses in Sorghum to Drought Stress. III. Free Proline Accumulation and Drought Resistance. *Crop Sci.* **1976,** *16*, 428–431

19. Bonilla, P.; Mackell, D.; Deal, K.; Gregorio, G. RFLP and SSLP Mapping of Salinity Tolerance Genes in Chromosome 1 of Rice (*Oryza sativa L.*) using Recombinant Inbred Lines. *Philippine Agric. Sci. (Philippines),* **2002,** *85*, 68–76.

20. Chartzoulakis, K. S.; Loupassaki, M. H. Effects of NaCl Salinity on Germination, Growth, Gas Exchange and Yield of Greenhouse Eggplant. *Agric. Water Manage.* **1997,** *32*, 215–225.

21. Cheeseman, J. M. Mechanism of Salt Tolerance in Plants. *Plant Physiol.* **1988,** *87*, 547–550.

22. Christian, J. H. The Influence of Nutrition on the Water Relations of *Salmonella oranienburg. Austr. J. Biol. Sci.* **1955,** *8*, 75–82.

23. Cicek, N.; Cakirlar, H. The Effect of Salinity on Some Physiological Parameters in Two Maize Cultivars. *Bulg. J. Plant Physiol.* **2002,** *28*, 66–74.

24. Cokkizgin, A. Salinity Stress in Common Bean (*Phaseolus vulgaris* L.) Seed Germination. *Notulae Botanicae Horti Agrobotanici Cluj-napoca,* **2012,** *40*, 177–182.

25. Dash, M.; Panda, S. K. Salt Stress Induced Changes in Growth and Enzyme Activities in Germinating *Phaseolus mungo* Seeds. *Biol. Plantarum* **2001,** *44*, 587–589.

26. Demir, M.; Arif, I. Effects of Different Soil Salinity Levels on Germination and Seedling Growth of Safflower (*Carthamus tinctorius* L.). *Turkish J. Agric.* **2003,** *27*, 221–227.

27. Demiral, T. I.; Turkan, K. Comparative Lipid Peroxidation, Antioxidant Defence Systems and Proline Content in Roots of Two Rice Cultivars Differing in Salt Tolerance. *Env. Exp. Bot.* **2005,** *53*, 247–257.

28. Dionisio-Sese, M. L.; Tobita, S. Antioxidant Responses of Rice Seedlings to Salinity Stress. *Plant Sci.* **1998,** *135*, 1–9.

29. Dubey, R. S. Photosynthesis in Plants Under Stressful Conditions. In *Handbook of Photosynthesis*; Pessarakli, M. Ed.; Marcel Dekker: New York, 1997; pp 859–875.

30. El-Kadi, D. A.; Afiah, S. A.; Aly, M. A.; Bardan, A. E. Bulked Segregant Analysis to Develop Molecular Markers for Salt Tolerance in Egyptian Cotton. *Arab. J. Biotech.* **2006,** *9*, 129–142.

31. Epstein, E.; Rains, D. W. Advances in Salt Tolerance. *Plant Soil* **1987,** *99*, 17–29.

32. FAO (Food and Agricultural Organization). *How to feed the World by 2050: A High-level Expert Forum.* FAO, Rome; 12–13 October, 2009; p 2.

33. Flowers, T. J. Preface. *J. Exp. Bot.* **2006,** *57*, i–iv.

34. Flowers, T. J.; Teo, A. R. Variability in the Resistance of NaCl Salinity Within Rice (*Oriza sativa L.*) Varieties. *New Phytol.* **1981,** *88*, 363–373.

35. Flowers, T. J.; Yeo, A. R. Breeding for Salinity Resistance in Crop Plants: Where Next? *Aust. J. Plant Physiol.* **1995**, *22*, 875–884.

36. Gama, P. B. S.; Inanaga, S.; Tanaka, K.; Nakazawa, R. Physiological Response of Common Bean (*Phaseolus vulgaris L.*) Seedlings to Salinity Stress. *Afr. J. Biotechnol.* **2007**, *6*, 79–88.

37. Ghafoor, A.; Ahmad, Z.; Qureshi, A. S.; Bashir, M. Genetic Relationship in *Vigna mungo* (L.) Hepper and *Vigna radiata* (L.) R. Wilczek based on Morphological Traits and SDS-PAGE. *Euphytica* **2002**, *123*, 367–378.

38. Gill, K. S. Stage Dependent Differential Effect of Saline Water Irrigation on Growth, Yield and Chemical Composition of Bajra. *Indian J. Agric. Sci.* **1988**, *88*, 210–212.

39. Grant, J. J.; Loake, J. G. Role of Reactive Oxygen Intermediate and Cognate Redox Signaling in Distance Resistance. *Plant Physiol.* **2000**, *124*, 21–29.

40. Greenway H.; Munns, R. Mechanisms of Salt Tolerance in Non-halophytes. *Annu. Rev. Plant Physiol.* **1980**, *31*, 149–190.

41. Gregorio, G. B. *Tagging Salinity Tolerance Genes in Rice using Amplified Fragment Length Polymorphism (AFLP).* Ph.D. dissertation; College, Laguna, Philippines: University of the Philippines Los Banõs, Laguna, 1997; p 118.

42. Gregorio, G. B., Senadhira, D.; Mendoza, R. D. Screening Rice for Salinity Tolerance. *IRRI Discussion Paper Series No. 22*; Philippines, 1997; p 13.

43. Gregorio, G. B.; Senadhira, D.; Mendoza, R. D.; Manigbas, N. L.; Roxas, J. P.; Guerta, C. Q. Progress in Breeding for Salinity Tolerance and Associated Abiotic Stresses in Rice. *Field Crops Res.* **2002**, *76*, 91–101.

44. Guevarra, E.; Loresto, E. C.; Jackson, M. T. Use of Conserved Rice Germplasm. *Plant Genet. Resour. Newsl.* **2001**, *124*, 51–56.

45. Gupta, S. K.; Gupta, I. C. *Crop Production in Waterlogged Saline Soils.* Scientific Publisher, Jodhpur, 1997; p 239.

46. Gupta, S. K.; Sharma, S. K. Response of Crops to High Exchangeable Sodium Percentage. *Irrig. Sci.*, **1990**, *11*, 173–179.

47. Hanson, A. D.; Burnet, M. Evolution and Metabolic Engineering of Osmoprotectant Accumulation in Higher Plants. In *Biochemical and Cellular Mechanisms of Stress Tolerance in Plants*; Cherry, J. H. Ed.; Springer-Verlag: Berlin, 1994; pp 291–301.

48. Hasegawa, P. M.; Bressan, R. A.; Zhung, J. K.; Bohnert, H. J. Plant Cellular and Molecular Responses to High Salinity. *Plant Mol. Biol.* **2000**, *51*, 463–499.

49. Hernandez, J. A.; Corpas, F. J.; Gomez, M.; Del Rio, L. A.; Sevilla, F. Salt-induced Oxidative Stress Mediated by Activated Oxygen Species in Pea Leaf Mitochondria. *Physiol. Plant.* **1993**, *89*, 103–110.

50. Hernandez, J. A.; Del Rio, L. A.; Sevilla, L. A. Salt Stress-induced Changes in Superoxide Dismutase Isozymes in Leaves and Mesophyll Protoplasts from *Vigna unguiculata* (L.) Walp. *New Phytol.* **1994**, *126*, 37–44.

51. Hernandez, J. A.; Mullineaux, P.; Sevilla, F. Tolerance of pea (*Pisum sativum* L.) to Long Term Stress is Associated with Induction of Antioxidant Defences. *Plant Cell Environ.* **2000**, *23*, 853–862.

52. Hernandez, J. A.; Olmos, E.; Corpas, F. J.; Sevilla, F.; Del Rio, L. A. Salt-induced Oxidative Stress in Chloroplasts of Pea Plants. *Plant Sci.* **1995**, *105*, 151–167.

53. Hoisington, D.; Khairallah, M.; Reeves, T.; Ribant J. M.; Skovand, B.; Taba, S.; Warburton, M. Plant Genetic Resources: What Can They Contribute Toward Increased Crop Productivity? *Proc. Natl. Acad. Sci. U.S.A.* **1999**, *96*, 5937–5943.

54. Huang, J.; Reddman, R. E. Salt Tolerance of Hordeum and Brassica Species During Germination and Early Seedling Growth. *Can. J. Plant Sci.* **1995**, *75*, 815–819.
55. Hurkman, W. J.; Fornari, C. S.; Tanaka, C. K. A Comparison of the Effect of Salt on Polypeptide and Translatable mRNA in Roots of a Salt Tolerant and Salt Sensitive Cultivar of Barley. *Plant Physiol.* **1989**, *90*, 1444–1456.
56. Hurkman, W. J.; Rao, H. P.; Tanaka, C. K. Germin-like Polypeptides Increase in Barley Roots During Salt Stress. *Plant Physiol.* **1991**, *97*, 366–374.
57. IRRI (International Rice Research Institute). *Annual Report*. International Rice Research Institute, Manila (Philippines), 1967; p 308.
58. Islam, M. M. *Mapping Salinity Tolerance Genes in Rice (Oryza sativa L.) at Reproductive Stage*. Ph.D. Dissertation; University of the Philippines Los Baños College, Laguna, Philippines, 2004; p 149.
59. Jaaska, V.; Jaaska, V. Isozyme Variation in Asian Bean. *Bot. Acta* **1990**, *103*, 223–322.
60. Jacoby, B. Mechanisms Involved in Salt Tolerance by Plants. In *Handbook of Plants and Crops Stress*; Pessarakli, M., Ed.; Marcel Dekker: New York, 1994; pp 97–123.
61. Jones, M. P. Genetic Analysis of Salt Tolerance in Mangrove Swamp Rice. *Proceedings of the Rice Genetics*; International Rice Research Institute, Manila, Philippines, 1986; pp 411–422.
62. Jones, R. G.; Wyn, S. R. Salt Stress and Comparative Physiology in the Gramineae, IV: Comparison of Salt Stress in *Spartina* × *townsendii* and Three Barley Cultivars. *Aust. J. Plant Physiol.* **1978**, *5*, 839–850.
63. Jones, R. G.; Wyn, S. R. Salt Tolerance. In *Physiological Processes Limiting Plant Productivity*; Johnson C. B., Ed.; Butterworths: London, 1981; pp 271–292.
64. Jones, W.; Gorham, R. C.; Mc Donnell, J. Organic and Inorganic Solute Contents in the *Triticeae*. In *Salinity Tolerance in Plants;* Staples, R. C., Tonniessen, G. H., Eds.; Wiley: New York, 1984; pp 189–203.
65. Kalir, A.; Omri, G.; Poljakoff-Mayber, A. Peroxidase and Catalase Activity in Leaves of *Halimione portulacoides* exposed to salinity. *Physiol. Plant.* **1984**, *62*, 238–244.
66. Karp, A.; Seberg, O.; Buiatti, M. Molecular Techniques in the Assessment of Botanical Diversity. *Ann. Bot.* (London), **1996**, *78*, 143–149.
67. Kaymakanova, M.; Stoeva, N. Physiological Reaction of Bean Plants (*Phaseolus vulg.* L.) to Salt Stress. *Gen. Appl. Plant Physiol. Spec.* **2008**, *34*, 3–4.
68. Khan, M. H.; Panda, S. K. Alterations in Root Lipid Peroxidation and Antioxidative Responses in Two Rice Cultivars Under NaCl-salinity Stress. *Acta Physyol. Plant.* **2008**, *30*, 91–89.
69. Koyama, M. L.; Levesley, A.; Koebner, R. M.; Flowers, T. J.; Yeo, A. R. Quantitative Trait Loci for Component Physiological Traits Determining Salt Tolerance in rice. *Plant Physiol.* 2001. *125*, 406–422.
70. Lambridg, C. J.; Godwin, I. D. Mung Bean. In *Genome Mapping and Molecular Breeding in Plants, Volume 3: Pulses, Sugar, and Tuber Crops*; Kole C., Ed.; Springer Verlag.: Heidelberg, 2007; pp 69–90.
71. Lang, N. T.; Buu, B. C.; Ismail, A. Molecular Mapping and Marker-assisted Selection for Salt Tolerance in Rice (*Oryza sativa L.*). *Omon Rice* **2008**, *16*, 50–60.
72. Lang, N. T.; Yanagihara, S.; Buu. B. C. Quantitative Trait Loci for Salt Tolerance in Rice via Molecular Markers. *Omon Rice* **2000**, *8*, 37–48.
73. Läuchli, A.; Epstein, E. Plant Responses to Saline and Sodic Conditions. In *Agricultural salinity Assessment and Management*. ASCE Manuals and Reports on Engineering Practice No. 71; ASCE: New York, 1990; pp 113–137.

74. Lauchli, A.; Grattan, S. R. Plant Growth and Development Under Salinity Stress (Chapter 1). In *Advance in Molecular Breeding towards Salt and Drought Tolerant Crop*; Jenks, M. A., Hasegawa, P. M., Jain, S. M., Eds.; Springer, 2007; pp 1–32.

75. Lee, T. M.; Lin, Y. H. Changes in Soluble and Cell Wall Bound Peroxidase Activities with Growth in Anoxia-treated Rice (*Oryza sativa* L.) Coleoptiles and Roots. *Plant Sci.* **1995,** *106*, 1–7.

76. Liener, I. E. Antitryptic and Other Anti-nutritional Factors in Legumes. In *Nutritional Improvement of Food Legumes by Breeding*; Milner, M., Ed.; John Wiley and Sons: New York, 1975; pp 239–258.

77. Lin, H. X.; Zhu, M. Z.; Yano, M.; Gao, J. P.; Liang, Z. W.; Su, W. A.; Chao, D. Y. QTLs for Na+ and K+ Uptake of The Shoots and Roots Controlling Rice Salt Tolerance. *Theor Appl Genet*, **2004,** *108*, 253–260.

78. Lone, M. I.; Kueh, J. S. H.; Wyn Jones, R. G.; Bright, S. W. J. Influence of Proline and Glycinebetaine on Salt Tolerance of Cultured Barley Embryos. *J Exp Bot.* **1987,** *38*, 479–490.

79. Lopez, F.; Vansuyt, G.; Fourcroy, P.; Case-Delbart, F. Accumulation of a 22-kDa Protein and its mRNA in the Leaves of *Raphanussativus* in Response to Salt Stress or Water Stress. *Physiol. Plant.* **1994,** *91*, 605–614.

80. Lutts, S.; Kinet, J. M.; Bouharmont, J. Effects of Salt Stress on Growth, Mineral Nutrition and Proline Accumulation in Relation to Osmotic Adjustment in Rice (*Oryza sativa* L.) Cultivars Differing in Salinity Tolerance. *Plant Growth Regul.* **1996,** *19*, 207–218.

81. Ma, H.; Yang, R.; Wang, Z.; Yu, T.; Jia, Y.; Gu, H.; Wand, X.; Ma, H. Screening of Salinity Tolerant Jute (*Corchorus capsularis & C. olitorius*) Genotypes via Phenotypic and Physiology-assisted Procedures. *Pak. J. Bot.* **2011,** *143*, 2655–2660.

82. Maas, E. V. Salt Tolerance of Plants. *App. Agric. Res.* **1986,** *1*, 12–26.

83. Maas E. V.; Hoffman G. J. Crop Salt Tolerance-current Assessment. ASCE *J. Irrig. Drain. Eng.* **1977,** *103*, 115–134.

84. MacAdam, J. W.; Nelson, C. I.; Sharp, R. E. Peroxidase Activity in the Leaf Elongation Zone of Tall Fescue. I. Spatial Distribution of Ionically Bound Peroxidase Activity in Genotypes Differing in Length of the Elongation Zone. *Plant Physiol.* **1992,** *99*, 872–878.

85. Maggio, A.; Miyazaki, S.; Veronese, P.; Fujita, T.; Ibeas, J. I.; Damsz, B.; Narasimhan, M. L.; Hasegawa, P. M.; Joly, R. J.; Bressan, R. A. Does Proline Accumulation Play an Active Role in Stress Induced Growth Reduction? *Plant J.* **2002,** *31*,699–712.

86. Mandhania, S.; Madan, S.; Sawhney, V. Antioxidant Defense Mechanism Under Salt Stress in Wheat Seedlings. *Biol. Plant.* **2006,** *50*, 227–231.

87. Mansour, M. M. F. Nitrogen Containing Compounds and Adaptation of Plants to Salinity Stress. *Biol. Plant.* **2000,** *43*, 491–500.

88. Mansuri, S. M.; Jelodar, N. B.; Bagheri, N. Evaluation of Rice Genotypes for Salt Stress in Different Growth Stages via Phenotypic and Random Amplified Polymorphic DNA (RAPD) Marker Assisted Selection. *Afr. J. Biotechnol.* **2012,** *11*, 9362–9372.

89. Martinez-Beltran, J.; Manzur, C. L. Overview of Salinity Problems in the World and FAO Strategies to Address the Problem. In *Proceedings of the International Salinity Forum*; Riverside, California, 2005; pp 311–313.

90. Mishra, B.; Gill, K. S. Selection of Sugar Beet Varieties for Sodic Soils. *Curr. Agric.* **1978,** *2*, 31–34.

91. Misra, K. L.; Porwal, A. K.; Gupta, S. K. Assessment of Crop Tolerance to Saline Irrigation Waters. *Bhu Jal News,* **1990,** *54*, 35–40.

92. Mittal, R.; Dubey, R. S. Behavior of Peroxidases in Rice: Changes in Enzyme Activity and Isoforms in Relation to Salt Tolerance. *Plant Physiol. Biochem.* **1991,** *29*, 31–40.

93. Mittler, R. Oxidative Stress, Antioxidants and Stress Tolerance. *Trends Plant Sci.* **2002,** *7*, 405–410.

94. Mohammadi-Nejad, G.; Arzani, A.; Rezai, A. M.; Singh, R. K.; Gregorio, G. B. Assessment of Rice Genotypes for Salt Tolerance Using Microsatellite Markers Associated With the Saltol QTL. *Afr. J. Biotechnol.* **2008,** *7*, 730–736.

95. Moose, S. P.; Munns, R. H. Molecular Plant Breeding as the Foundation for 21st Century Crop Improvement. *Plant Physiol.* **2008,** *147*, 969–977.

96. Muhammad, A.; Majid, R. F. Crop Breeding for Salt Tolerance in the Era of Molecular Markers and Marker-assisted Selection. *Plant Breed.* **2013,** *132*, 10–20.

97. Munns, R. Physiological Processes Limiting Plant Growth in Saline Soils: Some Dogmas and Hypotheses. *Plant Cell Environ.* **1993,** *16*, 15–24.

98. Munns, R. Comparative Physiology of Salt and Water Stress. *Plant Cell Environ.* **2002,** *25*, 239–250.

99. Munns R. Genes and Salt Tolerance: Bringing Them Together. *New Phytol.* **2005,** *167*, 645–663.

100. Munns, R.; James, R. A. Screening Methods for Salinity Tolerance: A Case Study with Tetraploid Wheat. *Plant Soil* **2003,** *253*, 201–218.

101. Munns, R.; James, R. A.; Läuchli, A. Approaches to Increasing the Salt Tolerance of Wheat and Other Cereals. *J. Exp. Bot.* **2006,** *57*, 1025–1043.

102. Munns, R.; Termaat, A. Whole-plant Responses to Salinity. *Aust. J. Plant Physiol.* **1986,** *13*, 143–160.

103. Murakeozy, E. P.; Nagy, Z.; Duhaze, C.; Bouchereau, A.; Tuba, Z. Seasonal Changes in the Levels of Compatible Osmolytes in Three Halophytic Species of Inland Saline Vegetation in Hungary. *J. Plant Physiol.* **2003,** *160*, 395–401.

104. Murty, B. R.; Arunachalam, V. The Nature of Divergence in Relation to Breeding in Crop Plants. *Indian J. Genet.* **1966,** *26*(A),188–198.

105. Natarajan, S. K.; Ganapathy, M.; Krishnakumar, S.; Dhanalakshmi, R.; Saliha, B. B. Grouping of Rice Genotypes for Salinity Tolerance based upon Grain Yield and Na:K Ratio Under Coastal Environment. *Res. J. Agric. Biol. Sci.* **2005,** *1*, 162–165.

106. Negrel, J.; Lherminier, J. Peroxidase-mediated Integration of Tyramine into Xylem Cell Walls of Tobacco Leaves. *Planta* **1987,** *172*, 494–501.

107. Niones, J. M. *Fine Mapping of the Salinity Tolerance Gene on Chromosome 1 of Rice (Oryza sativa L.) using Near-isogenic Lines.* M. Sc. Dissertation; University of the Philippines Los Baños College, Laguna, Philippines, 2000; p 78.

108. Noble, C. L.; Shannon, M. C. In *Salt Tolerance Selection of Forage Legumes Using Physiological Criteria.* Proceedings of International Congress of Plant Physiology; Sinha, S. K. et al., Eds.; Society of Plant Physiology and Biochemistry: New Delhi, India, 1988; Vol. 2, pp 989–994.

109. Palfi, G.; Juhasz, J. The Theoretical Basis and Practical Application of a New Method of Selection for Determining Water Deficiency in Plants. *Plant Soil* **1971,** *34*, 503–507.

110. Pareek, A.; Singla, S. L.; Grover, A. Salt Responsive Proteins/Genes in Crop Plants. In *Strategies for Improving Salt Tolerance in Higher Plants*; Jaiwal, P. K., Singh, R. P., Gulati, A., Eds.; Oxford and IBH Publication Co.: New Delhi, 1997; pp 365–391.

111. Pearson, G. A.; Ayers, S. D.; Eberhard, D. L. Relative Salt Tolerance of Rice During Germination and Early Seedling Development. *Soil Sci.* **1966,** *102*, 151–156.

112. Pitman, M. G.; Läuchli, A. Global Impact of Salinity and Agricultural Ecosystems. *Salinity: Environment—Plants—Molecules*, Läuchli, A.; Lüttge, U., Eds.; Kluwer Academic Publishers, Dordrecht, 2002; pp 3–20.

113. Poehlman, J. M. *The Mungbean.* Oxford & IBH Publ. Co., New Delhi, 1991; pp 27–30.

114. Polle, A.; Otter, T.; Seifert, F. Apoplastic Peroxidases and Lignification in Needles of Norway Spruce (*Piceaabies* L.). *Plant Physiol.* **1994,** *106*, 53–60.

115. Popp, M.; Smirnoff, N. Polyol Accumulation and Metabolism During Water Deficit. In *Environment and Plant Metabolism: Flexibility and Acclimation*; Smirnoff, N., Ed.; Bios Scientific: Oxford, 1995; pp 199–215.

116. Prasad, S. R.; Bagali, P. G.; Hittalmani, S.; Shashidhar, H. E. Molecular Mapping of Quantitative Trait Loci Associated with Seedling Tolerance to Salt (*Oryza sativa L.*). *Curr. Sci.* **2000,** *78*, 162–164.

117. Rains, D. W. Plant Tissue and Protoplast Culture: Application to Stress Physiology and Biochemistry. In *Plants Under Stresses. Biochemistry, Physiology and Ecology and Their Application to Plant Improvement*; Jones, H. G., Flowers, T. J., Jones, M. B., Eds.; Cambridge University Press: Cambridge, 1989; pp 181–196.

118. Ren Z. H.; Gao, J. P.; Li, G. L.; Cai, X. L.; Huang, W.; Chao, D. Y.; Zhu, M. Z.; Wang, Z. Y.; Luan, S.; Lin, H. X. A Rice Quantitative Trait Locus for Salt Tolerance Encodes a Sodium Transporter. *Nat. Genet.* **2005,** *37*, 1141–1146.

119. Rengasamy, P. World Salinization with Emphasis on Australia. *J. Exp. Bot.* **2006,** *57*, 1017–1023.

120. Roar, R.; Vaglio, M. D.; D'urzo, M. P.; Monti, L. Identification of *Vigna sp.* through Specific Seed Storage Polypeptides. *Euphytica* **1992,** *62*, 39–43.

121. Rudolph, A. S.; Crowe, J. H.; Crowe, L. M. Effect of three Stabilizing Agents—Proline, Betaine and Trehalose, on Membrane Phospholipids. *Arch. Biochem. Biophys.* **1986,** *245*, 134–143.

122. Sabouri, H.; Biabani, A. Toward the Mapping of Agronomic Characters on a Rice Genetic Map: Quantitative Trait Loci Analysis Under Saline Condition. *Biotechnology* **2009,** *8*, 144–149.

123. Sahi, C.; Singh, A.; Kumar, K.; Blumwald, E.; Grover, A. Salt Stress Response in Rice: Genetics, Molecular Biology, and Comparative Genomics. *Functi. Integr. Genomics* **2006,** *6*, 263–284.

124. Salama, S.; Trivedi, S.; Busheva, M.; Arafa, A.; Garab, G.; Erdei, L. Effects of NaCl Salinity on Growth, Cation Accumulation, Chloroplast Structure and Function in Wheat Cultivars Differing in Salt Tolerance. *J. Plant Physiol.* **1994,** *144*, 241–247.

125. Saradhi, A.; Saradhi, P. P. Proline Accumulation Under Heavy Metal Stress. *J. Plant Physiol.* **1991,** *138*, 554–558.

126. Schobert, B.; Tschesche, H. Unusual Solution Properties of Proline and Its Interaction with Proteins. *Biochem. Biophys. Acta* **1978,** *541*, 271–277.

127. Senguttuvel, P.; Ravindran, M.; Vijayalakshmi, C.; Thiyagarjan, K.; KannanBapu, J. R.; Viraktamath, B. C. Molecular Mechanism of Salt Tolerance for Genetic Diversity Analysed in Association with Na^+/K^+ Ratio Through SSR Markers in Rice. *Int. J. Agric. Res.* **2010,** *5*, 708–719.

128. Shannon, M. C. Breeding, Selection and the Genetics of Salt Tolerance. In *Salt Tolerance in Plants: Strategies for Crop Improvement*; Staples, R. C.; Toenniessen, G. H. Eds.; Wiley: New York, 1984; pp 300–308.

129. Shannon, M. C. Adaptation of Plants to Salinity. *Adv. Agron.* **1997**, *60*, 75–120.

130. Shannon, M. C.; Grieve, C. M. Tolerance of Vegetable Crops to Salinity. *Sci. Hortic.* **1999**, *78*, 5–38.

131. Sharma, D. K.; Singh, A. In *Greening the Degraded Lands: Achievements and Future Perspectives in Salinity Management in Agriculture.* Proceedings of Natural Resource Management in Arid and Semi-arid Ecosystem for Climate Resilient Agriculture; Pareek, N. K.; Arora, S., Eds.; Soil Conservation Society of India: New Delhi, 2016; pp 17–30.

132. Sheoran, I. S.; Garg, O. P. Quantitative and Qualitative Changes in Peroxidase During Germination of *mung* Bean Under Salt Stress. *Physiol. Plant.* **1979**, *46*, 147–150.

133. Singh, A. K.; Ansari, M. W.; Pareek, A.; Singla-Pareek, S. L. Raising Salinity Tolerant Rice: Recent Progress and Future Perspectives. *Physiol. Mol. Biol. Plants* **2008**, *14*, 137–154.

134. Singh, K. N. Crop Improvement Approaches for Salinity and Sodicity Tolerance (Chapter 19). In *Management of High RSC and Saline Waters*; Sharma, D. R.; Bhargava, G. P.; Kumar, C., Eds.; Central Soil Salinity Research Institute, Karnal, 2004; pp 115–120.

135. Singh, N. K.; Bracken, C. A.; Hasegawa, P. M.; Handa, A. K.; Buckel, S.; Hermodson, M. A.; Pfankoch, F.; Regnier, F. E.; Bressan, R. A. Characterization of Osmotin. A Thaumatin-like Protein associated with Osmotic Adjustment in Plant Cells. *Plant Physiol.* **1987**, *85*, 529–536.

136. Singha, S.; Choudhuri, M. A. Effect of Salinity (NaCl) Stress on H_2O_2 Metabolism in *Vigna* and *Oryza* seedlings. *Biochem. Physiol. Pflanzen.* **1990**, *186*, 69–74.

137. Smirnoff, N.; Cumbes, Q. J. Hydroxyl Radical Scavenging Activity of Compatible Solutes. *Phytochemistry* **1989**, *28*, 1057–1060.

138. Srivastava, U.; Mahajan, R. K.; Gangopadhyay, K. K.; Singh, M.; Dhillon, B. S. *Minimal Descriptors of Agri-horticultural Crops,* Part I; NBPGR: New Delhi, 2001; p 262.

139. Storey, R.; Ahmad, N.; Jones, R. G.; Wyn. Taxonomic and Ecological Aspects of the Distribution of Glycinebetaine and Related Compounds in Plants. *Oecologia* **1977**, *27*, 319–322.

140. Thomas, J. C.; Bohnert, H. J. Salt Stress Perception and Plant Growth Regulators in the Halophyte *Mesembryanthemum crystallinum. Plant Physiol.* **1993**, *103*, 1299–1304.

141. Thomson, M. J.; de Ocampo, M.; Egdane, J.; Rahman, M. A.; Sajise, A. G.; Adorada, D. L.; Gregorio, G. B. Characterizing the Saltol Quantitative Trait Locus for Salinity Tolerance in Rice. *Rice* **2010**, *3*, 148–160.

142. Uma, S.; Prasad, T. G.; Kumar, M. U. Genetic Variability in Recovery Growth and Synthesis of Stress Proteins in Response to Polyethylene Glycol and Salt Stress in Finger Millet. *Ann. Bot.* **1995**, *76*, 43–49.

143. Weimberg, R. Growth and Solute Accumulation in 3-week-old Seedlings of *Agropyron elongatum* Stressed with Sodium and Potassium Salts. *Physiol. Plant.* **1986**, *67*, 129–135.

144. Weimberg, R. Solute Adjustments in Leaves of Two Species of Wheat at Two Different Stages of Growth in Response to Salinity. *Physiol. Plant.* **1987**, *70*, 381–388.

145. Welbaum, G. E.; Tissaoui, T.; Bradford, K. J. Water Relations of Seed Development and Germination in Muskmelon (*Cucumismelo* L.) Z. Sensitivity of Germination to Water Potential and Abscisic Acid During Development. *Plant Physiol.* **1990**, *92*, 1029–1037.

146. Wicke, S.; Schneeweiss, G. M.; de Pamphilis, C. W.; Müller, K. F.; Quandt, D. The Evolution of the Plastid Chromosome in Land Plants: Gene Content, Gene Order, Gene Function. *Plant Mol. Biol.* **2011**, *76*, 273–297.

147. Yadav, S.; Rana, P.; Saini, N.; Jain, S.; Jain, R. K. Assessment of Genetic Diversity Among Rice Genotypes with Differential Adaptations to Salinity Using Physio-morphological and Molecular Markers. *J. Plant Biochem. Biotechnol.* **2008,** *17,* 1–8.

148. Yamaya, T.; Matsumoto, H. *Accumulation of Asparagine in NaCl-stressed Barley Seedlings. Berichte des Ohara Institut für landwirtschaftliche, Okayama Universität* **1989,** *19,* 181–188.

149. Yeo, A. R.; Caporn, S. J. M.; Flowers, T. J. The Effect of Salinity Upon Photosynthesis in Rice (*Oryza sativa* L.): Gas Exchange by Individual Leaves in Relation to Their Salt Content. *J. Exp. Bot.* **1985,** *36,* 1240–1248.

150. Yeo, A. R.; Flowers, T. J. Salinity Resistance in Rice (*Oryza sativa L.*) and a Pyramiding Approach to Breeding Varieties for Saline Soils. In *Effect of Drought on Plant Growth. Salts in Soils*; Turner, N. C., Passioura, J. B., Eds.; CSIRO: Melbourne, Australia, 1986; pp 161–173.

151. Yeo, A. R.; Yeo, M. E.; Flowers, S. E.; Flowers, T. J. Screening of Rice (*Oryza sativa L.*) Genotypes for Physiological Characters Contributing to Salinity Resistance and Their Relationship to Overall Performance. *Theor. Appl. Genet.* **1990,** *79,* 377–384.

152. Yokoi, S.; Bressan, R. A.; Hassegawa, P. M. *Salt Stress Tolerance of Plant. JIRCUS* Working Report, 2002; pp 25–33.

153. Yoshiba, Y.; Kiyosue, T.; Nakashima, K.; Yamaguchi-Shinozaki, K. Y.; Shinozaki, K. Regulation of Levels of Proline as an Osmolyte in Plants Under Water Stress. *Plant Cell Physiol.* **1997,** *38,* 1095–1102.

154. Zeng, Y.; Zhang, H.; Li, Z.; Shen, S.; Sun, J.; Wang, M.; Liao, D.; Liu, X.; Wang, X.; Xiao, F.; Wen, G. Evaluation of Genetic Diversity of Rice Landraces *(Oryza sativa)* in Yunan, China. *Breed. Sci.* **2007,** *57,* 91–99.

155. Zheng, X.; van Huystee, R. B. Peroxidase-regulated Elongation of Segments from Peanut Hypocotyls. *Plant Sci.* **1992,** *81,* 47–56.

156. Zhang, G. Y.; Guo, Y.; Chen, S. L.; Chen, S. Y. RFLP Tagging of a Salt Tolerance Gene in Rice. *Plant Sci.* **1995,** *110,* 227–234.

PLANTS UNDER WATERLOGGED CONDITIONS: AN OVERVIEW

ANUJ KUMAR SINGH, PANDURANGAM VIJAI, and
J. P. SRIVASTAVA

ABSTRACT

Waterlogging is considered as one of the major abiotic stresses that adversely affects crop growth, development, and productivity. It impedes diffusion of gases and gaseous exchange leading to conditions of hypoxic followed by anoxic which in turn affects the plant metabolic processes including photosynthesis, respiration, induces production of reactive species oxygen, synthesis and release of ethylene, etc. Redox potential and pH of waterlogged soils are altered resulting in derangement in nutrient availability to plants causing either their deficiency or toxicity. Plants tolerant to waterlogging exhibit adaptations at morphological, anatomical, biochemical, and molecular levels. Morphological adaptations like formation of adventitious root and rapid elongation of above-ground parts of plants help in tolerating submergence stress. Anatomical adaptations include development of aerenchyma to facilitate gaseous exchange, and lignification and suberization of cell walls preventing radial loss of oxygen. Sensing and signaling of low-oxygen levels in rhizosphere induces transcription of hypoxia-responsive genes and accumulation of anaerobic proteins. Group VII ethylene responsive factors are involved in low-oxygen signaling and they act as transcription factors inducing genes required for survival under anaerobic conditions. Shortage of carbohydrate availability increases the perception and transduction of secondary signaling molecules such as sucrose nonfermenting 1-related kinase 1 and target of rapamycin. Transcription factors, namely, basic leucine zipper family are also involved in inducing the expression of enzymes that utilize alternative carbon sources such as protein, structural carbohydrate, and lipids for energy. These anatomical, morphological, and biochemical adaptations in combination help the plants to sustain growth under waterlogged condition. Genes and gene products involved in waterlogging tolerance that have been identified provide opportunities for improving flooding tolerance in crop species.

8.1   INTRODUCTION

Waterlogging is a state of soil when all pore spaces of soil are fully saturated with water or having moisture over and above the saturation point. Such condition may prevail either for short duration (few days to months) or for long duration (permanently). Waterlogging results from water stagnation on the cropped land or it may also result when the water table of the ground water is too shallow to permit agriculture. It is one of the major abiotic stresses affecting growth and productivity of plants. Waterlogging is experienced from unpredicted excess precipitation (Fig. 8.1, top[75]), faulty irrigation practices, and/or poor drainage.[75] It is also commonly encountered on lands that are improperly leveled (Fig. 8.1, bottom).[75] Surface stagnation alone annually affects over 13 million hectare (M ha) of rice area in India, 3 M ha in Bangladesh, 5 M ha in Indonesia, and 1 M ha in Thailand during the monsoon season.[127,200,204] The frequency of floods is expected to increase in Southeast Asia, southern parts of India, East Africa, Siberia, and Northern parts of South America during this century mainly driven by climatic aberrations.[4,83] Climate change is likely to be the main cause of waterlogging of large agricultural lands in the coming few decades.[54]

Waterlogging adversely impacts crop productivity by influencing the physicochemical, biological, and thermal environment of the root zone, resulting in profound changes in physiology of plants. Factors such as anoxia, limited organic matter oxidation, imbalanced uptake of ions, ion toxicity, root damage, leaf epinasty, and other derangements either singly or in combination impact plant growth and development.

The yield reduction from waterlogging has been assessed to range from 15% to 80%, depending on the plant species, soil type, duration, and degree of the stress and the time at which plants are stressed.[216] An assessment revealed that yield losses in wheat were from 8% for a water stagnation of 1 day to 39% when water stagnated for 6 days in a semi-reclaimed alkali land.[77] The wheat yield losses were as high as 75% when the water table came within 70 cm of the soil surface as against the normal value of 120–150 cm.[76]

This chapter discusses and reviews the effects of waterlogging on soil characteristics, and plant growth and development. Mechanisms of adaptation of plants to waterlogging and molecular pathways of low oxygen sensing are also included to cover whole gamut of issues related to plants response to waterlogging.

FIGURE 8.1 **(See color insert.)** Water stagnation in mustard fields due to heavy rainfall in February (top) and nonuniform growth and yellowing of wheat crop in improperly leveled lands (bottom).

8.2 UNDESIRABLE EFFECTS OF WATERLOGGING ON SOIL, PLANTS, AND AIR

8.2.1 WATERLOGGING AND COMPOSITION OF SOIL AIR

Oxygen concentration, which is around 20.95% in the air, ranges from 1% to 7% in well-aerated roots, stem, tubers, and developing seeds.[17] Oxygen being moderately soluble in water, its maximum possible concentration at the plant surface may reach up to 0.25 mol m^{-3} that is well below the 8.31 mol m^{-3} present in the free air (at 25°C). The situation for CO_2 is somewhat different. Concentration of CO_2 in air is approximately 410 ppm (v/v). Although CO_2 is much more soluble in water than O_2, yet at equilibrium as compared to O_2, the maximum amount of dissolved CO_2 concentration in water is about 20 times lower than that of O_2. The waterlogged soil has altogether altered composition of soil air in the root zone with enhanced carbon dioxide concentration and reduced oxygen concentration.

If waterlogging persists longer, it gradually leads to hypoxia (low oxygen) followed by anoxia (absence of oxygen) in the soil because waterlogging prevents the diffusive escape of carbon dioxide and/or oxidative breakdown of gases such as ethylene[13] that are produced by roots and soil microorganisms. External water layer reduces the diffusion of CO_2 and ethylene 10,000 times than in a non-waterlogged condition,[11] as a result these gases accumulate in rhizosphere as well as in plant tissues and damage normal functions of plants.[10,27,71] The process is measured through oxygen diffusion rate, the optimum value required for good aeration being 30×10^{-8} g cm^{-2} min^{-1}.[180] If the value goes below, plants are likely to suffer from low oxygen and/or excess of CO_2. It is seen that waterlogging stress remains operative even when the water from the land has disappeared (Fig. 8.2).[76]

Improper mix of CO_2 and O_2 in the root zone is detrimental to the plant metabolic processes including photosynthesis and respiration with some serious consequences for root growth.[27,71] Although rice and other wetland plants can tolerate high concentrations of CO_2, yet other mesophytes suffer rapid damage.[71] Accumulation of carbon dioxide in the soil severely damages roots of certain species, for example, soybean (*Glycine max*) but not rice.[28]

Anoxic conditions also set a chain of processes that include denitrification, reduction of iron, manganese, and sulfate, and changing soil pH and soil redox potential (Eh). For example, in a typical series of reductions, NO_3^- is reduced to N_2, Mn^{4+} to Mn^{2+} and Fe^{3+} to Fe^{2+}. Similarly, SO_4^{2-} is reduced to H_2S, S^{2+}, or HS^- (depending upon the soil pH) and accumulations of acetic and butyric acids take place that are produced by microbial metabolism.[61,145]

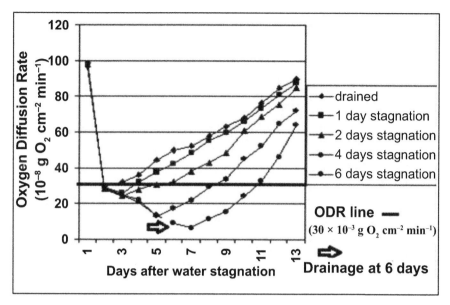

FIGURE 8.2 Oxygen diffusion rate (ODR) as affected by duration of water stagnation.
Note: Low oxygen diffusion rate (ODR) values even after the stress are removed especially for the case of 6 days of water stagnation. (Adapted from Ref. 76.)

8.2.2 PHYSICAL PROPERTIES

Color of the soil can often indicate the status of soil drainage or waterlogging. Gleyed/gray/green colored soils are often associated with poor drainage or waterlogging. The lack of air in these soils provides conditions for iron and manganese to form compounds that give these soils such colors.

Soil consistency, the force of cohesion and adhesion that are holding the peds together, is controlled to a great extent by the soil moisture content. Soil particles move easily in a wet soil, and the soil strength decreases with increasing moisture content above an optimum. As such, soils get compacted and become denser if heavy machinery is used to till the land when it is wet. Normal cultivation operations of tillage and plowing are delayed resulting in missing optimum sowing dates resulting in reduced yields. The bulk density of the waterlogged soils is somewhat higher that limits seed germination, water transmission and aeration.

Profound changes in soil structure have been reported with a decrease in the content of the silt-size aggregates and an increase in the content of the coarse blocky aggregates. Surface crusting is formed when waterlogged soil structure breaks down to form a dense layer on the soil surface which clogs pores, reduces infiltration of water and air, and has the potential to impede the emergence of seedlings.

Waterlogged soils have higher specific heat than dry soils and as such are cooler than non-waterlogged soils or slow to warm up. As a result, seeds may germinate poorly, plant root systems may be stunted and plants growth subdued in waterlogged soils. If plants are submerged with water that is turbid, deleterious effects become more pronounced due to shading effects.

8.2.3   CHEMICAL ENVIRONMENT

Under anaerobiosis, significant chemical changes occur in the root zone mainly because of the altered soil air constitution. Under waterlogging, roots produce excess of protons that acidify cytoplasm and vacuoles, which is suicidal for roots of many plants. It is observed that roots of pea, black-eyed peas and navy beans, die quickly under anoxic conditions as their cytoplasm becomes acidic more rapidly than those of maize, soybean, or pumpkin root tips which survive for longer periods under anaerobic conditions. Nutrient availability to plants is decreased as a result of denitrification of the organic as well as inorganic soil N. Besides, availability of micronutrients such as Mo is inhibited; several elements such as Mn and Fe are also released in the root environment that is toxic to plants impacting production and productivity under waterlogged condition.[124]

8.2.3.1   WATERLOGGING AND NUTRIENT AVAILABILITY

Changes in soil chemistry may alter the nutrient status due to microbial utilization of inorganic source of nutrients, leading to derangement in availability and uptake of nutrients.[144] Reduced soil redox potential due to development of anaerobic condition, favors facultative anaerobic microorganisms to reduce nitrate into nitrite, to nitrous oxide and finally to nitrogen gas (denitrification), resulting in significant reduction in availability of N in nitrate form to plants.[46] Volatilization of ammonia, ammonium ion fixation by clay minerals and leaching of nitrate, etc. adversely impact the plant growth. In waterlogged soils, the main form of plant-available nitrogen is NH_4^+ and, therefore, under prolonged waterlogging plants need to adapt to NH_4^+ nutrition, although it may help to conserve energy because as compared to NO_3^-, NH_4^+ requires lesser energy for further assimilation.[104] Under waterlogged condition organic matters oxidize very slowly, therefore, they are accumulated and hinder the availability of many macro and micro nutrients.

8.2.3.2   WATERLOGGING AND TOXIC ELEMENTS

During the succession of anaerobic oxidation processes, the redox potential (Eh) of flooded soils decreases due to formation of the reduced products. Approximate value for redox potential at which oxygen disappears is +330 mV (within few hours), while methane appears at an Eh of −250 mV (few weeks). Since the toxicity, solubility, mobility, and bioavailability of a given element or compound is mainly influenced by physicochemical properties of soil, therefore redox potential and pH of flooded soil play important role in regulating the mobility of trace metals, nutrients, and minerals. Low redox potential of soil induces activities of obligate anaerobic microorganisms to induce oxidative conversion of Mn^{4+} and Fe^{3+} to highly soluble Mn^{2+} and Fe^{2+} forms,[107] causing Mn^{2+} and Fe^{2+} toxicity[99,124] in plants. Ferrous ion toxicity can be a particular problem for rice grown on acidic soils. Iron toxicity is also reported in rice grown on nutrient poor and waterlogged soils.[101] In such conditions, therefore internal tolerance mechanism for detoxification of Fe toxicity is required.[53] Solubility of other metals is also increased which may be damaging to plants in flooded soils.[39,102,167]

Living organisms are important fraction of the soil, and their presence is encouraged by high organic matter levels, adequate soil moisture, drainage, and aeration. Under prolonged waterlogging, anaerobic bacteria convert SO_4^2 to H_2S, a poison to respiratory enzymes and non-respiratory oxidases. Soils that are low in iron are likely to contain free and undissociated H_2S, especially acidic soils.[144] Prolonged waterlogging also results in a drift toward neutrality, causing changes in nutrient solubility and uptake.

Another possible toxin in waterlogged soils is acetaldehyde. In alcoholic fermentation, activity of the enzyme alcohol dehydrogenase (ADH) that converts acetaldehyde to ethanol usually exceeds the activity of enzyme pyruvate decarboxylase (PDC) that promotes acetaldehyde production from pyruvic acid. Thus, it ensures low sub-toxic concentrations of acetaldehyde in anoxic cells. However, when such tissues return to aerobic condition, this control is sometimes lost and plant tissues typically produce excess of acetaldehyde that proves damaging.[25]

Nitric oxide, a gaseous signaling molecule that can be formed by the action of nitrate reductase (coded for by an anaerobically inducible gene) and possesses the ability to kill cells, is also a toxic compound produced in waterlogged soils.[47] However, the role of this molecule is still unclear, and it has even been suggested that the beneficial impact of nitrate on survival during anoxia may be an outcome of increase in nitric oxide arising from the reduction of nitrate to nitrite.[179]

8.2.4 WATERLOGGING AND PHYSIOLOGICAL PROCESSES

Poor supply of adenosine triphosphate (ATP) is one of the important factors for degeneration of growing root tips under waterlogged condition. Under waterlogging or submergence plants encounter oxygen shortage (hypoxia) that results in drop of respiration ceasing oxidative phosphorylation in mitochondria.[48] Although fermentation provides *ATP* by substrate-level phosphorylation, yet only 2–4 mole of ATP are generated per mole of glucose as compared to 36 mole of ATP per mole of glucose in oxidative phosphorylation. It is assessed that fermentation generates only 6–12% of the ATP as compared to mitochondria-based aerobic respiration. The modest ATP generation capability of glycolysis/fermentation depends on the supply of glucose and its precursors. Therefore, under waterlogged condition there will be high requirement of glucose and oxidized nicotinamide adenine dinucleotide (NAD^+) to carryon fermentation. Unless metabolic processes that consume ATP are simultaneously suppressed, the small yield of ATP in anaerobic cells is insufficient for the survival beyond a few hours.

Roots can respire in water equilibrated with air, but it is not possible for the submerged leaves to photosynthesize normally under similar conditions. In submerged plants, photosynthesis is at higher risk than respiration because photosynthetic influx of CO_2 is of the order of 350 μmol kg^{-1} s^{-1} and respiratory demands of roots for O_2 is about 50 μmol kg^{-1} s^{-1}.[12,79,146] Affinity of ribulose 1,5-bisphosphate carboxylase oxygenase (RuBISCO) for CO_2 is lower (Km of RuBISCO for CO_2 is 10 mmol m^{-3}) than the affinity of cytochrome oxidase for O_2 (Km of cytochrome oxidase for O_2 is 0.14 mmol m^{-3}). Therefore under submerged conditions, artificial supplement of CO_2 is required to achieve good photosynthetic activity. However, respiration continues.[168] Setter et al.[168] could grow rice plants to full size over many weeks when completely submerged but only when the water was highly enriched with CO_2 [about 10% (v/v) CO_2].

High concentration of CO_2 in rhizosphere may also lead to high concentrations of HCO_3^- in the cytoplasm that hampers the metabolic regulation of cells. Ethylene can also accumulate to phytotoxic concentrations in wet soils as a result of microbial action.[49,167]

8.2.4.1   CARBOHYDRATE SHORTAGE

Root carbohydrate reserves are considered a hallmark in conferring waterlogging resistance in plants.[16,19,63,148] Plants having low storage of carbohydrates are unable to meet the demand of glycolysis which lowers the duration of survival of plants under waterlogging condition. During hypoxic or anoxic conditions, photosynthesis is decreased and starch reserves in the roots are rapidly used up in fermentation. Maintenance of root growth and functioning becomes difficult or impossible. In in vitro studies of anaerobic roots, hexose feeding is a prerequisite for long periods of anaerobic survival of the cultures. Over shorter time scales too (hours or several days), seedling roots have been found to survive longer and ferment more vigorously when given external glucose.[188,203]

8.2.4.2   PRODUCTION OF REACTIVE OXYGEN SPECIES

The reactive oxygen species (ROS) are lethal to plant cells and can damage them by causing peroxidation of biomembrane lipids, DNA damage, protein denaturation, carbohydrate oxidation, pigment breakdown, and impairing enzymatic activities.[134,177] Hypoxic conditions favor generation of ROS such as superoxide radical ($O_2^{\cdot-}$), hydroxyl radical ($OH^{\cdot}$), and hydrogen peroxide (H_2O_2). Although H_2O_2 is less reactive than the other two ROS, in the presence of reduced transition metals such as Fe^{2+} (abundant in waterlogged soils), the formation of $OH^{\cdot}$ can occur in the Fenton reaction.[24] The major sources of ROS generation under waterlogged condition are electron transport chain (ETC) in mitochondria and oxidation reaction.[24]

The extent of the ROS-induced damage to cells depends on duration and severity of stress. Short-term O_2 deprivation results in a limited accumulation of ROS and lipid peroxidation. Constitutive endogenous antioxidants are responsible in regulating rate of ROS formation and the extent of lipid peroxidation during short-term oxidative stress.[24] In addition, hypoxia induces increased activities of antioxidant systems. Prolonged deprivation of O_2 is reported to diminish or even abolish synthesis, transport and turnover of antioxidants. As a consequence of the depleted antioxidants and associated enzymes, cells are unable to cope with the ROS and lipid peroxidation becomes severe, particularly during reoxygenation or drainage of water. In addition to causing nonspecific increases in membrane permeability resulting from lipid peroxidation, both H_2O_2 and $OH^{\cdot}$ have also been shown to directly control activity of Ca^{2+} and K^+ permeable plasma membrane ion channels.[42,43] Perturbations in intracellular ionic homeostasis may initiate programmed cell death (PCD).[42]

8.2.5 WATERLOGGING AND EMISSIONS OF GREENHOUSE GASES

Flooded soils are dynamic ecosystems that play an important role in biogeochemical cycling and in the production of greenhouse gases (GHGs). Methane (CH_4^+) and nitrous oxide (N_2O) are produced as by-products of anaerobic metabolism in the low-redox zones soils. The wetland rice produces much higher GHGs than the irrigated upland rice. Carbon dioxide (CO_2) may be produced at the interface of anaerobic–aerobic zones through the consumption of methane gas. However, the conversion of methane gas to carbon dioxide essentially reduces the global warming potential of methane. Apparently, impact of flooded soils on the global climate is to be ascertained.

8.2.6 WATERLOGGING AND ROOT GROWTH

Low levels of oxygen in the root zone trigger the adverse effects of water-logging on plant growth. Slowing of oxygen influx is the principal cause of injury to roots, and the shoots they support.[195] Anoxia is lethal for growing root tips; root growth ceasing with total arrest of O_2 influx.[23] Arrest of root growth and roots death principally arise because of (1) demand for ATP exceeds the supply and (2) self-poisoning by products of anaerobic metabolism. Nevertheless, even small amounts of external oxygen (e.g., 0.006–0.01 mol m^{-3} in solution) are observed to keep them alive.

Significant accumulation of organic substances (e.g., ethanol, acetaldehyde, and various short-chain fatty acids and phenolics) is observed in waterlogged soils.[8,49] These are secondary metabolites, produced as a result of anaerobic metabolism in plants[49,119] as well as by rhizospheric microorganisms.[49] Allelopathic effects of these secondary metabolites on plant growth are well known,[163] for example, accumulated ethylene slowdowns root growth.

8.2.7 LEAF EPINASTY

Ethylene triggers epinasty in flood sensitive plant species. In epinasty, more cell growth occurs on adaxial surface of leaf than on abaxial surface; as a result leaf drops from a horizontal to a more vertical position. In waterlogged tomato roots, it is reported that transcription of 1-aminocyclopropane-1-carboxylic acid (ACC) synthase (ACS) and ACC oxidase (ACO) is increased and there is increased synthesis of ACC. Nevertheless, conversion of ACC into ethylene by ACC oxidase (ACO) is decreased as this step is oxygen dependent. ACC thus formed is transported to aerial tissues where in the presence of oxygen ACO converts ACC into ethylene. This cause increased ethylene production in shoot tissues leading to leaf epinasty.[30] Downward bending of leaves reduces interception of incident light. As such, plants experience decreased transpiration rate and reduced absorption of water by the roots under waterlogged condition.

8.3 ADAPTATION OF PLANTS UNDER WATERLOGGED CONDITIONS

Plant species, on the basis of their flooding tolerance or adaptation, are classified as wetland species, flood tolerant, and flood sensitive species.[30] Genotypic differences are also reported in a number of species (Fig. 8.3) including maize[169] and pigeonpea,[22,173] etc.

Flood tolerant species utilize different traits in the adaptation mechanisms. Two major strategies involved are avoidance that prevents plants to escape from exposure to the stress while the other involves tolerance under which plants endure the stress conditions. On the basis of physiological and molecular mechanisms, survival tactics under low oxygen are described as low oxygen escape strategy (LOES) and low oxygen quiescent strategy (LOQS) under which energy saving is made through slowing the metabolism. LOES traits are:

- Increases in the growth rate of shoot organs, such as petioles and stems, so as to emerge above floodwater
- Initiate the development of aerenchyma to facilitate internal gas diffusion.

LOQS traits are characterized by:

- Conservation of energy and carbohydrates by reducing the underwater growth rate
- An increase of molecular components that prepare shoot and root organs for future conditions with low O_2 and production of protective

molecules that counteract harmful cellular changes associated with flooding, such as production of ROS.[16,17,200]

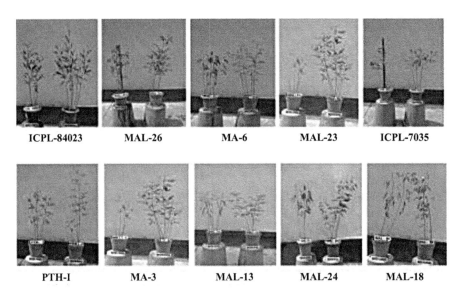

FIGURE 8.3   (**See color insert.**) Effects of waterlogging stress on pigeonpea genotypes (waterlogged plants are to the left and nonwaterlogged plants to the right for each genotype). Stress was imposed after 40 days of sowing and photographs were taken after 20 days of imposing waterlogging stress during 2007–2008.
Source: Adapted from Ref. [22].

8.3.1   MORPHOLOGICAL AND ANATOMICAL MECHANISMS

8.3.1.1   AERENCHYMA FORMATION

Aerenchyma is a continuous column of internal gas spaces from aerial part to roots.[95] In many plant species adapted to wetland, aerenchyma is constitutively present. Formation of aerenchymatous cells is also shown to be inducible upon flooding stress in both wetland and many mesophytic plant species.[36,181] It develops in existing root and shoot tissues, secondary tissues, or newly formed organs such as adventitious roots (ARs). Two types of aerenchyma are identified. The first type is primary aerenchyma, also called cortical aerenchyma, that is found in the roots of rice, maize, barley, and wheat.[11,56,91,133] The second type is secondary aerenchyma which is formed in plant parts like stem, hypocotyl, tap root, ARs as well as root nodules of legumes such as soybean, wild soybean (*Glycine soja*), new dhaincha (*Sesbania rostrata*), and native broom (*Viminaria juncea*) when they are grown under flooded conditions.[6,126,160] Secondary aerenchyma is characterized with a white, spongy tissue filled with gas spaces known as secondary tissue.[5,57] In roots, lysigenous aerenchyma is formed in cortex[56] while in shoot it is formed in cortex and pith.[11] Aerenchyma are formed in three ways:

- By programmed cell death (PCD) in the root cortex, termed as lysigenous aerenchyma,[149,158]
- By separation of cells without lysis, termed as schizogenous aerenchyma,[5] and
- By cell division and cell expansion without separation or lysis, called expansigenous aerenchyma.[56,166,176]

Lysigenous aerenchyma are observed in barley,[7] wheat,[192] rice,[96] maize,[73,80,93] and pigeonpea.[20] Schizogeneous aerenchyma is common in wetland species like *Rumex*.

Formation of inducible aerenchyma is triggered by ethylene as calcium-dependent and other signaling pathways are activated by ethylene accumulation. Under oxygen deprivation, Ca^{2+} is released from mitochondria into the cytoplasm.[183] The elevated cytosolic Ca^{2+} induces subsequent activation of kinases and phosphatases during aerenchyma formation.[182] Ca^{2+}-dependent plasma membrane-localized respiratory burst oxidase homologs (RBOHs) serves as a source of reactive oxygen that promotes production of apoplastic superoxides which in turn help to amplify ROS-mediated signaling. Genes responsible for aerenchyma formation are induced by these Ca^{2+} dependent signaling pathways.[51,182]

The accumulation of the gaseous plant growth regulator, namely, ethylene is induced in maize under waterlogged condition on account of increased activities of enzymes such as ACS and ACO that are involved in ethylene synthesis.[80] After ACS and ACO induction, RBOH gene expression occurs.[175,191] The ROS production through RBOH activity leads to the production and accumulation of apoplastic superoxide, which is further converted, either enzymatically or spontaneously, into hydrogen peroxide.[175] Signaling pathways through G-protein, phospholipase C, inositol-1,4,5-trisphosphate, and Ca^{2+} are involved in the formation of lysigenous aerenchyma in maize roots.[81] Gene encoding ROS scavenging protein, metallothionein, is significantly down regulated in the cortex.[149]

Aerenchyma formation in stems of deep water rice (*Oryza sativa*) has also been associated with an elevation in ROS and decreased expression of metallothionein 2b mRNA.[176] This further reduces scavenging of ROS, thereby leading to its accumulation in higher quantities, which activates PCD and lysis of the cortical cells and results in aerenchyma formation. Roots of wheat (*Triticum aestivum*), when pretreated with the ethylene precursor ACC, resulted in induction of high levels of RBOH transcripts, which has been reported to be responsible for more aerenchyma formation upon waterlogging, but this developmental change can be blocked by the NADPH oxidase inhibitor; diphenyleneiodonium.[210]

The cell wall is degraded enzymatically during the final stages of lysigenous aerenchyma formation. During cell death changes in esterified and de-esterified pectins in the cell wall of maize cortex are also observed.[73,74] Cortical cell wall is degraded by the combined action of cellulase (CEL), xyloglucan endotransglycosylase (XET), pectinase, and xylanase.[56,91] Treatment with ethylene, and reagents that increase intracellular Ca^{2+} also increases the activity of CEL and treatment with K252a, an inhibitor of Ca^{2+} decreases cellulose activity.[81] XET gene expression is observed to be induced in maize roots under flooding, and treatment with an ethylene biosynthesis inhibitor inhibits this induction.[157] Also treatment with an ethylene biosynthesis inhibitor prevents the formation of lysigenous aerenchyma. Hence, the induction of XET production in response to ethylene correlates with induction in aerenchyma formation through cell wall modification.[157]

Aerenchyma formed in waterlogged plants helps in the diffusion of oxygen into the roots and also outward diffusion of carbon dioxide and ethylene that have been generated within roots. It also helps in the diffusion and escape of methane generated in the soil.[33,108,170] Aerenchyma helps the plant to sustain the oxidative phosphorylation and ATP formation.[52] Ethylene removal prevents growth reduction.[199] Lysigenous aerenchyma forms continuous column of gas space and they also reduce the number of oxygen-consuming cells in plant, improving the oxygen balance in a dual fashion.

Aerenchyma formation in response to flooding facilitates relatively high O_2 concentrations in roots. Very steep radial O_2 diffusion gradient between root interior and the surrounding anaerobic soil can cause the loss of O_2 by outward diffusion into the soil (i.e., radial oxygen loss—ROL). Resistance offered by the roots for gaseous diffusion determines the rate of radial diffusion. But when the radial loss of oxygen is prevented by certain apoplastic barrier, higher O_2 concentrations in roots even in longer roots can be maintained.[33,34] The indispensable requirement of the apoplastic barrier has been demonstrated in wheat (*T. aestivum*).[122] *Hordeum marinum* (a wild barley) has suberized lamellae in the exodermal/hypodermal spaces near the root tips and lignified schlerenchyma/epidermal cells which imposes resistance to radial gaseous losses from roots. When these characteristics were introduced from *Hordeum marinum* into wheat, it conferred waterlogging tolerance in wheat.[122] These structures also constitute an efficient ROL barrier as well as an apoplastic blockade between living cells and the anaerobic toxic (i.e., saline or highly reduced) soil environment.[34,171,202] These modifications have also been correlated with minimizing ROL in other waterlogging tolerant species like teosinte (*Zea nicaraguensis*), a wild relative of maize and tolerant to waterlogging.[1] The exterior cell layers in roots of teosinte have larger deposition of lignin and suberin that increases gas impermeability and reduces ROL.[29] Similarly in paddy roots, suberization and lignification of the exodermis serve as a restriction for ROL.[1,150] Detoxification of phytotoxins in the rhizosphere is also brought about by the oxidation of the rhizosphere which is in turn facilitated by the increased ROL.[9,110]

Accumulation of malic acid and very long chain fatty acids have been observed during the metabolite profiling of longitudinal sections of ARs of rice (*O. sativa*). These rice plants were grown under barrier-forming stagnant conditions. The concentrations of these metabolites also increased from the root apex to the base, coinciding with the development of the barrier.[106] Malic acid is a substrate for fatty acid biosynthesis. It also serves as a precursor for suberin formation. ATP-binding cassette transporter (RCN1/OsABCG5) was identified from the short and shallow root phenotype of waterlogged reduced culm number 1 (rcn1) mutants of rice. These transporters facilitate the export of very long chain fatty acids and/or their derivates across the hypodermal plasma membrane into the apoplast where they serve as major components of suberin synthesis.[172] The roots of rcn1-2 mutants fail to develop effective suberized lamellae or a ROL barrier under deoxygenated conditions when compared to the wild type. Higher ethylene and CO_2 or low O_2 were not found to be essential for barrier formation, but root exudates or cellular degradation products are suggested to be important.[37,62]

8.3.1.2 FORMATION OF ADVENTITIOUS ROOTS

Plants tolerant to waterlogging produce adventitious roots (ARs) to compensate for the loss and damage of soil roots. They function to absorb nutrients and to anchor plants.[169] In many plant species, under waterlogged conditions ARs develop from the basal region of the stem or from the hypocotyls[162,199] replacing the existing but deteriorating primary root system. ARs that are formed under waterlogged conditions have porosities much higher than those that are produced under well aerated conditions[198] or the primary roots.[115] ARs grow in topsoil layers that are better aerated or may also float in flood waters.[40] Some ARs exhibit the capacity to develop chloroplasts, and they can serve as an additional source of carbohydrates and even oxygen to plants.[153,154]

In semi-aquatic rice varieties, the ARs develop from nodes on stem and help in reducing the distance over which oxygen has to be supplied. Also, the central cylinder of a root is often oxygen deficient as the endodermis with its Casparian strips that poses thick suberized cell walls, hinders easy diffusion of gases.[3] But the ARs, which mostly remain under water, have a poorly developed endodermis that helps in gaseous diffusion to the stele[31] exhibiting developmental changes in differentiation of endodermis.[2]

Accumulation of ethylene in submerged rice, maize, *Rumex palustris*, tomato, and hypoxic *Arabidopsis thaliana* is increased due to the induction of ethylene synthesis both at the level ACS and ACO gene expression and protein stability.[64,141,155,159,162,193] In tomato, ethylene promotes the formation of ARs but inhibits lateral root formation promoting shoot-borne root system.[132,197] Interactions of two growth regulators, namely, ethylene and auxin also control root branching in *Arabidopsis*.[90] de novo genesis of AR primordia is promoted by ethylene through both increased auxin flux toward the submerged stem[115,197] and also increased auxin activity.[26,215] Treatment with auxin efflux inhibitor naphthylphthalamic acid also inhibits AR formation.

During the formation and development of ARs under flooded conditions the plant growth regulators ethylene and auxin are integral at the cellular level,[199] whereas in the case of ARs preexisting primordia of stem nodes, it has been found that it is the ethylene and not the auxin, that signals activation of the cell cycle, followed by formation of ROS.[178] Programmed cell death (PCD) of the overlying epidermal cells is mediated by ethylene-promoted ROS production is involved in the emergence of delicate AR primordia. It occurs in a remarkably spatially specific manner and the location is determined by the force exerted by the outgrowing meristems. The reductions in the metallothionein 2b mRNA also participate in the nodal AR emergence, as a mutant of this ROS ameliorating protein showed enhanced force-induced epidermal PCD.

8.3.1.3   ENHANCED ELONGATION OF AERIAL PARTS

Rice is the only major crop species with seeds capable of germination under anaerobic conditions.[89,121] It is an example of escape mechanism based on organ elongation that is promoted by the absence of oxygen. Elongation growth response in deep water and lowland rice cultivars after 7 days of partial submergence showed distinctions in internode elongation.[78] *Arabidopsis* (Columbia-0 ecotype) after 3 days of complete submergence in darkness showed petiole elongation.[131] Elongation growth can exhaust energy reserves and cause death when the flooding depth is deep and the flooding period is long.[18,92] Aeration between shoot and root is more efficient when the shoots elongate and emerge above the floodwater.[82,154] Due to this reason, certain plants adapted to flood-prone environments have evolved the ability to elongate their porous shoots under submerged condition. This facilitates LOES and supplements the aerenchyma system and/or increases uprightness of submerged leaves.[38]

The resulting changes improve access to aerial or dissolved oxygen, or to light for the generation of photosynthetic oxygen. The carbon costs involved are high, and hence this trait is found to be restricted to species or accessions/landraces from environments characterized by shallow, but relatively prolonged floods.[72,201] In species examined to date, a hormonal hierarchy involving ethylene as a trigger, abscisic acid (ABA) as a repressor and gibberellic acid (GA)/auxin as promoters are associated with underwater elongation growth modification. Ethylene, gibberellic acid, and ABA are considered to regulate shoot elongation in waterlogged plants. In wild rice, it is reported that due to reduced gaseous diffusion, ethylene produced by the plant is not allowed to diffuse readily. As a result it accumulates in relatively higher concentration in plant tissues.[161] In such plants, increased ethylene level causes reduction in ABA content and induces internode elongation.[98]

The submergence tolerance gene SUB1A (submergence 1A) was identified in a landrace grown in lowlands of eastern India that are flood-prone.[58,209] Three quantitative trait loci (QTLs) snorkel1 and 2 (SK1/2) and two uncharacterized loci on chromosomes 1 and 3 (QTL 1 and 3) have been found to be responsible for underwater elongation in deep water rice.[78] Ethylene insensitive 3 binds the promoters of SK and promotes ethylene induced transcription. SUB1A and SK1/2 are found to be the key regulators of shoot elongation.[165] SUB1A mainly limits elongation growth, whereas SK1/2 promotes elongation from sixth internode that has been submerged underwater. All these are ethylene-responsive transcription factors (TFs) belonging to subfamily group VII (ERFVII). SUB1A limits elongation through multiple mechanisms including increased accumulation of the GA-response transcriptional inhibitors slender rice 1 and slender rice-like 1, and inhibition of ethylene biosynthesis, and restriction of chlorophyll degradation.[58,60,165] SUB1A positively controls a subset of hypoxia-induced genes encoding PDC and alcohol hedydrogenase (ADH), as well as ABA responsive genes.[59,60,94]

8.4   BIOCHEMICAL ADAPTATIONS

8.4.1   PHOTOSYNTHESIS UNDER SUBMERGED CONDITIONS

Waterlogging reduces carbon assimilation and transpiration rates.[14,104,143] Stomatal and root hydraulic conductances are also decreased under water-logged condition. Leaves under submerged conditions have enhanced boundary layer resistance due to decrease in diffusion coefficient by about 10,000 times. Boundary layer resistance to CO_2 and O_2 exchange is the major limiting factor to photosynthesis and respiration in leaves of submerged plants.[139] In waterlogged pigeonpea[20] and maize[174] leaves, carbon exchange rate, stomatal conductance, intercellular CO_2 concentration and transpiration rate are reported to reduce. Another factor that limits photosynthesis, particularly in submersed plants, is the exponential decrease in light intensity with depth.[37] Consequently, net photosynthesis of submerged leaves in mesophytes is often significantly reduced compared with aerial leaves. Traits that reduce diffusion resistance for CO_2 to leaves tend to increase photosynthesis under submerged conditions, as leaves that are newly developed are characterized by a higher specific leaf area, chloroplasts reorientation toward the epidermis of the leaf and thinner cuticles and cell walls, tend to photosynthesize relatively higher than mature leaves when submersed.[128]

Other relevant traits are the development of dissected leaves under water and the maintenance of gas films on leaf surface.[37] *Rumex palustris* develops new leaves under water, and acclimation in these leaves had led to a 38-fold decrease in CO_2 diffusion resistance when under water.[129] Most of the terrestrial plants in general have water-repellent (hydrophobic) leaf surfaces that help the leaves to retain a thin layer of air (gas film) when submerged. This phenomenon helps in increasing the interface of gas and water and allows faster CO_2 diffusion within the thin layer of air so that stomata can stay open. This was reported to result in a 1.5 to 6-fold increase in the rate of photosynthesis underwater when compared with leaves where the gas film was removed.[37,139,205,206,207] There is evidence of variation for this trait in rice. Submergence tolerant landrace FR13A had higher net photosynthesis underwater and also possessed longer leaf gas film retention even over a submergence period of 12 days.[207]

Although the trait involving persistence of gas film was correlated with better maintenance of carbohydrates during submergence in FR13A, yet the duration of retention of the gas film was lesser in Swarna-Sub1, indicating that there might be other genes which are involved in gas film formation or underwater photosynthesis. These morphological and anatomical leaf acclimations improve both the inward diffusion of CO_2 in light, and also the inward diffusion of O_2 at night or in turbid waters.[140,196] It has been found that leaf gas films adhere to the surface of leaves of many semi-aquatic species.[205] In *Melilotus siculus* (annual legume), it has been observed that gas films are formed on both the surfaces of leaves when plant is submerged in saline waters, and this not only improves gas exchange but also prevents salt intrusion into leaves.[189]

8.4.2   WATERLOGGING AND PLANT NUTRIENT UPTAKE

Flooding causes immediate cessation of root growth[67] and anaerobic respiration decreases the energy availability. As oxygen is depleted, soil microbes switch from using O_2 as an electron acceptor to NO_3^-, Fe^{3+}, and Mn^{4+} leading to highly reduced conditions in rhizosphere. More reduced conditions can lead to a lower availability of some plant nutrients, specifically nitrogen.[211] Also the plasma membrane proton pump (H^+-ATPase) requires ATP and the proton motive force generated is used to drive symporter-mediated ion uptake. All anions (e.g., NO_3^-) enter root cells via H^+-anion symporters. Furthermore, the H^+-ATPase maintains the negative membrane potential, essential to create electrochemical gradients allowing channel-mediated uptake of cations (e.g., K^+ uptake).[137]

Reduced ATP supply reduces active nutrient uptake considerably. Mineral elements uptake by roots is also greatly reduced on a per root weight basis[35,55] due to the low ATP availability. Not only is K^+ uptake significantly reduced but roots also loose substantial amounts of K^+ through depolarization-activated channels. Therefore, waterlogged plants often exhibit acute K^+ deficiency.[142] The organic acids present in waterlogged soils, from anaerobic microbial metabolism, also leads to membrane depolarization of root cells and reduced ion uptake. Anaerobic soil conditions increase oxalate-soluble P and Fe in soils.[214] Trapped CO_2 may form bicarbonate ions that can balance the effect of liming, leading to iron unavailability and chlorosis.

Soil flooding improves the bioavailability of P, Fe, and Mn to rice, which is adapted to waterlogging.[48,61] Therefore, waterlogged plants may suffer from Fe or Mn toxicity.[84] Drew[48] found that waterlogging often induced an increase of Mn and Fe in shoots of wheat and barley and significantly decreased N, P, K, Mg, Zn, and Mn concentrations in wheat and barley shoots.[130] On flooded acid soils, even rice may suffer from Fe toxicity.[41] Graven et al.[69] reported that the Mn concentration was increased to a toxic level in alfalfa shoots by soil flooding.

Upon waterlogging, root growth is generally arrested immediately whereas shoots may continue to grow. Increased shoot:root ratio causes an imbalance between shoot nutrient demands and supply by roots. Shoot and root nutrient content decreases under waterlogged conditions.[14,100,105,136] Controlled pot experiments have demonstrated that nutrient uptake is not only reduced through limited root growth, but that the uptake is also reduced on a per unit weight basis.[93,105,206] The diminished capacity for ion uptake, together with initial "dilution" of shoot nutrient concentrations by continued shoot growth relative to roots, explains a range of nutrient deficiencies observed in leaves of sensitive plants under waterlogged conditions.

In tolerant species, aerenchyma formation facilitates continuous O_2 supply to roots which helps in sustaining root respiration. Respiration supplies ATP to plants which in turn are used for active nutrient uptake. Adequate O_2 and energy used helps in deeper root penetration. Barriers to ROL in basal zones, also enables an aerobic rhizosphere at the root tips, which also benefit nutrient uptake by the roots.

Several studies have, however, shown that the anoxia-induced inhibition of nutrient uptake does not correlate with a reduced ATP synthesis[88] or with insufficient proton motive force.[213] The results indicate that inhibition of ion uptake is the result of down-regulation of transport systems quiescence strategy. Down regulation of transport system is not due to less ATP but to conserve the available energy under stress condition.

8.4.3   ROOT RESPIRATION AND ANAEROBIC METABOLISM

Prolonged anoxia in waterlogged soils is tolerated by rhizomes, tubers, and some shoot organs of wetland species, and by germinating seeds in rice and some paddy weeds (e.g., barnyard grass). Anoxia is not universal in plant tissues, it may occur in portions of the plant body (e.g., roots) or parts of tissues within roots. "Anoxic cores," coexistence of an anoxic stele and aerobic cortex, were demonstrated in maize roots under hypoxic conditions, using O_2-microelectrodes and biochemical indicators of fermentative metabolism.[71]

Oxidative phosphorylation yields 36 ATP units per hexose unit, while glycolysis yields only 2 ATP units. Therefore, glycolytic activity must be strongly upregulated under hypoxic conditions to generate sufficient ATP. This phenomenon requires the efficient recycling of NAD$^+$ from NADH; otherwise, glycolysis will be limited by the unavailability of NAD$^+$.[48,50,185,186] Therefore, the ethanol and lactate fermentation pathways are induced by activating the expression of the enzymes pyruvate decorboxylase (PDC) and alcohol dehydrogenase (ADH) (for the ethanol pathway) and lactate dehydrogenase (LDH) (for the lactate pathway), both of which use pyruvate and NADH as substrates. LDH/PDC pH-stat hypothesis of Davies Roberts confers that under O$_2$ deprivation pyruvate is converted into lactate, which lowers the pH of cytosol. At lowers cytosolic pH, there is inhibition of LDH but PDC activity is stimulated, which promotes the ethanol production.[30]

The PDC/pyruvate dehydrogenase-stat hypothesis confers that concentration of pyruvate stimulates the activity of PDC. An evaluation study with two rice genotypes: one sensitive and other tolerant to flooding during germination, along with two well-known paddy field weeds barnyard grasses, that is, *Echinochloa crusgalli* and *Echinochloa colona* revealed that rice was more flood-tolerant than either of the barnyard grass weeds and difference was associated with the stronger expression of the two main fermentative enzymes, that is, PDC and ADH.[85]

Lakshmanan et al.[109] identified pathways that are differentially regulated during germination when oxygen is limiting. They reported that genes linked to sucrose catabolism and fermentation pathways are being upregulated under anaerobic conditions, while others controlling central processes that run aerobically, such as oxidative phosphorylation, tricarboxylic acid (TCA) cycle and pentose phosphate pathway (PPP), are being down regulated. Roots in drained soil respire by catabolizing carbohydrates in the TCA cycle and the "reducing power" produced is used in the electron transport chain (ETC) with O$_2$ as the terminal electron acceptor.[187] Comparison of two rice varieties differing in anoxia tolerance suggested that fermentative activity is controlled by mutual regulation of the glycolytic enzymes phosphofructokinase (PFK) and pyrophosphate:fructose-6-phosphate-1-phophotransferase in an anoxia tolerant variety, whereas sensitive variety showed no correlation between PFK activity and fermentation rate.[66]

Anoxia tolerance lies in integration of energy production via anaerobic carbohydrate catabolism and energy consumption in reactions essential for survival. Accumulating evidence suggests two modes of tolerance based on slow and rapid rates of fermentation.[70] As an example of the "slow fermentation mode," lettuce seeds appear to survive anoxia by slowing carbohydrate catabolism in anoxia to less than 35% of the rate in air.[151] Other plant tissues, which survive but do not grow in anoxia, produce an initial burst of fermentative activity over 6–24 h before settling to slower fermentation rates. This two-phase pattern presumably provides the higher ATP required as cells acclimate to anoxia, but then the lower rates of fermentation would conserve carbohydrates for long-term survival.[212]

8.4.4   ANAEROBIC PROTEINS

During anoxia, normal protein synthesis is replaced by the selective transcription and translation of a set of proteins called "anaerobic proteins." In maize roots, there are 20–22 such proteins which include fermentative enzymes (e.g., PDC and ADH),[148] enzymes involved in anaerobic carbohydrate catabolism (e.g., sucrose synthase) and enzymes responsible for the reversible breakdown of sucrose as well as several glycolytic enzymes (e.g., aldolase). Other anaerobic proteins, especially antioxidants like superoxide dismutase (SOD), peroxidase, catalase, ascorbate peroxidase, polyphenol oxidase, etc., have been reported to be elevated under waterlogging in pigeonpea root.[20,117] Anaerobic proteins are also formed in rice embryos as those described for maize. In rice, an additional anaerobic protein, that is, tonoplast H^+-pyrophosphatase is also synthesized under waterlogging.

Elevated levels of transcripts (mRNAs) that encode for enzymes involved in an anaerobic metabolism module have been reported in plants that are raised in environments low in oxygen. These enzymes include amylases (starch consumption), sucrose synthase (sucrose catabolism), phosphofructokinase (PFK) (glycolysis), and pyruvat decorboxylase (PDC) (pyruvate metabolism), alcohol dehydrogenase, or lactate dehydrogenase, as well as alanine aminotransferase (alanine metabolism), glutamate decarboxylase (gamma aminobutyric acid), succinate, and several glucogenic amino acids. It is also suggested that when the electron transport in the mitochondria is hindered due to limited availability of oxygen, it results in bifurcation of the TCA cycle that causes enhanced production of ATP by succinyl-CoA ligase.[19,20,156,187]

Greenway and Gibbs[70] indicated that in waterlogged plants early cell death can only be avoided if the small amounts of available energy are successfully redirected to permit synthesis of certain critical "anaerobic" proteins (e.g., alcoholic fermentation enzymes) that support glycolysis and fermentation and help to prevent excessive acidification of the cytoplasm and vacuole, and maintain membrane integrity. This is illustrated by the ability of rice seeds to germinate without oxygen. Such ability is due, in part, to its possession of an anaerobically inducible gene coding for α-amylase, the enzyme principally responsible for degrading starch to a range of sugars.[116]

8.4.5 FUTILE NITRIC OXIDE CYCLE

An alternative pathway to recycle NAD^+ from NADH under low-oxygen conditions is via a futile nitric oxide (NO) cycle. Under hypoxic conditions, nitrite reduction by nitrate reductase increases, leading to NO production.[114] NO synthase is also involved in NO synthesis, which utilizes NADH. NO is oxidized to nitrate again by class-1 non-symbiotic hemoglobins that are specifically expressed when the oxygen concentration is low.[45,86] This cycle requires NADH as a cosubstrate and plays a role in balancing the antioxidant status of the cytosol.[87]

Plants that produce large amounts of NO upon hypoxia leads to decreased lactate accumulation,[135] supporting the hypothesis that the futile NO cycle alleviates fermentation. According to Manai et al.,[123] exogenous NO is involved in prevention of Na^+ accumulation and increase in K^+ concentrations. NO also influences Ca^{++} absorption and increases nitrate uptake. Exogenous application of sodium nitroprusside as a donor of NO in the flooding water has been observed to mitigate the deleterious effect of waterlogging in maize.[93] However, the effective concentration varied for different parameters and the different genotypes.

8.5 LOW OXYGEN SENSING IN PLANTS

During flooding, it is a common phenomenon that oxygen levels are low in the root tissues. Declining oxygen levels are sensed by the plants that help in rapid acclimatization to the prevailing conditions. Plants sense the lowering oxygen levels by the N-end rule pathway of protein degradation.[65,67,113] Group VII ethylene response factors (ERFs) are involved early in the signaling pathway during acclimation to hypoxia or anaerobic conditions. Five ERFVIIs genes have been identified in *Arabidopsis thaliana*. Three of these, namely, related to AP2.12 (RAP2.12), RAP2.2, and RAP2.3, are constitutively expressed and further upregulated by darkness or ethylene. The remaining two, namely, hypoxia responsive ERF1/2 are highly induced at transcriptional and translational levels by O_2 deprivation at different developmental stages.[68] All these five members of ERFs possess a specific motif at their N terminal. This characteristic motif at the N terminal has amino acids methionine followed by a cysteine.

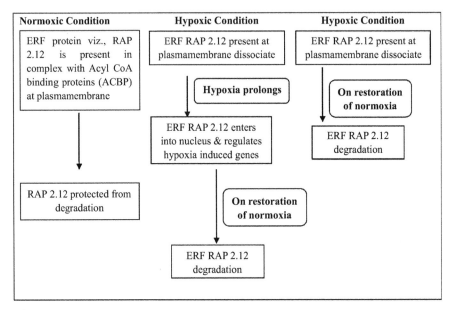

FIGURE 8.4   Status and function of ERFs (RAP2.12) under aerobic (normoxic) and hypoxic conditions.

Under aerobic conditions, the ERF proteins, for example, RAP2.12 (Fig.8.4) is bound to Acyl-CoA binding proteins in the plasma membrane and protected from degradation. Under hypoxic or anoxic conditions, there is shortage of oxygen and the ERFs dissociate and function as transcription factors inducing the genes required by the plants for survival. During this phase restoration of normoxic conditions (when flood water recedes/ is withdrawn), ERFs are degraded by the N-end rule pathway of protein degradation in the presence of oxygen, down regulating those genes which were earlier induced.

In the presence of oxygen (normoxic), the methionine is cleaved by the constitutive activity of enzyme methionine amino peptidase exposing cysteine residue for oxidation. When cysteine is oxidized, an arginine residue is added by the action of arginyl-tRNA-transferase. Now this protein (modified ERFs) is recognized by E3 ligase proteolysis 6 leading to ubiquitination and 26S proteosome-mediated protein degradation. As a result the ERFs are unable to perform their functions. However, under hypoxia or anoxia when

O_2 is limited as found in many organs of flooded plants, degradation of ERFs is inhibited as a consequence of a lack of cysteine oxidation. Also it has been found that ERF-VIIs participate both in homeostatic O_2^- and NO sensing mechanism as its turnover occurs when both O_2 and NO are available.[65,112,113]

8.5.1 SECONDARY SIGNALING PATHWAY ASSOCIATED WITH LOW OXYGEN

Flooding increases glycolysis pathway which, accelerates the normal carbohydrate consumption to a higher level, leading to shortage of carbohydrate supply.[59,147,194] In addition to basic model of low O_2 sensing many secondary molecules are involved in sensing. Under carbohydrate shortage energy conserving molecules such as sucrose non-fermenting 1-related kinase 1 (SnRK1) and target of rapamycin (TOR), perception and transduction have been found to increase.[15,190] In spinach, SnRK1 activation via phosphorylation occurs when high AMP:ATP ratios are prevalent,[184] whereas in *Arabidopsis* SnRK1 orthologs AKIN10 and AKIN11 activate members of the S1 class of the basic leucine zipper (bZIP) transcription factor family (bZIP1, bZIP11, and bZIP53) under carbohydrate shortage.[15,117]

These transcriptional regulators, which are themselves induced by carbohydrate shortage, activate the expression of enzymes which utilizes alternative carbon sources such as protein, structural carbohydrate, and lipid.[44,97] Phosphorylated SnRK1 is also able to affect the activities of enzymes involved in sterol synthesis (hydroxymethylglutaryl-CoA synthase), in sucrose synthesis (sucrose-phosphate synthase) and also in nitrogen assimilation (nitrate reductase).[120,125] In rice, the role of SnRK1 has been explored in the frame of starch utilization which can lead to fuel the process of glycolysis under flooding conditions.[111] It is suggested that the signal of sugar starvation activates calcineurin B-like interacting protein kinase 15. This in turn initiates a phosphorylation cascade, acting through SnRK1, and ultimately leads to induction of the hydrolytic enzyme α-amylase by the MYB transcription factor MYBS1.[118]

Germination of rice seeds under anoxic conditions also occur through the same mechanism to sustain postembryonic development.[138] With regard to TOR protein complex, the function of this complex in plants seems to be antagonistic to that of SnRK1s. It has been found to stimulate glycolysis and the TCA cycle, synthesis of amino acid and protein, and cell wall synthesis and down regulates proteolysis, amino acid catabolism, and autophagy.[190] TOR also exerts multilevel control over mRNA translation. It regulates ribosomal protein synthesis as well as rRNA gene expression and maintains translation reinitiation through phosphorylated eukaryotic translation initiation factor 3 subunit H.[152,164,208] As discussed above, low-oxygen and low-energy stresses also overlap in inhibiting protein synthesis. Together with oligouridylate binding protein 1C, mRNAs that are poorly translated associate into stress granules, where they are protected from nucleolysis until translation-favorable conditions return.

8.5.2   *LOW OXYGEN SENSING IN PLANT ORGANELLES*

Under hypoxic conditions, the plasma membrane-localized transcription factor RAP2.12 is released from the plasma membrane and its degradation is inhibited due to the low oxygen concentration in the cells. The transcription factor, therefore, accumulates in the nucleus and simultaneously induces genes involved in anaerobic metabolism and those that attenuate RAP2.12 activity. Furthermore, inhibition of the mitochondrial ETC causes transient ROS accumulation that act as signal, transduced by the mitogen-activated protein kinases (MPKs), MPK3, MPK4, and MPK6, to activate genes involved in ROS scavenging.[32] Moreover, inhibition of photosynthetic activity caused by hypoxia generates a signal of an unknown nature that induces genes involved in maintaining plastid membrane integrity.[103]

8.6   CONCLUSION AND FUTURE PROSPECTS

Adaptation of plants under waterlogging follows two strategies to escape from the stress: (1) increase in shoot growth rate to enable their emergence from water and/or (2) facilitating gaseous diffusion in submersed organs by aerenchyma formation. Nevertheless, some plants, termed as quiescent plants, remain alive under submersed condition. Such plants survive under submergence by conserving energy and maintain relatively higher carbohydrate levels in tissues, and/or by regulation at genetic and molecular level to enable plants to resist waterlogging with minimum harmful effects on plant process, for example, activation of ROS scavenging system, maintenance of membrane stability and cytosolic pH, alteration in metabolic processes and hormonal metabolism, etc. Hence adaptations can be either escape from hypoxia or anoxia by means of preexisting aerenchyma or induction in aerenchyma formation when stressed. The adaptations can also be in terms of regulating the internal metabolism to such low levels to meet the shortage of energy and promote survival (resilience through quiescence). The transcriptional regulation of the key processes include sensing shortage of oxygen, induction or repression of hypoxia-responsive genes, enhancing or decreasing cell elongation and the formation of lysigenous aerenchyma by means of spatially directed programmed death of cells.

Identifying genes and gene–product interactions involved in adaptive events provide opportunities for genetically manipulating the intracellular pathways that improve flooding tolerance in crop species. Much work is required to effectively elucidate the mechanism at molecular level. Various combinations of adaptations are to be developed by plants as the waterlogging/flooding stress imposes different sets of problems based on the severity of stress, duration or period of stress, and more importantly the age and stage of plant development. For example, when waterlogging of the soil damages roots through oxygen shortage, it also initiates shoot damage. In this case, root survival is expected to be favored by the development and presence of an extensive aerenchyma, as in the case of rice.

However to be successful, it is also required that it must be allied to other mechanisms that also minimize radial loss of internal oxygen by development of barriers. It also requires capacity or ability of combining hypoxia-induced quiescence of root apex which can grow with renewed vigor once fully reaerated, as found in aerenchymatous willow (*Salix alba*). The knowledge gained in efforts of overcoming these challenges can also be especially useful in the even more complex but relevant question of improving tolerance of crops to the combined stresses of flooding and salinity. The increasing threat from coastal flooding and salinity interactions with waterlogging of much fertile agricultural land as sea levels continue to rise make such work a priority for the future.

8.7 SUMMARY

Waterlogging is one of the major abiotic stresses affecting growth, development, and productivity in mesophytes. Waterlogging causes hypoxia (low oxygen) in soil that gradually leads to anoxia (absence of oxygen) if waterlogging persists for long. In addition to oxygen shortage, flooding also prevents the diffusive escape and/or oxidative breakdown of gases. Changes in concentrations of CO_2, O_2, and other gases in rhizosphere hamper physical, chemical, biological, and thermal properties of the soil and alter metabolic processes in plants. Ethylene, a phytohormone, is also accumulated to phytotoxic concentrations in wet soils. Under anaerobiosis roots produce excess of protons that acidify cytoplasm and vacuoles, which is suicidal for roots. Overall metabolism of plants is altered under waterlogged conditions.

Reactive oxygen species (ROS) are synthesized in mitochondria due to low oxygen hampering plant processes. Aerenchyma formation, elongation of aerial parts, reduction in oxygen losses from root surface through suberin deposition on epidermal layers are some of the adaptations of plants to survive under waterlogged condition. Aerenchyma in waterlogged plants enables them to transport oxygen and other gases between aerial parts and roots efficiently. Under waterlogged condition aerobic respiration ceases and synthesis of ethanol and lactate via fermentation is increased in plants. Several anaerobic proteins, namely, pyruvate decorboxylase, alcohol dehydrogenase, lactate dehydrogenase, superoxide dismutase, peroxidase, catalase, ascorbate peroxidase, polyphenol oxidase, glutamate decarboxylase, etc. are over expressed under waterlogged condition.

With advancement of biotechnological tools, it is possible to understand the molecular basis of waterlogging resistance in crop plants more precisely. In this review, attempt has been made to compile current status on effect of waterlogging on morpho-physiological and biochemical processes of plants vis-à-vis molecular mechanism of waterlogging resistance in mesophytes.

KEYWORDS

- aerenchyma
- anaerobiosis
- anoxia
- ethylene
- hypoxia
- waterlogging tolerance

REFERENCES

1. Abiko, T.; Kotula, L.; Shiono, K.; Malik, A. I.; Colmer, T. D.; Nakazono, M. Enhanced Formation of Aerenchyma and Induction of a Barrier to Radial Oxygen Loss in Adventitious Roots of *Zea nicaraguensis* Contribute to Its Waterlogging Tolerance as Compared with Maize (*Zea mays* ssp. *mays*). *Plant Cell Environ.* **2012**, *35*, 1618–1630.

2. Alassimone, J.; Naseer, S.; Geldner, N. A Developmental Framework for Endodermal Differentiation and Polarity. *Proc. Natl. Acad. Sci. U.S.A.* **2010**, *107*, 5214–5219.

3. Alassimone, J.; Roppolo, D.; Geldner, N.; Vermeer, J. E. M. The Endodermis-development and Differentiation of the Plant's Inner Skin. *Protoplasma* **2012**, *249*, 433–443.

4. IPCC. *Managing the Risks of Extreme Events and Disasters to Advance Climate Change Adaptation: Summary for Policy makers. A special report of Working Groups I and II of the Intergovernmental Panel on Climate Change (IPCC)*; Cambridge University Press, UK, 2012; pp 582.

5. Arber, A. *Water Plants: A Study of Aquatic Angiosperms*; Cambridge University Press: Cambridge, 1920; pp 421.

6. Arikado, H. Different Responses of Soybean Plants to an Excess of Water with Special Reference to Anatomical Observations. *Proc. Crop Sci. Soc. Jpn.* **1954**, *23*, 28–36.

7. Arikado, H.; Adachi, Y. Anatomical and Ecological Responses of Barley and Some Forage Crops to the Flooding Treatment. *Bull. Fac. Agric. Mie Univ.* **1955**, *11*, 1–29.

8. Armstrong, J.; Armstrong, W. Phragmites Die-back: Toxic Effects of Propionic, Butyric and Caproic Acids in Relation to pH. *New Phytol.* **1999**, *142*, 201–217.

9. Armstrong, J.; Armstrong, W. Rice: Sulfide-induced Barriers to Root Radial Oxygen Loss, Fe^{2+} and Water Uptake, and Lateral Root Emergence. *Ann. Bot.* **2005**, *96*, 625–638.

10. Armstrong, J.; Armstrong, W.; Armstrong, I. B.; Pittaway, G. R. Senescence, and Phytotoxin, Insect, Fungal and Mechanical Damage: Factors Reducing Convective Gas-Flows in *Phragmites australis*. *Aquat. Bot.* **1996**, *54*, 211–226.

11. Armstrong, W. Aeration in Higher Plants. In *Advances in Botanical Research*; Woolhouse, H. W., Ed.; Academic Press: London, 1979; Vol. 7, pp 225–332.

12. Armstrong, W.; Beckett, P. M.; Justin, S. H. F. W.; Lythe, S. Modeling, and Other Aspects of Root Aeration by Diffusion. In *Plant Life Under Oxygen Deprivation: Ecology, Physiology and Biochemistry*; Jackson, M. B., Davies, D. D., Lambers, H., Eds.; SPB Academic: The Hague, 1991; pp 267–282.

13. Arshad, M.; Frankenberger, W. T. J. Production and Stability of Ethylene in Soil. *Biol. Fertil. Soils*. **1990**, *10*, 29–34.

14. Ashraf, M.; Rehman, H. Interactive Effects of Nitrate and Long Term Waterlogging on Growth, Water Relation and Gaseous Exchange Properties of Maize (*Zea mays* L.). *Plant Sci.* **1999**, *144*, 35–43.

15. Baena-Gonzalez, E.; Rolland, F.; Thevelein, J. M.; Sheen, J. A Central Integrator of Transcription Reprogramming Networks in Plant Stress and Energy Signaling, *Nature*. **2007**, *448*, 938–942.

16. Bailey-Serres, J.; Voesenek, L. A. C. J. Flooding Stress: Acclimations and Genetic Diversity. *Annu. Rev. Plant Biol.* **2008**, *59*, 313–339.

17. Bailey-Serres, J.; Voesenek, L. A. C. J. Life in the Balance: A Signaling Network Controlling Survival of Flooding. *Curr. Opin. Plant Biol.* **2010**, *13*, 489–494.

18. Bailey-Serres, J.; Fukao, T.; Ronald, P.; Ismail, A.; Heuer, S.; Mackill, D. Submergence Tolerant Rice: *SUB1*'s Journey from Landrace to Modern Cultivar. *Rice* **2010**, *3*, 138–147.

19. Bailey-Serres, J.; Fukao, T.; Gibbs, D. J.; Holdsworth, M. J.; Lee, S. C.; Licausi, F.; Perata, P.; Voesenek, L. A. C. J.; van Dongen, J. T. Making Sense of Low Oxygen Sensing. *Trends Plant Sci.* **2012**, *17*, 129–138.

20. Bansal, R. Diurnal and Temporal Changes in Biochemical and Physiological Parameters in Pigeonpea Genotype during Early Stages of Waterlogging Stress. Ph. D. Thesis, Banaras Hindu University, Varanasi, India, 2010; pp 150.

21. Bansal, R.; Srivastava, J. P. Antioxidative Defense System in Pigeon Pea Roots Under Waterlogging Stress. *Acta Physiol. Plant* **2012**, *34*, 515–522.

22. Bansal, R.; Srivastava, J. P. Effect of Waterlogging on Photosynthetic and Biochemical Parameters in Pigeon Pea. *Russ. J. Plant Physiol.* **2015**, *62*, 322–327.

23. Blackwell, P. S.; Wells, E. A. Limiting Oxygen Flux Densities for Oat Root Extension. *Plant Soil* **1983**, *73*, 129–139.

24. Blokhina, O.; Virolainen, E.; Fagerstedt, K. V. Antioxidants, Oxidative Damage and Oxygen Deprivation Stress: A Review. *Ann. Bot.* **2003**, *91*, 179–194.

25. Boamfa, E. I.; Ram, P. C.; Jackson, M. B.; Reuss, J.; Harren, F. J. M. Dynamic Aspects of Alcoholic Fermentation of Rice Seedlings in Response to Anaerobiosis and to Complete Submergence: Relationship to Submergence Tolerance. *Ann. Bot.* **2003**, *91*, 279–290.

26. Boerjan, W.; Cervera, M. T.; Delarue, M.; Beeckman, T.; Dewitte, W.; Bellini, C.; Caboche, M.; Van Onckelen, H.; Van Montagu, M.; Inzé D. Super Root, a Recessive Mutation in *Arabidopsis*, Confers Auxin Overproduction. *Plant Cell* **1995**, *7*, 1405–1419.

27. Boru, G.; Van Ginkel, M.; Trethowan, R. M.; Boersma, L.; Kronstad, W. E. Oxygen Use from Solution by Wheat Genotypes Differing in Tolerance to Waterlogging. *Euphytica* **2003**, *132*, 151–158.

28. Boru, G.; Vantoai, T.; Alves, J.; Hua, D.; Knee, M. Responses of Soybean to Oxygen Deficiency and Elevated Root-zone Carbon Dioxide Concentration. *Ann. Bot.* **2003**, *91*, 447–453.

29. Bramley, H.; Turner, N. C.; Turner, D. W.; Tyermann, S. D. The Contrasting Influence of Short-term Hypoxia on the Hydraulic Properties of Cells and Roots of Wheat and Lupine. *Funct. Plant Biol.* **2010**, *37*, 183–193.

30. Buchanan, B.; Gruissem, W.; Jones, R. Responses to Abiotic Stresses. In *Biochemistry and Molecular Biology of Plants*; American Society of Plant Physiologists; John Wiley & Sons: Rockville, Maryland, 2000; pp 1280.

31. Calvo-Polanco, M.; Senorans, J.; Zwiazek, J. J. Role of Adventitious Roots in Water Relations of Tamarack (*Larix laricina*) Seedlings Exposed to Flooding. *BMC Plant Biol.* **2012**, *12*, 9. http://www.biomedcentral.com/1471-2229/12/99 (accessed May 16, 2017).

32. Chang, R.; Jang, C. H.; Branco-Price, C.; Nghiem, P.; Bailey-Serres, J. Transient MPK6 Activation in Response to Oxygen Deprivation and Reoxygenation is Mediated by Mitochondria and Aids Seedling Survival in *Arabidopsis*. *Plant Mol. Biol.* **2012**, *78*, 109–122.

33. Colmer, T. D. Aerenchyma and an Inducible Barrier to Radial Oxygen Loss Facilitate Root Aeration in Upland, Paddy and Deep-water Rice (*Oryza sativa* L.). *Ann. Bot.* **2003**, *91*, 301–309.

34. Colmer, T. D. Long-distance Transport of Gases in Plants: A Perspective on Internal Aeration and Radial Oxygen Loss from Roots. *Plant Cell Environ.* **2003**, *26*, 17–36.

35. Colmer, T. D.; Greenway, H. Ion Transport in Seminal and Adventitious Roots of Cereals During O_2 Deficiency. *J. Exp. Bot.* **2011**, *62*, 39–57.

36. Colmer, T. D.; Voesenek, L. A. C. J. Flooding Tolerance: Suites of Plant Traits in Variable Environments. *Funct. Plant Biol.* **2009**, *36*, 665–681.

37. Colmer, T. D.; Winkel, A.; Pedersen, O. A Perspective on Underwater Photosynthesis in Submerged Terrestrial Wetland Plants. *AoB Plants* **2011**, *2011*, plr030. DOI: 10.1093/aobpla/plr030. https://www.ncbi.nlm.nih.gov/pmc/articles/PMC3249690/ (accessed May 16, 2017).

38. Cox, M. C. H.; Millenaar, F. F.; de Jong van Berkel, Y. E. M.; Peeters, A. J. M.; Voesenek, L. A. C. J. Plant Movement: Submergence-induced Petiole Elongation in *Rumex paulstris* Depends on Hyponastic Growth. *Plant Physiol.* **2003,** *132,* 282–291.

39. Das, K. K.; Panda, D.; Sarkar, R. K.; Reddy, J. M.; Ismail, A. M. Submergence Tolerance in Relation to Variable Floodwater Conditions in Rice. *Environ. Exp. Bot.* **2009,** *66,* 425–434.

40. Dawood, T.; Rieu, I.; Wolters-Arts, M.; Derksen, E. B.; Mariani, C.; Visser, E. J. W. Rapid Flooding-induced Adventitious Root Development from Preformed Primordia in *Solanum dulcamara. Ann. Bot.* **2014,** *6,* plt058. DOI: 10.1093/aobpla/plt058.

41. DeDatta, S. K.; Buresh, R. J.; Mamaril, C. P. Increasing Nutrient Use Efficiency in Rice with Changing Needs. *Fertil. Res.* **1990,** *26,* 157–167.

42. Demidchik, V.; Cuin, T. A.; Svistunenko, D.; Smith, S. J.; Miller, A. J.; Shabala, S.; Sokolik, A.; Yurin, V. *Arabidopsis* Root K^+-efflux Conductance Activated by Hydroxyl Radicals: Single-channel Properties, Genetic Basis and Involvement in Stress-induced Cell Death. *J Cell Sci.* **2010,** *123,* 1468–1479.

43. Demidchik, V.; Shabala, S. N.; Davies, J. M. Spatial Variation in H_2O_2 Response of *Arabidopsis thaliana* Root Epidermal Ca^{2+} Flux and Plasma Membrane Ca^{2+} Channels. *Plant J.* **2007,** *49,* 377–386.

44. Dietrich, K.; Weltmeier, F.; Ehlert, A.; Weiste, C.; Stahl, M.; Harter, K.; Dröge-Laser, W. Heterodimers of the *Arabidopsis* Transcription Factors bZIP1 and bZIP53 Reprogram Amino Acid Metabolism During Low Energy Stress. *Plant Cell* **2011,** *23,* 381–395.

45. Dordas, C. Non Symbiotic Hemoglobins and Stress Tolerance in Plants. *Plant Sci.* **2009,** *176,* 433–440.

46. Dordas, C.; Hasinoff, B. B.; Rivoal, J.; Hill, R. D. Class-1 Hemoglobin, Nitrate and NO Levels in Anoxic Maize Cell-suspension Cultures. *Planta* **2004,** *219,* 66–72.

47. Dordas, C.; Hasinoff, B. B.; Igamberdiev, A. U.; Manach, N.; Rivoal, J.; Hill, R. D. Expression of a Stress Induced Hemoglobin Affects NO Levels Produced by Alfalfa Root Cultures Under Hypoxic Stress. *Plant J.* **2003,** *35,* 763–770.

48. Drew, M. C. Effects of Flooding and Oxygen Deficiency on Plant Mineral Nutrition. *Adv. Plant Nutr.* **1988,** *3,* 115–159.

49. Drew, M. C.; Lynch, J. M. Soil Anaerobiosis, Microorganisms, and Root Function. *Annu. Rev. Phytopathol.* **1980,** *18,* 37–66.

50. Drew, M. C.; Cobb, B. G.; Johnson, J. R.; Andrews, D.; Morgan, P. W.; Jordan, W.; Jiu, H. C. Metabolic Acclimation of Root Tips to Oxygen Deficiency. *Ann. Bot.* **1994,** *74,* 281–286.

51. Drew, M. C.; He, C. J.; Morgan, P. W. Programmed Cell Death and Aerenchyma Formation in Roots. *Trends Plant Sci.* **2000,** *5,* 123–127.

52. Drew, M. C.; Saglio, P. H.; Pradet, A. Larger Adenylate Energy Charge and ATP/ADP Ratios in Aerenchymatous Roots of *Zea mays* in Anaerobic Media as a Consequence of Improved Internal Oxygen Transport. *Planta* **1985,** *165,* 51–58.

53. Dufey, I.; Hakizimana, P.; Draye, X.; Lutts, S.; Bertin, P. QTL Mapping for Biomass and Physiological Parameters Linked to Resistance Mechanisms to Ferrous Iron Toxicity in Rice. *Euphytica* **2009,** *167,* 143–160.

54. Durack, P. J.; Wijffels, S. E.; Matear, R. J. Ocean Salinities Reveal Strong Global Water Cycle Intensification During 1950 to 2000. *Science* **2012,** *336,* 455–458.

55. Elzenga, J. T. M.; van Veen, H. Waterlogging and Plant Nutrient Uptake. In *Waterlogging Signaling and Tolerance in Plants*; Mancuso, S., Shabala, S., Eds.; Springer-Verlag: Heidelberg, 2010; pp 23–36.

56. Evans, D. E. Aerenchyma Formation. *New Phytol.* **2003,** *161,* 35–49.

57. Fraser, L. The Reaction of *Viminaria denudata* to Increased Water Content of the Soil. *J. Proc. Linn. Soc.* **1931,** *56,* 391–406.

58. Fukao, T.; Bailey-Serres, J. Submergence Tolerance Conferred by *Sub1A* is Mediated by *SLR1* and *SLRL1* Restriction of Gibberellin Responses in Rice. *Proc. Natl. Acad. Sci. U.S.A.* **2008,** *105,* 16814–16819.

59. Fukao, T.; Xu, K.; Ronald, P. C.; Bailey-Serres, J. A Variable Cluster of Ethylene Response Factor-like Genes Regulates Metabolic and Developmental Acclimation Responses to Submergence in Rice. *Plant Cell* **2006,** *18,* 2021–2034.

60. Fukao, T.; Yeung, E.; Bailey-Serres, J. The Submergence Tolerance Regulator *SUB1A* Mediates Crosstalk between Submergence and Drought Tolerance in Rice. *Plant Cell* **2011,** *23,* 412–427.

61. Gambrell, R. P.; Patrick, W. H. Chemical and Microbiological Properties of Anaerobic Soils and Sediments. In *Plant Life in Anaerobic Habitats*; Hook, D. D., Crawford, R. M. M., Eds.; MI, Ann Arbor Science Publishers: Ann Arbor, USA, **1978;** pp 375–423.

62. Garthwaite, A. J.; von Bothmer, R.; Colmer, T. D. Diversity in Root Aeration Traits Associated with Waterlogging Tolerance in the Genus *Hordeum. Funct. Plant Biol.* **2003,** *30,* 875–889.

63. Gerard, B.; Alaoui-Sosse, B.; Badot, B. M. Flooding Effects on Starch Partitioning During Early Growth of Two Oak Species. *Trees Struct. Funct.* **2009,** *23,* 373–380.

64. Geisler-Lee, J.; Caldwell, J.; Gallie, D. R. Expression of the Ethylene Biosynthetic Machinery in Maize Roots is Regulated in Response to Hypoxia. *J. Exp. Bot.* **2010,** *61,* 857–871.

65. Gibbs, D. J.; Lee, S. C.; Isa, N. M.; Gramuglia, S.; Fukao, T.; Bassel, G. W.; Correia, C. S.; Corbineau, F.; Theodoulou, F. L.; Bailey-Serres, J.; Holdsworth, M. J. Homeostatic Response to Hypoxia is Regulated by the N-end Rule Pathway in Plants. *Nature* **2011,** *479,* 415–418.

66. Gibbs, J.; Morrell, S.; Valdez, A.; Setter, T. L.; Greenway, H. Regulation of Alcoholic Fermentation in Coleoptiles of Two Rice Cultivars Differing in Tolerance to Anoxia. *J. Exp. Bot.* **2000,** *51,* 785–796.

67. Gibbs, J.; Turner, D. W.; Armstrong, W.; Darwent, M. J.; Greenway, H. Response to Oxygen Deficiency in Primary Maize Roots. I. Development of Oxygen Deficiency in the Stele Reduces Radial Solute Transport to the Xylem. *Aust. J. Plant Physiol.* **1998,** *25,* 745–758.

68. Giuntoli, B.; Lee, S. C.; Licausi, F.; Kosmacz, M.; Oosumi, T.; Van Dongen, J. T.; Bailey-Serres, J.; Perata, P. Trihelix DNA Binding Protein Counterbalances Hypoxia-responsive Transcriptional Activation in *Arabidopsis. PLOS Biol.* **2014,** 12 (9), e1001950. DOI: 10.1371/journal.pbio.1001950.

69. Graven, E. H.; Attoe, O. J.; Smith, D. Effect of Liming and Flooding on Manganese Toxicity in Alfalfa Soil. *Soil Sci. Soc. Am. J.* **1965,** *56,* 166–172.

70. Greenway, H.; Gibbs, J. Mechanism of Anoxia Tolerance in Plants. II. Energy Requirements for Maintenance and Energy Distribution to Essential Processes. *Funct. Plant Biol.* **2003,** *30,* 999–1036.

71. Greenway, H.; Armstrong, W.; Colmer, T. D. Conditions Leading to High CO_2 (>5 kPa) in Waterlogged-flooded Soils and Possible Effects on Root Growth and Metabolism. *Ann. Bot.* **2006,** *98,* 9–32.

72. Groeneveld, H. W.; Voesenek, L. A. C. J. Submergence-induced Petiole Elongation in *Rumex palustris* is Controlled by Developmental Stage and Storage Compounds. *Plant Soil* **2003**, *253*, 115–123.

73. Gunawardena, A.; Pearce, D. M.; Jackson, M. B.; Hawes, C. R.; Evans D. E. Characterization of Programmed Cell Death During Aerenchyma Formation induced by Ethylene or Hypoxia in Roots of Maize (*Zea mays* L.). *Planta* **2001**, *212*, 205–214.

74. Gunawardena, A.; Pearce, D. M. E.; Jackson, M. B.; Hawes, C. R.; Evans, D. E. Rapid Changes in Cell Wall Pectic Polysaccharides are Closely Associated with Early Stages of Aerenchyma Formation, a Spatially Localized form of Programmed Cell Death in Roots of Maize (*Zea mays* L.) Promoted by Ethylene. *Plant Cell Environ.* **2001**, *24*, 1369–1375.

75. Gupta, S. K. *Subsurface Drainage for Waterlogged Saline Lands*. Report Submitted to Agricultural Education Division, New Delhi; Central Soil Salinity Research Institute: Karnal, 2011; pp 353 (Unpublished).

76. Gupta, S. K.; Gupta, I. C. *Salt Affected Soils: Reclamation and Management*; Scientific Publishers: Jodhpur, India, 2014; pp 316.

77. Gupta, S. K.; Sharma, D. P.; Swarup, A. Relative Tolerance of Crops to Surface Stagnation. *J. Agric. Eng.* **2004**, *41*, 44–48.

78. Hattori, Y.; Nagai, K.; Furukawa, S.; Song, X. J.; Kawano, R.; Sakakibara, H.; Wu, J.; Matsumoto, T.; Yoshimura, A.; Kitano, H.; Matsuok, M.; Mori, H.; Ashikari, M. The Ethylene Response Factors SNORKEL1 and SNORKEL2 Allow Rice to Adapt to Deep Water. *Nature* **2009**, *460*, 1026–1030.

79. Haupt-Herting, S.; Fock, H. P. Oxygen Exchange in Relation to Carbon Assimilation in Water-stressed Leaves During Photosynthesis. *Ann. Bot.* **2002**, *89*, 851–859.

80. He, C. J.; Finlayson, S. A.; Drew, M. C.; Jordan, W. R.; Morgan, P. W. Ethylene Biosynthesis During Aerenchyma Formation in Roots of Maize Subjected to Mechanical Impedance and Hypoxia. *Plant Physiol.* **1996**, *112*, 1679–1685.

81. He, C. J.; Morgan, P. W.; Drew, M. C. Transduction of an Ethylene Signal is Required for Cell Death and Lysis in the Root Cortex of Maize During Aerenchyma Formation Induced by Hypoxia. *Plant Physiol.* **1996**, *112*, 463–472.

82. Herzog, M.; Pedersen, O. Partial Versus Complete Submergence: Snorkeling Aids Root Aeration in *Rumex palustris* but not in *R. acetosa*. *Plant Cell Environ.* **2014**, *37*, 2381–2390.

83. Hirabayashi, Y.; Mahendran, R.; Koirala, S.; Konoshima, L.; Yamazaki, D.; Watanabe, S.; Kim, H.; Kanae, S. Global Flood Risk Under Climate Change. *Nat. Clim. Change.* **2013**, *3*, 816–821.

84. Horst, W. J. The Physiology of Manganese Toxicity. In *Manganese in Soils and Plants*; Graham, R. D., Hannam, R. J., Uren, N. C., Eds.; Kluwer Academic Publishers: Dorrech, The Netherlands, 1988, pp 175–188.

85. Estioka, P.; Moro, B.; Baltazar, A. M.; Merca, F. E.; Ismail, A. M.; Johnson, D. E. Differences in Response to Flooding by Germinating Seeds of Two Contrasting Rice Cultivars and Two Species of Economically Important Grass Weeds. *AoB Plants* **2014**, *6*, plu064. DOI: 10.1093/aobpla/plu064.

86. Igamberdiev, A. U.; Kleczkowski, L. A. Magnesium and Cell Energetics in Plants Under Anoxia. *Biochem. J.* **2011**, *437*, 373–379.

87. Igamberdiev, A. U.; Stoimenova, M.; Seregélyes, C.; Hill, R. Class-1 Hemoglobin and Antioxidant Metabolism in Alfalfa Roots. *Planta* **2006**, *223*, 1041–1046.

88. Ishizawa, K.; Murakami, S.; Kawakami, Y.; Kuramochi, H. Growth and Energy Status of Arrowhead Tubers, Pondweed Turions and Rice Seedlings Under Anoxic Conditions. *Plant Cell Environ.* **1999**, *22*, 505–514.

89. Ismail, A. M.; Ella, E. S.; Vergara, G. V.; Mackill, D. J. Mechanisms Associated with Tolerance to Flooding During Germination and Early Seedling Growth in Rice (*Oryza sativa*). *Ann. Bot.* **2009**, *103*, 197–209.

90. Ivanchenko, M. G.; Muday, G. K.; Dubrovsky, J. G. Ethylene-auxin Interactions Regulate Lateral Root Initiation and Emergence in *Arabidopsis thaliana*. *Plant J.* **2008**, *55*, 335–347.

91. Jackson, M.; Armstrong, W. Formation of Aerenchyma and the Processes of Plant Ventilation in Relation to Soil Flooding and Submergence. *Plant Biol.* **1999**, *1*, 274–287.

92. Jackson, M. B.; Ram, P. C. Physiological and Molecular Basis of Susceptibility and Tolerance of Rice Plants to Complete Submergence. *Ann. Bot.* **2003**, *91*, 227–241.

93. Jaiswal, A.; Srivastava, J. P. Effect of Nitric Oxide on Some Morphological and Physiological Parameters in Maize Exposed to Waterlogging Stress. *Afr. J. Agric. Res.* **2015**, *10*, 3462–3471.

94. Jung, K. H.; Seo, Y. S.; Walia, H.; Cao, P. J.; Fukao, T.; Carnlas, P. E.; Amonpant, F.; Bailey-Serres, J.; Ronald, P. C. The Submergence Tolerance Regulator Sub 1 A Mediates Stress Response Expression of AP2/ERF Transcription Factors. *Plant Physiol.* **2010**, *152*, 1674–1692.

95. Justin, S. H. F. W.; Armstrong, W. The Anatomical Characteristics of Roots and Plant Response to Soil Flooding. *New Phytol.* **1987**, *106*, 465–495.

96. Justin, S. H. F. W.; Armstrong, W. Evidence for the Involvement of Ethenein Aerenchyma Formation in Adventitious Roots of Rice (*Oryza sativa*). *New Phytol.* **1991**, *118*, 49–62.

97. Kang, S. G.; Price, J.; Lin, P. C.; Hong, J. C.; Jang, J. C. The *Arabidopsis* bZIP1 Transcription Factor is Involved in Sugar Signaling, Protein Networking, and DNA Binding. *Mol. Plant.* **2010**, *3*, 361–373.

98. Kende, H.; Van der Knaap, E.; Cho, H. T. Deep Water Rice: A Model Plant to Study Stem Elongation. *Plant Physiol.* **1998**, *118*, 1105–1110.

99. Khabaz-Saberi, H.; Rengel, Z.; Wilson, R.; Setter, T. L. Variation of Tolerance to Manganese Toxicity in Australian Hexaploid Wheat. *J. Plant Nutr. Soil Sci.* **2010**, *173*, 103–112.

100. Khabaz-Saberi, H.; Setter, T. L.; Waters, I. Waterlogging Induces High to Toxic Concentrations of Iron, Aluminum, and Manganese in Wheat Varieties on Acidic Soil. *J. Plant Nutr.* **2006**, *29*, 899–911.

101. Kirk, G. J. D. *The Biogeochemistry of Submerged Soils*; Wiley: Chichester, UK, 2004; pp 283.

102. Kirk, G. J. D.; Greenway, H.; Atwell, B. J.; Ismail, A. M.; Colmer, T. D. Adaptation of Rice to Flooded Soils. In *Progress in Botany*; Lüttge, U., Beyschlag, W., Cushman, J., Eds.;Springer: Berlin, Heidelberg, 2014;Vol. 75, pp 215–253.

103. Klecker, M.; Gasch, P.; Peisker, H.; Dormann, P.; Schlicke, H.; Grimm, B.; Mustroph, A. A Shoot-specific Hypoxic Response of *Arabidopsis* Sheds Light on the Role of the Phosphate-responsive Transcription Factor *Phosphate-starvation Response1*. *Plant Physiol.* **2014**, *165*, 774–790.

104. Kronzucker, H. J.; Kirk, G. J. D.; Siddiqi, M. Y.; Glass, A. D. M. Effects of Hypoxia on NH_4^+ Fluxes in Rice Roots, Kinetics and Compartmental Analysis. *Plant Physiol.* **1998**, *116*, 581–587.

105. Kuiper, P. J. C.; Walton, C. S.; Greenway, H. Effects of Hypoxia on Ion Uptake by Nodal and Seminal Wheat Roots. *Plant Physiol. Biochem.* **1994,** *32*, 267–276.

106. Kulichikhin, K.; Yamauchi, T.; Watanabe, K.; Nakazono, M. Biochemical and Molecular Characterization of Rice (*Oryza sativa* L.) Roots Forming a Barrier to Radial Oxygen Loss. *Plant Cell Environ.* **2014,** *37, 2406–2420.*

107. Laanbroek, H. J. Bacterial Cycling of Minerals that Affect Plant Growth in Waterlogged Soils: A Review. *Aquat. Bot.* **1990,** *38*, 109–125.

108. Laanbroek, H. J. Methane Emission from Natural Wetlands: Interplay between Emergent Macrophytes and Soil Microbial Processes—A Mini-review. *Ann. Bot.* **2010,** *105*, 141–153.

109. Lakshmanan, M.; Mohanty, B.; Lim, S. H.; Ha, S. H.; Lee, D. Y. Metabolic and Transcriptional Regulatory Mechanisms Underlying the Anoxic Adaptation of Rice Coleoptile. *Ann. Bot.* **2014,** *6*, plu026. DOI: 10.1093/aobpla/plu026.

110. Lamers, L. P. M.; Govers, L. L.; Janssen, I. C. J. M.; Geurts, J. J. M.; Van der Welle, M. E. W.; Van Katwijk, M. M.; Van der Heide, T.; Roelofs, J. G. M.; Smolders, A. J. P. Sulfide as a Soil Phytotoxin—A Review. *Front. Plant Sci.* **2012,** *4,* 1–14.

111. Lee, K. W.; Chen, P. W.; Lu, C. A.; Chen, S.; Ho, T. H. D.; Yu, S. M. Coordinated Responses to Oxygen and Sugar Deficiency Allow Rice Seedlings to Tolerate Flooding. *Sci. Signaling.* **2009,** *2, ra61.*

112. Licausi, F.; Perata, P. Low Oxygen Signaling and Tolerance in Plants. *Adv. Bot. Res.* **2009,** *50*, 139–198.

113. Licausi, F.; Kosmacz, M.; Weits, D. A.; Giuntoli, B.; Giorgi, F. M.; Voesenek, L. A. C. J.; Perata. P.; van Dongen, J. T. Oxygen Sensing in Plants is Mediated by an N-end Rule Pathway for Protein Destabilization. *Nature* **2011,** *479*, 419–422.

114. Limami, A.; Diab, H.; Lothier, J. Nitrogen Metabolism in Plants Under Low Oxygen Stress. *Planta* **2014,** *239*, 531–541.

115. Liu, S.; Wang, J.; Wang, L.; Wang, X.; Xue, Y.; Wu, P.; Shou, H. Adventitious Root Formation in Rice Requires *OsGNOM1* and is Mediated by the *OsPINs* Family. *Cell Res.* **2009,** *19*, 1110–1119.

116. Loreti, E.; Novi, G.; Poggi, A.; Alpi, A.; Perata, P. A Genome-wide Analysis of the Effects of Sucrose on Gene Expression in *Arabidopsis* Seedlings Under Anoxia. *Plant Physiol.* **2005,** *137*, 1130–1138.

117. Loreti, E.; Poggi, A.; Novi, G.; Alpi, A.; Perata, P. A Genome-wide Analysis of the Effects of Sucrose on Gene Expression in *Arabidopsis* Seedlings Under Anoxia. *Plant Physiol.* **2005,** *137*, 1130–1138.

118. Lu, C. A.; Lin, C. C.; Lee, K. W.; Chen, J. L.; Huang, L. F.; Ho, S. L.; Liu, H. J.; Hsing, Y. I.; Yu, S. M. The *SnRK1A* Protein Kinase Plays a Key Role in Sugar Signaling During Germination and Seedling Growth of Rice. *Plant Cell* **2007,** *19*, 2484–2499.

119. Lynch, J. M. Phytotoxicity of Acetic-acid Produced in Anaerobic Decomposition of Wheat Straw. *J. Appl. Bacteriol.* **1977,** *42*, 81–87.

120. Mackintosh, R. W.; Davies, S. P.; Clarke, P. R.; Weekes, J.; Gillespie, J. G.; Gibb, B. J.; Hardie, D. G. Evidence for a Protein Kinase Cascade in Higher Plants. *Eur. J. Biochem.* **1992,** *209*, 923–931.

121. Magneschi, L.; Perata, P. Rice Germination and Seedling Growth in the Absence of Oxygen. *Ann. Bot.* **2009,** *103*, 181–196.

122. Malik, A. I.; Islam, A. K. M. R.; Colmer, T. D. Transfer of the Barrier to Radial Oxygen Loss in Roots of *Hordeum marinum* to Wheat (*Triticum aestivum*): Evaluation of Four *H. marinum*-wheat amphiploids. *New Phytol.* **2001,** *190*, 499–508.

123. Manai, J.; Chokri, Z.; Mohamed, D.; Houda, G. D.; Bouthour. Role of Nitric Oxide in Saline Stress: Implications on Nitrogen Metabolism. *Afr. J. Plant Sci.* **2012,** *6,* 376–382.

124. Marschner, H. Mechanisms of Adaptation of Plants to Acid Soils. *Plant Soil.* **1991,** *134,* 1–20.

125. McMichael, R. W. Jr.; Bachmann, M.; Huber, S. C. Spinach Leaf Sucrose-phosphate Synthase and Nitrate Reductase are Phosphorylated/inactivated by Multiple Protein Kinases *in vitro.* *Plant Physiol.* **1995,** *108,* 1077–1082.

126. Mochizuki, T.; Takahashi, U.; Shimamura, S.; Fukuyama, M. Secondary Aerenchyma Formation in Hypocotyl in Summer Leguminous Crops. *Jpn. J. Crop Sci.* **2000,** *69,* 69–73.

127. Mohanty, H. K. Breeding for Submergence Tolerance in Rice in India. In *Progress in Rain-fed Lowland Rice*; International Rice Research Institute: Los Baños, Philippines, 1987; pp 200–205.

128. Mommer, L.; Pons, T. L.; Visser, E. J. W. Photosynthetic Consequences of Phenotypic Plasticity in Response to Submergence: *Rumex palustris* as a Case Study. *J. Exp. Bot.* **2005,** *57,* 283–290.

129. Mommer, L.; Pons, T. L.; Wolters-Arts, M.; Venema, J. H.; Visser, E. J. W. Submergence-induced Morphological, Anatomical, and Biochemical Responses in a Terrestrial Species Affect Gas Diffusion Resistance and Photosynthetic Performance. *Plant Physiol.* **2005,** *139,* 497–508.

130. Morad, P.; Silvestre, J. Plant Injury Due to Oxygen Deficiency in the Root Environment of Soilless Culture: A Review. *Plant Soil.* **1996,** *184,* 243–254.

131. Mustroph, A.; Lee, S. C.; Oosumi, T.; Zanetti, M. E.; Yang, H.; Ma, K.; Yaghoubi-Masihi, A.; Fukao, T.; Bailey-Serres, J. Cross-kingdom Comparison of Transcriptomic Adjustments to Low-oxygen Stress Highlights Conserved and Plant-specific Responses. *Plant Physiol.* **2010,** *152,* 1484–1500.

132. Negi, S.; Sukumar, P.; Liu, X.; Cohen, J. D.; Muday, G. K. Genetic Dissection of the Role of Ethylene in Regulating Auxin-dependent Lateral and Adventitious Root Formation in Tomato. *Plant J.* **2010,** *61,* 3–15.

133. Nishiuchi, S.; Yamauchi, T.; Takahashi, H.; Kotula, L.; Nakazono, M. Mechanisms for Coping with Submergence and Waterlogging in Rice. *Rice.* **2012,** *5:*2, 1–14.

134. Noctor, G.; Foyer, C. H. Ascorbate and Glutathione: Keeping Active Oxygen Under Control. *Annu. Rev. Plant Physiol. Plant Mol. Biol.* **1998,** *49,* 249–279.

135. Oliveira, H. C.; Freschi, L.; Sodek, L. Nitrogen Metabolism and Translocation in Soybean Plants Subjected to Root Oxygen Deficiency. *Plant Physiol. Biochem.* **2013,** *66,* 141–149.

136. Pang, J.; Ross, J.; Zhou, M.; Mendham, N.; Shabala, S. Amelioration of Detrimental Effects of Waterlogging by Foliar Nutrient Spray in Barley. *Funct. Plant Biol.* **2007,** *34,* 221–227.

137. Pang, J.; Shabala, S. Membrane Transporters and Waterlogging Tolerance. In *Waterlogging Signaling and Tolerance in Plants*; Mancuso, S., Shabala, S., Eds.; Springer-Verlag: Heidelberg, 2010; pp 197–219.

138. Park, M.; Yim, H. K.; Park, H. G.; Lim, J.; Kim, S. H.; Hwang, Y. S. Interference with Oxidative Phosphorylation Enhances Anoxic Expression of Rice *α-amylase* Genes Through Abolishing Sugar Regulation. *J. Exp. Bot.* **2010,** *61,* 3235–3244.

139. Pedersen, O.; Colmer, T. D.; Sand-Jensen, K. Underwater Photosynthesis of Submerged Plants—Recent Advances and Methods. *Front. Plant Sci.* **2013,** *4, 140,* 1–19,

140. Pedersen, O.; Rich, S. M.; Colmer, T. D. Surviving Floods: Leaf Gas Films Improve O_2 and CO_2 Exchange, Root Aeration, and Growth of Completely Submerged Rice. *Plant J.* **2009,** *58*, 147–156.

141. Peng, H. P.; Lin, T. Y.; Wang, N. N.; Shih, M. C. Differential Expression of Genes Encoding 1-Aminocyclopropane-1-Carboxylate Synthase in *Arabidopsis* During Hypoxia. *Plant Mol. Biol.* **2005,** *58*, 15–25.

142. Petraglia, T.; Poole, R. J. Effect of Anoxia on ATP Levels and Ion Transport Rates in Red Beet. *Plant Physiol.* **1980,** *65*, 973–974.

143. Pezeshki, S. R. Wetland Plant Responses to Soil Flooding. *Environ. Exp. Bot.* **2001,** *46*, 299–312.

144. Ponnamperuma, F. N. The Chemistry of Submerged Soil. In *Advances in Agronomy*; 1972; Vol. 24, pp 29–95.

145. Ponnamperuma, F. N. Effects of Flooding on Soils. In *Flooding and Plant Growth*; Kozlowski, T. T., Ed.; Academic Press Inc.: Orlando, FL, USA, 1984; pp 1–44.

146. Poorter, H.; Vanderwerf, A.; Atkin, O. K.; Lambers, H. Respiratory Energy Requirements of Roots Vary with the Potential Growth Rate of a Plant. *Physiol. Plant* 1991, *83*, 469–475.

147. Porterfield, D. M.; Crispi, M. L.; Musgrave, M. E. Changes in Soluble Sugar, Starch, and Alcohol Dehydrogenase in *Arabidopsis thaliana* Exposed to N_2 Diluted Atmospheres. *Plant Cell Physiol.* **1997,** *38*, 1354–1358.

148. Rai, R. K.; Srivastava, J. P.; Shahi, J. P. Effect of Waterlogging on Some Biochemical Parameters During Early Growth Stages of Maize. *Indian J. Plant Physiol.* **2004,** *9*, 66–69.

149. Rajhi, I.; Yamauchi, T.; Takahashi, H.; Nishiuchi, S.; Shiono, K.; Watanabe, R.; Mliki, A.; Nagamura, Y.; Tsutsumi, N.; Nishizawa, N. K.; Nakazono, M. Identification of Genes Expressed in Maize Root Cortical Cells During Lysigenous Aerenchyma Formation Using Laser Micro-Dissection and Microarray Analyses. *New Phytol.* **2011,** *190*, 351–368.

150. Ranathunge, K.; Lin, J.; Steudle, E.; Schreiber, L. Stagnant Deoxygenated Growth Enhances Root Suberization and Lignifications, but Differentially Affects Water and NaCl Permeability in Rice (*Oryza sativa* L.) Roots. *Plant Cell Environ.* **2011,** *34*, 1223–1240.

151. Raymond, P.; Pradet, A. Stabilization of Adenine Nucleotide Ratios at Various Values by an Oxygen Limitation of Respiration in Germinating Lettuce (*Lactuca sativa*) Seeds. *Biochem. J.* **1980,** *190*, 39–44.

152. Ren, M.; Qiu, S.; Venglat, P.; Xiang, D.; Feng, L.; Selvaraj, G.; Datla, R. Target of Rapamycin Regulates Development and Ribosomal RNA Expression through Kinase Domain in *Arabidopsis*. *Plant Physiol.* **2011,** *155*, 1367–1382.

153. Rich, S. M.; Ludwig, M.; Colmer, T. D. Photosynthesis in Aquatic Adventitious Roots of the Halophyte Stem Succulent *Tecticornia pergranulata*. *Plant Cell Environ.* **2008,** *31*, 1007–1016.

154. Rich, S. M.; Pedersen, O.; Ludwig, M.; Colmer, T. D. Shoot Atmospheric Contact is of Little Importance to Aeration of Deeper Portions of the Wetland Plant *Meionectes brownii*; Submerged Organs Mainly Acquire O_2 from the Water Column or Produce it Endogenously in Underwater Photosynthesis. *Plant Cell Environ.* **2012,** *36*, 213–223.

155. Rieu, I.; Cristescu, S. M.; Harren, F. J. M.; Huibers, W.; Voesenek, L. A.; Mariani, C.; Vriezen, W. H. *RP-ACS1*, a Flooding-induced 1-Aminocyclopropane-1-Carboxylate

Synthase Gene of *Rumex palustris*, is Involved in Rhythmic Ethylene Production. *J. Exp. Bot.* **2005,** *56*, 841–849.

156. Rocha, M.; Sodek, L.; Licausi, F.; Hameed, M. W.; Dornelas, M. C.; van Dongen, J. T. Analysis of Alanine Aminotransferase in Various Organs of Soybean (*Glycine max*) and in Dependence of Different Nitrogen Fertilizers During Hypoxic Stress. *Amino Acids.* **2010,** *34*, 1043–1053.

157. Saab, I. N.; Sachs, M. M. A Flooding-induced Xyloglucanendo-transglycosylase Homolog in Maize is Responsive to Ethylene and Associated with Aerenchyma. *Plant Physiol.* **1996,** *112*, 385–391.

158. Sachs, J. A. *A Text Book of Botany*; Oxford University Press: Oxford, UK, 1882.

159. Santosa, I. E.; Ram, P. C.; Boamfa, E. I.; Laarhoven, L. J. J.; Reuss, J.; Jackson, M. B.; Harren, F. J. M. Patterns of Peroxidative Ethane Emission from Submerged Rice Seedlings Indicate that Damage from Reactive Oxygen Species Takes Place During Submergence and is not Necessarily a Post-anoxic Phenomenon. *Planta* **2007,** *226*, 193–202.

160. Saraswati, R.; Matoh, T.; Sekiya, J. Nitrogen Fixation of *Sesbania rostrata*: Contribution of Stem Nodules to Nitrogen Acquisition. *Soil Sci. Plant Nutr.* **1992,** *38*, 775–780.

161. Satler, S. O.; Kende, H. Ethylene and the Growth of Rice Seedlings. *Plant Physiol.* **1985,** *79*, 194–198.

162. Sauter, M. Root Responses to Flooding. *Curr. Opin. Plant Biol.* **2013,** *16*, 282–286.

163. Sauter, M.; Moffatt, B.; Saechao, M. C.; Hell, R.; Wirtz, M. Methionine Salvage and S-adenosylmethionine: Essential Links between Sulfur, Ethylene and Polyamine Biosynthesis. *Biochem. J.* **2013,** *451*:2, 145–154.

164. Schepetilnikov, M.; Dimitrova, M.; Mancera-Mart´ınez, E.; Geldreich, A.; Keller, M.; Ryabova, L. A. *TOR* and *S6K1* Promote Translation Reinitiation of *uORF* Containing mRNAs via Phosphorylation of *eIF3h*. *EMBO J.* 2013, *32*, 1087–1092.

165. Schmitz, A. J.; Folsom, J. J.; Jikamaru, Y.; Ronald, P.; Walia, H. *SUB1A*-mediated Submergence Tolerance Response in Rice Involves Differential Regulation of the Brassinosteroid Pathway. *New Phytol.* **2013,** *198*, 1060–1070.

166. Seago, J. L. Jr.; Marsh, L. C.; Stevens, K. J.; Soukup, A.; Votrubov, O.; Enstone, D. E. A Re-examination of the Root Cortex in Wetland Flowering Plants with Respect to Aerenchyma. *Ann. Bot.* **2005,** *96*, 565–579.

167. Setter, T. L.; Waters, I.; Sharma, S. K.; Singh, K. N.; Kulshreshtha, N.; Yaduvanshi, N. P. S.; Ram, P. C.; Singh, B. N.; Rane, J.; McDonald, G.; Khabaz-Saberi, H.; Biddulph, T. B.; Wilson, R.; Barclay, I.; McLean, R.; Cakir, M. Review of Wheat Improvement for Waterlogging Tolerance in Australia and India: The Importance of Anaerobiosis and Element Toxicities Associated with Different Soils. *Ann. Bot.* **2009,** *103*, 221–235.

168. Setter, T. L.; Waters, I.; Wallace, I.; Bhekasut, P.; Greenway, H. Submergence of Rice. I. Growth and Photosynthetic Response to CO_2 Enrichment of Floodwater. *Aust. J. Plant Physiol.* **1989,** *16*, 251–263.

169. Shah, N. A.; Srivastava J. P.; Teixeira da Silva, J. A.; Shahi, J. P. Morphological and Yield Responses of Maize (*Zea mays* L.) Genotypes Subjected to Root Zone Excess Soil Moisture Stress. *Plant Stress* **2012,** *6*, 59–72.

170. Shannon, R. D.; White, J. R.; Lawson, J. E.; Gilmour, B. S. Methane Efflux from Emergent Vegetation in Peat Lands. *J. Ecol.* **1996,** *84*, 239–246.

171. Shiono, K.; Ogawa, S.; Yamazaki, S.; Isoda, H.; Fujimura, T.; Nakazono, M.; Colmer, T. D. Contrasting Dynamics of Radial O_2^- Loss Barrier Induction and Aerenchyma Formation in Rice Roots of Two Lengths. *Ann. Bot.* **2011,** *107*, 89–99.

172. Shiono, K.; Yamauchi, T.; Yamazaki, S.; Mohanty, B.; Malik, A. I.; Nagamura, Y.; Nishizawa, N. K.; Tsutsumi, N.; Colmer, T. D.; Nakazono, M. Microarray Analysis of Laser Micro Dissected Tissues Indicates the Biosynthesis of Suberin in the Outer Part of Roots During Formation of a Barrier to Radial Oxygen Loss in Rice (*Oryza sativa*). *J. Exp. Bot.* **2014**, *65*, 4795–4806.

173. Singh, V. P. Physiological and Biochemical Changes in Pigeonpea [*Cajanus cajan* (Millsp.)] Genotypes to Waterlogging Stress at Early Stage. Ph. D. Thesis, Banaras Hindu University, Varanasi, India, 2010; pp 180.

174. Srivastava, J. P.; Singh, P.; Singh, V. P.; Bansal, R. Effect of Waterlogging on Carbon Exchange Rate, Stomatal Conductance and Mineral Nutrient Status in Maize and Pigeon Pea. *Plant Stress* **2010**, *4*(Special Issue 1), 95–99.

175. Steffens, B.; Sauter, M. Epidermal Cell Death in Rice is Confined to Cells with a Distinct Molecular Identity and is Mediated by Ethylene and H_2O_2 through an Auto Amplified Signal Pathway. *Plant Cell* 2009, *2*, 184–196.

176. Steffens, B.; Geske, T.; Sauter, M. Aerenchyma Formation in the Rice Stem and Its Promotion by H_2O_2. *New Phytol.* **2011**, *190*, 369–378.

177. Steffens, B.; Kovalev, A.; Gorb, S. N.; Sauter, M. Emerging Roots Alter Epidermal Cell Fate through Mechanical and Reactive Oxygen Species Signaling. *Plant Cell.* **2012**, *24*, 3296–3306.

178. Steffens, B.; Steffen-Heins, A.; Sauter, M. Reactive Oxygen Species Mediate Growth and Death in Submerged Plants. *Front. Plant Sci.* **2013**, *4:179,* 1–7.

179. Stoimenova, M.; Hansch, R.; Mendel, R.; Gimmler, H.; Kaiser, W. M. The Role of Nitrate Reduction in the Anoxic Metabolism of Roots I. Characterization of Root Morphology and Normoxic Metabolism of Wild Type Tobacco and a Transformant Lacking Root Nitrate Reductase. *Plant Soil* **2003**, *253*, 145–153.

180. Stolzy, L. H.; Letey, J. Measurement of Oxygen Diffusion Rates with the Platinum Micro-electrode, III: Correlation of Plant Response to Soil Oxygen Diffusion Rates. *Hilgardia* **1964**, *35*, 567–576.

181. Striker, G. G.; Insausti, P.; Grimoldi, A. A.; Vega, A. S. Trade-off between Root Porosity and Mechanical Strength in Species with Different Types of Aerenchyma. *Plant Cell Environ.* **2007**, *30*, 580–589.

182. Subbaiah, C. C.; Bush, D. S.; Sachs, M. M. Elevation of Cytosolic Calcium Precedes Anoxic Gene Expression in Maize Suspension-culture Cells. *Plant Cell* **1994**, *6*, 1747–1762.

183. Subbaiah, C. C.; Sachs, M. M. Molecular and Cellular Adaptations of Maize to Flooding Stress. *Ann. Bot.* **2003**, *90*, 119–127.

184. Sugden, C.; Crawford, R. M.; Halford, N. G.; Hardie, D. G. Regulation of Spinach *SNF1*-related (*SnRK1*) Kinases by Protein Kinases and Phosphatases is Associated with Phosphorylation of the T Loop and is Regulated by 5′-AMP. *Plant J.* **1999**, *19*, 433–439.

185. Summers, J. E.; Jackson, M. B. Anaerobic Conditions Strongly Promote Extension by Stem of Overwintering Tubers of *Potamogeton pectinatus* L. *J. Exp. Bot.* **1994**, *45*, 1309–1318.

186. Summers, J.; Ratcliffe, R.; Jackson, M. Anoxia Tolerance in the Aquatic Monocot *Potamogeton pectinatus*: Absence of Oxygen Stimulates Elongation in Association with an Unusually Large Pasteur Effect. *J. Exp. Bot.* **2002**, *51*, 1413–1422.

187. Sweetlove, L. J.; Beard, K. F. M.; Nunes-Nesi, A.; Fernie, A. R.; Ratcliffe, R. G. Not Just a Circle: Flux Modes in the Plant TCA Cycle. *Trends Plant Sci.* **2010**, *15*, 462–470.

188. Tadege, M.; Brandle, R.; Kuhlemeiei, C. Anoxia Tolerance in Tobacco Roots: Effect of Overexpression of Pyruvate Decarboxylase. *Plant J.* **1998**, *14*, 327–335.

189. Teakle, N. L.; Colmer, T. D.; Pedersen, O. Leaf Gas Films Delay Salt Entry and Enhance Underwater Photosynthesis and Internal Aeration of *Melilotus siculus* Submerged in Saline Water. *Plant Cell Environ.* **2014,** *37*, 2339–2349.

190. Tom´e, F.; Nägele, T.; Adamo, M.; Garg, A.; Marco-Llorca, C.; Nukarinen, E.; Pedrotti, L.; Peviani, A.; Simeunovic, A.; Tatkiewicz, A.; Tomar, M.; Gamm, M. The Low Energy Signaling Network. *Front. Plant Sci.* **2014,** *5:353,* 1–12.

191. Torres, M. A.; Dang, J. L. Functions of the Respiratory Burst Oxidase in Biotic Interactions, Abiotic Stress and Development. *Curr. Opin. Plant Biol.* **2005,** *8*, 397–403.

192. Trought, M. C. T.; Drew, M. C. The Development of Waterlogging Damage in Young Wheat Plants in Anaerobic Solution Cultures. *J. Exp. Bot.* **1980,** *31*, 1573–1580.

193. Van der Straeten, D.; Zhou, Z., Prinsen, E.; Van Onckelen, H. A.; van Montagu, M. C. A Comparative Molecular—Physiological Study of Submergence Response in Low Land and Deep Water Rice. *Plant Physiol.* **2001,** *125*, 955–968.

194. van Veen, H.; Mustroph, A.; Barding, G. A.; Vergeer-van Eijk, M.; Welschen-Evertman, R. A. M.; Pedersen, O.; Visser, E. J. W.; Larive, C. K.; Pierik, R.; Bailey-Serres, J., Voesenek, L. A.; Sasidharan, R. Two *Rumex* Species from Contrasting Hydrological Niches Regulate Flooding Tolerance through Distinct Mechanisms. *Plant Cell.* **2013,** *25*, 4691–4707.

195. Vartapetian, B. B.; Jackson, M. B. Plant Adaptations to Anaerobic Stress. *Ann. Bot.* **1997,** *79*, 3–20.

196. Verboven, P.; Pedersen, O.; Ho, Q. T.; Nicolai, B. M.; Colmer, T. D. The Mechanism of Improved Aeration Due to Gas Films on Leaves of Submerged Rice. *Plant Cell Environ.* **2014,** *37*, 2433–2452.

197. Vidoz, M. L.; Loreti, E.; Mensuali, A.; Alpi, A.; Perata, P. Hormonal Interplay During Adventitious Root Formation in Flooded Tomato Plants. *Plant J.* **2010,** *63*, 551–562.

198. Visser, E. J. W.; Colmer, T. D.; Blom, C. W. P. M.; Voesenek, L. A. C. J. Changes in Growth, Porosity, and Radial Oxygen Loss from Adventitious Roots of selected mono- and Dicotyledonous Wetland Species with Contrasting Types of Aerenchyma. *Plant Cell Environ.* **2000,** *23*, 1237–1245.

199. Visser, E. J. W.; Voesenek, L. A. C. J. Acclimation to Soil Flooding Sensing and Signal-transduction. *Plant Soil* **2004,** *254*, 197–214.

200. Voesenek, L. A. C. J.; Colmer, T. D.; Pierik, R.; Millenaar, F. F.; Peeters, A. J. How Plants Cope with Complete Submergence. *New Phytol.* **2006,** *170*, 213–226.

201. Voesenek, L. A. C. J.; Rijnders, J.; Peeters, A.; Van de Steeg, H. M.; De Kroon, H. Plant Hormones Regulate Fast Shoot Elongation Under Water: From Genes to Communities. *Ecology* **2004,** *85*, 16–27.

202. Watanabe, K.; Nishiuchi, S.; Kulichikhin, K.; Nakazono, M. Does Suberin Accumulation in Plant Roots Contribute to Waterlogging Tolerance? *Front. Plant Sci.* **2013,** *4:178,* 1–7.

203. Webb, T.; Armstrong, W. The Effects of Anoxia and Carbohydrates on the Growth and Viability of Rice, Pea and Pumpkin Roots. *J. Exp. Bot.* **1983,** *34*, 574–603.

204. Widjaya-Adhi, D. A. Management, Utilization and Development of Indonesia Tidal Swampy Area. In *Indonesia Land Resources and Its Utilization*; Adimihardjo, A., Ed.; Indonesia Center for Agricultural Land Resource Research and Development: Bogor, Indonesia, 2000; pp 127–164.

205. Winkel, A.; Colmer, T. D.; Pedersen, O. Leaf Gas Films of *Spartina anglica* Enhance Rhizome and Root Oxygen During Tidal Submergence. *Plant Cell Environ.* **2011**, *34*, 2083–2092.

206. Winkel, A.; Colmer, T. D.; Ismail, A. M.; Pedersen, O. Internal Aeration of Paddy Field Rice (*Oryza sativa*) During Complete Submergence—Importance of Light and Floodwater O_2. *New Phytol.* **2013**, *197*, 1193–1203.

207. Winkel, A.; Pedersen, O.; Ella, E.; Ismail, A. M.; Colmer, T. D. Gas Film Retention and Underwater Photosynthesis During Field Submergence of Four Contrasting Rice Genotypes. *J. Exp. Bot.* **2014**, *65*, 3225–3233.

208. Xiong, Y.; McCormack, M.; Li, L.; Hall, Q.; Xiang, C.; Sheen, J. Glucose-*TOR* Signaling Reprograms the Transcriptome and Activates Meristems. *Nature* **2013**, *496*, 181–186.

209. Xu, K.; Xu, X.; Fukao, T.; Canlas, P.; Maghirang-Rodriguez, R.; Heuer, S.; Ismail, A. M.; Bailey-Serres, J.; Ronald, P. C.; Mackill, D. J. *Sub1A* is an Ethylene Response—Factor-like Gene that Confers Submergence Tolerance to Rice. *Nature* **2006**, *442*, 705–708.

210. Yamauchi, T.; Shimamura, S.; Nakazono, M.; Mochizuki, T. Aerenchyma Formation in Crop Species: A Review. *Plant Soil.* **2013**, *370*, 447–460.

211. Zhang, B. G.; Puard, M.; Couchat, P. Effect of Hypoxia, Acidity and Nitrate on Inorganic Nutrition in Rice Plants. *Plant Physiol. Biochem.* **1990**, *28*, 655–661.

212. Zhang, H.; Greenway, H. Anoxia Tolerance and Anaerobic Catabolism of Aged Beet Root Storage Tissue. *J. Exp. Bot.* **1994**, *45*, 567–575.

213. Zhang, Q.; Greenway, H. Membrane Transport in Anoxic Rice Coleoptiles and Storage Tissues in Beetroot. *Aust. J. Plant Physiol.* **1995**, *22*, 965–975.

214. Zhang, Y.; Lin, X.; Werner, W. The Effect of Soil Flooding on the Transformation of Fe Oxides and the Adsorption/desorption Behavior of Phosphate. *J. Plant Nutr. Soil Sci.* **2003**, *166*, 68–75.

215. Zhao, Y.; Christensen, S. K.; Fankhauser, C.; Cashman, J. R.; Cohen, J. D.; Weigel, D.; Chory, J. A Role for Flavin monooxygenase-like Enzymes in Auxin Biosynthesis. *Science.* **2001**, *291*, 306–309.

216. Zhou, M. Z. Improvement of Plant Waterlogging Tolerance. In *Waterlogging Signaling and Tolerance in Plants*; Mancuso, S., Shabala, S., Eds.; Springer-Verlag: Heidelberg, Germany, 2010; pp 267–285.

PART IV

Non-Conventional and High-Value Crops for Salt-Affected Lands

POTENTIAL AND ROLE OF HALOPHYTE CROPS IN SALINE ENVIRONMENTS

ASHWANI KUMAR, ANITA MANN, ARVIND KUMAR, SARITA DEVI, and PRABODH CHANDER SHARMA

ABSTRACT

Salinity is one of the major abiotic stresses that hinder the performance of the crop plants. Besides osmotic stress, the toxic components of salinity, that is, Na^+ and Cl^- ions interfere with the normal physiological processes resulting in significantly reduced yields. Many research groups across the world are vigorously promoting halophytic plants as potential candidates for the productivity enhancements of salt-affected lands. Halophytes constitute only about 1% of the world's flora, but are characterized to survive in environments having salt concentration as high as 200 mM NaCl. Halophytes have evolved a number of adaptive traits which include ion compartmentalization, osmotic adjustment, succulence, selective transport and uptake of ions, enzymatic and nonenzymatic antioxidant response, maintenance of redox and energy status, salt inclusion/excretion, and genetic regulation. Besides, they are also being increasingly viewed as potential sources of novel salt tolerant genes/transcripts for use in the genetic improvement. Many such genes isolated from the halophytic species have been introgressed in model plants to test their efficacy in enhancing the salt tolerance. The elucidation of mechanisms and gene transfer from halophytes to glycophytes or vice versa is an important field to work on to develop designer crops for the challenging environments to which agriculture will be increasingly exposed in future.

9.1 INTRODUCTION

Salinity is one of the major environmental stresses that adversely impact soil health, environmental quality, and agricultural production. With intensification of agricultural practices and global climate change, problems of soil and water salinity are assuming serious dimensions.[121] According to a report published by the United Nations Environmental Program, "Globally ~1000 million hectare (M ha) of land in around 100 countries is salt-affected constituting about 20% of the total agricultural land and more than 6% of the world's total land area."[36] Although salt-affected soils (SAS) have been reported from various agro-ecological regions of the world, yet salinization of agricultural lands is particularly severe in irrigated tracts of arid and semiarid zones, where intensive water use, often at the neglect of the drainage, has dealt a severe blow to the sustainability of soil, water, and agro-biodiversity resources.[1,121] In India, where SAS occupy an estimated 6.73 M ha area, technologies such as subsurface drainage, use of chemical amendments, irrigation management, and agro-forestry techniques have made it possible to grow a wide range of crops even in the highly deteriorated saline and sodic soils. Nonetheless, many of these techniques suffer from inherent limitations that have thwarted their large-scale adoption by farmers in the salinity-affected regions. For example, inadequate supply, low product quality, and reduced subsidy resulting in high prices have proved impediments in the use of gypsum in reclaiming sodic soils.[121,122]

Studies conducted globally have revealed that salt induced crop losses are likely to increase substantially in the coming decades if appropriate corrective measures to tackle the intertwined menaces of salinity, waterlogging, and ions toxicities are not in place. Although tremendous success of different salinity mitigation technologies is evident in India and elsewhere, yet a range of social, policy, and monetary constraints have restricted the scope of these techniques in the soil restoration projects. In this backdrop, it is increasingly being realized that affordable and feasible solutions are urgently required to harness the productivity of SAS under a changing scenario characterized by the challenges like fresh water scarcity, alterations in the crop growing conditions and the dismal adaptive capacity of the farmers. Use of salt tolerant cultivars (STCs) represents an innovative and cost-effective strategy to sustain the crop production in saline environments. Several direct (e.g., reduced use of amendments, stable yields, etc.) and indirect (e.g., long-term improvements in soil quality, carbon sequestration, etc.) benefits that accrue from the use of STCs have captured the imagination of salinity researchers.[121]

Although a number of high yielding and salinity tolerant genotypes are available in different crops, yet they often fail to grow under extreme salinity necessitating the identification and development of alternative crops that can be produced under harsh conditions. In addition to the conventional genetic improvement techniques, marker-assisted breeding and genetic transformation can be of great help to improve the salt tolerance of crop plants. One of the prerequisites to enhance the salinity tolerance through these approaches is the screening of available germplasm to identify the genes linked to plant survival under extreme salt stress. In the last few decades, use of halophytes as economic crops and phytoremediation plants for SAS has gained the attention of researchers. Besides, they are also being increasingly viewed as potential sources of novel salt tolerant genes/transcripts for use in the genetic improvement. Many such genes isolated from the halophytic species have been introgressed in model plants to test their efficacy in enhancing the salt tolerance.

This chapter discusses various physiological and genetic bases of salt tolerance in halophytes; and explores the use of these indices in halophyte improvement and their potential applications in biosaline agriculture.

9.2 HALOPHYTES: CHARACTERISTICS AND CLASSIFICATION

Halophytes are halotolerant plants that grow under saline conditions but do not require high concentration of salt for their growth in contrast to "halophilic" (i.e., salt loving) or "halophile" organisms that require high salinity for survival and growth. In common parlance, plants growing exclusively on saline soil are referred to as halophytes.[113] The preceding definition may be a misnomer as many of the halophytes also thrive on ordinary soils devoid of salt. They are the plants which can tolerate salt concentrations over 0.5% at any stage of life.[125,127] They have also been defined as native flora of saline soils containing solutions with an osmotic potential of at least 3.3 bar, being equivalent to 70 mM monovalent salts.[52] Plants that cannot survive in these habitats are classified as non-halophytes. Unquestionably, these definitions are not quite adequate since there is a continuum from the least to the most salt tolerant species. Thus, broadly, halophytes can be defined as a group of plants, which survive and reproduce in environments with salt concentration of 200 mM or more, whereas the non-halophytes, to which most of the conventional crops belong, are sensitive to lower levels of salinity.[41]

Dicot halophytes are more salt tolerant as compared to monocot species and exhibit optimal growth at 100–200 mM of NaCl, while the latter show optimal growth at NaCl concentration of 50–100 mM.[12,41] Some of the halophytes can also grow when sprinkled with saline water as salt sprays being non detrimental to their survival. In comparison to glycophytes, which make an exaggerating with about 98% of the plant species listed, only ~2% of the plant species are categorized as halophytes. Glenn et al.[51] suggested the halophytic species as high as 6000, whereas the eHALOPH Halophyte database[40] currently identifies more than 1500 species as salt tolerant, albeit without labeling them as "halophyte." These are highly evolved and specialized plants/organisms with well-adapted morphological, anatomical, and physiological characteristics allowing them to proliferate in the soils possessing high salt concentrations.[41,44]

9.2.1 CLASSIFICATION OF HALOPHYTES

Halophytes have been classified based on their habitat, distribution, and the response to salinity. Based on ecological aspect, halophytes are classified as true halophytes, xerohalophytes, and marine phanerogams.

9.2.1.1 TRUE HALOPHYTES

These are the plants that can thrive in water having more than 0.5% of NaCl. A small number of plant lineages have evolved structural, phenological, physiological, and biochemical mechanisms for salt resistance and true halophytes have evolved convergently in many related families.

9.2.1.2 XEROHALOPHYTES

These are essentially the desert plant species classified as halophytes. Desert and coastal halophytes possess the same mechanisms to overcome the salt stress.

9.2.1.3 MARINE PHANEROGAMS

These are the seed-bearing plants that complete their life cycle under complete submergence in seawater.

9.2.2   *TRUE VERSUS FACULTATIVE HALOPHYTES*

Some halophytes, which consistently require a particular concentration of NaCl for growth, are called "obligate halophytes" or "true mangroves." In contrast, terms "facultative halophytes" or "mangrove associates" are used to describe the halophytes adapted to high saline conditions that can also grow well on normal soils devoid of salt. As halophytes grow profusely in the presence of salts, they have potential to augment the economic value of highly salinized lands otherwise unsuitable to grow salt sensitive crops.[140] For diverse habitats differing in salt content, plants are classified as[150]:

- Euhalophytes that grow in polyhaline habitat with salt range equaling or exceeding 1% of NaCl[150];
- Mesohalophytes growing in mesohaline habitats with salt content in the range of 0.1–1% of NaCl; and
- Oligohalophytes that grow in ligohaline habitat having salt content in the range of 0.01–0.1% NaCl.

This classification, however, is not exclusive in the sense that some halophytes from each of these groups may be found in more than one habitat. For example, halophytes adapted to both oligohaline and mesohaline habitats are called oligo-mesohalophytes, and those found growing in all the three habitats are called euryhalophytes.

9.2.3   *CLASSIFICATION ON THE BASIS OF PHYSIOLOGICAL RESPONSE*

On the basis of their response to salinity, halophytes are classified into four distinct types[140]:

- Obligatory halophytes are the plants requiring saline growth conditions throughout their life cycle.
- Preferential halophytes are those commonly found in saline habitats despite their occasional appearance in non-saline habitats.
- Supporting halophytes are the nonaggressive plants capable of growing in saline substrate.
- Accidental halophytes are the species which grow in marsh saline habitats only accidentally.

9.2.4   SUCCULENT, NON-SUCCULENT, AND ACCUMULATING TYPES

Salt marsh plants, which endure excess Cl⁻in the cell sap due to increased succulence (e.g., *Salicornia herbacea*), are called succulent halophytes. In contrast, non-succulent halophytes overcome the salt stress by desalinization of their tissues and secreting the excess salts through salt glands (e.g., *Spartina alterniflora*). Accumulating type halophytes lack special mechanism(s) of salt removal. Salt concentration in such plants increases continuously until the death of plants (e.g., *Juncus gerardii* and *Suaeda jructicosa*).

Although halophytes are found in many different saline ecosystems around the globe, yet coastal regions represent the highest halophyte diversity among such habitats. This fact, however, does not diminish the importance of other habitats as rich sources of halophyte diversity. For example, even in salt deserts (e.g., Rann of Kutchin Gujarat, India)[46, 53] salt tolerant halophytes form the predominant vegetation group. The halophytic plants comprise of diverse plant groups including the flowering plants, shrubs, climbers, herbs, trees, and grasses[60]; and supply fuel, fodder, and timber to the local communities.

9.3   SALINITY AND HALOPHYTE GROWTH

Salinized plants suffer initially from the osmotic stress and subsequently from the specific ion effects. Osmotic stress (which inhibits water uptake, cell expansion, and lateral bud development) is experienced by the plants as soon as salt concentration in the root zone begins to rise.[84] Although most of the plant species tend to adjust osmotically (i.e., lowering the water potential) in saline soils to prevent the loss of turgor, yet it often comes at the cost of slow growth afterwards. One of the deleterious effects of the high salt levels is manifested as nutrient imbalance.[117] For instance, the high soil Na^+ concentrations reduce the amounts of available K^+, Mg^{++}, and Ca^{++} to plants resulting in Na^+ toxicity on one hand and deficiencies of essential cations on the other.[34] A summary of salt stress induced responses is presented in Table 9.1.

The second phase of growth inhibition (i.e., "specific ion toxicity") commences with the accumulation of injurious concentrations of ions such as Na^+ and Cl^- in the plant cells. Higher plants growing in normal (non-saline) soils maintain 100–200 mM of K^+ and 1–10 mM of Na^+ in cell cytosol that ensures optimum enzyme activities. Key metabolic processes are affected when salt concentration exceeds the threshold limit; photosynthesis is inhibited due to excessive accumulation of Na^+ and/or Cl^- ions in the chloroplasts. Although photosynthetic electron transport appears to be relatively insensitive to salt stress, yet both carbon metabolism and photophosphorylation are affected to varying degrees. A high Na^+ level in plant leaves accelerates chlorosis and necrosis of the cells with a concurrent decrease in the activity of essential cellular components and metabolites.[86] Under certain conditions, both osmotic and salt stresses simultaneously suppress the growth and metabolism. An abnormally high Na^+ to K^+ ratio coupled with high concentrations of total salts inactivate enzymes and inhibit the protein synthesis.

TABLE 9.1 Salt Stress Induced Responses in Plants.[a]

Causes	What happens		Recovery/adaptation
	Salinity stress		
Osmotic stress (dehydration)	Inhibition of water uptake, cell elongation, and leaf bud development		Osmotic adjustment through ions/solutes and organic compounds
Ionic stress (ionic toxicity and deficiencies)	Leaf senescence, inhibition of photosynthesis, protein synthesis, and enzyme activity	Cell death	Ion homeostasis, ion extrusion, and ion compartmentation
Imbalance in ionic uptake	Nutritional problems, reduced availability of K^+, Mg^{++}, and Ca^{++}		Ion reabsorption

[a]Adapted and modified from Ref. [59].

Excess Na^+ displaces Ca^{2+} from the plasma membrane leading to changes in the plasma membrane composition, integrity and permeability.[23] Saline conditions often lead to the closure of stomata resulting in the reduced availability of carbon dioxide in leaves and the subsequent inhibition of carbon fixation.[89] It causes the exposure of chloroplast to excessive excitation energy which in turn could cause the generation of reactive oxygen species (ROS) manifested as oxidative stress.[90] As previously indicated, although a number of salt tolerant genotypes have been developed in different crops they give best results only in moderately salt-affected soils. Under these circumstances, it has become absolutely essential to identify the alternative crops adapted to extreme salinity such that vast tracts of abandoned lands can be put to productive use ensuring long-term improvements in environmental quality and better economic returns to the land owners.

9.3.1 MORPHOLOGICAL AND ANATOMICAL ADAPTATIONS IN HALOPHYTES TO SALT STRESS

In the course of evolution, halophytic plants have developed numerous adaptive mechanisms to sustain and grow under saline conditions[27] as discussed in this section.

9.3.1.1 ROOTS

In addition to normal roots, halophytes also develop many stilt or prop roots from the aerial parts for better anchorage in muddy or loose sandy soil. These roots grow downward and enter the deep and tough strata of the soil. The stilt roots of mangrove plants show normal features with periderm on the surface, aerenchymatous cortex containing sclereids, normal endodermis, secretory pericycle, radially arranged xylem and phloem, and extensively developed pith. Sometimes, a large number of adventitious root buttresses develop from the basal parts of tree trunks. These root buttresses provide sufficient support to the plants. In poorly drained and oxygen deficient soils of the coastal regions, the hydro halophytes develop special type of negatively geotropic roots referred to as pneumatophores or breathing roots to compensate the lack of oxygen in the plant root environment.

The pneumatophores usually develop from the underground roots and project in the air well above the surface of mud and water. They appear as peg-like structures. The tips of these respiratory roots may be pointed. Very young pneumatophores, however, show root features, that is, exarch xylem and radial arrangement of vascular tissues. Generally, the negatively geotropic breathing roots show features of stem and not of roots. They possess numerous lenticels or pneumathodes on their surface and prominent aerenchyma enclosing large air cavities internally. Highly developed air chambers are continuous with the stomata of leaves and with the cortex and primary phloem of the stems. The gaseous exchange takes place in these roots through the lenticels. The aerenchyma helps in the conduction of air down to the subterranean or submerged roots. Pneumatophores do not develop in some species of *Rhizophora*. In those cases, the upper aerial parts of descending stilt roots probably take up the respiratory activity.

9.3.1.2 STEM

Many halophyte species develop succulent stems with the degree of succulence depending on the ratio of absorbed to free ions in the plant cells rather than absolute amounts of sodium chloride or sulfate present.[8] Succulence is induced only after the accumulation of free ions in an organ exceeding the critical level. Salinity inhibits the cell division and stimulates cell elongation. Such effects cause decrease in the cell number and increase in cell size; typical of succulents. Succulence is directly correlated with salt tolerance of plants and the degree of their development can serve as an indicator of the ability of plants to survive in highly saline habitats.[104] The temperate halophytes are herbaceous, but the tropical ones are mostly bushy and show dense cymose branching. Submerged marine angiosperms are among the very few species of halophytes that do not become succulent.

9.3.1.3 LEAVES

The leaves in most of the halophytes are thick, entire, succulent, small in size, and often glassy in appearance. Some species are aphyllous. Stems and leaves of coastal aero-halophytes show additional mode of adaptation to their habitats. Their surfaces are densely covered with trichomes. Leaves of submerged marine halophytes are thin and have very poorly developed vascular system and frequently green epidermis. They are adapted to absorb water and nutrients from the medium directly.

9.3.1.4 FRUITS, SEEDS, AND THEIR DISPERSAL

The fruits and seeds in halophytes are generally light in weight. Fruit walls have a number of air chambers. The fruits, seeds, and seedlings can float on the water for a long time and are dispersed to the distant places by the water current. Mangrove vegetation of tropical sea-shores from Australia to East Africa includes approximately the same species of plants. Similarly, the mangroves of West Asia show considerable resemblances with those of East Asia and East Africa. It is due, in part, to the fact that medium and temperature remain uniform throughout and partly due to the efficient means of dispersal or migration of plants.

9.3.2   VIVIPAROUS MODE OF SEED GERMINATION

As high degree of salinity in the soil or water retards seed germination, many halophyte species (such as *Rhizophora, Aegiceras, Avicennia, Cassula,* and *Ranansatia vivipara*) have evolved an effective adaptation mechanism in the form of viviparous germination to avoid the retarding effects of salinity on seed germination. Such halophytes or mangrove plants exhibit vivipary, that is, germination of seeds inside the fruits still attached to the mother plants. In *Rhizophora* plants, when the embryo reaches advanced stage of development, the massive club-shaped hypocotyl and terminal radicle pointing downwardly emerge out of the fruit. When the hypocotyl attains a length of several centimeters (~50–80 cm), the seedling falls vertically down. Thus, the radicle and a part of hypocotyl become fixed in the mud and the remaining upper part of hypocotyl along with other embryonic parts, such as plumule and cotyledons, remain above the surface of mud or water. Within a few hours, the radicle develops a tuft of roots, and plumule also starts growing rapidly. Sometimes, seedlings fall in deep water and they float on the surface of water vertically with hypocotyl pointing downward. On reaching shallow water depths, the radicle becomes fixed in the soft mud and the plant starts growing rapidly.

9.3.3   ANATOMICAL FEATURES

The appearance and structures, which characterize certain groups of plants, contribute greatly to the ecological and physiological means of adaptation. Some of these specific traits include large cells and small intercellular spaces, highly elastic cell walls, well-developed water storing tissues, smaller relative surface area, tiny and fewer stomata, and low chlorophyll content. Similarly, a range of anatomical features, which impart xerophytic character to these plants, include, *inter alia*, presence of thick cuticle on foliage.[131] Leaves may be dorsiventral or isobilateral. They develop protected stomata which are not deeply sunken. Epidermal cells are thin-walled. The palisade consists of several layers of narrow cells with intercalated tannin and oil cells. The leaves and stems of coastal halophytes are abundantly covered with various types of simple and branched trichomes giving the plants a grayish appearance. The trichomes may exert a protective function in plants by affecting water economy, affecting the temperature of the leaves and preventing sea water droplets from reaching the live tissues of leaves.

9.4 PHYSIOLOGICAL ADAPTATIONS

Halophytes grow in saline habitats not because they are salt loving but because they tolerate high concentration of salts better than other plants growing in non-saline habitat. Physiological studies have shown that the halophytes may not experience difficulties in absorbing too saline water because (1) they show high rates of transpiration, (2) they show exudation of sap that contains dissolved salts, and (3) they develop many shallow absorbing roots. On the other hand, morphology and anatomy of the halophytes clearly show xeromorphic features in them. The question arises as to why xeromorphy develops in halophytes even when these plants are growing in an environment where water is available in abundance.

Saline conditions are not essentially "dry" for all plant species. Under saline conditions sometimes higher transpiration rates have been observed in halophytes than in neighboring salt hating plants.[18,29] It is, therefore, possible that halophytes show xeromorphism partly for withstanding high salinity of water and partly for absorbing the available water with ease. Active accumulation of salt increases the osmotic concentration of cell sap in these plants and thus enables them to absorb saline water. The significance of succulence, however, is not so clearly understood. Probably, it is induced by accumulation of salts in the cytoplasm. It seems reasonable because sodium salts, if present in the soil water, definitely stimulate succulence even in non-halophytes. On the other hand, characteristic succulence of some plants may disappear if they are grown on the soil lacking in these common salts. Excessive accumulation of sodium does not harm these plants.

9.4.1 ADAPTIVE MECHANISMS FOR SALT EXCLUSION

Halophytes have evolved a number of adaptive traits which allow them to germinate, grow, and complete their life cycle under harsh saline conditions.[44] The mechanisms enhancing abiotic stress tolerance in halophytes include ion compartmentalization, osmotic adjustment through osmolytes accumulation, succulence, selective transport, and uptake of ions, enzymatic, and non-enzymatic antioxidant response, and maintenance of redox and energy status, salt inclusion/excretion, and genetic regulation.[41,65] These processes help to achieve three interconnected aspects of plant activity: prevention of damage to the plant, reestablishment of the homeostatic condition and the resumption of growth. Understanding the mechanism of tolerance in halophytes at morphological, anatomical, physiological, biochemical, and molecular levels is thus crucial to improve the tolerance of the crop plants and their adoption under abiotic stress conditions to exploit such problem soils.

9.4.2 GENERAL STRATEGIES TO AVOID SALT INJURY

Plants minimize the salt injury by excluding salt from meristems, particularly from the actively growing, photosynthesizing shoots, and leaves. In plants, sensitivity or resistance to moderate levels of salinity in the soil depends, partly, on the ability of the roots to prevent potentially harmful ions from reaching the shoots.

9.4.3 SALT AVOIDANCE

Avoidance is an adaptive mechanism by which plants tend to keep the salt ions away from those parts where they may be toxic or harmful.[5] Salt avoidance may involve certain physiological and structural adaptations so as to minimize the salt concentrations of the cell through physiological exclusion by the root membranes. This may involve passive exclusion of ions through permeable membrane by means of a pump or dilution of ions in plant tissues.[5] Halophytes employ four principal mechanisms to avoid salt stress. They are salt exclusion, salt excretion, salt dilution, and ion compartmentation.

9.4.3.1 SALT EXCLUSION

Salt exclusion essentially involves filtration at the surface of the root. It is an efficient but complex way of preventing ion uptake so that salt accumulation in upper plant parts, especially in the transpiring organs, is avoided. Salt exclusion is based on lower root permeability for ions even under conditions of high external salinity. Root membranes prevent salt from entering while allowing the water to pass through. The red mangrove is an example of the salt excluding species. Restricted uptake by the roots, although important as the first line of defense, is not a very efficient mechanism and is often found in conjunction with internal exclusion mechanisms which involve sequestering salt ions in specialized tissues by removing them from the transpiration stream. One way plants achieve this is by exchanging K^+ ions for Na^+ ions as they pass through the xylem. However, the ion accumulation capacity of the xylem is limited and reduces the efficacy of this process. In case of mangroves, 99% of the salts are excluded through the roots.[129] Exclusion of salt at whole plant level occurs at the roots and the casparian strips may play role in salt exclusion from the inner tissues.[43] Some well-documented examples of the salt excluding species are *Rhizophora mucronata, Ceriops candolleana, Bruguiera gymnorrhiza*, and *Kandelia candel*.

9.4.3.2   SALT EXCRETION/EXTRUSION

Some plants get rid of salt by excreting it back into the environment. It is also a very efficient way of preventing excessive concentrations of salts accumulating in photosynthetic tissues. This mechanism is a characteristic feature of those species, which have developed special features, mostly localized at the leaf epidermis, known as salt glands and salt hairs (bladders). Plants can excrete salt through their roots, shoots, and leaves. One of the most obvious signs of salt excretion by the species having salt glands or salt hairs is the salt crust on leaves and shoots.[95] Some plants transport and accumulate salt to storage areas that are shed later. Specialized structures for excretion have evolved in the epidermis of some species. Bladder hairs are structures on leaf surfaces that consist of several stalk cells and a bladder cell (e.g., *Atriplex, Mesembryanthemum crystallinum*). The stalk cells transport ions into the vacuole of the bladder cell, which eventually dies and falls off the plant. Salt glands are used by some species to excrete salt. These specialized structures transport ions directly out of the plant through both roots and leaves. Salt glands are dump sites for the excess salt absorbed in water from the soil; help plants adapt to saline environments. Salt glands are usually not found in nonhalophytes.[100] Salt glands appear mostly in the mesophyll tissue of C_4 grasses. Salt excreters remove salt through glands or bladders or cuticle (e.g., *Tamarix*) located on each leaf.

9.4.3.3   SALT DILUTION/SUCCULENCE

One mode of defense against salt in plant tissues is to simply dilute the concentration of ions and hence maintaining succulence. To achieve this, plants increase their storage volume by developing thick, fleshy, and succulent structures. Succulence stands for a plant condition that involves increase in cell size, decrease in the growth extension, and decrease in surface area as well as higher water content per tissue volume.[39] Succulence is mainly a result of vacuoles of mesophyll cells filling with water and increasing in size. This mechanism is common in wet saline environments, like salt marshes, where water is not limiting. The mechanism is limited by the dilution capacity of plant tissues. Halophyte species exhibiting this trait include *Sonneratia apetala*, *Sonneratia acida*, *Sonneratia alba*, *Limnitzera racemosa*, *Excoecaria agallocha*, *Salvadora persica*, *Sesuvium portulacastrum*, and *Suaeda nudiflora*.

9.4.3.4   COMPARTMENTATION OF IONS

To avoid the risk of ion toxicity associated with salt stress, Na^+ and Cl^- are generally compartmentalized in the vacuole and/or in less sensitive tissues. However, when plants are growing in high salt concentrations, an adequate sequestration of ions in the vacuole can become a limiting factor, especially in the case of glycophytes. In this scenario, plants can accumulate excessive amount of Na^+ in the cytosol which negatively affects many aspects of cellular physiology. These ions can be compartmentalized either at organ level where high salts accumulate only in roots compared to shoots especially leaves or at cellular level where salts accumulate in vacuoles than cytoplasm thus protecting enzymes without toxic ion accumulation in the cytosol. The ability to compartmentalize ions within vacuoles, a key aspect of salt tolerance, is a process that involves ion transporters at the tonoplast (vacuolar membrane) and particularly vacuolar Na^+/H^+ exchangers (NHX). However, efficient ion compartmentation relies not only on transport across the tonoplast, but also on retention of ions within vacuoles.

In terms of metabolic energy, use of ions to balance tissue water potential in a saline environment clearly has a lower energy cost for the plant than use of carbohydrates or amino acids, the production of which has a significantly higher energy cost. On the other hand, high-ion concentrations are toxic to many cytosolic enzymes, so ions must be accumulated in the vacuole to minimize toxic concentrations in the cytosol.

9.4.3.5   ION HOMEOSTASIS AND COMPARTMENTALIZATION

Irrespective of their nature, both glycophytes and halophytes are unable to tolerate high salt concentration in their cytoplasm. Hence, the excess salt is either transported to the vacuole or sequestered in older tissues that eventually are sacrificed, thereby protecting the plant from salinity stress. One of the most detrimental effects of salinity stress is the accumulation of Na^+ and Cl^- ions in tissues of plants exposed to soils with high NaCl concentrations and focus of research is the study about the transport mechanism of Na^+ ion and its compartmentalization. Entry of both Na^+ and Cl^- into the cells causes severe ion imbalance and excess uptake might cause significant physiological disorder(s). High Na^+ and Cl^- concentration is deleterious to cellular systems.[116] Plant regulates the relative concentration of inorganic ions (mainly Na^+ and K^+) to adjust cell turgor, cell volume, intracellular pH value, ionic strength, and many other crucial physiological parameters. Halophytes rely heavily on the use of inorganic ions (Na^+, Cl^-, and K^+) to maintain shoot osmotic and turgor pressure under saline conditions,[44,50,126,127] while glycophytes achieve this predominantly by increased de novo synthesis of compatible solutes. These three inorganic ions account for 80–95% of the cell sap osmotic pressure in both halophytic grasses and dicots.[50,119]

Ion homeostasis, by keeping tissue Na^+ to low/tolerable ranges and selective uptake of essential nutrients like K^+ and Ca^{2+}, is an important aspect of salt tolerance that is not only crucial for normal plant growth but is also an essential process for growth during salt stress. The Na^+ ion that enters the cytoplasm is then transported to the vacuole via Na^+/H^+ antiporter. Two types of H^+ pumps are present in the vacuolar membrane: vacuolar type H^+-ATPase (V-ATPase) and the vacuolar pyrophosphatase (V-PPase).[33] Of these, V-ATPase is the most dominant H^+ pump present within the plant cell. Studies in hypocotyls of *Vigna unguiculata* seedlings revealed that the activity of VATPase pump was increased when exposed to salinity stress; however under similar conditions, activity of V-PPase was inhibited.[87] In the case of halophyte *Suaeda salsa* V-ATPase activity was upregulated and V-PPase played a minor role.[135]

At high salt levels, Na^+ exclusion mechanisms are insufficient to maintain low intracellular Na^+ levels; and Na^+ ions start accumulating in plant tissues. Ionic effect is continuous and long-term effect of cumulative nature increase intracellular salt ion levels with the duration of salinity stress. Therefore, it required to be compartmentalized internally via vacuolar sequestration. The major tonoplast proteins ensuring Na^+ vacuolar sequestration belong to vacuolar H^+-ATPases and H^+-pyrophosphatases creating a sufficient electrochemical potential for energizing tonoplast membrane and NHX Na^+/H^+ exchangers which provide Na^+ influx into vacuole in exchange with H^+. An increased intracellular Na^+ level (activity) also induces Ca^{2+}signaling leading to the activation of Na^+ active efflux from plant cells via salt overlay sensitive 1 (SOS1) and *SOS2/SOS3* pathway.[55]

First, Na^+ induced Ca^{2+} signaling activates *SOS3* which is a calcineurin B-type like calcium binding protein with a myristoyl anchor at the *N*-terminus. *SOS3* with bound calcium can attach plasma membrane and interact with *SOS2* which is a serine/threonine protein kinase. *SOS3/SOS2* complex then activates *SOS1* which is an adenosine triphosphate (ATP)-dependent Na^+/H^+ exchanger with 12 transmembrane helices and a long intracellular chain which can be phosphorylated by *SOS2/SOS3* complex. Active *SOS1* ensures Na^+ exclusion from plant cells via a mechanism of ion exchange (Na^+ is exchanged by H^+) at the cost of ATP. *SOS1* also seems to control Na^+ loading into xylem thus regulating Na^+ level in transpiration stream and its root-to-shoot transport. The upregulation of the SOS pathway proteins under salt stress leads to the salt tolerance. Proteomic studies supporting elevation in expression of SOS proteins were carried out under salinity[148] and ozone stress.[3]

9.4.4   OSMOTIC ADJUSTMENT

Through osmotic adjustment, many plants acclimatize to dry and saline conditions. Some salt tolerant plants control the accumulation of salt ions to counterbalance low water potentials in the saline soils. Salt ions are compartmentalized in vacuoles to protect proteins and membranes from ion toxicity. The active transport of ions requires energy, wherein a trade-off is established such that energy is allocated to tolerance rather than to growth and reproduction. Solute-rich vacuoles have a high osmotic potential that creates a gradient in which water moves from the cytoplasm into the vacuole. To counterbalance this gradient, plants produce osmotically active organic solutes called compatible solutes that do not interfere with plant physiological processes.

Osmolytes are the organic compounds that play role in maintaining fluid balance as well as cell volume. In situations where increased external osmotic pressure tends to rupture the plant cells, certain osmotic channels are switched on to allow the efflux of certain osmolytes. As these osmolytes move outside, they carry water with themselves preventing the cell from bursting out. Sugars, alcohols, amino acids, polyols, tertiary, and quaternary ammonium and sulphonium compounds are some examples of such osmolytes.[105] A variety of compounds such as amino acids and amides (e.g., proline), ammonium compounds (e.g., betaine), and soluble carbohydrates act as compatible solutes.

9.4.5 REACTIVE OXYGEN SPECIES SCAVENGING

The halophytic plants display a cascade of events upon exposure to environmental stresses leading to metabolic dysfunction. Such events include: physiological water-deficit; abscisic acid (ABA) regulated stomatal closure, limited CO_2 availability, over-reduction of electron transport chain in the chloroplast and mitochondria, and finally the generation of reactive oxygen species (ROS).[45] These ROSs are highly toxic and in the absence of protective mechanism in the plant can cause oxidative damage to proteins, nucleic acids and lipids.[80,81] Excessive ROS cause phytotoxic reactions such as lipid peroxidation, protein degradation and deoxyribonucleic acid (DNA) mutation.[78,93,134,136] The increased concentration of ROS damages the D1 protein of PS II leading to photo-inhibition. Additionally, this may also lead to alteration in the redox state resulting in further damage to the cell.[82]

Stress enhanced photorespiration and nicotineamide adenine dinucleotide phosphate activity also contributes to increase in H_2O_2 accumulation, which may inactivate enzymes by oxidizing their thiol groups. Within the group of ROS, hydrogen peroxide (H_2O_2) has been extensively studied on account of its relative stability and function as a signal molecule in plants.[79] It is essential to control an optimal amount of H_2O_2 in plant cells and cellular organelles in order to endure salt stress, implemented by various enzymatic and non-enzymatic antioxidants. To regulate the ROS levels, plant cells have evolved many complex enzymatic and non-enzymatic antioxidant defense mechanisms, which together help to control the cellular redox state under changing environmental conditions. Ascorbate peroxidase and catalase are key enzymatic antioxidants that detoxify excessive H_2O_2.[79,85] Superoxide dismutase (SOD) is also an important antioxidant enzyme and is the first line of defense against oxidative stress in plants. SOD causes dismutation of superoxide radicals at almost diffusion-limited rates to produce H_2O_2.[109] It plays an important part in determining the concentration of O_2^- and H_2O_2 in plants and hence performs a key role in the defense mechanism against free-radical toxicity.[16] Recent studies have shown that ROS plays a key role in plants as signal transduction molecules involved in mediating responses to pathogen infection, environmental stresses, programmed cell death, and developmental stimuli.[82,130]

9.4.6 ION TRANSPORT AND UPTAKE

Membranes along with their associated components play an integral role in maintaining ion concentration within the cytosol during the period of stress by regulating ion uptake and transport.[108] The transport phenomenon is carried out by different carrier proteins, channel proteins, antiporters, and symporters. A high K^+/Na^+ ratio in cytosol is essential for normal cellular functions of plants. Maintaining cellular Na^+/K^+ homeostasis is pivotal for plant survival in saline environments. Na^+ competes with K^+ uptake through Na^+/K^+ cotransporters and may also block K^+ specific transporters of root cells under salinity.[147] This leads to toxic levels of Na^+ and insufficient K^+ concentration for enzymes and osmotic adjustment, also an essential element for growth and development. It results in lower productivity and may even lead to death. Multiple transport mechanisms govern $K^+:Na^+$ selectivity and contribute to K^+ and Na^+ uptake in higher plants.[112] High-affinity potassium transporter (HKT1) (first K^+ carrier) encodes a high affinity K^+ transporter that functions as a Na^+ coupled K^+ transporter and is thought to contribute to sodium uptake in saline soil.[107] However, increased Na^+ concentrations caused changes in HKT1 behavior that have more physiological relevance for Na^+ uptake than K^+ uptake.[32]

Antisense expression of wheat HKT1 in transgenic wheat causes significantly less[22] Na uptake and enhances growth under salinity than in control plants.[68] This suggests that either inactivation of low affinity Na^+ transporter (HKT) activity or suppression of its expression can considerably improve salt tolerance. Intracellular NHX proteins are Na^+, K^+/H^+ antiporters involved in K^+ homeostasis, endosomal pH regulation, and salt tolerance. Barragán et al.[11] showed that tonoplast localized NHX proteins (NHX1 and NHX2: the two major tonoplast-localized NHX isoforms) are essential for active K^+ uptake at the tonoplast, for turgor regulation, and for stomatal function. In fact more such NHX isoforms have been identified and their roles in ion (Na^+, K^+, and H^+) homeostasis established from different plant species (e.g., LeNHX3 and LeNHX4 from tomato).[47] Two major genes, Nax1 and Nax2, for Na^+ exclusion have been found in durum wheat.[83] The proteins encoded by these genes are shown to increase retrieval of Na^+ from xylem into roots thereby reducing shoot Na^+ accumulation. In particular, Nax1 gene confers reduced rate of transport of Na^+ from root to shoot and retention of Na^+ in the leaf sheath thus maintaining a higher sheath-to-blade Na^+ concentration ratio. The second gene, Nax2 also confers a lower rate of transport of Na^+ from root to shoot and has a higher rate of K^+ transport, thus enhanced K^+ versus Na^+ discrimination in leaf.[61]

The knowledge, on how Na^+ is sensed, is still very limited in most cellular systems. Theoretically, Na^+ can be sensed either before or after entering the cell or both. Extracellular Na$^+$ may be sensed by a membrane receptor, whereas intracellular Na$^+$ may be sensed by membrane proteins or by any of many Na$^+$ sensitive enzymes in cytoplasm. In spite of molecular identity of Na$^+$ sensor(s) remaining elusive, the plasma membrane Na$^+$/H$^+$ antiporter *SOS1* is a possible candidate. It plays an important role in Na$^+$ extrusion and in controlling long-distance Na$^+$ transport from root to shoot.[126] Thisanti-porter forms one component in a mechanism based on sensing of salt stress that involves an increase of cytosolic Ca$^+$ protein interactions and revers-ible phosphorylation with SOS1 acting in concert with other two proteins named as SOS2 and SOS3. Both the protein kinase, SOS2 and its associated calcium-sensor unit, SOS3 are required for posttranslational activation of SOS1 Na$^+$/H$^+$ exchange activity. SOS4 and SOS5 have also been charac-terized,[74] which help in maintenance of cell wall integrity and architecture under salt stress.

9.4.7   HORMONE SYNTHESIS

ABA is well recognized as an important stress hormone. The concentration of ABA increases when water deficits occur, with its *de novo* synthesis beginning in the roots, in response to sensing an insufficient supply of water.[145] ABA stress hormone hardens plants against excess salts to trigger metabolic and physiological changes. In halophytes, which grow in conditions of "physiological drought," due to low water potential in the root medium, the lowest concentrations of ABA were found under salinity concentrations optimum for growth.[22] It was the case for the highly tolerant halophytic species *Suaeda maritima*, which exhibited the lowest seasonal range of ABA contents (from 649.4 ng g^{-1} to 835.6 ng g^{-1} dry weight) in comparison with several other species, where higher ABA concentrations were correlated with increased sodium content of the shoot.[26] Numerous genes involved in the acclimation/adaptation processes are up and/or down-regulated under stress conditions. Although not all of them are subjected to ABA regulation, expression of a large number of them is controlled by ABA. For the past several years, researchers have been trying to identify transcription factors that regulate the expression of ABA/stress-responsive genes via the consensus element, which is generally known as "abscisic acid response element."

9.5   APPLICATION OF HALOPHYTES IN SALINE ENVIRONMENT

9.5.1   ROLE OF HALOPHYTES IN SALINITY MANAGEMENT

Halophytes are naturally evolved salt tolerant plants. These species possess a range of highly efficient and complementary morphological, physiological, and anatomical characteristics to combat and even benefit from a saline environment.[41,44,118] As such, these have the ability to complete their life cycle in a NaCl rich environment where almost 99% of the salt sensitive species die. Thus, halophytes are regarded as a source of potential new crops[50,51] for saline environment. Halophytes can either be improved into new salt resistant crops or may be used as a source of genes to be introduced into conventional crop species that in general have their economical production decreased with increasing soil salt levels.

Halophytes have been utilized practically for managing the stressful environment and shown to be involved in increasing the economy of the developing countries in many parts of the world. Halophytes have shown their role in desalination of saline lands in arid and semiarid regions as well as the stabilization of saline lands along the coasts. Extensive application of halophytes in phytoremediation of heavy-metal-contaminated sites and wetland restoration and revegetation through introduction of variety of halophyte species has been made. This has led to developing agriculture on saline lands as alternatives sources of food, forage, fodder, medicine, ornamental, and services (e.g., improvement of soil structure and fertility, habitat for wildlife, source of biomass for the production of biodiesel) and important plant-based chemicals to ever-growing human population. While the following sections describe some of these applications, it is emerging that constructive strategies have to be developed and implemented for the protection of world's nonrenewable resources, wherein available halophyte diversity can be utilized as an important source.[6,10,64,97,98,142,144]

9.5.2   GREENING THE WASTELANDS

There is a great opportunity to exploit highly saline land and water resources through cultivation of halophytes. Based on his experiments in an Israel, the ecologist Hugo Boyko showed that several plant species can be cultivated even while irrigating with the sea water.[17] It stimulated a global interest in exploring the constraints posed by salinity and led to systematic identification of plant species that can grow under saline conditions to make the best use of such wastelands. Amongst conventional crops, beetroot (*Beta vulgaris*) and date palm (*Phoenix dactylifera*) are well known for their food value and are grown successfully in saline environment. A range of cultivation systems for the utilization of halophytes have also been developed especially in India, for the production of food, fuel, energy including biofuels, landscaping, ornamental uses, vegetables, and more. Seapurslane, *S. portulacastrum* (L.), is a pioneer, psammophytic facultative halophyte naturally growing in the subtropical, Mediterranean, coastal, and warmer zones of the world. It is an emerging source of secondary metabolite phytoecdysteroids used in sericulture industry. This underutilized plant can be cultivated on large scale for food as well as forages.[72]

The coastal badam (almond: *Terminalia catappa*) and species of *Pandanus* are known for their oils of industrial application. Fruits of Pandanus are staple food for coastal population of bay islands (India), and both of these plants are found naturally growing in tidal zones but can be cultivated successfully to provide livelihood to the coastal communities. Palmirah palm (*Borassus flabellifer*), widely used for toddy, jaggery, vinegar, beverage, juice for sugar and edible radicles and fruits, is found widely distributed all along Andhra coast, India. It needs to be genetically improved for wider cultivation. The young leaves and shoots of *Chenopodium album*, species of Amaranthus, *Portuleca oleracea*, *S. portulacastrum*, and many others are used as vegetable and salad in many parts of the country. Many of these are even cultivated.[25] *S. persica*, a facultative halophyte, appears to be a potentially valuable oilseed crop for saline and alkali soils. Its seed contains 40–45% of oil rich in industrially important lauric and myrestic acids.[101]

Rao et al.[101] studied the physiological mechanisms of its tolerance and developed cultivation practices for these halophytes for wider cultivation in saline and alkali vertisols of India. The oilseed halophyte, *Salicornia bigelovii*, can yield 2 t ha^{-1} of seed containing 28% oil and 31% protein being similar to soybean yield and seed quality. *S. bigelovii* has been evaluated as a source of vegetable oil and the cake as animal feed. It is grown in some areas of Gujarat and Rajasthan as it withstands high salinity of soil and water. New halophyte crops such as NyPa* forage and NyPa grain have the potential for utilizing salty land, stabilizing soil from wind and water erosion, providing pasturage, and helping less advantaged populations to feed themselves.[141]

The potential of halophytic grass species *Distichlis spicata* accessions as fodder has been investigated.[19] Although less salt tolerant than species of *Atriplex*, the ash content of tested *D. spicata* accessions never exceeded 11% of the dry matter, highlighting its potential as a fodder crop.[19] In Pakistan and India, halophytic species *Leptochloa fusca* (*Kallar* grass) has shown high productivity of about 20 t dry matter ha^{-1} from 4 to 5 cuts per year.[66,75,76]

Crithmum maritimum has the potential to become a multipurpose halophytic cash crop. While the aromatic, succulent leaves can be used as a salty vegetable its delicate inflorescences may be attractive for ornamental purposes.[13]

9.5.3   PHYTODESALINATION

Desalination is a process that connotes several processes such as reclamation of saline soils, sodic soils, and removal of minerals from saline water.[146] The use of biological materials for desalination is referred as phytodesalination. It is a biological approach that aims to rehabilitate saline, sodic, or saline sodic soils by Na-hyperaccumulating vegetation. Some halophytes have the potential as a source of "tolerant genes" to alter the salt tolerance of salt sensitive plants while others are suitable as cash crops under saline conditions simultaneously helping in desalination. This ability of vegetation could be important particularly in the arid and semiarid regions where insufficient precipitations and inappropriate irrigation systems are unable to reduce the salt burden in the rhizosphere of plants and suitable physicochemical methods are too expensive.[120] These hyperaccumulator plants can be harvested and dumped at point sink, or incinerated for further industrial uses.

Boyko[17] was one of the pioneers who first suggested that halophytic plants can be used to desalinate the soil and water. It was found that *S. salsa* produces biomass of about 20 t dry weight ha^{-1} containing 3–4 t of salt.[62] Native flora of saline waste lands in Pakistan was made productive by growing salt tolerant plants, namely, *Suaeda fruticosa, Kochia indica, Atriplex crassifolia, Sporobolus arabicus, Cynodon dactylon, Desmostachya bipinnata, Polypogon monspeliensis,*[75] *S. nudiflora, Salsola baryosma, Haloxylon recurvum,* and *Atriplex lentiformis.* These hyperaccumulators had high biomass. These plants had the potential of desalinization of saline soils from 16 dS m^{-1} to 2 dS m^{-1} in 4.9–6.1 years.[28,30,31] Eslamzadh[35] studied that high (30%) Na$^+$ accumulation corresponding to 31,500 µg g^{-1} dry weight in *Salicorni aeuropia.* Amounts of Ni, Cr, Cd, Pb, and Hg were also found to be high in this plant.

Hamidov et al.[57,58] showed that *C. album* produced 3.25 t ha^{-1} year^{-1} dry biomass removing 569.6 kg ha^{-1} salt ions from 0.3 m of the soil profile amounting to 1.47% of the soil salts. *Portulaca oleracea* also accumulated high amounts of salt (497 kg ha^{-1}) with biomass production of 3948 kg ha^{-1}. Cultivation of *S. portulacastrum* accumulated NaCl and reduced electrical conductivity of saturation extract (EC$_e$) from 4.9 to 2.5 dS m^{-1} in 120 days by accumulating NaCl.[102] Ravindran et al.[102] also observed that *S. maritime* and *S. portulacastrum* exhibited greater accumulation of salts in their tissue and higher reduction of salts from the saline land. It is estimated that these two halophytes could remove 504 and 474 kg of NaCl, respectively, from the saline land from 1 ha in 4 months.

Rabhi et al.[99] reported that *Arthrocnemum indicum*, *S. fruticosa*, and *S. portulacastrum* seedlings grown on a saline soil significantly reduced the soil salinity and electrical conductivity (EC) by absorbing soluble salts mainly sodium ions and they also reported that *S. portulacastrum* was able to accumulate nearly 30% of Na^+ content in shoot over a period of 170 days. Farah Al-Nasir[37] reported that different halophytic species showed decrease in soil salinity to various levels at the end of the experiment with salinity decreasing from an average EC_e of 84 to 5.46, 5.04 and 6.3 mS cm^{-1} in the 0–30-cm depth and from 49.6 to 5.46, 13.45 and 7.14 mS cm^{-1} for the lower soil depth (30–60 cm depth) for *Atriplex hallimus* L., *Atriplex numularia* L., and *Tamarix aphylla* L., respectively.

Zorrig et al.[149] estimated phytodesalination capacity of *Tunisian sabkha* at about 0.65 t Na^+ ha^{-1} in summer and 0.75 t Na^+ ha^{-1} in winter. Kumar et al.[67] found that by growing grass and nongrass halophytes (*Urochondrasetulosa*, *S. nudiflora*, *Sporobolus marginatus*, *Aleuropus lagopoides*), soil pH was reduced from 9.5 to 9.15 and from 10.0 to 9.6 in alkali/sodic soil and in saline microplots, EC_e was reduced from 15 to 3.2; 25 to 6.4, and 35 to 14.4 dS m^{-1}, respectively. On the other hand, in combined saline and sodic stress conditions, EC_e was reduced from 10 to 6; 15 to 7.2; 20 to 9.2 dS m^l and pH was reduced from 9 to 8.27, 8.33, 8.35, respectively. *Atriplex nummularia* (oldman salt bush) has been shown to achieve a biomass yield of 20–30 t ha^{-1} year^{-1} and accumulate between 20 and 40% NaCl in its dry matter when irrigated with saline water.[42,49,77,138]

Similarly, *Suaeda fruticose* (seablite) can remove > 2.5 t ha^{-1} in a single harvest of the aerial parts of the plant each year,[20] while *S. salsa* at a density 15 plants m^{-2} could potentially remove between 3 and 4 t of Na^+ per hectare, if the plants are harvested at the end of the growing season.[146] This ability to remove significant amounts of salt point out that remediated soils can attain the normal agricultural productivity making halophytes promising in phytodesalination. Few reports on salinity remediation through halophytes have been compiled in Table 9.2.

9.5.4 PHYTOREMEDIATION

Phytoremediation is visualized as benign and cost-effective plant-based technology that depends upon the remarkable ability of some plants to remove

TABLE 9.2 Remediation of Saline Soils Through Cultivation of Halophytes.[a]

Hyperaccumulator plant(s) used; site and nature of studies	Salient findings	References
Suaeda salsa; saline plots, Yu Cheng, East China	About 67% reduction in rhizospheric Na$^+$ at 20–30 cm depth due to hyperaccumulation in shoots at plant density of 30 plants m^{-2}	[146]
Suaeda salsaco cultivated on saline medium with tomatoes in "closed insulated pallet system," Oregon, USA	Initial NaCl concentration of 4 g L^{-1} in irrigation water had no adverse effect on tomato fruit yield when S. salsa was cocultivated with tomato	[4,7]
Atriplex prostrata, Hordeum jubatum, Spergularia marina, Suaeda calceoformis; brine contaminated saline sites in Ohio, USA	Compared with nonvegetated control plots, these species reduced soil salinity ranging from 4% in Hordeum jubatum to 17% in Atriplex prostrate	[63]
Salicornia europia collected from Maharloo salt lake, Farz Province, Iran	Exhibited high (30%) Na$^+$ accumulation corresponding to 31,500 µg g^{-1} dry weight	[35]
Apocynum lancifolium, Chenopodium album; Northwest Uzbekistan	C. album produced 3.25 t ha^{-1} year^{-1} dry biomass removing 569.6 kg ha^{-1} salt from 0.3 m of the soil profile	[57,58]
Sesuvium portulacastrum; Pichavaram mangrove forest, Tamil Nadu, India	Cultivation of S. portulacastrum reduced EC$_e$ from 4.9 to 2.5 dS m^{-1} in 120 days	[102]
Suaeda fruticosa, S. nudiflora, Salsola baryosma, Haloxylon recurvum, Atriplex lentiformis; saline microplots, Hisar, India	High biomass producing and salt hyperaccumulator plants which decreased EC$_e$ from 16 dS m^{-1} to 2 dS m^{-1} in about 6 years	[28,31]
Wild growing Tecticornia indica and Suaeda nudiflora; North-east Tunisia	Dry weight biomass and Na$^+$ removal (both in t ha^{-1}) were, respectively, 7.4 and 0.7 in T. indica; and 0.75 and 0.22 in S. nudiflora. Remediated soil had high microbial biomass	[88]
Urochondrasetulosa, Suaeda nudiflora, Sporobolus marginatus, Aleuropus lagopoides, sodic, saline, and saline-sodic microplots, Karnal, India	Soil pH$_s$ and EC$_e$ reduced considerably in sodic and saline treatments. In saline-sodic soils, EC$_e$ reduced from 10 to 6; 15 to 7.2; 20 to 9.2 dS m^{-1} and pH reduced from 9 to 8.27, 8.33, 8.35, respectively	[67]

[a]Adapted and modified from Ref. [7].

or neutralize various chemicals (organics and metal ions) from the soil, water, and air. It is eco-friendly, cost-effective, esthetically pleasing, noninvasive, and socially acceptable way to redress the removal of environmental contaminants.[9,92,96,110] This technique is potentially applicable to a variety of contaminants, including some of the most significant such as heavy metals (Cu, Cd, Zn, Mn, Fe, Pb, Hg, As, Cr, Se, Ur, etc.)[111], radionuclides, petroleum hydrocarbons, chlorinated solvents, pentachlorophenol, and polycyclic aromatic hydrocarbons, and it encompasses a number of different methods that can lead to contaminant removal through accumulation (phytoextraction and rhizofiltration) or dissipation (phytovolatilization), degradation (rhizodegradation and phytodegradation), or immobilization (phytostabilization).[38,92,94,114,132] Phytoremediation constitutes a group of strategies meant not only to reduce the metal load at the contaminated site but also to stabilize the site. These strategies are:

- *Phytodegradation/phytotransformation*: Enzymatic degradation of the pollutants in the plant tissue.
- *Phytoexcretion*: Excretion of heavy metals from the leaves.
- *Phytoextraction/phytoaccumulation*: Transfer of atmosphere pollutants from the soil and accumulation in the above ground parts of the plant.
- *Phytostabilization*: Stabilization of heavy metals in the soil/root surface and reduction of heavy metal mobility.
- *Phytovolatization*: Transfer of pollutants from the soil to the atmosphere.
- *Rhizofiltration*: Transfer of pollutants from the soil and accumulation in the roots of the plant.
- *Rhizosphere enhanced bioremediation or phytostimulation*: Enhancement of the microbial community and increase of biodegradation in the rhizosphere.

In *K. candel* seedlings, inhibition of leaf and root development was observed[21] only at the highest applied metal concentrations (400 mg kg^{-1} Cu and Zn). On the other hand, Pb (0–800 mg g^{-1}) had little negative effect on *Avicennia marina* seedlings.[73] Studies have demonstrated the accumulation of metals (Cu, Zn, Pb, Fe, Mn, and Cd) predominantly in root tissue, rather than in foliage, in numerous mangrove species grown in the field conditions, such as *Avicennia* sps., *Rhizophora* sps., and *Kandelia* spp.[91]

Eslamzadh[35] observed that in *S. europia*, amounts of Ni, Cr, Cd, Pb, and Hg were quite high. There were no adverse effects on the growth of *R. mucronata* and *Avicennia alba* seedlings treated with Zn (10–500 mg mL^{-1}) and Pb (50–250 mg mL^{-1}). On the basis of aforementioned reports, it is clear that several species are able to phytoremediate the contaminated soils efficiently and effectively to support meaningful land-use systems to meet the current challenges of global food security.

9.6   TOWARD GENETIC IMPROVEMENT: GENOMICS APPROACHES

Conventional breeding to improve salt tolerance of crops has worked on a limited scale and may not be able to provide major breakthroughs in future. A significant progress has been made to identify genes that play an important role in the plant system to overcome the unfavorable stresses.[24] Several published papers and reports document intensive works on genetic engineering for abiotic stress tolerance and development of transgenics using different candidate genes.[103,106,139] A number of genes have also been isolated and characterized from *Porteresia coarctata* that are related to the salt tolerance property of the plant, especially the genes operational in inositol metabolic pathway. Myoinositol phosphate synthase has been analyzed in details.[115] A total of 162 unique transcripts corresponding to possible salt-related genes were identified in *Puccinellia tenuiflora*.[137]

Transgenic approaches to salt tolerance are largely achieved through the overexpression of the genes involved in Na^+ exclusion from the root or leaves and Na^+ compartmentalization in the vacuoles.[14,136] In the past decade, vacuolar Na^+/H^+ antiporters have been the center of attention in transgenic studies for the alleviation of Na^+ induced toxicity in plants.[70,133,143] Several Na^+/H^+ antiporter genes have been characterized from salt tolerant grasses[124] and have been showed to impart salt tolerance in transgenic experiments. Several investigators have used candidate gene cDNAs from halophytic species for over-expression in *Arabidopsis thaliana* or other glycophyte hosts, and observed a significant alleviation of salt stress in the transgenic hosts, proving these genes do contribute to salt tolerance. A number of candidate genes encoding transcription factors, ion transporters, osmoprotectants, and antioxidants have been reported from salt tolerant *S. alterniflora*.[128] Success of glycophyte genes transfer into a halophyte through the gene gun technique[69] has given new hopes of vice versa transfer of halophyte genes in glycophytes.

Increased salinity tolerance have been achieved in a range of plant species over expressing *NHX* gene from the halophytes and nonhalophytes, which indicates that these genes are involved in Na^+ tolerance in plants.[54] Characterization of salt induced expressed sequence tags (ESTs) in salt tolerant *Aleuropus littoralis* revealed that about 20% of the ESTs bear no resemblance in the protein database and are thus novel.[150] The analysis of such ESTs could be important in the development of salt tolerant transgenic cereal crops.

To date, dehydration-responsive element binding (DREBs) transcriptions, which play a key role in plant stress signal transduction pathway, have been isolated from a wide variety of plants,[2] however, only a few reports are available from the halophytic plants like *Physcomitrella patens*[71] and *Atriplex hortensis*.[123] DREB SbDREB2A has been isolated from the halophyte *Salicornia brachiata* and its role in abiotic stress was studied in *Escherichia coli* BL21 (DE3). SbDREB2A, an A-2 type DREB transcription factor did confer abiotic stress tolerance in *E. coli* and exhibited better growth in basal lysogeny *broth* (LB) medium as well as in supplemented with NaCl, polyethylene glycol (PEG), and mannitol.[56] Genes for salinity tolerance have been identified in *A. marina* and have been transferred to rice and other crops and patented. AmMYB1, a single-repeat myeloblastosis (MYB) transcription factor, isolated from the salt-tolerant mangrove tree *A. marina* has been used to derive transgenic tobacco plants constitutively expressing the AmMYB1 transcription factor.[48] The transgenic tobacco showed better tolerance to NaCl stress.

The genetic mechanisms may be simple or they may be inordinately complex precluding easy transfer to conventional crops. In some halophytes, over 1400 genes may be involved. Moreover, evidences show that these genes do not operate in solo but in concert with other enzyme systems. To complicate matters further, it may be that each species of halophyte has a specific "orchestra" to match its particular enzyme system. Thus, even if one is able to identify the whole system, the system configuration may be different in different halophytes. In spite of these difficulties, it will still be interesting to test the effect of single-gene additions or alterations on glycophytes, if the relevant halophyte genes can be cloned and transferred to glycophytes.[15] The elucidation of mechanisms and gene transfer from halophytes to glycophytes or vice versa is an important field to work on to develop designer crops for the challenging environments to which agriculture will be increasingly exposed in future.

9.7 CONCLUSION AND WAY FORWARD

A number of techniques have been suggested to revive the productivity of salt-affected soils, but in many situations, these interventions fail to provide adequate relief, especially under permanent water inundation and extreme levels of root zone salinity. Under such conditions, halophytes provide an opportunity to ensure economic returns and gradual improvements in soil quality. Physiological and molecular approaches can enhance the salt tolerance and economic usage of the halophytes. Genetic improvement and novel industrial applications of such halophytic crops are needed to make them an attractive proposition for extremely saline and sodic soils where it is not possible to grow the traditional crops.

Glenn et al. proposed *greening the world's deserts and transforming lands with brackish water tables into sites of commercial agriculture in 1998 yet the dream has so far not been realized as such ideas takes much longer to ripe.*[52] Nonetheless, the idea and the pace with which it is being taken forward have been quite satisfactory. Moreover, it is picking up momentum especially in the light of envisaged climate change. The halophytes, naturally occurring salt tolerant plants, can bridge the gap between the demand and supply of food for humans and forage/fodder for livestock, besides providing compounds for industrial or pharmaceutical use. However, the greatest impediment to use halophytes as cultivated plants is due to many undesirable characteristics such as uneven germination, lack of seed retention, and toxic substances in tissues. The agronomy of many of these crops is not as well developed as that of the cultivated crops resulting in low yields. Marketing including postharvest value addition technologies also pose problems. Additional steps toward domestication will require identifying the correct plant species for the target environment and application and eventually improving wild plants though crop breeding. It appears that halophytic crops should undergo the same processes that have undergone in conventional agricultural crops so that during short time spans economically profitable and consumer-acceptable products can be attained.

Besides, we need to understand the salt tolerant mechanisms in halophytes as these will play an increasingly important role both as models for generating tolerance in cultivated crops and as "niche crops" in their own right for landscapes with saline soils and saline water including sea water. We believe that it should be possible to improve crop performance in biosaline agriculture integrating halophytes and conventional crops.

9.8 SUMMARY

Salinity is one of the major abiotic stresses that hinder the performance of the crop plants. Besides osmotic stress, the toxic components of salinity, that is, Na^+ and Cl^- ions interfere with the normal physiological processes resulting in significantly reduced yields. Many research groups across the world are vigorously promoting halophyte plants as potential candidates for the productivity enhancements of salt-affected lands. Halophytes constitute about 1% of the world's flora, but are able to survive in environments where the salt concentration is high, around 200 mM NaCl. They are "salt-plant" and might just be the answer to a question, namely, how can we put our planet's practically infinite volumes of saltwater and large acreage of saline lands to good use?

The general physiology of halophytes has been reviewed with reference to salinity tolerance. Briefly, the halophytes manage toxicity of ions through ion compartmentalization, osmotic adjustment through osmolytes accumulation, succulence, selective transport and uptake of ions, enzymatic and non-enzymatic antioxidant response, maintenance of redox and energetic status, salt inclusion/excretion, etc. which have been explained in the present chapter. This explanation provides an insight into understanding of the mechanism of salinity tolerance in halophytes.

The current chapter also discusses the use of these plants as a means of greening the waste lands, phytoremediation of soils and waste water and a source of genes to genetically engineer plants that can sustain their growth and productivity in saline environments. The later application requires exploring gene pool of wild relatives of crop plants with high salt tolerance. Due to their close affinity to the cereal crops, the success of transformation of genes from halophytes in cereal crops is expected to be very pronounced. It is believed that improved knowledge of halophytes is of important to understand our natural world and to enable the use of some of these plants in land revegetation, forages for livestock, fuel, energy including biofuels, landscaping, ornamental uses, vegetables, and develop salt-tolerant crops.

KEYWORDS

- abiotic stresses
- adaptive mechanisms
- ion homeostasis
- reactive oxygen species (scavenging)
- salt exclusion
- specific ion toxicity

REFERENCES

1. Abrol, I. P.; Yadav, J. S. P.; Massoud, F. I. *Salt Affected Soils and Their Management*; FAO Soils Bulletin. Food and Agriculture Organization: Rome, Italy, 1988; pp 39.
2. Agarwal, P. K.; Agarwal, P.; Reddy, M. K.; Sopory, S. K. Role of DREB Transcription Factors in Abiotic and Biotic Stress Tolerance in Plants. *Plant Cell Rep.* **2006**, *25*, 1263–1274.
3. Agrawal, G. K.; Rakwal, R.; Yonekura, M.; Kubo, A.; Saji, H. Proteome Analysis of Differentially Displayed Proteins as a Tool for Investigating Ozone Stress in Rice (*Oryza sativa* L.) Seedlings. *Proteomics* **2002**, *2*, 947–959.
4. Albaho, M. S.; Green, J. L. *Suaeda salsa*, a Desalinating Companion Plant for Greenhouse Tomato. *Hort. Sci.* **2000**, *35*, 620–623.
5. Allen, J. A.; Cambers, J. L.; Stine, M. Prospects for Increasing Salt Tolerance of Forest Trees: A Review. *Tree Physiol.* **1994**, *14*, 843–853.
6. Al-Oudat, M.; Qadir, M. *The Halophytic Flora of Syria*; International Center for Agricultural Research in the Dry Areas: Aleppo, Syria, 2011; pp 186.
7. Angrish, R.; Devi, S. Potential of Salt Hyper-accumulator Plants in Salinity Phytoremediation. In: *Advances in Plant Physiology: An International Treatise Series*; Hemantaranjan, A., Ed.; Scientific Publishers: Jodhpur, India, 2014; Vol. 15, pp 307–323.
8. Arnold, A. Die Bedeutung der Chlorionen für die Pflanze. In: *Botanische Studien Hrgb*; Troll, W., Guttenberg, H. V., Eds.; Heft 2, Gustav Fischer Verlag: Jena, 1955; pp 148.
9. Arthur, E. L.; Rice, P. J.; Rice, P. J.; Anderson, T. A.; Baladi, S. M.; Henderson, K. L. D.; Coats, J. R. Phytoremediation—An Overview. *Crit. Rev. Plant Sci.* **2005**, *24*, 109–122.
10. Barnes, R. F.; Baylor, J. E. Forages in a Changing World. In:*Forages: An Introduction to Grassland Agriculture*; State University Press: New York, USA, 1995; pp 3–13.
11. Barragán, V.; Leidi, E. O.; Andrés, Z.; Rubio, L.; De Luca, A.; Fernández, J. A.; Cubero, B.; Pardo, J. M. Ion Exchangers NHX1 and NHX2 Mediate Active Potassium Uptake into Vacuoles to Regulate Cell Turgor and Stomatal Function in *Arabidopsis. Plant Cell* **2012**, *24*, 1127–1142.
12. Bell, H. L.; O'Leary, J. W. Effects of Salinity on Growth and Cation Accumulation of *Sporobolus virginicus* (Poaceae). *Am. J. Bot.* **2003**, *90*, 1416–1424.
13. Hamed, B. K.; Youssef, B. N.; Ranieri, A.; Zarrouk, M.; Abdelly, C. Changes in Content and Fatty Acid Profiles of Total Lipids and Sulfolipids in the Halophyte *Crithmum maritimum* Under Salt Stress. *J. Plant Physiol.* **2005**, *162*, 599–602.
14. Bhatnagar-Mathur, P.; Vadez, V.; Sharma, K. K. Transgenic Approaches for Abiotic Stress Tolerance in Plants: Retrospect and Prospects. *Plant Cell Rep.* **2008**, *27*, 411–424.
15. Bohnert, H.; Jensen, R. Metabolic Engineering for Increased Salt Tolerance—the Next Step. *Aust. J. Plant Physiol.* **1996**, *23*, 661–667.
16. Bowler, C.; Van Montagu, M.; Inzé, D. Superoxide Dismutase and Stress Tolerance. *Annu. Rev. Plant Physiol. Plant Mol. Biol.* **1992**, *43*, 83–116.
17. Boyko, H. Basic Ecological Principles of Plant Growing by Irrigation with High Saline Seawater. In: *Salinity and Aridity*; Boyko, H., Ed.; D. W. Junk Publisher: The Hague, 1966; pp 131.
18. Braun-Blanquet, J. *Plant Sociology: A Study of Plant Communities*; McGraw-Hill Book Company, Inc.: New York, 1932; pp 439.
19. Bustan, A.; Pasternak, D.; Pirogova, I. Evaluation of Salt grass as a Fodder Crop for Livestock. *J. Sci. Food Agric.* **2005**, *85*, 2077–2084.

20. Chaudhri, I. I.; Shah, B. H.; Naqvi, N.; Mallick, I. A. Investigations on the Role of *Suaeda fruticosa* Forsk in the Reclamation of Saline and Alkaline Soils in West Pakistan Plains. *Plant Soil* **1964**, *21*, 1–7.

21. Chiu, C. Y.; Hsiu, F. S.; Chen, S. S.; Chou, C. H. Reduced Toxicity of Cu and Zn to Mangrove Seedlings in Saline Environments. *Bot. Bull. Acad. Sin.* **1995**, *36*, 19–24.

22. Clipson, N. J. W.; Lachno, D. R.; Flowers, T. J. Salt Tolerance in the Halophyte *Suaeda maritima* (L.) Dum.: Abscisic Acid Concentration in the Response to Constant and Altered Salinity. *J. Exp. Bot.* **1988**, *39*, 1381–1388.

23. Cramer, G. R.; Lauchli, A.; Polito, V. S. Displacement of Ca^{2+} by Na^+ from the Plasma Lemma of Root Cells. A Primary Response to Salt Stress? *Plant Physiol.* **1985**, *79*, 207–211.

24. Cushman, J. C.; Bohnert, H. J. Genomic Approaches to Plant Stress Tolerance. *Curr. Opin. Plant Biol.* **2000**, *3*, 117–124.

25. Dagar, J. C. Ecology, Management and Utilization of Halophytes. *Bull. Nat. Inst. Ecol.* **2005**, *15*, 81–97.

26. Dajic, Z.; Pekic, S.; Mrfat-Vukelic, S. Salinity Problem and Adaptations of Plants Grown in Arid and Semi-arid Regions. In: *Drought and Plant Production*; Jeftic, S., Pekic, S., Eds.; Agricultural Research Institute Serbia: Belgrade, 1997; pp 355–362.

27. Dansereau, P. *Biogeography: An Ecological Perspective*; Ronald Press: New York, USA, 1957; pp 394.

28. Datta, K. S.; Angrish, R. Selection, Characterization and Quantification of Plant Species for Phytoremediation of Saline Soils. In: *Final Progress Report,* Ministry of Environment and Forests, Government of India, New Delhi; 2006; pp 166.

29. Delf, E. M. Transpiration and Behavior of Stomata in Halophytes. *Ann. Bot.* **1911**, *25*, 485–505.

30. Devi, S.; Nandwal, A. S.; Angrish, R.; Arya, S. S.; Kumar, N.; Sharma, S. K. Phytoremediation Potential of Some Halophytic Species for Soil Salinity. *Int. J. Phytorem.* **2016**, *18*, 693–696.

31. Devi, S.; Rani, C.; Datta, K. S.; Bishnoi, S. K.; Mahala, S. C.; Angrish, R. Phytoremediation of Soil Salinity Using Salt Hyper-accumulator Plants. *Indian J. Plant Physiol.* **2008**, *4*, 347–356.

32. Diatloff, E.; Kumar, R.; Schachtman, D. P. Site Directed Mutagenesis Reduces the Na+ Affinity of HKT1, a Na^+ Energized High Affinity K^+ Transporter. *FEBS Lett.* **1998**, *432*, 31–36.

33. Dietz, K. J.; Tavakoli, N.; Kluge, C.; Mimura, T.; Sharma, S. S.; Harris, G. C., et al. Significance of the V-type ATPase for the Adaptation to Stressful Growth Conditions and its Regulation on the Molecular and Biochemical Level. *J. Exp. Bot.* **2001**, *52*, 1969–1980.

34. Epstein, E. *Mineral Nutrition of Plants: Principles and Perspectives*; John Wiley: New York; 1972; pp 600.

35. Eslamzadh, T. *Salicorni aeuropea*, a Bioaccumulator in Maharloo Salt Lake Region. *Int. J. Soil Sci.* **2006**, *1*, 75–80.

36. F. A. O. *FAO Agristat*; 2007. www.fao.org (accessed May 15, 2016).

37. Al-Nasir, F. Bio-reclamation of a Saline Sodic Soil in a Semi-arid Region/Jordan. *Am. Eurasian J. Agric. Environ. Sci.* **2009**, *5*, 701–706.

38. Fletcher, J. Phytoremediation and Rhizoremediation. In: *Focus on Biotechnology*; Hoffmann, M., Anné, J., Eds.; Springer: Dordrecht, the Netherlands; 2006; pp 198.

39. Flowers, T. J. *eHALOPH Halophytes Database*; 2014. http://www.sussex.ac.uk/ affiliates/halophytes (accessed Dec 15, 2016).

40. Flowers, T. J.; Colmer, T. D. Salinity Tolerance in Halophytes. *New Phytol.* **2008,** *179,* 945–963.

41. Flowers, T. J.; Yeo, A. R. Ion Relations of Plants under Drought and Salinity. *Aust. J. Plant Physiol.* **1986,** *13,* 75–91.

42. Flowers, T. J.; Galal, H. K.; Bromham, L. Evolution of Halophytes: Multiple Origins of Salt Tolerance in Land Plant. *Funct. Plant Biol.* **2010,** *37,* 604–612.

43. Flowers, T. J.; Hajibagheri, M. A.; Clipson, N. J. W. Halophytes. *Q. Rev. Biol.* **1986,** *61,* 313–335.

44. Flowers, T. J.; Troke, P. F.; Yeo, A. R. The Mechanism of Salt Tolerance in Halophytes. *Annu. Rev. Plant Physiol.* **1977,** *28,* 89–121.

45. Foyer, C. H.; Noctor, G. Redox Sensing and Signaling Associated with Reactive Oxygen in Chloroplasts, Peroxisomes and Mitochondria. *Physiol. Plant* **2003,** *119,* 355–364.

46. G. T. D. *Gujarat Tourism Document*; 2008. www.gujarattourism.com (accessed Nov 04, 2016).

47. Galvani, A. The Challenge of the Food Sufficiency through Salt Tolerant Crops. *Rev. Environ. Sci. Biotechnol.* **2007,** *6,* 3–16.

48. Galvez, F. J.; Baghour, M.; Hao, G.; Cagnac, O.; Rodrıguez Rosales, M. P.; Venema, K. Expression of LeNHX Isoforms in Response to Salt Stress in Salt Sensitive and Salt Tolerant Tomato Species. *Plant Physiol. Biochem.* **2012,** *51,* 109–115.

49. Ganesan, G.; Sankararamasubramanian, H. M.; Harikrishnan, M.; Ashwin, G.; Parida, A. A MYB Transcription Factor from the Grey Mangrove is Induced by Stress and Confers NaCl Tolerance in Tobacco. *J. Exp. Bot.* **2012,** *63,* 4549–4561.

50. Ghnaya, T.; Nouairi, I.; Slama, I.; Messedi, D.; Grignon, C.; Adbelly, C.; Ghorbel, M. H. Cadmium Effects on Growth and Mineral Nutrition of Two Halophytes: *Sesuvium portulacastrum* and *Mesembryanthemum crystallinum. J. Plant Physiol.* **2005,** *162,* 1133–1140.

51. Glenn, E. P.; Anday, T.; Chaturvedi, R., et al. Three Halophytes for Saline-water Agriculture: An Oilseed, Forage and a Grain Crop. *Environ. Exp. Bot.* **2013,** *92,* 110–121.

52. Glenn, E. P.; Brown, J. J.; Blumwald, E. Salt Tolerance and Crop Potential of Halophytes. *Crit. Rev. Plant Sci.* **1999,** *18,* 227–255.

53. Greenway, H.; Munns, R. Mechanisms of Salt Tolerance in Non-halophytes. *Annu. Rev. Plant Physiol.* **1980,** *31,* 149–190.

54. Guo, S. L.; Yin, H. B.; Zhang, X.; Zhao, F. Y.; Li, P. H.; Chen, S. H.; Zhao, Y. X.; Zhang, H. Molecular Cloning and Characterization of a Vacuolar H$^+$-pyrophosphatase Gene, SsVP, from the Halophyte *Suaeda salsa* and its Overexpression Increases Salt and Drought Tolerance of *Arabidopsis. Plant Mol. Biol.* **2006,** *60,* 41–50.

55. Gupta, B.; Huang, B. Mechanism of Salinity Tolerance in Plants: Physiological, Biochemical, and Molecular Characterization. *Int. J. Genomics* **2014,** *2014,* 1–18.

56. Gupta, K.; Agarwal, P. K.; Reddy, M. K.; Jha, B. SbDREB2A, an A-2 Type DREB Transcription Factor from Extreme Halophyte *Salicornia brachiata* Confers Abiotic Stress Tolerance in *Escherichia coli. Plant Cell Rep.* **2010,** *29,* 1131–1137.

57. Hamidov, A.; Beltrao, J.; Neves, A.; Khaydarova, V.; Khamidov M. *Apocynum lancifolium* and *Chenopodium album*—Potential Species to Remediate Saline Soils. *WSEAS Trans. Environ. Dev.* **2007,** *3,* 123–128.

58. Hamidov, A.; Khaydarova, V.; Khamidov, M.; Neves, A.; Beltrao, J. In *Remediation of Saline Soils Using Apocyanum lancifolium and Chenopodium album*. Proceedings of the 3rd IASME/WSEAS International Conference on Energy, Environment; Agios Nikolas: Greece, July 24–26, 2007, pp 157–164.

59. Horie, T.; Karahara, I.; Katsuhara, M. Salinity Tolerance Mechanisms in Glycophytes: An Overview with the Central Focus on Rice Plants. *Rice* **2012**, *5*, 1–18.

60. Ishnava, K.; Ramarao, V.; Mohan, J.; Kothari, I. Ecologically Important and Life Supporting Plants of Little Rann of Kachchh. *Gujarat J. Ecol. Nat. Environ.* **2011**, *3*, 33–38.

61. James, R. A.; Davenport, R. J.; Munns, R. Physiological Characterization of Two Genes for Na^+ Exclusion in Durum Wheat Nax1 and Nax2. *Plant Physiol.* **2006**, *142*, 1537–1547.

62. Ke-Fu, Z. Desalinization of Saline Soils by *Suaeda salsa*. *Plant Soil* **1991**, *135*, 303–305.

63. Keiffer, C. H.; Ungar, I. A. Germination and Establishment of Halophytes on Brine Affected Soils. *J. Appl. Ecol.* **2002**, *39*, 402–415.

64. Khan, M. A.; Qaiser, M. Halophytes of Pakistan: Distribution, Ecology, and Economic Importance. In: *Sabkha Ecosystems*; Khan, M. A., Kust, G. C., Barth, H. J., Boer, B., Eds.; Springer: Drodrecht, 2006; Vol. II, pp 129–153.

65. Koyro, H. W.; Geißler, N.; Hussin, S.; Debez, A.; Huchzermeyer, B. Strategies of Halophytes to Survive in a Salty Environment. In: *Abiotic Stress and Plant Responses*, 1st ed.; I. K. International Publishing House: New Delhi, India, 2008; pp 83–104.

66. Kumar, A.; Abrol, I. P. *Grasses in Alkali Soils*; Central Soil Salinity Research Institute:Karnal, India, 1986; pp 95; Bulletin No. 11.

67. Kumar, A.; Kumar, A.; Lata, C.; Kumar, S. Eco-physiological Responses of *Aeluropus lagopoides* (Grass Halophyte) and *Suaeda nudiflora* (Non-grass Halophyte) under Individual and Interactive Sodic and Salt Stress. *S. Afr. J. Bot.* **2016**, *105*, 36–44.

68. Laurie, S.; Feeney, K. A.; Maathuis, F. J. M.; Heard, P. J.; Brown, S. J.; Leigh, R. A. A Role for HKT1 in Sodium Uptake by Wheat Roots. *Plant J.* **2002**, *32*, 139–149.

69. Li, X. G.; Gallagher, J. L. Expression of Foreign Genes, GUS and Hygromycin Resistance, in the Halophyte *Kosteletzkya virginica* in Response to Bombardment with Particle Inflow Gun. *J. Exp. Bot.* **1996**, *47*, 1437–1447.

70. Liu, D.; Li, T. Q.; Jin, X. F.; Yang, X. E.; Islam, E.; Mahmood, Q. Lead Induced Changes in the Growth and Antioxidant Metabolism of the Lead Accumulating and Non-accumulating Ecotypes of *Sedum alfredii*. *J. Integr. Plant Biol.* **2008**, *50*, 129–140.

71. Liu, N.; Zhong, N. Q.; Wang, G. L.; Li, L. J.; Liu, X. L.; He, Y. K.; Xia, G. X. Cloning and Functional Characterization of PpDBF1 Gene Encoding a DRE-binding Transcription Factor from *Physcomitrella patens*. *Planta* **2007**, *226*, 827–838.

72. Lokhande, V. H.; Nikam, T. D.; Suprasanna, P. *Sesuvium portulacastrum* (L.), a Promising Halophyte: Cultivation, Utilization and Distribution in India. *Genet. Resour. Crop Evol.* **2009**, *56*, 741–747.

73. MacFarlane, G. R.; Burchett, M. D. Toxicity, Growth and Accumulation Relationship of Copper, Lead and Zinc in the Grey Mangrove, *Avicennia marina* (Forsk.) Vierh: Biological Indication Potential. *Environ. Pollut.* **2002**, *123*, 139–151.

74. Mahajan, S.; Pandey, G. K.; Tuteja, N. Calcium-and Salt-stress Signalling in Plants: Shedding Light on SOS Pathway. *Arch. Biochem. Biophys.* **2008**, *471*, 146–158.

75. Mahmood, H.; Malik, K. A.; Lodhi, M. A. K.; Sheikh, K. H. Seed Germination and Salinity Tolerance in Plant Species Growing on Saline Wastelands. *Biol. Plant* **1996**, *38*, 309–315.

76. Mahmood, K.; Malik, K. A.; Lodhi, M. A. K.; Sheikh, K. K. Soil–Plant Relationships in Saline Wastelands: Vegetation, Soils, and Successional Changes, During Biological Amelioration. *Environ. Conserv.* **1994,** *21,* 236–241.

77. Manousaki, E.; Kalogerakis, N. Halophytes—An Emerging Trend in Phytoremediation. *Int. J. Phytorem.* **2011,** *13,* 959–969.

78. McCord, J. M. The Evolution of Free Radicals and Oxidative Stress. *Am. J. Med.* **2000,** *108,* 652–659.

79. Mhamdi, A.; Queval, G.; Chaouch, S.; Vanderauwera, S.; Breusegem, F. V.; Noctor, G. Catalase Function in Plants: A Focus on Arabidopsis Mutants as Stress-mimic Models. *J. Exp. Bot.* **2010,** *61,* 4197–4220.

80. Miller, G.; Suzuki, N.; Ciftci-Yilmaz, S.; Mittler, R. Reactive Oxygen Species Homeostasis and Signaling During Drought and Salinity Stresses. *Plant Cell Environ.* **2010,** *33,* 453–467.

81. Mittler, M. Oxidative Stress, Antioxidants and Stress Tolerance. *Trends Plant Sci.* **2002,** *7,* 405–410.

82. Mittler, R.; Vanderauwera, S.; Gollery, M.; Breusegem, F. V. Reactive Oxygen Gene Network of Plants. *Trends Plant Sci.* **2004,** *9,* 490–498.

83. Munns, R.; Tester, M. Mechanisms of Salinity Tolerance. *Annu. Rev. Plant Biol.* **2008,** *59,* 651–681.

84. Munns, R.; Rebetzke, G. J.; Husain, S.; James, R. A.; Hare, R. A. Genetic Control of Sodium Exclusion in Durum Wheat. *Aust. J. Agric. Res.* **2003,** *54,* 627–635.

85. Nakano, Y.; Asada, K. Hydrogen Peroxide is Scavenged by Ascorbate-specific Peroxidase in Spinach Chloroplasts. *Plant Cell Physiol.* **1981,** *22,* 867–880.

86. Niu, X.; Bressan, R. A.; Hasegawa, P. M.; Pardo, J. M. Ion Homeostasis in NaCl Stress Environments. *Plant Physiol.* **1995,** *109,* 735–742.

87. Otoch, M. L. O.; Sobreira, A. C. M.; Aragão, M. E. F.; Orellano, E. G.; Lima, M. G. S.; de Melo, D. F. Salt Modulation of Vacuolar H$^+$-ATPase and H$^+$-Pyrophosphatase Activities in *Vigna unguiculata. J. Plant Physiol.* **2001,** *158,* 545–551.

88. Ouni, Y.; Lakhdar, A.; Rabi, M.; Aoui, A. S.; Maria, A. R.; Chedly, A. Effects of the Halophytes *Tecticornia indica* and *Suaeda fruticosa* on Soil Enzyme Activities in a Mediterranean Sabkha. *Int. J. Phytorem.* **2013,** *15,* 188–197.

89. Parida, A.; Das, A. B. Salt Tolerance and Salinity Effects on Plants: A Review. *Ecotoxicol. Environ. Saf.* **2005,** *60,* 324–349.

90. Parvaiz, A.; Satyawati, S. Salt Stress and Phytobiochemical Responses of Plants. *Plant Soil Environ.* **2008,** *54,* 89–99.

91. Peters, E. C.; Gassman, N. J.; Firman, J. C.; Richmond, R. H.; Power, E. A. Ecotoxicology of Tropical Marine Ecosystems. *Environ. Toxicol. Chem.* **1997,** *16,* 12–40.

92. Pilon-Smits, E. Phytoremediation. *Annu. Rev. Plant Biol.* **2005,** *56,* 15–39.

93. Pitzschke, A.; Forzani, C.; Hirt, H. Reactive Oxygen Species Signaling in Plants. *Antioxid. Redox Signaling.* **2006,** *8,* 1757–1764.

94. Pletsch, M. Plants and the Environment/Phytoremediation. In: *Encyclopedia of Applied Plant Sciences*; Thomas, A., Murphy, D. J., Murray, B. G., Eds.; Elsevier: London (UK), 2003; pp 781–786.

95. Popp, M. Salt Resistance in Herbaceous Halophytes and Mangroves. *Prog. Bot.* **1995,** *56,* 415–429.

96. Prasad, M. N. V. Phytoremediation of Metals in the Environment for Sustainable Development. *Proc. Indian Nat. Sci. Acad.* **2004,** *70,* 71–98.

97. Rabhi, M.; Ferchichi, S.; Jouini, J., et al. Phytodesalination of a Salt-affected Soil with the Halophyte *Sesuvium portulacastrum* L. to Arrange in Advance the Requirements for the Successful Growth of a Glycophytic Crop. *Bioresour. Technol.* **2010**, *101*, 6822–6828.

98. Rabhi, M.; Karray-Bouraoui, N.; Medini, R.; Attia, H.; Athar, H.; Lachaâl, M.; Abdelly, C.; Smaoui, A. Seasonal Variations in the Capacities of Phytodesalination of a Salt-affected Soil in Two Perennial Halophytes in Their Natural Biotope. *J. Biol. Res.* **2010**, *14*, 181–189.

99. Rabhi, M.; Talbi, O.; Atia, A.; Chedly, A.; Smaoui, A. Selection of Halophyte That Could be Used in the Bio Reclamation of Salt Affected Soils in Arid and Semi-arid Regions. In: *Biosaline Agriculture and High Salinity Tolerance*; Abdelly, C., Öztürk, M., Ashraf, M., Grignon, C., Eds.; Birkhäuser Verlag AG: Berlin; 2008; pp 242–246.

100. Ramadan, T.; Flowers, T. J. Effects of Salinity and Benzyl Adenine on Development and Function of Microhairs of *Zea mays* L. *Planta* **2004**, *219*, 639–648.

101. Rao, G. G.; Nayak, A. K.; Chinchmalatpure, A. R.; Nath, A.; Babu, V. P. Growth and Yield of *Salvadora persica*—A Facultative Halophyte Grown on Saline Black Soil. *Arid. Land Res. Manage.* **2004**, *18*, 165–168.

102. Ravindran, K. C.; Venkatesan, K.; Balakrishnan, V.; Chellappan, K. P.; Balasubramanian, T. Restoration of Saline Land by Halophytes for Indian Soils. *Soil Biol. Biochem.* **2007**, *39*, 2661–2664.

103. Reguera, M.; Peleg, Z.; Blumwald, E. Targeting Metabolic Pathways for Genetic Engineering Abiotic Stress-tolerance in Crops. *BBA* **2012**, *1819*, 186–194.

104. Repp, G. The Salt Tolerance of Plants; Basic Research and Tests. *UNESCO Arid. Zone Res.* **1961**, *14*, 153–161.

105. Rhodes, D.; Hanson, A. D. Quarternary Ammonium and Tertiary Sulphonium Compounds of Higher Plants. *Annu. Rev. Plant Physiol. Plant Mol. Biol.* **1993**, *44*, 357–384.

106. Roy, S. J.; Tucker, E. J.; Tester, M. Genetic Analysis of Abiotic Stress Tolerance in Crops. *Curr. Opin. Plant Biol.* **2011**, *14*, 232–239.

107. Rubio, F.; Gassmann, W.; Schroeder, J. I. High-affinity Potassium Uptake in Plants Response. *Science* **1996**, *273*, 978–979.

108. Sairam, R. K.; Tyagi, A. Physiology and Molecular Biology of Salinity Stress Tolerance in Plants. *Curr. Sci.* **2004**, *86*, 407–421.

109. Salin, M. L. Toxic Oxygen Species and Protective Systems of the Chloroplast. *Physiol. Plant* **1987**, *72*, 681–689.

110. Salt, D. E.; Smith, R. D.; Raskin, I. Phytoremediation. *Annu. Rev. Plant Biol.* **1998**, *49*, 643–668.

111. Sarma, H. Metal Hyper-accumulation in Plants: A Review Focusing on Phytoremediation Technology. *J. Environ. Sci. Technol.* **2011**, *4*, 118–138.

112. Schachtman, D. P. Molecular Insights into the Structure and Function of Plant K^+ Transport Mechanisms. *BBA* **2000**, *1465*, 127–139.

113. Schimper, A. F. W. *Plant Geography upon a Physiological Basis*; Clarendon Press: Oxford, UK; 1903; pp 839.

114. Schnoor, J. L. *Phytoremediation: Technical Evaluation Report for Groundwater Remediation Technologies Analysis Center Pittsburgh*; The University of Iowa, Iowa, USA, 1997; pp 37.

115. Sengupta, S.; Majumder, A. L. Insight Into the Salt Tolerance Factors of a Wild Halophytic Rice, *Porteresia coarctata*: A Physiological and Proteomic Approach. *Planta* **2009**, *229*, 911–929.

116. Serrano, R.; Marquez, J. A.; Rios, G. Crucial Factors in Salt Stress Tolerance. In: *Yeast Stress Responses*; Hohmann, S., Mager, W. H., Eds.; R. G. Lande Company: Austin, TX; 1997; pp 147–169.

117. Serrano, R.; Mulet, J. M.; Rios, G.; Marquez, J. A.; de Larrinoa, I. F.; Leube, M. P.; Mendizabal, I.; Pascual-Ahuir, A.; Proft, M.; Ros, R.; Montesinos, C. A Glimpse of the Mechanism of Ion Homeostasis during Salt Stress. *J. Exp. Bot.* **1999**, *50*, 1023–1036.

118. Shabala, S. Learning from Halophytes: Physiological Basis and Strategies to Improve Abiotic Stress Tolerance in Crops. *Ann. Bot.* **2013**, *112*, 1209–1221.

119. Shabala, S.; Mackay, A. Ion Transport in Halophytes. *Adv. Bot. Res.* **2011**, *57*, 151–199.

120. Shahid, S. A. In: *New Technologies for Soil Reclamation and Desert Greenery*. Proceedings of the Joint KISR—PEC Symposium Yazd, Iran; Nader, M. A., Faisal, K. T., Eds.; 2002; pp 308–329.

121. Sharma, D. K.; Singh, A. Salinity Research in India-achievements, Challenges and Future Prospects. *Water Energ. Int.* **2015**, *58*, 35–45.

122. Sharma, D. K.; Singh, A. Greening the Degraded Lands: Achievements and Future Perspectives in Salinity Management in Agriculture. In: *Natural Resource Management in Arid and Semi-arid Ecosystem for Climate Resilient Agriculture*; Pareek, N. K., Arora, S., Eds.; Soil Conservation Society of India: New Delhi, 2016; pp 17–30.

123. Shen, Y. G.; Zhang, W. K.; Yan, D. Q.; Du, B. X.; Zhang, J. S.; Liu, Q.; Chen, S. Y. Characterization of a DRE-binding Transcription Factor from a Halophyte *Atriplex hortensis*. *Theor. Appl. Genet.* **2003**, *107*, 155–161.

124. Shi, H.; Quintero, F. J.; Prado, J. M.; Zhu, J. K. The Putative Plasma Membrane Na^+/H^+ Antiporter SOS1 Controls Long-distance Na^+ Transport in Plants. *Plant Cell* **2002**, *14*, 465–477.

125. Stocker, O. D. Halophyte Problem. In: *Ergebnisse der Biologie*; Frisch, K. V., Goldschmidt, R., Ruhland, W., Winterstein, H., Eds.; Springer: Berlin, Germany, 1928; pp 266–353.

126. Storey, R. Salt Tolerance, Ion Relations and the Effect of Root Medium on the Response of Citrus to Salinity. *Aust. J. Plant Physiol.* **1995**, *22*, 101–114.

127. Storey, R.; Wyn Jones, R. G. Salt Stress and Comparative Physiology in the *Gramineae*. III. Effect of Salinity upon Ion Relations and Glycine Betaine and Proline Levels in *Spartina × townsendii*. *Aust. J. Plant Physiol.* **1979**, *5*, 831–838.

128. Subudhi, P. K.; Baisakh, N. *Spartina alterniflora* Loisel., a Halophyte Grass Model to Dissect Salt Stress Tolerance. *In Vitro Cell Dev. Biol. Plant* **2011**, *47*, 441–457.

129. Tomlinson, P. B. *The Botany of Mangroves*; Cambridge University Press: Cambridge, London; 1986; pp 413.

130. Torres, M. A.; Dangl, J. L. Functions of the Respiratory Burst Oxidase in Biotic Interactions, Abiotic Stress and Development. *Curr. Opin. Plant Biol.* **2005**, *8*, 397–403.

131. USEPA (US Environmental Protection Agency). *Phytoremediation of Contaminated Soil and Ground Water at Hazardous Waste Sites. Ground Water Issue, Office of Solid Waste and Emergency Response*; US Environmental Protection Agency: Washington, DC, 2001; pp 36; EPA/540/S-01/500.

132. Uphof, J. C. T. Halophytes. *Bot. Rev.* **1941**, *7*, 1–57.

133. Vera Estrella, R.; Barkla, B. J.; Garcia-Ramirez, L.; Pantoja, O. Salt Stress in *Thellungiella halophila* Activates Na^+ Transport Mechanisms Required for Salinity Tolerance. *Plant Physiol.* **2005**, *139*, 1507–1517.

134. Vinocur, B.; Altman, A. Recent Advances in Engineering Plant Tolerance to Abiotic Stress: Achievements and Limitations. *Curr. Opin. Biotechnol.* **2005**, *16*, 123–132.

135. Wang, B.; Lüttge, U.; Ratajczak, R. Effects of Salt Treatment and Osmotic Stress on V-ATPase and V-PPase in Leaves of the Halophyte *Suaeda salsa*. *J. Exp. Bot.* **2001,** *52,* 2355–2365.

136. Wang, W.; Vinocur, B.; Altman, A. Plant Responses to Drought, Salinity and Extreme Temperatures: Towards Genetic Engineering for Stress Tolerance. *Planta* **2003,** *218,* 1–14.

137. Wang, Y.; Chu, Y.; Liu, G.; Wang, M. H.; Jiang, J.; Hou, Y.; Qu, G.; Yang, C. Identification of Expressed Sequence Tags in an Alkali Grass (*Puccinellia tenuiflora*) cDNA Library. *J. Plant Physiol.* **2007,** *164,* 78–89.

138. Watson, M. C.; O'Leary, J. W. Performance of *Atriplex* Species in the San Joaquin Valley, California, under Irrigation and with Mechanical Harvests. *Agric. Ecosyst. Environ.* **1993,** *43,* 255–266.

139. Wu, L.; Fan, Z.; Guo, L.; Li, Y.; Chen, Z. L.; Qu, L. J. Over-expression of the Bacterial *nhaA* Gene in Rice Enhances Salt and Drought Tolerance. *Plant Sci.* **2005,** *168,* 297–302.

140. www.biologidiscussion.com (accessed Dec 15, 2016).

141. Yensen, N. P. *Preface: Halophyte Data Base: Salt Tolerant Plants and Their Uses*; United States Department of Agriculture, Agricultural Research Service; Washington DC, 1999. http://www.ussl.ars.usda.gov/pls/caliche/Halophyte.query (accessed Oct 10, 2016).

142. Zaier, H.; Ghnaya, T.; Lakhdar, A.; Baioui, R.; Ghabriche, R.; Mnasri, M.; Sghair, S.; Lutts, S.; Abdelly, C. Comparative Study of Pb-phytoextraction Potential in *Sesuvium portulacastrum* and *Brassica juncea*: Tolerance and Accumulation. *J. Hazard. Mater.* **2010,** *183,* 609–615.

143. Zhang, H. X.; Blumwald, E. Transgenic Salt-tolerant Tomato Plants Accumulate Salt in Foliage but not in Fruit. *Nat. Biotechnol.* **2001,** *19,* 765–768.

144. Zhang, J. Y.; Wang, Z. Y.; Jenks, M. A.; Hasegawa, P. M.; Jain, S. M. Recent Advances in Molecular Breeding of Forage Crops for Improved Drought and Salt Stress Tolerance. In: *Advances in Molecular Breeding towards Salinity and Drought Tolerance*; Jenks, M. A., Hasegawa, P. M., Mohan Jain, S., Eds.; Springer: Netherlands, 2007; pp 797–817.

145. Zhang, J.; Gowing, D. J.; Davies, W. J. ABA as a Root Signal in Root to Shoot Communication of Soil Drying. In: *Importance of Root to Shoot Communication in the Responses to Environmental Stress*; Davies, W. J., Jeffcoat, B., Eds.; British Society for Plant Growth Regulation: Bristol, 1989; pp 163–174.

146. Zhao, K. F. Desalinization of Saline Soils by *Suaeda salsa*. *Plant Soil* **1991,** *135,* 303–305.

147. Zhu, J. K. Regulation of Ion Homeostasis under Salt Stress. *Curr. Opin. Plant Biol.* **2003,** *6,* 441–445.

148. Zorb, C.; Schmitt, S.; Neeb, A.; Karl, S.; Linder, M.; Schubert, S. The Biochemical Reaction of Maize (*Zea mays* L.) to Salt Stress is Characterized by a Mitigation of Symptoms and not by a Specific Adaptation. *Plant Sci.* **2004,** *167,* 91–100.

149. Zorrig, W.; Rabhi, M.; Ferchichi, S.; Smaoui, A.; Abdelly, C. Phytodesalination: A Solution for Salt-affected Soils in Arid and Semi-arid Regions. *J. Arid. Land Stud.* **2012,** *22,* 299–302.

150. Zouari, N.; Saad, R. B.; Legavre, T.; Azaza, J.; Sabau, X.; Jaoua, M.; Masmoudi, K.; Hassairi, A. Identification and Sequencing of ESTs from the Halophyte Grass *Aeluropus littoralis*. *Gene* **2007,** *404,* 61–69.

CHAPTER 10

APPROACHES FOR ENHANCING SALT TOLERANCE IN SEED SPICES

ARVIND KUMAR VERMA, ANSHUMAN SINGH,
RAMESHWAR LAL MEENA, and BALRAJ SINGH

ABSTRACT

India continues to be the world leader in production of seed spices that are used to impart flavor and aroma to the food products. Seed spice crops, extensively grown in arid and semiarid tracts, have to face the compounded impacts of salinity, water scarcity, and climate change. Seed spices are moderately sensitive to salt stress; excess salt concentrations in the root zone often debilitate crop growth resulting in moderate to heavy reductions in economic yield. Salt tolerance of a plant describes its capacity to endure salinity stress without appreciable reductions in plant growth and yield. An understanding of the physiological mechanisms, morphological traits, and genetic mechanisms that impart salt tolerance may help develop management practices to maximize crop output under saline conditions. This chapter describes several conventional and improved techniques for salinity management in seed spices. Beginning with the selection of salt-tolerant planting material, the role of different techniques such as seed priming, nutrient management, microbial inoculants, plant growth regulators, and bio-stimulants in mitigating the salt hazard is discussed. It is concluded that a well thought out ensemble of agronomic manipulations can help realize high seed spice yields in salty soils.

10.1 INTRODUCTION

Seed spices comprise of a wide variety of plants that produce volatile and nonvolatile food additives. They are used to impart flavor and taste to the food and also for their health promoting benefits.[64] Seed spices have also been traditionally used as preservatives, colorants, and flavor enhancers. In many countries, they are used as the ingredients of traditional medicine. Both *in vitro* and *in vivo* studies have also revealed their antioxidant, digestion stimulant, antibacterial, antiinflammatory, antiviral, and anticarcinogenic activities.[113] In India, out of 63 spices grown in different parts, 20 are classified as seed spices with cumin (*Cuminum cyminum* L.), coriander (*Coriandrum sativum* L.), fenugreek (*Trigonella foenum-graecum* L., *Trigonella cornicu-lata* L.), and fennel (*Foeniculum vulgare* Mill) being cultivated in a sizeable area. Similarly, celery (*Apium graveolens* L.), nigella (*Nigella sativa* L.), dill (*Anethum graveolens* L., *Anethum sowa* Kurz), ajwain (*Trachyspermum ammi* Sprague), anise (*Pimpenella anisum* L.), and caraway (*Carum carvi* L.) are the minor seed spices grown in India and other parts of the world (Table 10.1).

TABLE 10.1   Important Seed Spices Cultivated in Different Parts of the World.

English name	Hindi name	Botanical name
Aniseed	Vilayati sounf	*Pimpinella anisum* L.
Caraway	Shahi jeera	*Carum carvi* L.
Carom seed	Ajwain	*Trachyspermum ammi* L.
Celery	Ajmoda	*Apium graveolens* L.
Coriander	Dhania	*Coriandrum sativum* L.
Cumin	Jeera	*Cuminum cyminum* L.
Dill	Sowa	*Anethum graveolens* L.
Fennel	Saunf	*Foeniculum vulgare* Mill.
Fenugreek	Methi	*Trigonella foenum-graecum* L.
Nigella	Kalaunji	*Nigella sativa* L.

Note: The English names given here are used throughout the text except in case of Carom seed for which the Hindi name *Ajwain* is used. For the minor crops not mentioned here, English names along with botanical name (only at the first mention) are given.

Most of the major seed spices included in this chapter are native to the land regions in vicinity of the Mediterranean Sea. Similarly, most of them are umbelliferous plants (i.e., aromatic flowering plants) belonging to the family Apiaceae with the exception of fenugreek (family Fabaceae) and nigella (family Ranunculaceae). In India, five major seed spices (i.e., cumin, coriander, fennel, fenugreek, and carom) occupy an area of about 1.45 million hectare (M ha) with total production of about 1.0 million tons (MT) and productivity of about 0.9 t ha^{-1}.[81] Historically, India has always been recognized as the "land of spices" and continues to be their largest producer, exporter and consumer in the world. Besides India, other major seed spice producers are Iran, Turkey, Egypt, Morocco, Canada, Pakistan, Romania, Soviet Union, Israel, China, Burma, and Thailand.

In India, Rajasthan and Gujarat states are known as the "Seed Spices Bowl" together contributing to more than 80% of the total seed spices produced in India. Other states where seed spices are commonly grown are Bihar, West Bengal, Uttar Pradesh, Madhya Pradesh, Odisha, Punjab, Karnataka, and Tamil Nadu. Over the years, the global demand for Indian spices has consistently increased so that exports were increased by about 29% in coriander, 70% in cumin, 3.1% in celery, 58% in fennel, 49% in fenugreek, and 97% in others with an overall increase of 62% in the recent past. Despite significant surge in the export volume in recent years, there is still a huge demand for Indian seed spices in the global market.[96] The growing predominance of India in global seed spice industry is evident by the fact that out of total global demand of seed spices (>100 thousand tons) about 60% are supplied by India alone.

Salinity is a major environmental stress that reduces soil's capability to produce food. As a generic term, salinity is used to refer to soils containing either excess soluble salts (i.e., saline) or excess exchangeable Na^+ (sodic) or both (saline-sodic). Often, saline and sodic soils are also underlain with marginal quality saline or sodic ground water unfit for irrigation. Predominance of excess soluble salts raises the saturation paste salinity of soil (EC_e > 4 dS m^{-1}) leading to osmotic stress and specific ion toxicities (Na^+ and Cl^-) in plants.

In sodic soils, more than 15% of the exchange sites are occupied by Na^+ ions giving rise to high exchangeable sodium percentage (ESP > 15) conditions leading to soil structure loss and poor water transmission through the soil profile. Globally, the problem of salinization is especially severe in arid and semiarid regions of the world where faulty and over irrigation coupled with the neglect of land drainage are major drivers of excessive salt accumulation in the crop root zone. On the contrary, salinization of agricultural lands in dry and rain-fed regions is primarily ascribed to the removal of perennial vegetation to grow the annual crops. Problems of irrigation-induced as well as dry land salinities arise largely due to inappropriate anthropogenic interventions and unscientific land management to augment the crop productivity at the expense of soil and environmental quality.[102]

Traces of soluble salts and harmful ions such as Na^+ are naturally found in almost all cultivated soils. In some regions of the world, certain natural factors result in the excessive concentration of such geogenic salts; and this condition is referred to as primary salinity. However, in most of the cases, primary salinity affected lands remain in a manageable condition and pose limited constraints to crop production. In contrast as previously mentioned, unscientific on-farm water management often leads to secondary soil salinity characterized by heavy salt accumulation and subsoil waterlogging. Currently, over 800 M ha of global agricultural land are affected by primary salinity. In addition, human-induced secondary salinization affects the productivity of about 77 M ha of world's crop lands. Of the total secondary salinity affected area, about 45.4 M ha are in irrigated commands of arid regions and the rest are distributed in rain-fed areas.[124] In India, salt-affected soils cover approximately 6.73 M ha area which is projected to increase to 16.20 M ha by 2050 pointing to the fact that rate of salinization will surpass the reclamation efforts.[53]

Besides salt-induced land degradation, several other factors such as population growth, urban and industrial expansion and pervasive land use have, over time, led to reduced availability of productive lands for crop production. Fresh water, a key input in agricultural production, is not only becoming a scarce resource but the problem of poor quality water has also significantly increased in the last few decades. Climate change impacts are likely to further reduce the availability and/or quality of land and fresh water for agriculture in the coming decades necessitating agricultural expansion into marginal environments.[102] In so far as salt affected lands are concerned, a suit of measures can be successfully employed to harness their productivity.

This chapter discusses the response of seed spices to salt stress, various adverse effects of salinity on plant physiological relations and productivity of these crops. Besides, this chapter also discusses some conventional and the advanced techniques of salinity management in seed spices to augment the productivity under saline conditions.

10.2   PLANT RESPONSES TO SALT STRESS

A majority of the cultivated crops, referred to as glycophytes, are sensitive to salinity and fail to survive when salt concentration exceeds the threshold value. In contrast, plants which not only profusely grow but also produce viable economic yields in hyper-saline environments are described as halophytes. One of the basic differences between halophyte and glycophytes is that halophytes have the ability to survive under severe salt shock[87] by modulating the physiological activities at cellular, tissue and whole plant levels. Apparently, a plant is adjudged to be salt tolerant if it exhibits the ability to withstand high salt concentrations in the root zone without any appreciable reduction in the plant growth and yield.[101]

Under high salinity conditions, salt-induced hyper-ionic and hyperosmotic effects often severely affect the plant growth and in certain cases may result in complete crop loss; especially in salt sensitive genotypes. Adverse effects of salinity are often severe in arid regions compared to both semiarid and humid climates apparently due to scanty rainfall and very high evapotranspiration demand in arid zones. Poor soil and water management practices often accentuate salt stress to the levels that crop production virtually becomes impossible in salt affected arid lands.[35] A set of factors including high osmotic stress due to low soil water potential, Na^+ and/or Cl^- toxicities and imbalanced nutrition account for most of the adverse effects on plant growth under saline conditions.

Elevated salt levels raise the osmotic pressure of the soil solution and create the physiological drought, that is, though water remains available, plants fail to extract it from the soil.[77,102] Nutritional imbalance and specific ion effects are mainly due to high accumulation of Na^+ and Cl^- hampering the availability of other essential elements like K^+, Ca^+, NO_3^- or P that in turn affects the activities of key enzymes.[13]

10.2.1 MECHANISMS OF SALINITY TOLERANCE

Salt tolerance describes the ability of plants to endure salinity stress without any appreciable reductions in growth, yield, and produce quality. Different morphological, physiological, and genetic mechanisms contribute to salt tolerance. For example at the cellular level, salt tolerance depends on the ability of growing plant cells either to exclude the toxic ions or an increased tolerance to minimize the adverse effects on cellular processes. Some of such mechanisms are briefly summarized in this section.

10.2.1.1 OSMOTIC ADJUSTMENT

Generally, osmotic stress suppresses the plant growth in salt-stressed plants regardless of their capacity to exclude the salt. Partitioning of salt from the cytoplasm into the vacuole creates a strong osmotic gradient across the vacuolar membrane. Salinized plants tend to balance this gradient by increasing accumulation of solute molecules in the cytoplasm, a process known as osmotic adjustment,[68] which helps the plants to adapt to salinity by maintaining the cell turgor and volume. Different kinds of compatible solutes have been identified from the plants exposed to salinity. They exhibit more or less similar properties including a low polar charge, high solubility, and large hydration shell.[84] Besides maintaining the cell turgor, compatible solutes are also implicated in stabilizing the active conformation of cytoplasmic enzymes and thus protecting them against inactivation by inorganic ions.[105] Compatible solutes include compounds such as proline, glycine-betaine, and other related quaternary ammonium compounds, pinitol, mannitol, and sorbitol. Notwithstanding the advantages offered by such cell benign molecules, their adequate production comes at the cost of metabolic expense as plants have to spend a significant portion of energy otherwise to be used to support the active growth.[46] As a consequence, plants tend to accumulate high amounts of inorganic ion from the growing substrate to prevent the energy expenditure as comparatively far lesser energy is required for the osmotic adjustment through inorganic ions than that conferred by organic molecules synthesized in the cell.[116] Excessive ion accumulation, however, tends to debilitate the plants by altering the vital cell processes.[47]

10.2.1.2 SALT INCLUSION VERSUS EXCLUSION

Most of the glycophytes fail to utilize the salt transported to the leaves from the root leading to progressive increase in salt concentration, slower growth and the eventual leaf death. The genotypes having ability to exclude the salt, even partially, exhibit greater salt tolerance than those lacking this mechanism. Compared to the glycophytes, vast majority of halophytes employ salt inclusion as a strategy to achieve osmotic balance with the external medium.[111]

10.2.1.3 Na^+/K^+ DISCRIMINATION

Besides salt exclusion, selective ion uptake from the growing medium greatly affects the ability of plants to withstand the salt damage. Salt tolerant genotypes often show the preferential accumulation of K^+ over Na^+. Such genotypes possess the ability to discriminate Na^+ from K^+ and thus succeed in substituting Na^+ for K^+ for root uptake.[97] Maintenance of adequate levels of K^+ in the young expanding tissues is often linked to salt tolerance even in salt sensitive genotypes.[106]

10.3   RESPONSES OF SEED SPICES TO SALINITY

The adverse effects of salt on plant growth and development vary with the magnitude and duration of salinity as well as the crop growth stage. Owing to the complexity of the trait, the genetic and physiological bases of salt tolerance are not yet fully understood in most of the crop plants. Often, several genes controlling salinity tolerance in a given plant species interact strongly with environmental factors. It is due to this reason that genetic variation can only be demonstrated indirectly by measuring the responses of different genotypes. In practice, measurements on growth parameters and economic yield are widely used to estimate the plant response to salinity, especially at moderate salinities. In most of the annual crops, salt-induced reduction in biomass production (which usually directly correlates with the yield) compared with the control plants over a period of time is considered to be a reliable estimate of the relative impact of salinity on different crops/genotypes. Similarly, in many perennial crops, percent reduction in plant survival is often adjudged to be a good indicator of salt-induced damage.[76,77,102] Different effects of salinity on physiological activities, metabolism, growth, and economic yield in seed spices are presented in this section.

10.3.1   PLANT PHYSIOLOGY UNDER SALINE CONDITIONS

The detrimental effects of salt on plant growth and development are direct result of the changes at cellular level which impact key physiological activities. Initially, soil water deficit caused by the osmotic stress followed by high salt concentrations inside the plant lead to an array of changes in plant metabolism. It then affects seed germination, seedling establishment, biomass production, yield, and quality of the final product.

10.3.1.1   CELL MEMBRANE STABILITY

In most of the glycophytes, reduced availability of CO_2 in chloroplast cells and accumulation of toxic ions such as Cl^- lead to overproduction of reactive oxygen species (ROS) and lipid peroxidation destabilizing the cell membranes. Cell membrane stability (CMS) is a routinely used measure to distinguish the salt tolerant and salt susceptible genotypes in crops as higher CMS seems to confer salt tolerance. Electrolyte leakage (EL) often provides an indirect, but reasonable estimate of the relative CMS under stress conditions.[107] In fenugreek, leaf membrane permeability was gradually increased with addition of NaCl (0, 60, 120, and 180 mM) into irrigation water.[78] Increase in salinity from 0 to 25 mM NaCl led to almost two-fold increase in EL in chili (*Capsicum annuum* L.) plants. At 50 mM NaCl level, plants registered about two-and-half fold increase in EL over control.[98] In salinized fenugreek plants (100 or 200 mM NaCl), membrane permeability, lipid peroxidation and chlorophyll loss eventually caused the leaf senescence.[29] Salt stress induced a noticeable increase in malondialdehyde (MDA: a measure of lipid peroxidation) content in fennel. MDA level was increased by 18% at 50 mM NaCl and 29% at 100 mM NaCl compared with control indicating salt-induced cell membrane damage.[43] Hydrogen peroxide and MDA contents were increased in the leaves of salinized coriander plants.[59]

10.3.1.2   PLANT WATER STATUS

Salt-stressed plants show reduced water uptake and content. The water potential of plants thus decreases with increasing salinity.[47] A genotype exhibiting minimum reductions in leaf turgidity will be better able to support turgor dependent processes such as growth and stomatal activity in stressful environments.[75] To overcome the water shortage, plants undergo osmotic regulation by increasing the negativity of the osmotic potential of the leaf sap. Relative water content (RWC) of a leaf is a commonly used measure of its hydration status (i.e., actual water content) under stress conditions. By providing a measurement of the "water deficit" of the leaf, RWC indicates the degree of stress experienced by the plants. RWC combines the effects of both leaf water potential and osmotic adjustment as an index of plant water status.

Ajwain plants did not show appreciable reduction in RWC up to 4 dS m^{-1} salinity compared to control. However, RWC was declined by about 33% and 40% at 8 and 12 dS m^{-1} salinity levels, respectively, than the control.[66] Salt-induced decrease in plant water consumption suppressed plant biomass production and seed yield of fennel.[99] Fennel cv. RMt-1 showed decline in RWC with increasing salinity (50–200 mM NaCl) levels with the decrease in RWC becoming more severe with the increasing duration of the salt exposure.[58] Lower soil osmotic potential decreased the water uptake by salinized sweet basil (*Ocimum basilicum* L.) plants resulting in marked reductions in growth and biomass production.[92] Water content of seedling tissues was 44.84 and 46.01%, respectively, in cumin and fennel at 10 dS m^{-1} salinity compared to about 53% in the seedlings grown in normal soils in both the crops.[50]

10.3.1.3 LEAF CHLOROPHYLL

Leaf chlorophyll levels (*a*, *b*, and total chlorophyll), in conjunction with other physiological parameters, are frequently assessed to understand the extent of salt injury in plants. Although leaf chlorophyll levels tend to drop in salt-stressed plants, low-to-moderate salinity may sometimes enhance its synthesis in certain genotypes. Again, two types of chlorophyll (*a* and *b*) may be differentially affected by salinity. For example, NaCl stress reduced chlorophyll-*b* but increased chlorophyll-*a* content in summer savory (*Satureja hortensis* L.).[4] Decline in leaf chlorophyll levels under saline conditions is ascribed to the breakdown of chlorophyll pigments and the instability of the pigment-protein complex. Salt ions also interrupt with *de novo* synthesis of structural proteins of the chlorophyll molecule.[54] Nitrogen deficiency, insufficient leaf turgor and increased activity of chlorophylase enzyme also account for the low chlorophyll levels in salinized plants.[3] Salinity-induced changes in leaf chlorophyll levels in seed spices (Table 10.2) indicate that factors such as growing season, degree of salinity and presence of chemicals other than salts also influence the extent of chlorophyll loss.

TABLE 10.2 Salt-induced Changes in Leaf Chlorophyll Concentration in Seed Spices.

Crop	Findings	References
Dill	Total leaf chlorophyll significantly increased in the autumn regardless of salinity. In contrast, in spring, it increased at low salinity (2–4 dS m^{-1}) but decreased at higher salinity (6–8 dS m^{-1})	[109]
Fennel	Salt treatment (50 and 75 mM NaCl) caused significant reductions in fresh and dry weights, and chlorophyll (*a* and *b*) and β-carotene contents in seedlings	[82]
Fenugreek	Saline irrigation (0, 60, 120, and 180 mM NaCl) decreased plant dry matter and chlorophyll production with the highest reductions observed at 180 mM NaCl than control. Application of silicon (0.2 mM Na$_2$SiO$_3$) partially mitigated the salt injury	[78]
Nigella	NaCl (150 mM) stressed plants showed decrease in leaf RWC, leaf area and chlorophyll content resulting in the lower P_N and dry matter production. Kinetin (10 μM) spray appreciably reduced most of these adverse effects and upregulated antioxidant enzyme activities leading to higher yield compared to the untreated stressed plants	[100]

10.3.1.4   CARBON ASSIMILATION

Reduced availability of photo-assimilates is one of the major causes of plant growth reduction in saline soils. Depending on plant species, salinity may either decrease the leaf surface area for photosynthesis or may result in a reduced rate of photosynthesis.[108] In so far as decrease in photosynthetic rate is concerned, either stomatal or non-stomatal factors may be responsible.

Reduced stomatal activity, restricted CO_2 diffusion to chloroplast and alterations in photosynthetic metabolism are the stomatal factors which hamper carbon assimilation in salinized plants. Similarly, oxidative stress may also contribute to the lower photosynthetic rate.[22] In general, a decrease in stomatal activity is recognized as a major cause of salt-induced photosynthetic decline in crop plants. Salt stress may also affect the ability of a plant to transport the photo-assimilates to the growing leaves.[114] In saline soils, plants tend to reduce the stomatal conductance (gs) to arrest the transpiration rate (E) for efficient water use as well as reduced salt loading via transpiration stream. The fact that reduced E limits salt loading into the foliage may be an efficient adaptive trait, especially when salt stress lasts only for a limited period of time. Although plants resort to stomatal closure mainly to curtail the water loss, yet it comes at the cost of net photosynthesis (P_N) as CO_2 diffusion to chloroplast cells is substantially lowered.

Stomatal limitations appeared to be responsible for decreased P_N in NaCl stressed (40 and 80 mM) coriander plants as salinity reduced gs, E and internal CO_2 concentration.[28] Low salinity (12.5 mM Na_2SO_4) improved salt and drought tolerance in several accessions of coriander due to decrease in E and the consequent increase in water use efficiency.[18] Despite significant reductions in plant growth and seed yield, increasing NaCl levels (40, 80, 120, and 160 mM) did not affect P_N and E rates indicating a negative correlation between gas exchange parameters and salt tolerance in ajwain.[10] Saline irrigation (10 dS m^{-1}) differentially affected leaf metabolism in cumin genotypes with genotype "UC-198" exhibiting the highest decline (14.8%) in starch concentration followed by "RZ-209" (11.1%), while it was least in the tolerant genotype "RZ-19" probably due to higher reductions in P_N in the former ones.[39]

Low NaCl (25 mM) had negligible effect on carbon assimilation in celery, but intermediate salinity (100 mM NaCl) decreased gs, while high NaCl levels (300 mM) diminished the carboxylation capacity and thus adversely affecting P_N.[32] Although sufficient data are not available to reach to conclusive evidence regarding the gas exchange characteristics in seed spices under salinity, yet it appears that relative impacts of salt on photosynthesis and water use may vary with the agro-climatic conditions, species/genotype as well as the level and duration of salinity.

10.3.1.5   MINERAL UPTAKE AND ASSIMILATION

Plants require different macro- and micronutrients for metabolic activities such as photosynthesis and the production of secondary metabolites. Salinity-induced decrease in soil osmotic potential and increased concentrations of toxic ions (Na^+ and Cl^-) restrain the uptake of water and essential nutrients by plants. Na^+ and Cl^- concentrations consistently increased in roots, stems and leaves of marjoram (*Origanum majorana* L.) plants in the presence of NaCl (0, 50, 100, and 150 mM). Although root Na^+ decreased, leaf and stem Na^+ was considerably increased with increasing NaCl levels; yet, Cl^- content was invariably increased in all plant parts albeit at a far lower rate than that of Na^+. Salt treatments resulted in decreased K^+ concentration in roots at 100 mM NaCl and in stems at 150 mM NaCl. The K^+ content in leaves was not affected by salt. Ca^{2+} content decreased in roots and stems but was not much affected in leaves. Data suggested a higher allocation of K^+ and Ca^{2+} in leaves than in stems and roots at 100 mM and higher NaCl levels.[15]

In Bishop's weed (*Ammi majus* L.), Na^+ and Cl^- were increased in both shoots and roots while K^+ and Ca^{2+} were decreased with increase in salinity. Plants maintained markedly higher K^+/Na^+ ratio in the shoots than in roots, and the ratio remained >1 even at the highest (160 mM NaCl) salt level.[10] Similar results have been reported in ajwain where Na^+ and Cl^- shoots and roots increased, whereas K^+ and Ca^{2+} consistently decreased with increase in salt level. Plants showed favorable K^+/Na^+ and Ca^{2+}/Na^+ ratios in shoots than in roots with K^+/Na^+ ratio remaining above 1 even at 120 mmol L^{-1} NaCl.[11] Irrigation water salinity above 4 dS m^{-1} significantly enhanced leaf Na^+ content in fennel. The highest and the lowest leaf Mg^{2+} levels were recorded in normal and highest salinity levels (12 dS m^{-1}), respectively. Leaf K^+ and Ca^{2+} contents were not significantly affected by salinity.[99] Increasing salinity (2, 6, and 10 dS m^{-1}) had little influence on plant growth, water relations and the macronutrient levels in celery plants. However, salt stress markedly enhanced the accumulation of Na^+ and Cl^- in the mature leaves but to a much lesser extent in the young leaves.[86]

The data presented here are illustrative: salt effects on mineral nutrition are complex and the results often significantly vary with the crop and experimental conditions. In most of the cases, Na^+ and Cl^- ions tend to increase resulting in the reduced availability of other essential ions such as K^+ and Ca^{2+}. Ion accumulation patterns in different plant parts may be different. Some species often maintain adequate K^+ and Ca^{2+} levels in the foliar and root tissues to counteract adverse effects of Na^+ and Cl^-, especially at low to moderate salt levels. Salinity levels beyond tolerance threshold, however, lead to the breakdown of salt exclusion and avoidance mechanisms resulting in abrupt increases in Na^+ and Cl^- concentrations and a range of salt injury symptoms.

Salinity almost invariably leads to reduced N availability in plants.[36] Similarly, in most of the cases, salt-stressed plants show deficient P levels with adverse consequences for photosynthesis and other energy-dependent growth processes.[83] Elevated Na^+ levels in growing medium usually suppress K^+ supply to plants. K^+ not only acts as an essential cofactor for many enzymes but also plays critical roles in cellular osmotic balance and stomatal regulation.[76] Available evidence suggests that some crop genotypes tend to preferentially accumulate K^+ to partly overcome Na^+ toxicity.[15] In majority of the crop plants, salinity lowers Ca^{2+} levels leading to membrane permeability, EL and other harmful effects.[10] In saline soils, availability of micronutrients such as Fe, Zn, Mn, and Cu may either increase or decrease.[71]

10.3.1.6   ANTIOXIDANT DEFENSE SYSTEM

As previously mentioned, salt stress leads to increased levels of harmful ROS such as superoxide (O_2^-), hydrogen peroxide (H_2O_2), and hydroxyl radicals (OH) which, by oxidizing lipids, proteins and nucleic acids, impair the cell structure and functions. To overcome the oxidative damage, plants activate different enzymatic [superoxide dismutase (SOD), catalase (CAT), ascorbate peroxidase (APX), glutathione reductase (GR), and peroxidase (PER)] and nonenzymatic antioxidants (e.g., ascorbic acid, α-tocopherol, glutathione, and carotenoids) to scavenge the free radicals.[31] Coriander plants sprayed with 10 μM triacontanol showed higher salt tolerance due to enhanced levels of SOD, CAT, APX, and PER.[59] Inoculation with mycorrhizal strain *Glomus intraradices* improved the concentration and activity of SOD, CAT, PX, APX, and GR enzymes in salt treated (0, 50, 100, and 200 mM NaCl) fenugreek plants and thus reduced oxidative damage compared to nonmycorrhizal plants.[31]

The activity of antioxidant enzymes was significantly increased in response to NaCl stress (150 mM) in Indian mustard (*Brassica juncea* L. Czern.). Exogenous application of salicylic acid (SA) further enhanced the antioxidant enzyme levels.[118] Salinity upregulated CAT levels in fenugreek plants. Seed treatment with ascorbic acid, on the other hand, improved PER and esterase (Est) levels but decreased the CAT activity.[19] The activity of PER and polyphenol oxidase enzymes was increased in response to NaCl-induced salinity (0, 50, 100, 150, and 200 mM) enhancing the tolerance of fenugreek plants against oxidative damage.[88] Certain phenolic compounds also show antioxidant activities and thus prevent the damage caused by ROS, which are inevitably produced when plant metabolism is impaired by environmental stresses. Water deficit enhanced the phenolic content in cumin seeds.[90]

10.3.1.7   OSMOREGULATION

Proline and glycine betaine (GB) are the two major osmoprotectants synthesized by salinized plants to maintain the cellular osmotic balance. Proline is also implicated in removing the ROS, stabilizing the cell organelles and in buffering the cellular redox potential under stress conditions. Salinity stress responsive genes, whose promoters contain proline responsive elements (ACTCAT), are also known to be induced by proline.[23] While soluble and insoluble carbohydrates and proline contents were increased, and other free amino acids were declined with increasing NaCl concentrations (up to 300 mM) in black cumin.[49] NaCl stress (160, 200, 240, and 280 mM) markedly increased the proline and other amino acids in anise and coriander. In contrast, only proline increased but other amino acids decreased in caraway and cumin plants.[123] Shoot proline content significantly increased at 100 mM NaCl level in fennel.[9] Although leaf proline did not increase up to 80 m mol L^{-1}, yet it significantly increased at the highest NaCl level (120 m mol L^{-1}) suggesting a positive role of proline in salt stress adaptation in ajwain.[11] Both proline and GB consistently increased in NaCl treated ajwain plants. Proline and GB levels were about 3.5-fold and 2-fold higher at 150 mM NaCl than control.[117]

Treatment of Gamma ray (0, 25, 50, 100, and 150 Gy) irradiated fenugreek seeds with GB (50 mM) significantly improved the levels of nucleic acids in plants compared to untreated (irradiated) control indicating its protective role against oxidative stress.[73] Mannitol, a compatible solute, increased in NaCl (25, 100, and 300 mM) treated celery plants due to increased activity of mannose-6-phosphate reductase (a key enzyme in mannitol biosynthesis); especially in the young and fully expanded leaves.[32] Soluble sugars increased in the leaves of salinized chili pepper (*Capsicum frutescens* L.) plants. Tolerance of cultivars "Awlad Haffouzz" and "Korba" to the highest NaCl (12 g L^{-1}) level was linked to the corresponding higher concentrations of soluble sugars.[121]

10.3.1.8 HORMONAL CHANGES

Plant hormones, also referred to as phytohormones, comprise of diverse organic compounds which modulate different plant physiological responses. They are produced in one part and are translocated to other parts in trace amounts; they exert a modifying influence on different physiological activities. Chemically synthesized compounds mimicking hormonal action when applied externally are called plant growth regulators (PGRs). Phytohormones may either exhibit a growth promoting (auxins, gibberellins, and cytokinin) or growth inhibiting effect [abscisic acid (ABA) and ethylene].[62] It is seen that the levels of growth promoting phytohormones commonly decline while those of growth inhibitors increase in response to salinity. As a consequence, different growth promoting PGRs are often externally applied to alleviate the salt stress in plants. For example, exogenous application of ABA reduces the ethylene-induced leaf abscission probably by decreasing the accumulation of toxic Cl$^-$ ions in leaves.[55] Several other PGRs have also been successfully used to enhance the salt tolerance in crop plants (see Section 10.4.6) in this chapter.

Salt treatment leads to increase in ABA levels which, in most cases, are positively correlated with leaf and/or soil water potential implying that elevated ABA levels are mainly due to water deficit. It has been shown that only slight increase in ABA concentration enhances the plant adaptation to stresses as higher concentrations may prove inhibitory to plant growth.[120] Salt-stressed ABA to overcome adverse effects of osmotic stress on photosynthesis, growth and plants tend to synthesize translocation of assimilates that also plays a critical role in the expression of salt-induced genes.[112]

ABA accumulation may also favor higher uptake of Ca^{2+} and reduced absorption of Cl$^-$ ions resulting in better CMS and low ethylene-induced leaf abscission, respectively, in salt-stressed plants.[93,120] Although little evidence is available to support the salt stress protective role of ABA in seed spices, yet some experimental findings have established the plant growth promoting effects of ABA application in crops such as coriander possibly due to tangible improvements in stress tolerance.[94] In most of the cases, accumulated ABA rapidly disappears subsequent to stress release to allow the normal plant metabolism.[120]

10.3.2   EFFECTS ON PLANT GROWTH

Salinity adversely affects plant growth and yield by decreasing the seed germination, plant stand, biomass production, flowering, and seed formation in seed spices. Considerable variation is noted with respect to genotypes, crop growth stages as well as the level and duration of the salt treatment.

10.3.2.1   SEED GERMINATION AND SEEDLING ESTABLISHMENT

In most crop plants, seed germination, seedling emergence, and plant survival are particularly sensitive to salinity stress. High levels of salt may either partially or completely inhibit the seed germination. Salt stress lowers the seed germination primarily by decreasing the osmotic potential of the soil solution which in turn retards the water absorption by seeds. Excessive accumulation of Na^+ and Cl^- ions proves toxic to the seed embryo. In four umbelliferous seed spices exposed to different salt levels, 50% reduction in seed germination (G_{50}) was seen at 120 mM NaCl in anise, at 150 mM NaCl in coriander, and at 200 mM NaCl in caraway and cumin each.[122] Seedling dry weight of anise and coriander was decreased with increasing salinity, but seedling growth of caraway and cumin appeared to be stimulated by NaCl concentrations up to 80 mM.[123] Increasing salinity reduced the seed germination and vegetative growth in coriander and fennel crops. Germination and plant height were not affected up to 5.0 dS m^{-1} salinity. Fennel showed higher salt tolerance than coriander under high salinity levels.[67] Salinity significantly decreased percent seed germination, rate of germination and the root and shoot lengths in fenugreek seedlings. Root length was more affected than shoot length.[57]

Although seed germination was not significantly affected up to 200 mM NaCl, yet further increase to 300 mM NaCl decreased germination by about 70% compared to control.[8] Different accessions of fenugreek exposed to varying NaCl levels (0, 4, 6, 8, and 10 dS m^{-1}) germinated and grew at low salinity, but germination was delayed as NaCl concentration increased and germination was almost completely inhibited at 10 dS m^{-1} NaCl.[7] Salinity significantly reduced germination percentage and the growth of shoots and roots in ajwain. Seed treatment with 0.2% chitosan solution, however, improved the salt tolerance in ajwain as evident from increased shoot and root lengths, shoot dry weight, and RWC in salinized plants (4–12 dS m^{-1}).[66] Salt stress severely reduced seed germination in ajwain with virtually no germination observed at very high salinity (18 dS m^{-1}).[74]

Seed germination and seedling emergence rate, seedling length and weight and radical length were unaffected up to 9 dS m^{-1} salinity in nigella. While seed germination rate was very high (94.8%) in normal soils, it was reduced to about 90% at 15 dS m^{-1} and consistently decreased with increasing salinity. No seed germination was seen at a salinity level of 36 dS m^{-1}.[34]

10.3.2.2   VEGETATIVE GROWTH AND BIOMASS PRODUCTION

As with other crops, salt stress invariably causes significant yield reduction unless appropriate interventions are applied to overcome the salt hazard. Under field conditions, salt-stressed crops show stunted and sparse growth resulting in lower productivity (Fig. 10.1).

FIGURE 10.1   (See color insert.) Sparse and stunted plant growth in fenugreek in a saline water irrigated field at Hisar, Haryana, India (Courtesy: Rameshwar Lal Meena).

Saline irrigated (0, 60, 120, and 180 mM NaCl) fenugreek plants showed decline in plant dry matter and chlorophyll production with the highest reductions observed at 180 mM NaCl level than control. Application of silicon (0.2 mM Na_2SiO_3) partially mitigated the salt injury.[78] Increasing salinity in irrigation water (0.25, 1, 2, 4, 6, 8, 10, and 12 dS m^{-1}) resulted in decreased water consumption, plant height, plant fresh and dry weights, and seed yield in fennel. Based on these observations, threshold salt tolerance for fennel was found to be 2.64 dS m^{-1} (salinity at which yield starts to decrease) with slope of 4.5% (yield decline with per unit increase in electrical conductivity) pointing to the moderately salt sensitive nature of fennel.[99]

NaCl (1500 ppm) treated thyme (*Thymus vulgaris* L.) showed significant decrease in plant height, number of branches and fresh and dry plant mass.[27] NaCl-induced salinity (0, 3, and 6 dS m^{-1}) hampered shoot length and fresh and dry shoot weight in nigella but root length increased with increasing salinity. In control plants, the mean shoot length was 12.58 cm, while at 3 and 6 dS m^{-1} salinity levels, it reduced to 8.4 and 6.05 cm, respectively.[52] Salinized (12 dS m^{-1}) fennel plants showed considerable decrease in plant height, leaf weight, and bulb weight which declined by 33%, 49%, and 71%, respectively, compared to control.[24]

Increasing salt concentrations (0, 2, 4, 6, 8, and 10 dS m^{-1}) significantly decreased the seedling height, shoot and root lengths, seed germination percentage, germination rate, fresh, and dry seedling weight and seed vigor index in fennel and cumin.[50] Pot experiments on tolerance of seed spices revealed that decrease in the seed and biomass yield of coriander decreased 6% and 28% and 3% and 23%, respectively. Similar results have been obtained in case of fennel although it emerges that under similar conditions fennel is relatively more tolerant to saline water irrigation then coriander.[115] The experiments also established benefits of conjunctive use as a management option to use highly saline waters.

Similar results for the fennel crop were later reported by Meena et al.[69] additionally revealing the benefits of organic manures in mitigating the adverse effects of saline irrigation. In sodic soils, emergence of secondary branches and seed setting are considerably depressed in coriander resulting in low seed yield than yield obtained in normal soils. The cation composition of stover revealed Na^+ inclusion mechanism with narrow K^+/Na^+ and Ca^{2+}/Na^+ ratios. The study further revealed that the fennel crop could tolerate medium level of sodicity.[40]

10.3.2.3   SEED YIELD

Increasing NaCl concentrations (0, 40, 80, 120, and 160 mM) caused significant reductions in the fresh and dry weights of shoots and roots as well as seed yield in Bishop's weed.[10] Although NaCl stress (0, 40, 80, and 120 mmol L^{-1}) significantly reduced both plant dry mass and seed yield in ajwain, yet the seed yield was relatively more adversely affected. While shoot dry biomass decreased by about 27%, reduction in seed yield was almost 50% than control at 120 mmol L^{-1} NaCl.[11] Salt stress (0, 40, and 80 mM NaCl) significantly decreased the fresh and dry plant weight, umbels per plant, 1000 seed weight and seed yield in fennel. Addition of sodium silicate (0.5 and 1 mM) into saline solution mitigated these adverse effects.[89] Salinized plants (75 mM NaCl) of coriander showed about 36% decrease in seed yield than control.[80] With increase in salinity from 0.3 to 9 dS m^{-1}, the mean biological and seed yields decreased from 550.2 to 268.6 g m^{-2} and 105.5 to 40.2 g m^{-2}, respectively, in black cumin.[34]

Experiments conducted on dill on saline black soils of different unirrigated sites in Khanpur, Warsada, and Bamangam villages of Anand district showed that the *in-situ* salinity was negatively correlated with seed yield of dill. The study revealed that the dill can profitably be grown on saline black soils (up to 4–6 dS m^{-1} salinity) without any irrigation during winter season.[48] With a unit increase in salinity beyond this, the yield reduction was 0.043 t ha^{-1}. It is also revealed that the water of submarginal quality (salinity ~4 dS m^{-1}) can be used without any significant yield reduction.[48] Chauhan[21] revealed that while seed and stover yields of fennel decreased with increasing salinity of irrigation water, 1000 seed weight was not significantly affected even up to 8 dS m^{-1} salinity (Table 10.3). Salinity–induced reduction in plant growth and seed yield in cumin crop are given in Figure 10.2. Table 10.4 indicates effects of salinity levels on yield attributing characteristics and yield of fennel crop.

TABLE 10.3   Reduction in Seed and Total Biomass Yield (%) of Coriander and Fennel Under Different Modes of Saline Water Irrigation.

Irrigation water salinity (dS m^{-1})	Coriander (t ha^{-1})		Fennel (t ha^{-1})	
	Biomass	Seed yield	Biomass	Seed yield
3.7	–	–	–	–
3.7 and 8.7[a]	3	6	2	5
8.7	23	28	15	21

"–": Assumed no reduction in yield.

[a]Alternate irrigation beginning with 3.7 dS m^{-1}.

FIGURE 10.2 (See color insert.) Adverse effects of salinity on vegetative growth (left) and seed formation (right) in cumin crop (Jaisalmer, Rajasthan, India).

TABLE 10.4 Effect of Salinity Levels on Yield Attributing Characteristics and Yield of Fennel Crop.

Salinity of irrigation water	Weight of 1000 seeds			Seed yield			Stover yield		
	2010–2011	2011–2012	Mean	2010–2011	2011–2012	Mean	2010–2011	2011–2012	Mean
dS m^{-1}		g				t ha^{-1}			
BAW	6.06	6.04	6.05	1.17	1.11	1.14	5.52	5.35	5.44
4	6.01	6.00	6.00	1.09	1.05	1.07	5.43	5.30	5.37
6	6.01	6.00	6.00	1.03	1.01	1.02	5.42	5.28	5.35
8	5.92	5.90	5.91	0.92	0.89	0.91	4.67	4.61	4.64
CD at 5%	NS	NS	–	0.117	0.2.10	–	0.89	0.87	–

BAW, best available water.

10.3.2.4 SEED OIL YIELD AND QUALITY

Although increasing salinity (0–120 mmol L^{-1} NaCl) adversely affected the seed yield in ajwain, yet seed oil concentration was not affected.[11] Seed oil yield in fennel consistently decreased with the increasing salinity (0–100 mM NaCl).[9] Seed essential oil production decreased with increasing NaCl levels (0–160 mM) in Bishop's weed.[10] Despite a significant reduction in seed yield, salinized coriander plants showed increase in essential oil yield with increasing NaCl levels. Essential oil yield increased by 77% and 84% at 50 and 75 mM NaCl, respectively, than control. Again, the major constituents of oil (linalool and camphor) also increased with increasing salinity.[80]

Dill plants grown under saline conditions (0, 4, 8, and 12 dS m^{-1}) showed significant reductions in leaf dry weight per plant. Yet, flower and seed dry weights did not significantly decrease and essential oil yield even increased with increasing salinity.[44] NaCl stress (60 mM) decreased the seed yield of nigella by about 58% than control. Essential oil yield increased by 0.53, 0.56, and 0.72% at 20, 40, and 60 mM NaCl level, respectively, over control. Salinity enhanced the linoleic acid percentage but did not affect the unsaturated degree of the fatty acids pool and thus oil quality.[20] In spite of almost 50% decrease in seed yield, salinity did not significantly decrease the seed oil percentage in black cumin.[34] Moderate sodicity levels (ESP—20 and 30) slightly suppressed the seed yield in coriander and fennel but seed oil yield was not affected in both the crops.[104] These data tend to show that salt-stress-induced reductions in seed yield are compensated, to a great extent, by increased essential oil levels in different crops.

10.4   SALINITY MANAGEMENT OPTIONS

A range of solutions have been proposed to enhance the salt tolerance in crop plants or, alternatively, by modifying the growing conditions to get higher yields. In seed spices, enhanced adaptation to salt stress can be achieved by developing the high yielding and salt tolerant genotypes, presowing seed treatments, exogenous applications of chemicals and PGRs, supplemental nutrition and the use of microbial inoculants. Use of salt tolerant cultivars (STCs) is an economically viable and environment-friendly approach to obtain stable yields in saline soils. Cultivation of plants having inherent ability to endure the high salt levels implies reduced dependence on chemical soil ameliorants to make the root zone conditions favorable to plant growth and yield. Such measures are briefly discussed in the succeeding sections.

10.4.1   GENETIC IMPROVEMENT FOR SALT TOLERANCE

Conventional breeding methods as well as advanced biotechnological techniques have been used with varying degree of successes to develop STCs in crop plants. Although traditional breeding approaches have led to the identification of a few STCs in different crops, yet the rate of success through conventional improvement is rather slow due to polygenic inheritance of the salt tolerance. Owing to their cross pollinated nature (except fenugreek), seed spices exhibit high degree of heterozygosity and polymorphism for different traits. Nonetheless, low genetic variability has been reported for salinity tolerance in most of such species. To overcome this problem, mass screening of seed spice crops is often the method of choice to identify salt tolerant lines for commercial cultivation as well as for use as parents in breeding programs. Over the years, such screening programs have led to the identification of a handful of STCs in different seed spice crops (Table 10.5).

TABLE 10.5   Genotypic Differences in Seed Spices Under Salt Stress.

Crop	Finding	References
Cumin	Genotype RZ-19 showed higher salt tolerance than UC-198 and RZ-209. Adverse effects of salinity (10 dS m^{-1}) on total chlorophyll, K$^+$/Na$^+$ ratio, soluble protein, free amino acids, starch, reducing sugars, and nitrate reductase activity were considerably less in RZ-19 than sensitive genotypes	[39]
Coriander	On the basis of seed germination and seedling growth under NaCl stress (25, 50, 75, and 100 mM), cultivars PD-21 and Kalmi were found to be highly and moderately salt tolerant, respectively while Pant Haritma was categorized as salt sensitive	[60]
Fenugreek	Among the eight cultivars exposed to varying NaCl levels (0, 4, 6, 8, and 10 g L^{-1}), Berber was found to be the most salt tolerant and Damar I as the most salt sensitive	[38]
	Only two accessions (F-3 and F-98) showed about 40% germination stress index, while it was low/very low (5–34%) in other six accessions at 10 dS m^{-1} salinity	[7]
Fennel	Among the four cultivars (HF-107, Local BRS, NDF-7, NDF-9) grown in moderately sodic soils (ESP ~ 30), HF-107 and NDF-7 produced the highest (1956 kg ha^{-1}) and the lowest (1078 kg ha^{-1}) seed yields	[40]
Sweet fennel	Cultivars Dulce, ZefaFino, and Selma differentially responded to saline irrigation. ZefaFino recorded the highest plant length, leaf number, plant fresh weight, leaf yield, and essential oil content compared to other cultivars	[119]
Chili pepper	Higher proline accumulation in leaves was linked to high salt tolerance (NaCl 12 g L^{-1}) of cultivars Awlad Haffouzz and Korba in comparison to cultivar Souk jedid	[121]

Induced mutagenesis may also be tried to create genetic variability for identifying the salt tolerant mutants. Dimethyl sulfate (DMS; 0, 1000, and 2000 ppm) treatment of fennel seeds exposed to saline irrigation (0, 50, 100 mM NaCl) resulted in two salt tolerant mutants in the second generation. While mutant 1 was obtained from the 1000 ppm DMS treatment, mutant 2 was isolated from 2000 ppm DMS treated seeds grown under 100 mM

NaCl. Both the mutants were superior to the wild type (control plants) with respect to plant height, branching, leaf production, stem diameter, total seeds weight, oil percentage, and seed oil yield.[43]

Advances in molecular marker technology in the recent past have made it possible to dissect the quantitative trait loci (QTLs) linked to salt tolerance. QTLs represent those genomic regions in the plants where the major genes controlling salt tolerance are located. Mapping of such genomic stretches helps improve the selection efficiency by differentiating the candidate genes from the other genes. QTL mapping may also be helpful in identifying the candidate genes responsible for modulating plant salt tolerance during different growth stages such as seed germination, vegetative growth, and seed setting.[37]

Genetic transformation technology has also made rapid strides in the recent times and has significantly contributed to crop improvement for desirable traits. The *SbNHX1* gene, cloned from halophyte *Salicornia brachiata* was introgressed in cumin using *Agrobacterium*-mediated transformation method. *SbNHX1* gene encodes a vacuolar Na^+/H^+ antiporter and is involved in the compartmentalization of excess Na^+ ions into the vacuole and maintenance of ion homeostasis. Transgenic lines (L3, L5, L10, and L13), overexpressing the *SbNHX1* gene, showed higher photosynthetic pigments (chlorophyll *a*, *b*, and carotenoids) and lower electrolytic leakage, lipid peroxidation, and proline contents compared to wild type plants under salinity stress.[85]

10.4.2 SELECTION OF SALT-TOLERANT CROPS AND CULTIVARS

Similar to other economically important plants, seed spice crops vary with each other in their relative salt tolerance. Some of the findings presented in Table 10.6 point to the distinct salt tolerance threshold levels ranging from salt sensitive to highly salt tolerant in different species.

Different seed spice crops employ varied mechanisms to overcome the salt injury. High salinity tolerance in celery is attributed to the higher accumulation of mannitol which acts as an efficient osmo-protectant[32] and the partitioning of toxic Na^+ and Cl^- ions into older leaves so that young and actively photo-synthesizing leaves are protected from the salt ions.[86] Although a few studies have indicated moderate salt sensitivity in fennel,[2,99] Graifenberg et al.[45] found markedly lower salt tolerance threshold ($EC_e \sim 1.5$ dS m^{-1}) in the tested culti-vars (Monte Bianco and Everest) which lacked the ion partitioning mechanism to prevent Na^+ and Cl^- accumulation in the young leaves.[45] It implies that salt stress adaptation traits may vary across genotypes in a particular species.

Elevated proline levels and the maintenance of high shoot K^+/Na^+ ratios contribute to the high salt tolerance in nigella[49] and ajwain,[10] respectively.

TABLE 10.6 Relative Salt Tolerance in Seed Spices.

Crop	Salt tolerance	References
Ajwain	Moderately salt tolerant; salt stress adaptation is linked to the maintenance of high K^+/Na^+ and Ca^{2+}/Na^+ ratios in shoots and roots	[11]
Bishop's weed	Moderately salt tolerant; salt tolerance is achieved by the maintenance of high shoot K^+/Na^+ ratio and proline accumulation in shoots	[10]
Celery	Salt tolerant species; mannitol accumulation and partitioning of Na^+ and Cl^- ions in the older leaves contribute to salt tolerance	[32,86]
Dill	Highly salt tolerant; NaCl-induced salinity up to 12 dS m^{-1} does not affect seed and oil yield	[44]
Fennel	Moderately salt sensitive; threshold salinity at which yield starts to decline is around 2.5 dS m^{-1}	[99]

10.4.3 SEED PRIMING

Different types of seed priming methods (e.g., hydro-priming, osmo-priming, and priming with chemicals and PGRs) have proved effective in overcoming the salt-induced decrease in seed germination and seedling establishment in seed spices. Coriander seeds primed with aerated solutions of 0.13 M NaCl and $CaCl_2$ showed improved salt tolerance as evident by higher seed germination, seedling emergence, better growth and improved levels of K^+ and Ca^{2+} than unprimed seeds.[14] The beneficial effects of hydration–dehydration seed treatment on germination of many seed spice crops have been demonstrated.[95] Seed priming with distilled water (hydro-priming) and Zn_2SO_4 shortened the mean germination time in cumin under NaCl stress. However, hydro-primed seeds recorded significantly higher final germination rate as well as better seedling growth suggesting toxicity caused by Zn_2SO_4 to the seed embryo.[79]

Hydro-priming for 6 h followed by matrix priming with synthetic soil for 72 h hastened the seed germination in cumin; over 90% of the primed seeds germinated on 4th day after inoculation compared to delayed germination (on 8th day) in untreated seeds. Genotype "GC-4" was found to be more responsive to priming than "RZ-209".[95] Different methods of seed priming [SA, gibberellic acid (GA), and hydro-priming] improved the length and dry weights of plumule and radicle in NaCl stressed (50 and 100 mM) fenugreek. However, both SA and GA treatments gave better results as evident from increased chlorophyll and proline levels and decrease in MDA content than salinized (both unprimed and hydro-primed) seedlings.[33] SA (0.00001 mM) primed ajwain seeds showed improved germination percentage and rate, radicle and plumule lengths and seed vigor up to 12 dS m⁻¹ salinity level.[74] Seed priming with SA (0.2 mM) and KNO₃ (3%) improved different germination traits under osmotic stress in black cumin.[17]

10.4.4   NUTRIENT MANAGEMENT

Organic manure application enhanced the salt tolerance of fennel cultivar "Dulce." Vegetative growth parameters such as plant height, leaf number, plant fresh and dry weights, leaf yield, macronutrient levels (N, P, and K), K^+/ Na^+ ratio and proline contents increased by organic manuring.[2] Application of magnetite iron and sheep manure improved the plant height, branching, plant fresh and dry weights, umbels per plant, fruit yield per plant, essential oil percentage in fruits, and essential oil yield per plant in saline irrigated fennel.[70] Increased levels of nitrogen and phosphorus fertilizers enhanced the seed and stover yields in nigella under sodic conditions.[41] Application of 90 kg ha⁻¹ N and 50 kg ha⁻¹ P resulted in 37% higher seed yield in coriander while 80 kg ha⁻¹ N and 25 kg ha⁻¹ P resulted in significantly higher (67%) seed yield in fennel over unfertilized control plants in moderately sodic soils (ESP ~ 20 and 30).[42] Plant growth enhancement due to application of organic inputs in fennel crop is shown in Figure 10.3.

FIGURE 10.3   Organic manures improved plant growth in saline irrigated (~3 dS m⁻¹, right) fennel compared to control (saline irrigated but untreated, left) under arid conditions of Hisar, Haryana, India.

10.4.5 MICROBIAL INOCULANTS

It has been shown that arbuscular mycorrhizal fungi (AMFs) improve the plant growth under stress conditions by increasing the nutrient availability, maintaining a favorable K^+/Na^+ ratio and osmotic adjustment by accumulation of compatible solutes such as proline or soluble sugars. AMF-treated plants exhibit higher photosynthesis and water use efficiency under salt stress compared to stressed but nonmycorrhizal plants.[6]

Inoculation with AMF *Glomus intraradices* significantly decreased the extent of salt-induced (0, 50, 100, and 200 mM NaCl) ultrastructural alterations such as protoplasm shrinkage, grana disorganization, thylakoid swelling, chloroplast membrane disintegration, and aggregation of chromatin in nucleus in fenugreek.[103] Lesser damage in AMF-inoculated plants may be attributed to higher osmolyte (GB and sugars) and polyamines concentrations, and more and bigger plastoglobules.[30] *Glomus mosseae* inoculation enhanced shoot and root dry weights, leaf area, photosynthetic pigments, and soluble sugars in NaCl-stressed (0, 25, 50, and 100 mM) chili (cv. Zhongjiao 105) plants.[65] Plant growth-promoting rhizobacteria (PGPR) is a group of bacteria present in the rhizosphere. PGPR protect the plants against various biotic and abiotic stresses. They alleviate salt stress in plants by enhancing the biosynthesis of phytohormones and enzyme 1-aminocyclopropane-1-carboxylate (ACC) deaminase, which suppresses the endogenous levels of ACC—a precursor of ethylene. Decrease in ACC availability hampers ethylene production and thus lessens the extent of ethylene-induced leaf abscission and other harmful effects.[26]

Seed treatment with biofertilizers containing different bacterial and fungal isolates improved the salt tolerance in coriander. Nitroxin (including *Azospirillium* and *Azotobacter*) and Super nitroplus (including *Aspirillium*, *Bacillus subtilis*, *Pseudomonas fluorescens*) applications resulted in better plant growth and higher levels of chlorophyll pigments (*a*, *b*, and *a+b*); especially up to 60 mM NaCl stress.[91] Seed inoculation with *Rhizobium meliloti* strain "FRS-7" gave the best results with regard to shoot length, shoot dry weight, shoot total nitrogen, root length, root dry weight, root total nitrogen, seed yield, 1000 grain weight, number of root nodules, and nodule fresh and dry weights in fenugreek crop grown under semiarid conditions. The performance of this strain was even better than 20 kg N ha^{-1}. Seed yields obtained with FRS-7 during two consecutive years were about 36.8% and 45.9% higher over control.[102]

PGPR (*Paenibacillus polymyxa* and *Azospirillum lipoferum*) treatment reduced salt stress and resulted in significant improvements in plant height, number of branches, plant fresh and dry weights, essential oil percentage, and essential oil yield in sweet basil.[1] Use of PGPR (*Azospirillum lipoferum* and/or *Bacillus megaterium*) significantly improved the plant growth, photosynthetic pigments and yield in saline irrigated (~3 and 6 dS m⁻¹) ajwain plants.[63]

10.4.6   PLANT GROWTH REGULATORS AND BIOSTIMULANTS

As the levels of plant-growth-promoting phytohormones decline in salt-stressed plants, several studies have been conducted to assess the effect of synthetic PGRs to overcome the salt injury in different crops. SA, a plant bioregulator, has been extensively used to alleviate drought and salt stresses in crop plants. Application of gibberellic acid (GA_3) has been found to counteract some of the adverse effects of salinity in plants.[110] Jasmonates can also play an important role in plant salt tolerance. Polyamines, known to elicit diverse physiological activities in plants such as cell division, tuber formation, root initiation, flower development, and fruit ripening, are also implicated in improving abiotic stress tolerance.[51] Brassinosteroids have received attention due to their beneficial influence in improving the salt tolerance.[5] Foliar application of indole acetic acid and kinetin alleviated the adverse effects of salinity such as decrease in plant biomass production, chlorophyll content, and RWC in maize plants exposed to 100 mM NaCl.[61]

Application of SA improved antioxidant enzyme levels, leaf water potential, RWC, leaf pigments and osmolytes, and seed essential oil content in fennel genotypes exposed to drought stress.[12] SA (5 and 10 μM) treatment also increased the drought tolerance of nigella seedlings as evident from negligible injury symptoms in the pretreated plants.[56] Fenugreek cultivars "Deli Kabul" and "Kasuri" showed significant reductions in plant biomass in saline (100 mM NaCl) than normal soils. However, biomass production was less reduced in "Deli Kabul" as evident from higher shoot fresh weight compared to "Kasuri." Foliar spray of SA (100 ppm) overcame salt-induced growth reduction in both the cultivars.[16]

Coriander plants sprayed with 10 μM triacontanol showed higher salt tolerance due to enhanced levels of antioxidant enzymes (SOD, CAT, APX, and PER).[59] Application of a nonionic surfactant (3 ppm) enhanced the photosynthetic activity in fenugreek plants at moderate and high salinities

(6–10 dS m^{-1}).[25] Soil drenching with humic acid (3 g L^{-1}) and foliar spray of dry yeast (20 and 25 g L^{-1}) effectively alleviated salt stress in fennel by increasing K$^+$ accumulation and lowering Na$^+$ uptake by the plants.[72]

10.5 CONCLUSION

Critical review of the adverse effects of soil/water salinity on germination and seedling establishment, vegetative growth and biomass production, seed yield and seed oil yield, and quality of seed spice crops reveal that not only crops but varieties within the crops respond differently to salt stress. Apparently, selection of salt tolerant crops and cultivars is suggested as one of the major intervention to improve production and productivity of seed spices cultivated in saline environments. A case is build-up to suggest genetic improvement of these crops. Until then, some nonstructural interventions such as seed priming, nutrient management, use of microbial inoculants, and application of PGRs and biostimulants can be pursued to blunt the adverse effects of salts on crop productivity.

10.6 SUMMARY

The introductory part of this chapter highlights the role of seed spices in human diets and healthcare followed by a brief account of the recent trends in production and export of seed spices with special reference to India. Introduction section also briefly deals with expanding salinity problem and the need to overcome the salt stress for higher crop productivity. While discussing the plant responses to salt stress, various adverse effects of salinity on plant physiological relations with special reference to seed spices have been dealt with.

Continuing with plant response to salinity, adverse effects on seed germination and seedling establishment, vegetative growth and biomass production, seed yield and seed oil yield and quality of seed spice crops are critically analyzed. Toward the end of this chapter, conventional approaches and the advanced techniques of salinity management in seed spices are thoroughly discussed. Some of the options discussed include selection of salt tolerant crops and cultivars, seed priming, nutrient management, microbial inoculants and application of PGRs and biostimulants. Genetic improvement for salt tolerance is included as it may be prove to be the most eco-friendly approach to manage saline environments in the future.

KEYWORDS

- abiotic stresses
- osmoregulation
- physiological mechanisms
- salinization
- seed spices
- stomatal activity
- water potential

REFERENCES

1. Abo-Kora, H. A.; Mohsen, M. A. Reducing Effect of Soil Salinity Through Using Some Strains of Nitrogen Fixers Bacteria and Compost on Sweet Basil Plant. *Int. J. Pharm. Tech. Res.* **2016,** *9*, 187–214.

2. Abou El-Magd, M. M.; Zaki, M. F.; Habou Hussein, S. D. Effect of Organic Manure and Different Levels of Saline Irrigation Water on Growth, Green Yield and Chemical Content of Sweet Fennel. *Aust. J. Basic Appl. Sci.* **2008,** *2*, 90–98.

3. Aggrwal, K. B.; Ranjan, J. K.; Rathore, S. S.; Saxena, S. N.; Mishra, B. K. Changes in Physical and Biochemical Properties of Fenugreek (*Trigonella* sp. L.) Leaf During Different Growth Stages. *Int. J. Seed Spices* **2013,** *3*, 31–35.

4. Akbari, S.; Kordi, S.; Fatahi, S.; Ghanbari, F. Physiological Responses of Summer Savory (*Satureja hortensis* L.) Under Salinity Stress. *Int. J. Agric. Crop Sci.* **2013,** *5*, 1702–1708.

5. Ali, B.; Hayat, S.; Ahmed, A. 24-epibrassinolide Ameliorates the Saline Stress in Chickpea (*Cicer arientum* L). *Environ. Exp. Bot.* **2007,** *59*, 217–223.

6. Aliabadi, F. H.; Lebaschi, M. H.; Hamidi, A. Effects of *Arbuscular Mycorrhizal* Fungi, Phosphorus and Water Stress on Quantity and Quality Characteristics of Coriander. *J. Adv. Natl. Appl. Sci.* **2008,** *2*, 55–59.

7. Al-Saady, N. A.; Khan, A. J.; Rajesh L.; Esechie, H. A. Effect of Salt Stress on Germination, Proline Metabolism and Chlorophyll Content of Fenugreek (*Trigonella foenum gracium* L.). *J. Plant Sci.* **2012,** *7*, 176–185.

8. Asaadi, A. M. Investigation of Salinity Stress on Seed Germination of *Trigonella foenum-graecum*. *Res. J. Biol. Sci.* **2009,** *4*, 1152–1155.

9. Ashraf, M.; Akhtar, N. Influence of Salt Stress on Growth, Ion Accumulation and Seed Oil Content in Sweet Fennel. *Biol. Plant* **2004,** *48*, 461–464.

10. Ashraf, M.; Mukhtar, N.; Rehman, S.; Rha, E. S. Salt-induced Changes in Photosynthetic Activity and Growth in a Potential Medicinal Plant Bishop's Weed (*Ammi majus* L). *Photosynthetica* **2004,** *42*, 543–550.

11. Ashraf, M.; Orooj, A. Salt Stress Effects on Growth, Ion Accumulation and Seed Oil Concentration in an Arid Zone Traditional Medicinal Plant Ajwain [*Trachyspermum ammi* (L.) Sprague]. *J. Arid Environ.* **2006,** *64*, 209–220.

12. Askari, E.; Ehsanzadeh, P. Drought Stress Mitigation by Foliar Application of Salicylic Acid and Their Interactive Effects on Physiological Characteristics of Fennel (*Foeniculum vulgare* Mill.) Genotypes. *Acta Physiol. Plant* **2015**, *37*, 1–14.

13. Assem, K. S.; Hussein, H. A. E.; Hussein, A. H.; Awaly, B. S. Transformation of the Salt Tolerance Gene BI-GST into Egyptian Maize Inbred Lines. *Arab J. Biotechnol.* **2009**, *13*, 99–114.

14. Aymen, E. M.; Cherif, H. Influence of Seed Priming on Emergence and Growth of Coriander (*Coriandrum sativum* L.) Seedlings Grown Under Salt Stress. *Acta Agric. Slovenica* **2013**, *1*, 41–47.

15. Baatour, O.; Kaddour, R.; Wannes, W. A.; Lachaâl, M.; Marzouk, B. Salt Effects on the Growth, Mineral Nutrition, Essential Oil Yield and Composition of Marjoram (*Origanum majorana*). *Acta Physiol. Plant.* **2010**, *32*, 45–51.

16. Babar, S.; Siddiqi, E. H.; Hussain, I.; Hayat Bhatti, K.; Rasheed, R. Mitigating the Effects of Salinity by Foliar Application of Salicylic Acid in Fenugreek. *Physiol. J.* **2014**, *2014*, 1–6.

17. Balouchi, H.; Dehkordi, S. A.; Ehnavi, M. M.; Behzadi, B. Effect of Priming Types on Germination of *Nigella sativa* Under Osmotic Stress. *South Western J. Horticult., Biol. Environ.* **2015**, *6*, 1–20.

18. Bashtanova, U. B.; Flowers, T. J. Effect of Low Salinity on Ion Accumulation, Gas Exchange and Postharvest Drought Resistance and Habit of *Coriandrum sativum* L. *Plant Soil* **2012**, *355*, 199–214.

19. Behairy, R. T.; El-Danasoury, M.; Craker, L. Impact of Ascorbic Acid on Seed Germination, Seedling Growth, and Enzyme Activity of Salt-stressed Fenugreek. *J. Med. Active Plants* **2012**, *1*, 106–113.

20. Bourgou, S.; Bettaieb, I.; Saidani, M.; Marzouk, B. Fatty Acids, Essential Oil and Phenolics Modifications of Black Cumin Fruit Under NaCl Stress Conditions. *J. Agric. Food Chem.* **2010**, *58*, 12,399–12,406.

21. Chauhan, S. K. Effect of Saline Water Irrigation in Fennel (*Foeniculum vulgare* Mill.) Grown in Semi-arid Condition. *TECHNOFAME*, **2016**, *5*, 141–143.

22. Chaves, M. M.; Flexas, J.; Pinheiro, C. Photosynthesis Under Drought and Salt Stress: Regulation Mechanisms from Whole Plant to Cell. *Ann. Bot.* **2009**, *103*, 551–560.

23. Chinnusamy, V.; Jagendorf, A.; Zhu, J. K. Understanding and Improving Salt Tolerance in Plants. *Crop Sci.* **2005**, *45*, 437–448.

24. Cucci, G.; Lacolla, G.; Boari, F.; Cantore, V. Yield Response of Fennel (*Foeniculum vulgare* Mill.) to Irrigation with Saline Water. *Acta Agric. Scand. Sect. B—Soil Plant Sci.* **2014**, *64*, 129–134.

25. Dadresan, M.; Luthe, D. S.; Reddivari, L.; Chaichi, M. R.; Yazdani, D. Effect of Salinity Stress and Surfactant Treatment on Physiological Traits and Nutrient Absorption of Fenugreek Plant. *Commun. Soil Sci. Plant Anal.* **2015**, *46*, 2807–2820.

26. Egamberdieva, D.; Lugtenberg, B. Use of Plant Growth-promoting Rhizobacteria to Alleviate Salinity Stress in Plants. In *Use of Microbes for the Alleviation of Soil Stresses;* Miransari, M, Ed.; Springer: New York, 2014; pp 73–96.

27. El-Din, A. E.; Aziz, E. E.; Hendawy, S. F.; Omer, E. A. Response of *Thymus vulgaris* L. to salt stress and alar (B9) in newly reclaimed soil. *J. Appl. Sci. Res.* **2009**, *5*, 2165–2170.

28. Elhindi, K. M.; El-Hendawy, S.; Abdel-Salam, E.; Schmidhalter, U.; Rahman, S.; Hassan, A. A. Foliar Application of Potassium Nitrate Affects the Growth and Photosynthesis in Coriander (*Coriander sativum* L.) Plants Under Salinity. *Progr. Nutr.* **2016**, *18*, 63–73.

29. Evelin, H.; Giri, B.; Kapoor, R. Contribution of Glomus Intraradices Inoculation to Nutrient Acquisition and Mitigation of Ionic Imbalance in NaCl-stressed *Trigonella foenum-graecum*. *Mycorrhiza* **2012,** *22,* 203–217.

30. Evelin, H.; Giri, B.; Kapoor, R. Ultrastructural Evidence for AMF Mediated Salt Stress Mitigation in *Trigonella foenum-graecum*. *Mycorrhiza* **2013,** *23,* 71–86.

31. Evelin, H.; Kapoor, R. Arbuscular Mycorrhizal Symbiosis Modulates Antioxidant Response in Salt-stressed *Trigonella foenum-graecum* Plants. *Mycorrhiza* **2014,** *24,* 197–208.

32. Everard, J. D.; Gucci, R.; Kann, S. C.; Flore, J. A.; Loescher, W. H. Gas Exchange and Carbon Partitioning in the Leaves of Celery (*Apium graveolens* L.) at Various Levels of Root Zone Salinity. *Plant Physiol.* **1994,** *106,* 281–292.

33. Farahmandfar, E.; BagheriShirvan, M.; AzimiSooran, S.; Hoseinzadeh, D. Effect of Seed Priming on Morphological and Physiological Parameters of Fenugreek Seedlings Under Salt Stress. *Int. J. Agric. Crop Sci.* **2013,** *5,* 811–815.

34. Faravani, M.; Davazdehemami, S.; Gholami, B. A. The Effect of Salinity on Germination, Emergence, Seed Yield and Biomass of Black Cumin. *J. Agric. Sci.* **2013,** *58,* 41–49.

35. Farooq, M.; Hussain, M.; Wakeel, A.; Siddique, K. H. M. Salt Stress in Maize: Effects, Resistance Mechanisms, and Management. A Review. *Agron. Sustainable Dev.* **2015,** *35,* 461–481.

36. Feigin, A. Fertilization Management of Crops Irrigated with Saline Water. *Plant Soil* **1985,** *89,* 285–299.

37. Foolad, M. R. Comparison of Salt Tolerance During Seed Germination and Vegetative Growth in Tomato by QTL Mapping. *Genome* **1999,** *42,* 727–734.

38. Forawi, H. A. S.; Elsheikh, E. A. Response of Fenugreek to Inoculation as Influenced by Salinity, Soil Texture, Chicken Manure and Nitrogen. *Univ. Khartoum J. Agric. Sci.* **1995,** *3,* 77–89.

39. Garg, B. K.; Burman, U.; Kathju, S. Genotypic Differences in Seed Yield, K:Na Ratio and Leaf Metabolism of Cumin (*Cuminum cyminum* L.) Under Salt Stress. *J. Spices Aromatic Crops* **2015,** *12,* 113–119.

40. Garg, V. K. Sodicity Affects Growth, Yield and Cation Composition of Fennel (*Foeniculum vulgare* Mill.). *J. Spices Aromatic Crops* **2012,** *21,* 130–135.

41. Garg, V. K.; Malhotra, S. Response of *Nigella sativa* L. to Fertilizers Under Sodic Soil Conditions. *J. Med. Aromatic Plant Sci.* **2008,** *30,* 122–125.

42. Garg, V. K.; Singh, P. K.; Katiyar, R. S. Yield, Mineral Composition and Quality of Coriander (*Coriandrum sativum*) and Fennel (*Foeniculum vulgare*) Grown in Sodic Soil. *Indian J. Agric. Sci.* **2004,** *74,* 221–223.

43. Gehan, M. G.; Alhamd, M. F. A. Induction of Salt Tolerant Mutants of *Foeniculum vulgare* by Dimethyl Sulphate and Their Identification Using Protein Pattern and ISSR Markers. *Alexandria J. Agric. Res.* **2015,** *60,* 95–109.

44. Ghassemi-Golezani, K.; Zehtab-Salmasi, S.; Dastborhan, S. Changes in Essential Oil Content of Dill (*Anethum graveolens*) Organs Under Salinity Stress. *J. Med. Plants Res.* **2011,** *5,* 3142–3145.

45. Graifenberg, A.; Botrini, L.; Giustiniani, L.; Lipucci di Paola, M. Salinity Affects Growth Yield and Elemental Concentration of Fennel. *Hortic. Sci.* **1996,** *31,* 1131–1134.

46. Greenway, H.; Munns, R. Mechanisms of Salt Tolerance in Non-halophytes. *Annu. Rev. Plant Physiol.* **1980,** *31,* 149–190.

47. Gupta, B.; Huang, B. Mechanism of Salinity Tolerance in Plants: Physiological, Biochemical, and Molecular Characterization. *Int. J. Genomics* **2014,** *2014,* 1–18.

48. Gururaja Rao, G.; Nayak, A. K.; Chinchmalatpure, A. R.; Singh, R.; Tyagi, N. K. *Resource Characterization and Management Options for Salt Affected Black Soils of Agro—Ecological Region V of Gujarat State*. Central Soil Salinity Research Institute, Regional Research Station, Anand 388001, Gujarat, India; Technical Bulletin 1/2001, 2001; p 83.

49. Hajar, A. S.; Zidan, M. A.; Al-Zahrani, H. S. Effect of Salinity Stress on the Germination, Growth and Some Physiological Activities of Black Cumin (*Nigella sativa* L.). *Arab Gulf J. Sci. Res.* **1996**, *14*, 445–454.

50. Hokmalipour, S. Effect of Salinity and Temperature on Seed Germination and Seed Vigor Index of Chicory (*Chichoriumintynus* L.), Cumin (*Cuminiumcyminium* L.) and Fennel (*Foeniculum vulgare*). *Indian J. Sci. Technol.* **2105**, *8*, 1–9.

51. Hopkins, W. G. *Introduction to Plant Physiology*, 2nd ed.; Wiley: New York, 1999; p 512.

52. Hussain, K.; Majeed, A.; Nawaz, K.; Hayat, K. B.; Nisar, M. F. Effect of Different Levels of Salinity on Growth and Ion Contents of Black Seeds (*Nigella sativa* L). *Curr. Res. J. Biol. Sci.* **2009**, *1*, 135–138.

53. ICAR-CSSRI. *ICAR-Central Soil Salinity Research Institute Vision 2050*. Indian Council of Agricultural Research, New Delhi, 2015; p 31.

54. Jaleel, C. A.; Sankar, B.; Sridharan, R.; Panneerselvam, R. Soil Salinity Alters Growth, Chlorophyll Content, and Secondary Metabolite Accumulation in *Catharanthus roseus*. *Turkish J. Biol.* **2008**, *32*, 79–83.

55. Javid, M. G.; Sorooshzadeh, A.; Moradi, F.; Sanavy, S. A. M. M.; Allahdadi, I. The Role of Phytohormones in Alleviating Salt Stress in Crop Plants. *Aust. J. Crop Sci.* **2011**, *5*, 726–734.

56. Kabiri, R.; Nasibi, F.; Farahbakhsh, H. Effect of Exogenous Salicylic Acid on Some Physiological Parameters and Alleviation of Drought Stress in *Nigella sativa* Plant Under Hydroponic Culture. *Plant Protect. Sci.* **2014**, *50*, 43–51.

57. Kapoor, N.; Arif, M.; Pande, V. Antioxidative Defense to Salt Stress *in Trigonellafoenum graecum* L. *Curr. Discov.* **2013**, *2*, 123–127.

58. Kapoor, N.; Pande, V. Effect of Salt Stress on Growth Parameters, Moisture Content, Relative Water Content and Photosynthetic Pigments of Fenugreek Variety RMt-1. *J. Plant Sci.* **2015**, *10*, 210–221.

59. Karam, E. A.; Keramat, B. Foliar Spray of Triacontanol Improves Growth by Alleviating Oxidative Damage in Coriander Under Salinity. *Indian J. Plant Physiol.* **2017**, *22*, 120–124.

60. Kaur, G.; Kumar, A. Influence of Salinity Stress on Germination and Early Seedling Growth of Coriander (*Coriandrum sativum* L.) Cultivars. *Int. J. Sci. Res.* **2017**, *5*, 564–568.

61. Kaya, C.; Tuna, A. L.; Dikilitas, M.; Cullu, M. A. Response of Some Enzymes and Key Growth Parameters of Salt-stressed Maize Plants to Foliar and Seed Application of Kinetin and Indole Acetic Acid. *J. Plant Nutr.* **2010**, *33*, 405–422.

62. Kaya, C.; Tuna, A. L.; Yokaş, I. The Role of Plant Hormones in Plants Under Salinity Stress. In *Salinity and Water Stress*; Springer: Netherlands, 2009; pp 45–50.

63. Khalil, S. E. Alleviating Salt Stress in *Thymus capitatus* Plant Using Plant Growth-promoting Bacteria (PGPR). *Int. J. ChemTech Res.* **2016**, *9*, 140–155.

64. Kochhar, K. P. Dietary Spices in Health and Diseases. *Indian J. Physiol. Pharmacol.* **2008**, *52*, 106–122.

65. Latef, A. A. H. A.; Chaoxing, H. Does Inoculation with Glomusmosseae Improve Salt Tolerance in Pepper Plants? *J. Plant Growth Regul.* **2014,** *33,* 644–653.

66. Mahdavi, B.; Rahimi, A. Seed Priming with Chitosan Improves the Germination and Growth Performance of Ajowan (*Carum copticum*) Under Salt Stress. *EurAsian J. BioSci.* **2013,** *7,* 69–76.

67. Mangal, J. L.; Yadava, A.; Singh, G. P. Effects of Different Levels of Soil Salinity on Germination, Growth, Yield and Quality of Coriander and Fennel. *South Indian Hortic.* **1986,** *34,* 26–31.

68. McCue, K. F.; Hanson, A. D. Drought and Salt Tolerance: Towards an Understanding and Application. *Trends Biotechnol.* **1990,** *8,* 358–362.

69. Meena, R. L.; Ambast, S. K.; Gupta, S. K.; Chinchmalatpure, A. R.; Sharma, D. K. Performance of Fennel (*Foeniculum vulgare* Mill) as Influenced by Saline Water Irrigation and Organic Input Management in Semi-arid Conditions. *J. Soil Salinity Water Qual.* **2014,** *6,* 52–58.

70. Mohsen, M. M. A.; AbeerKassem, H. M. Influence of Magnetic Iron and Organic Manure on Fennel Plant Tolerance Saline Water Irrigation. *Int. J. PharmTech Res.* **2016,** *9,* 86–102.

71. Moreno, D. A.; Pulgar, G.; Romero, L. Yield Improvement in Zucchini Under Salt Stress: Determining Micronutrient Balance. *Sci. Hortic.* **2000,** *86,* 175–183.

72. Mostafa, G. G. Improving the Growth of Fennel Plant Grown Under Salinity Stress Using Some Biostimulants. *Am. J. Plant Physiol.* **2015,** *10,* 77–83.

73. Moussa, H. R.; Jaleel, C. A. Physiological Effects of Glycine Betaine on Gamma-irradiated Stressed Fenugreek Plants. *Acta Physiol. Plant.* **2011,** *33,* 1135–1140.

74. Movaghatian, A.; Khorsandi, F. Germination of *Carum copticum* Under Salinity Stress as Affected by Salicylic Acid Application. *Ann. Biol. Res.* **2014,** *5,* 105–110.

75. Mullan, D.; Pietragalla, J. Leaf Relative Water Content. In *Physiological Breeding II: A Field Guide to Wheat Phenotyping*; CIMMYT, Mexico, 2012; pp. 25–27.

76. Munns, R. Genes and Salt Tolerance: Bringing them Together. *New Phytol.* **2005,** *167,* 645–663.

77. Munns, R.; James, R. A.; Läuchli, A. Approaches to Increasing the Salt Tolerance of Wheat and Other Cereals. *J. Exp. Bot.* **2006,** *57,* 1025–1043.

78. Nasseri, M.; Arouiee, H.; Kafi, M.; Neamati, H. Effect of Silicon on Growth and Physiological Parameters in Fenugreek (*Trigonellafoenum graceum* L.) Under Salt Stress. *Int. J. Agric. Crop Sci.* **2012,** *21,* 1554–1558.

79. Neamatollahi, E.; Bannayan, M.; Darban, S. A.; Ghanbari, A. Hydropriming and Osmopriming Effects on Cumin (*Cuminum cyminum* L.) Seeds Germination. *World Acad. Sci. Eng. Technol.* **2009,** *33,* 526–529.

80. Neffati, M.; Sriti, J.; Hamdaoui, G.; Kchouk, M. E.; Marzouk, B. Salinity Impact on Fruit Yield, Essential Oil Composition and Antioxidant Activities of *Coriandrum sativum* Fruit Extracts. *Food Chem.* **2010,** *124,* 221–225.

81. NHB. *Indian Horticulture Database.* National Horticulture Board (NHB), Govt. Of India, Gurugram, India, 2014; p 286.

82. Nourimand, M.; Mohsenzadeh, S.; Teixeira da Silva, J. A. Physiological Responses of Fennel Seedling to Four Environmental Stresses. *Iranian J. Sci. Technol.* **2012,** *1,* 37–46.

83. Overlach, S.; Diekmann, W.; Raschke, K. Phosphate Translocator of Isolated Guard-cell Chloroplasts from *Pisum sativum* L. Transports Glucose-6-phosphate. *Plant Physiol.* **1993,** *101,* 1201–1207.

84. Paleg, L. G.; Stewart, G. R.; Starr, R. The Effect of Compatible Solutes on Proteins. *Plant Soil* **1985**, *89*, 46–57.

85. Pandey, S.; Patel, M. K.; Mishra, A.; Jha, B. In Planta Transformed Cumin (*Cuminum cyminum* L.) Plants, Overexpressing the SbNHX1 Gene Showed Enhanced Salt Endurance. *PLoS One* **2016**, *11*, e0159349.

86. Pardossi, A.; Bagnoli, G.; Malorgio, F.; Campiotti, C. A.; Tognoni, F. NaCl Effects on Celery (*Apium graveolens* L.) Grown in NFT. *Sci. Hortic.* **1999**, *81*, 229–242.

87. Parida, A. K.; Das, A. B. Salt Tolerance and Salinity Effects on Plants: A Review. *Ecotoxicol. Environ. Saf.* **2005**, *60*, 324–349.

88. Pour, A. P.; Farahbakhsh, H.; Saffari, M.; Keramat, B. Response of Fenugreek Plants to Short-term Salinity Stress in Relation to Photosynthetic Pigments and Antioxidant Activity. *Int. J. Agric.* **2013**, *3*, 80–86.

89. Rahimi, R.; Mohammakhani, A.; Roohi, V.; Armand, N. Effects of Salt Stress and Silicon Nutrition on Chlorophyll Content, Yield and Yield Components in Fennel (*Foeniculum vulgare* Mill.). *Int. J. Agric. Crop Sci.* **2012**, *4*, 1591–1595.

90. Rebey, I. B.; Jabri-Karoui, I.; Hamrouni-Sellami, I.; Bourgou, S.; Limam, F.; Marzouk, B. Effect of Drought on the Biochemical Composition and Antioxidant Activities of Cumin (*Cuminum cyminum* L.) Seeds. *Ind. Crops Prod.* **2012**, *36*, 238–245.

91. Rabiei, Z.; Pirdashti, H.; Rahdari, P. The Effect of Plant-growth Promoting Rhizobacteria (PGPR) on Chlorophyll and Carotenoid Contents of the Coriander (*Coriandrum sativum* L.) Under Salinity Stress. *Plant Ecosyst.* **2013**, *8*, 61–74.

92. Said-Al Ahl, H. A. H.; Mahmoud, A. A. Effect of Zinc and/or Iron Foliar Application on Growth and Essential Oil of Sweet Basil (*Ocimum basilicum* L.) Under Salt Stress. *Ozean J. Appl. Sci.* **2010**, *3*, 97–111.

93. Said-Al Ahl, H. A. H.; Omer, E. A. Medicinal and Aromatic Plants Production Under Salt Stress. A Review. *Herbapolonica* **2011**, *57*, 72–87.

94. Saxena, S. N.; Kakani, R. K.; Rathore, S. S.; Singh, B. Use of Plant Growth Regulators for Yield Improvement in Coriander (*Coriandrum sativum* L.). *J. Spices Aromatic Crops* **2014**, *23*, 192–199.

95. Saxena, S. N.; Kakani, R. K.; Sharma, L. K.; Agrawal, D.; Rathore, S. S. Usefulness of Hydro-matrix Seed Priming in Cumin (*Cuminum cyminum* L.) for Hastening Germination. *Int. J. Seed Spices* **2015**, *5*, 24–28.

96. SBI. *Annual Report 2014–2015.* Spice Board India (SBI); Cochin, Govt. of India; 2015; p 393.

97. Schroeder, J. I.; Ward, J. M.; Gassmann, W. Perspectives on the Physiology and Structure of Inward-rectifying K$^+$ Channels in Higher Plants: Biophysical Implications for K$^+$ Uptake. *Annu. Rev. Biophys. Biomol. Struct.* **1994**, *23*, 441–471.

98. Selvakumar, G.; Thamizhiniyan, P. The Effect of the Arbuscular Mycorrhizal (AM) Fungus Glomus Intraradices on the Growth and Yield of Chilli (*Capsicum annuum* L.) Under Salinity Stress. *World Appl. Sci. J.* **2011**, *14*, 1209–1214.

99. Semiz, G. D.; Ünlukara, A.; Yurtseven, E.; Suarez, D. L.; Telci, I. Salinity Impact on Yield, Water Use, Mineral and Essential Oil Content of Fennel (*Foeniculum vulgare* Mill.). *J. Agric. Sci.* **2012**, *18*, 177–186.

100. Shah, S. H. Kinetin Improves Photosynthetic and Antioxidant Responses of *Nigella sativa* to Counteract Salt Stress. *Russian J. Plant Physiol.* **2011**, *58*, 454–459.

101. Shannon, M. C.; Grieve, C. M. Salt Tolerance of Vegetable Crops to Salinity. *Sci. Hortic.* **1999**, *78*, 5–38.

102. Sharma, D. K.; Singh, A. Salinity Research in India: Achievements, Challenges and Future Prospects. *Water Energy Int.* **2015**, *58*, 35–45.
103. Singh, N. K.; Patel, D. B. Performance of Fenugreek Bioinoculated with Rhizobium Meliloti Strains Under Semi-arid Condition. *J. Environ. Biol.* **2016**, *37*, 31–35.
104. Singh, P. K.; Chowdhury, A. R.; Garg, V. K. Yield and Analysis of Essential Oil of Some Spice Crops Grown in Sodic Soils. *Indian Perfumer* **2002**, *46*, 35–40.
105. Smirnoff, C.; Thonke, B.; Popp, M. The Compatibility of D-pinitol and 1D-1-*O*-Methyl-mucoinositol with Malate Dehydrogenase Activity. *Bot. Acta* **1990**, *103*, 270–273.
106. Storey, R.; Gorham, J.; Pitman, M. C.; Hanson, M. G.; Gage, D. Response of *Melanthera biflora* to Salinity and Water Stress. *J. Exp. Bot.* **1993**, *44*, 1551–1561.
107. Sudhakar, C.; Lakshmi, A.; Giridarakumar, S. Changes in the Antioxidant Enzyme Efficacy in Two High Yielding Genotypes of Mulberry (*Morus alba* L.) Under NaCl Salinity. *Plant Sci.* **2001**, *161*, 613–619.
108. Terry, N.; Waldron, L. J. Salinity, Photosynthesis, and Leaf Growth. *California Agric.* **1984**, *38*, 38–39.
109. Tsamaidi, D.; Daferera, D.; Karapanos, I. C.; Passam, H. C. The Effect of Water Deficiency and Salinity on the Growth and Quality of Fresh Dill (*Anethum graveolens* L.) During Autumn and Spring Cultivation. *Int. J. Plant Product.* **2017**, *11*, 33–46.
110. Tuna, A. L.; Kaya, C.; Higgs, D.; Murillo-Amador, B.; Aydemir, S.; Girgin, A. R. Silicon Improves Salinity Tolerance in Wheat Plants. *Environ. Exp. Bot.* **2008**, *62*, 10–16.
111. Ungar, I. A. *Ecophysiology of Vascular Halophytes*. CRC Press, Boca Raton, FL.; 1991; p 209.
112. Vaidyanathan, R.; Kuruvilla, S.; Thomas, G. Characterization and Expression Pattern of an Abscisic Acid and Osmotic Stress Responsive Gene from Rice. *Plant Sci.* **1999**, *140*, 21–30.
113. Viuda-Martos, M.; Ruiz-Navajas, Y.; Fernández-López, J.; Pérez-Alvarez, J. A. Spices as Functional Foods. *Crit. Rev. Food Sci. Nutr.* **2010**, *51*, 13–28.
114. Xiong, L.; Zhu, J. K. Molecular and Genetic Aspects of Plant Responses to Osmotic Stress. *Plant Cell Environ.* **2002**, *25*, 131–139.
115. Yadav, R. K.; Meena, R. L.; Sethi, M.; Dagar, J. C. Evaluation of Salinity Tolerance of Coriander and Fennel Seed Spices. *Salinity News*. Central Soil Salinity Research Institute: Karnal, 2011; Vol. 17; p 2.
116. Yeo, A. R. Salinity Resistance: Physiologies and Prices. *Physiol. Plant.* **1983**, *58*, 214–222.
117. Yogita, R.; Nikam, T. D.; Dhumal, K. N. Seed Germination and Seedling Physiology of Ajowan (*Trachyspermum ammi* L.) Under Chloride Salinity. *J. Spices Aromatic Crops* **2014**, *23*, 102–105.
118. Yusuf, M.; Fariduddin, Q.; Varshney, P.; Ahmad, A. Salicylic Acid Minimizes Nickel and/or Salinity-induced Toxicity in Indian Mustard (*Brassica juncea*) Through an Improved Antioxidant System. *Environ. Sci. Pollut. Res.* **2012**, *19*, 8–18.
119. Zaki, M. F.; Aboul-Hussein, S. D.; Abou El-Magd, M. M.; El-Abagy, H. M. H. Evaluation of Some Sweet Fennel Cultivars Under Saline Irrigation Water. *Eur. J. Sci. Res.* **2009**, *30*, 67–78.
120. Zhang, J.; Jia, W.; Yang, J.; Ismail, A. M. Role of ABA in Integrating Plant Responses to Drought and Salt Stresses. *Field Crops Res.* **2006**, *97*, 111–119.
121. Zhani, K.; Hermans, N.; Ahmad, R.; Hannachi, C. Evaluation of Salt Tolerance (NaCl) in Tunisian Chili Pepper (*Capsicum frutescens* L.) on Growth, Mineral Analysis and Solutes Synthesis. *J. Stress Physiol. Biochem.* **2013**, *9*, 209–228.

122. Zidan, A. M. A.; Elewa, M. A. Effect of NaCl Salinity on the Rate of Germination, Seedling Growth, and Some Metabolic Changes in Four Plant Species (Umbelliferae). *J. Agric. Trop. Subtrop.* **1994,** *95*, 87–97.

123. Zidan, M. A.; Elewa, M. A. Effect of Salinity on Germination, Seedling Growth and Some Metabolic Changes in Four Plant Species (Umbelliferae). *Indian J. Plant Physiol.* **1995,** *38*, 57–61.

124. Zinck, J. A.; Metternicht, G. Soil Salinity and Salinization Hazard. In *Remote Sensing of Soil Salinization: Impact and Land* Management; CRC Press: Boca Raton, FL, 2009; pp 3–20.

INDEX

A

Adenosine triphosphate (ATP), 292
Adventitious roots (ARs), 299–300
Aerenchyma formation, 296–299
Anaerobic
    metabolism, 304–306
    proteins, 306–307
Antioxidant enzymatic activity, 169–170
Arbuscular mycorrhizal (AM), 66
Ascorbate peroxidase (APX), 55

B

Biochemical adaptations
    anaerobic metabolism, 304–306
    anaerobic proteins, 306–307
    futile nitric oxide (NO) cycle, 307
    photosynthesis, 302–303
    plant nutrient uptake, 303–304
    root respiration, 304–306
Biochemical approaches
    salt tolerance, 163
        antioxidant enzymatic activity,
            169–170
        genetic and genomic approaches,
            176–180
        genetic engineering in plants, 185–187
        hormone-mediated regulation, 172–174
        ION homeostasis, 163–167
        ION transporters, 184, 188
        multilevel signal transduction, 174
        nitric oxide (NO), 171
        osmoprotectants, 168–169
        plats, epigenetics, 180–183
        polyamines (PAs), 170–171
    soil salinity
        plant response, 161
    stress avoidance, 161–162
    tolerance, 161–162
    transgenic plants
        compatible organic solutes, 188
        enhanced antioxidant production,
            188–189

C

Challenges of fruit crops
    farm yard manure (FYM), 57
    genetic transformation, 61–62
    in vitro screening, 62
    marker-assisted selection, 60–61
    quantitative trait loci mapping, 60–61
    rootstocks, development of, 59–60
    salt tolerance, genetic improvement for,
        58–59
    transgrafting, 62
Chemical environment
    nutrient availability, 290–291
    toxic elements, 291–292
Conservation agriculture (CA), 87–88,
    92–95
    environmental conservation, 95–96
    experimental evidence, 103–106
    framework of, 91–92
    higher economics, 95–96
    IGP, 94
    Indian IGP
        management practices in, 105
        mitigating soil salinity role, 102
        uptake, 94
    potential benefits, 95
    progressive global, 94
    salt tolerant varieties, compatibility, 103
    soil organic carbon (SOC), 102–103
    zero tillage (ZT), 93

D

Deficit irrigation (DI), 70

E

Enzmatic and nonenzymatic responses
    ascorbate peroxidase (APX), 55
    stress protection, 55

F

Farm yard manure (FYM), 57
Fruit crops
  agricultural water management, 41
  agronomic practices
    arbuscular mycorrhizal (AM), 66
    fruit-based agro-forestry systems,
      65–66
    mycorrhizal inoculants, use, 66–67
    nutrient management, 66
    paclobutrazol (PBZ), 66–67
    plant growth substances, application,
      67–68
    planting techniques, 63–65
    salicylic acid (SA), 68
    salt tolerant cultivars, selection of, 63
  antioxidant defense system under salinity,
    56
  botanical names, 85
  challenges
    farm yard manure (FYM), 57
    genetic transformation, 61–62
    in vitro screening, 62
    marker-assisted selection, 60–61
    quantitative trait loci mapping, 60–61
    rootstocks, development of, 59–60
    salt tolerance, genetic improvement
      for, 58–59
    transgrafting, 62
  development
    fruit yield, 54–55
    quality, 54–55
    root, 52–53
    seed germination, 51–52
    seedling establishment, 51–52
    shoot growth inhibition, 52–53
    visible injury symptoms, 53–54
  english names, 85
  enzymatic responses
    ascorbate peroxidase (APX), 55
    stress protection, 55
    superoxide dismutase (SOD), 55
  irrigation projects, 40
  limitations
    farm yard manure (FYM), 57
    genetic transformation, 61–62
    in vitro screening, 62
    marker-assisted selection, 60–61
    quantitative trait loci mapping, 60–61

  rootstocks, development of, 59–60
  salt tolerance, genetic improvement
    for, 58–59
  transgrafting, 62
local names, 85
nonenzymatic responses
  ascorbate peroxidase (APX), 55
  stress protection, 55
  superoxide dismutase (SOD), 55
osmo-protection
  proline accumulation for, 56
perennial trees and pastures, 40
plant growth, 42
salinity, 39, 42–43
salinity alleviation
  arbuscular mycorrhizal (AM), 66
  fruit-based agro-forestry systems,
    65–66
  mycorrhizal inoculants, use, 66–67
  nutrient management, 66
  paclobutrazol (PBZ), 66–67
  plant growth substances, application,
    67–68
  planting techniques, 63–65
  salicylic acid (SA), 68
  salt tolerant cultivars, selection of, 63
salt stress, effects of
  cell membrane damage, 44–45
  gas exchange characteristics, 47–49
  ion toxicity, 49–51
  leaf pigments, loss of, 46–47
  leaf water relations, alterations in,
    45–46
  osmotic stress, 49–51
  oxidative stress, 45
salt-affected soils, irrigation management
  in
  deficit irrigation (DI), 70
  drip irrigation, 68–69
  partial root-zone drying (PRD), 70–71
  subsurface drip irrigation (SSDI), 69
scope
  farm yard manure (FYM), 57
  genetic transformation, 61–62
  in vitro screening, 62
  marker-assisted selection, 60–61
  quantitative trait loci mapping, 60–61
  rootstocks, development of, 59–60

salt tolerance, genetic improvement
for, 58–59
transgrafting, 62
sodicity, 42–43
soil salinity, 42
tree growth
fruit yield, 54–55
quality, 54–55
root, 52–53
seed germination, 51–52
seedling establishment, 51–52
shoot growth inhibition, 52–53
visible injury symptoms, 53–54
Futile nitric oxide (NO) cycle, 307

G

Gypsum, 5, 17–22

H

Halophyte crops
characteristics and classification, 331
accumulating types, 334
marine phanerogams, 332
non-succulent, 334
physiological response, 333
succulent, 334
true halophytes, 332
true *versus* facultative halophytes, 333
xerohalophytes, 332
genetic improvement
genomics approaches, 354–355
greening wastelands, 348–350
growth, 334–335
anatomical features, 338
fruits, 337
leaves, 337
roots, 336
seeds and dispersal, 337
stem, 337
physiological adaptations
compartmentalization, 342–344
homeostasis, 342–344
hormone synthesis, 347
IONS, compartmentation, 342
osmotic adjustment, 344
reactive oxygen species scavenging, 345
salt dilution/succulence, 341–342

salt exclusion, 340–341
salt exclusion, adaptive mechanisms, 339–340
salt excretion/extrusion, 341
salt injury, 340
transport and uptake, 345–347
phytodesalination, 350–351
phytoremediation, 351–354
remediation of saline soils, 352
role, 348
saline environment
greening wastelands, 348–350
phytodesalination, 350–351
phytoremediation, 351–354
role, 348
salinity, 334–335
anatomical features, 338
fruits, 337
leaves, 337
roots, 336
seeds and dispersal, 337
stem, 337
seed germination
viviparous mode of, 338
Hormonal signaling
abscisic acid (ABA), 140
ethylene, 140–141
salicylic acid (SA), 139

I

Indian IGP
nature and properties of soils, 96–97
CSSRI projections, 99
salinity, 97–98
salt-affected soils, 99
water logging problems, 97–98
Indo-Gangetic Plain (IGP), 87–88
ION homoestasis
efflux, 133–134
halophytes *versus* glycophytes, 132–133
influx, 133–134
ions to vacuoles, compartmentation, 134–135
stelar cells, 135
xylem, 135

K

Kandelia candel, 119

L

Land degradation, 87
Limitations of fruit crops
  farm yard manure (FYM), 57
  genetic transformation, 61–62
  in vitro screening, 62
  marker-assisted selection, 60–61
  quantitative trait loci mapping, 60–61
  rootstocks, development of, 59–60
  salt tolerance, genetic improvement for,
    58–59
  transgrafting, 62
Low oxygen sensing, 307–308
  organelles, 310
  signaling pathway, 309–310

M

Marker-assisted selection (MAS)
  wheat breading, 236–237
Mitigating soil salinity, 88–90
  conservation agriculture (CA), 92–95
    environmental conservation, 95–96
    experimental evidence, 103–106
    framework of, 91–92
    higher economics, 95–96
    IGP, 94
    Indian IGP uptake, 94
    Indian IGP, mitigating soil salinity
      role, 102
    management practices in Indian IGP,
      105
    potential benefits, 95
    progressive global, 94
    salt tolerant varieties, compatibility,
      103
    soil organic carbon (SOC), 102–103
    zero tillage (ZT), 93
  Indian IGP, nature and properties of soils,
    96–97
    CSSRI projections, 99
    salinity, 97–98
    salt-affected soils, 99
    water logging problems, 97–98
  salt-affected soils, reclamation/
    management
    flushing, 100
    saline soils, 99–100
    sodic soils, 99–100

transformation of agriculture systems
  IGP, 92–95
  zero tillage (ZT), 93
Molecular approaches
  salt tolerance, 163
    antioxidant enzymatic activity,
      169–170
    genetic and genomic approaches,
      176–180
    genetic engineering in plants, 185–187
    hormone-mediated regulation, 172–174
    ION homeostasis, 163–167
    ION transporters, 184, 188
    multilevel signal transduction, 174
    nitric oxide (NO), 171
    osmoprotectants, 168–169
    plats, epigenetics, 180–183
    polyamines (PAs), 170–171
  soil salinity
    plant response, 161
  stress avoidance, 161–162
  tolerance, 161–162
  transgenic plants
    compatible organic solutes, 188
    enhanced antioxidant production,
      188–189
Morpho-biochemical and molecular markers
  morphological descriptors for screening,
    262
    germination traits, 263–264
  salinity and plant response
    definition, 258–259
    mitigate salinity stress, 260
    soil salinity, 259–260
  salt tolerance, screening of germplasm,
    260
    descriptors for, 262
  screening
    molecular markers, 271–273
    physiological and biochemical
      parameters for, 268–271
  soil salinity
    crop tolerance, 267
  threshold EC and slope, 268
  whole plant and crop yield, 264
    salinity and relative yield, 266–267

N

Nitric oxide (NO), 171

Nonenzymatic antioxidants
  α-Tocopherol, 129
  ascorbate, 128
  carotenoids and anthocyanins, 129
  glutathione, 128–129
  polyamines, 130

O

OMICS-based approaches
  ionome, 230–231
  metabolomics, 229–230
  phenomics, 231–232
  proteomics, 224–229
  system biology, 232–233
  transcriptome profiling, 222–223
Osmolyte metabolism mechanism, 119
  abscisic acid (ABA), 121
  Arabidopsis thaliana, 121
  glycine betaine, 121–123
  induce salt tolerance, exogenous
    application, 123–124
  proline synthesis, 120–121

P

Paclobutrazol (PBZ), 66–67
Partial root-zone drying (PRD), 70–71
Physiological adaptations
  halophyte crops
    compartmentalization, 342–344
    homeostasis, 342–344
    hormone synthesis, 347
    IONS, compartmentation, 342
    osmotic adjustment, 344
    reactive oxygen species scavenging,
      345
    salt dilution/succulence, 341–342
    salt exclusion, 340–341
    salt exclusion, adaptive mechanisms,
      339–340
    salt excretion/extrusion, 341
    salt injury, 340
    transport and uptake, 345–347
Physiological and biochemical changes in
  plants
  cell wall changes, 130–132
  hormonal signaling
    abscisic acid (ABA), 140
    ethylene, 140–141

salicylic acid (SA), 139
ION homeostasis
  efflux, 133–134
  halophytes *versus* glycophytes,
    132–133
  influx, 133–134
  ions to vacuoles, compartmentation,
    134–135
  stelar cells, 135
  xylem, 135
Kandelia candel, 119
metabolomics, 118–119
nonenzymatic antioxidants
  ascorbate, 128–130
osmolyte metabolism mechanism, 119
  abscisic acid (ABA), 121
  Arabidopsis thaliana, 121
  glycine betaine, 121–123
  induce salt tolerance, exogenous
    application, 123–124
  proline synthesis, 120–121
phenolic metabolism, 130–132
proteomics, 118–119
ROS scavenging and antioxidant
  machinery
  antioxidant enzymes, 124–126
  detoxification enzymes, 126–128
salinity, 118
signal transduction
  calcium signaling, 136–137
  no signaling, 137–139
  ROS, 137–139
transcriptomic, 118–119
Physiological approaches
salt tolerance, 163
  antioxidant enzymatic activity,
    169–170
  genetic and genomic approaches,
    176–180
  genetic engineering in plants, 185–187
  hormone-mediated regulation, 172–174
  ION homeostasis, 163–167
  ION transporters, 184, 188
  multilevel signal transduction, 174
  nitric oxide (NO), 171
  osmoprotectants, 168–169
  plats, epigenetics, 180–183
  polyamines (PAs), 170–171
soil salinity

plant response, 161
stress avoidance, 161–162
tolerance, 161–162
transgenic plants
  compatible organic solutes, 188
  enhanced antioxidant production,
    188–189
Plant growth
  biomass production, 382–383
  quality, 385–386
  seed germination, 381–382
  seed oil yield, 385–386
  seed yield, 384
  seedling establishment, 381–382
  vegetative growth, 382–383
Polyamines (PAs), 170–171
Programmed cell death (PCD), 294

Q

Quantitative trait locus
  mapping, 213–214, 217
  studies performed, 215–216

R

Reactive oxygen species (ROS), 117,
  293–294
  scavenging and antioxidant machinery
    antioxidant enzymes, 124–126
    detoxification enzymes, 126–128
Rice–wheat (RW) cropping system, 88

S

Salicylic acid (SA), 68
Saline environment
  halophyte crops
    greening wastelands, 348–350
    phytodesalination, 350–351
    phytoremediation, 351–354
    role, 348
Salinity
  management
    biostimulants, 392–393
    cultivars, selection of, 388–389
    microbial inoculants, 391–392
    nutrient management, 390
    plant growth regulators, 392–393
    salt tolerance, genetic improvement
      for, 386–388

salt tolerant crops, 388–389
  seed priming methods, 389–390
responses of
  antioxidant defense system, 378–379
  assimilation, 377–378
  carbon assimilation, 376–377
  cell membrane stability, 373–374
  hormonal changes, 380–381
  leaf chlorophyll levels, 375
  mineral uptake, 377–378
  osmoregulation, 379–380
  plant water status, 374
Salt stress
  effects of
    cell membrane damage, 44–45
    gas exchange characteristics, 47–49
    ion toxicity, 49–51
    leaf pigments, loss of, 46–47
    leaf water relations, alterations in,
      45–46
    osmotic stress, 49–51
    oxidative stress, 45
Salt tolerance, 163
  antioxidant enzymatic activity, 169–170
  genetic and genomic approaches,
    176–180
  genetic engineering in plants, 185–187
  genomic selection, 237
  genomics technologies
    wheat to salinity stress, physiological
      responses, 211–212
  genomics-based applications
    genes, 236
    osmolytes/compatible solutes, 235
    sodium transporters, 234–235
    transcription factors, 233–234
  hormone-mediated regulation, 172–174
  ION homeostasis, 163–167
  ION transporters, 184, 188
  marker-assisted selection (MAS)
    wheat breding, 236–237
  mining wheat germplasm collections
    novel, genes/alleles, 238
  multilevel signal transduction, 174
  nitric oxide (NO), 171
  OMICS-based approaches
    ionome, 230–231
    metabolomics, 229–230
    phenomics, 231–232

proteomics, 224–229
system biology, 232–233
transcriptome profiling, 222–223
osmoprotectants, 168–169
plats, epigenetics, 180–183
polyamines (PAs), 170–171
related wheat species
genomic-based exploration of,
238–240
screening of germplasm, 260
unravel mechanisms
genome sequence, 212–213
genome wide association (GWA),
217–218
microRNA (miRNA), 218–220
molecular marker resources, 212–213
QTL mapping, 213–214, 217
wild
genomic-based exploration of,
238–240
wild and related wheat species
genomic-based exploration of,
238–240
Salt-affected soils, 4
classification of, 12
extent and distribution, 14–15
irrigation management in
deficit irrigation (DI), 70
drip irrigation, 68–69
partial root-zone drying (PRD), 70–71
subsurface drip irrigation (SSDI), 69
reclamation/management
flushing, 100
saline soils, 99–100
sodic soils, 99–100
Salt-affected soils (SAS), 210, 330
Scope of fruit crops
farm yard manure (FYM), 57
genetic transformation, 61–62
in vitro screening, 62
marker-assisted selection, 60–61
quantitative trait loci mapping, 60–61
rootstocks
development of, 59–60
salt tolerance
genetic improvement for, 58–59
transgrafting, 62
Screening
molecular markers, 271–273

physiological and biochemical parameters
for, 268–271
Seed germination
viviparous mode of, 338
Seed spices, 367–368
plant growth
biomass production, 382–383
quality, 385–386
seed germination, 381–382
seed oil yield, 385–386
seed yield, 384
seedling establishment, 381–382
vegetative growth, 382–383
salinity management
biostimulants, 392–393
cultivars, selection of, 388–389
microbial inoculants, 391–392
nutrient management, 390
plant growth regulators, 392–393
salt tolerance, genetic improvement
for, 386–388
salt tolerant crops, selection of,
388–389
seed priming methods, 389–390
salinity, responses of
antioxidant defense system, 378–379
assimilation, 377–378
carbon assimilation, 376–377
cell membrane stability, 373–374
hormonal changes, 380–381
leaf chlorophyll levels, 375
mineral uptake, 377–378
osmoregulation, 379–380
plant water status, 374
salt stress, plant responses, 370
Na^+/K^+ discrimination, 372
osmotic adjustment, 370–372
salt inclusion versus exclusion, 372
Signal transduction
calcium signaling, 136–137
no signaling, 137–139
ROS, 137–139
Sodic (alkali) soils, nomenclature and
reclamation
characteristics, 15
chemical factors, 16
microbiological factors, 16
physical factors, 16
chemical amendments, 16

gypsum, 17–22
  phosphogypsum, 22–24
exchangeable sodium percentage (ESP),
    6, 9
  sodic soils, 6, 9, 13–14
experiments, 5
  gypsum, 5
  reclamation technology, 24–25
  resodification, 29–30
  salt-affected soils
    classification of, 12
    extent and distribution, 14–15
  semireclaimed sodic lands, 29–30
  technology package, 25
    management package, 26–27
    stakeholders' participation, 28–29
  terminology
    black alkali soils, 6
    Central Soil Salinity Research Institute
      (CSSRI), 9
    electrical conductivity (EC), 7
    management issues, 12
    productivity, 8
    sodic soils, 7
    Soil Salinity Laboratory of United
      States, 7
    United States Department of
      Agriculture (USDA), 8
  Zarifa Viran Series
    physicochemical characteristics of,
      10–11
Soil organic carbon (SOC), 102–103
Soil salinity
  crop tolerance, 267
  plant response, 161
Subsurface drip irrigation (SSDI), 69
Superoxide dismutase (SOD), 55

T

Terminology of sodic (alkali) soils
  nomenclature and reclamation
    black alkali soils, 6
    Central Soil Salinity Research Institute
      (CSSRI), 9
    electrical conductivity (EC), 7
    management issues, 12
    productivity, 8
    sodic soils, 7

    Soil Salinity Laboratory of United
      States, 7
    United States Department of
      Agriculture (USDA), 8
Transformation of agriculture systems
  IGP, 92–95
  zero tillage (ZT), 93
Transgenic plants
  compatible organic solutes, 188
  enhanced antioxidant production,
    188–189

U

Unravel mechanisms
  genome sequence, 212–213
  genome wide association (GWA),
    217–218
  microRNA (miRNA), 218–220
  molecular marker resources, 212–213
  QTL mapping, 213–214, 217

W

Waterlogged conditions, plants, 286
  adaptation of
    low oxygen escape strategy (LOES),
      295
    low oxygen quiescent strategy
      (LOQS), 295
  anatomical mechanisms
    adventitious roots (ARs), 299–300
    aerenchyma formation, 296–299
    aerial parts, enhanced elongation of,
      300–302
  biochemical adaptations
    anaerobic metabolism, 304–306
    anaerobic proteins, 306–307
    futile nitric oxide (NO) cycle, 307
    photosynthesis, 302–303
    plant nutrient uptake, 303–304
    root respiration, 304–306
  chemical environment
    nutrient availability, 290–291
    toxic elements, 291–292
  greenhouse gases, emissions of, 294
  leaf epinasty, 294–295
  low oxygen sensing, 307–308
    organelles, 310

signaling pathway, 309–310
morphological
    adventitious roots (ARs), 299–300
    aerenchyma formation, 296–299
    aerial parts, enhanced elongation of,
        300–302
physiological processes
    adenosine triphosphate (ATP), 292
    carbohydrate shortage, 293
    programmed cell death (PCD), 294

reactive oxygen species (ROS),
    293–294
root growth, 294
soil air, composition of, 288
    physical properties, 289–290

Z

Zarifa Viran Series
    physicochemical characteristics of, 10–11
Zero tillage (ZT), 93

Printed and bound by CPI Group (UK) Ltd, Croydon, CR0 4YY

23/10/2024

01777702-0015